Seismic Exploration of the Deep Continental Crust

Methods and Concepts of DEKORP and Accompanying Projects

Edited by
Dirk Gajewski
Wolfgang Rabbel

1999

Springer Basel AG

Reprint from Pure and Applied Geophysics
(PAGEOPH), Volume 156 (1999), No. 1/2

Editors:

Dirk Gajewski
Institute for Geophysics
University of Hamburg
Hamburg
Germany
email: gajewski@dkrz.de

Wolfgang Rabbel
Institute of Geosciences
University of Kiel
Kiel
Germany
e-mail: wrabbel@geophysik.uni-kiel.de

A CIP catalogue record for this book is available from the Library of Congress, Washington D.C., USA

Deutsche Bibliothek **Cataloging-in-Publication Data**

Seismic exploration of the deep continental crust : methods and concepts of DEKORP and accompanying projects / ed. by Dirk Gajewski ; Wolfgang Rabbel. - Basel ; Boston ; Berlin : Birkhäuser, 1999
  (Pageoph topical volumes)
  ISBN 978-3-7643-6210-2   ISBN 978-3-0348-8670-3 (eBook)
  DOI 10.1007/978-3-0348-8670-3

© 1999 Springer Basel AG
Originally published by Birkhäuser Verlag in 1999
Printed on acid-free paper produced from chlorine-free pulp

ISBN 978-3-7643-6210-2

9 8 7 6 5 4 3 2 1

Contents

Pure appl. geophys. 156 (1999) 1–6
0033–4553/99/010001–06 $ 1.50 + 0.20/0

Foreword

Seismic Exploration of the Deep Continental Crust

The DEKORP Project

When DEKORP, the German continental reflection seismic program, started its first deep seismic survey in 1984, no one would have expected that this project would run for 14 consecutive years. During this time it became the major focus of deep seismic research in Germany. Its open organisational structure, comprising separate executive and steering committees, admitted the participation of many geoscientific institutions and university institutes in the planning and performing of research. Scientifically, the DEKORP activities were inspired by similar foreign research programs already in existence at that time, such as COCORP and BIRPS. DEKORP was funded by the German Federal Ministry of Research (BMBF) and complemented by special funding programs of the German science foundation (DFG), dedicated to the lower crust and the Variscan orogenic belt.

In the 1980s DEKORP focussed on deep seismic reflection profiling of the crystalline crust of the Variscan mountain belt located between the North German Basin and the Alps. The seismic sections provided fundamental new insight into the structure of the European continent and the dynamics of continental formation: "from collision to collapse." They formed the basis for worldwide comparative studies of orogenic structure. Many of the findings of DEKORP and other reflection programs have become common geoscientific knowledge today, such as thin skinned thrusts and detachment zones, deeply extending fracture zones, laminated lower crust and missing mountain roots in the central European Variscides, evolution models of upper and lower crust, among many others.

In 1989, in support of the Continental Deep Drilling Program (KTB), DEKORP performed the first 3-D survey of deep crystalline crust worldwide imaging the profound complexity of tectonics at the KTB drilling site in the Oberpfalz (S. Germany). The DEKORP group broadened its spectrum of research targets in the 1990s to approach further fundamental questions of continental evolution, such as the formation of intracontinental basins and subduction. The suite of DEKORP-profiles was finalized, partly in international cooperation, by spectacular depth images of the Urals, including superdeep reflections from the earth's mantle, of the Central Andes and the North German/Danish Basin.

From the outset of DEKORP near vertical reflection profiling was complemented by wide-angle reflection and refraction investigations to better determine velocity depth functions, to improve the imaging of reflections, and to render petrological interpretation feasible. In the processing of seismic reflection data, DEKORP basically relied on the CMP-method approved for sedimentary data but applied with special care to the complex crystalline targets by the processing centres in Clausthal-Zellerfeld and Potsdam.

The complicated signature of the reflections from the deep crust often resembled scattering patterns rather than echos from continuous interfaces. This seismic structure as well as low impedance contrasts and low signal-to-noise ratios indicated that new processing and interpretation techniques must be considered to better image the crystalline crust and to take into account its particular characteristics in wave propagation. Therefore, seismic data acquisition and geological interpretation were complemented by research projects aiming at the improvement of existing and the development of new seismic methods and interpretation concepts, such as pre-stack migration, 3-D imaging, or the use of shear waves and seismic anisotropy. Most of these accompanying projects were directly conducted under the roof of the DEKORP organisation, others closely related to it but funded by the DFG. The results of some of these projects are presented in this volume of Pure and Applied Geophysics. The seismic reflection sections and interpretation of the DEKORP data are published in a number of Journal articles and most of the reflection data gathered before 1989 can be viewed in the DEKORP-Atlas (Meissner, R. and Bortfeld, R.K. (eds.): DEKORP-Atlas, Springer, Berlin, 1990). The first article of the present volume by Meissner and Rabbel is a review of major results of the DEKORP project shown in the context of continental reflection profiling worldwide.

This Volume

The articles of this special issue are organized in three topical groups:

(a) Seismic Structure of the Deep Crust and Upper Mantle: Reflectivity, Anisotropy and Scaling Effects

– *Meissner and Rabbel* provide a review of reflection images associated with different tectonical processes and compare these observations against each other.

– *Tittgemeyer et al.* show that different scale lengths of seismic heterogeneity exist at different depth levels of the crust and mantle and they demonstrate how scale length can be assessed quantitatively by seismic profiling and modelling.

– *Lüschen* emphasizes the need for shear-wave investigations as a lithological discriminator in deep crustal exploration and discusses problems of *S*-wave data acquisition.

– *Weiss et al.* present a review of the anisotropy of the lower continental crust where laboratory investigations of rock samples are considered and compared to numerical models using the determined elastic parameters.

– In a second article, *Weiss et al.* discuss the petrology of the upper mantle beneath southern Germany and its relation to seismic velocity, anisotropy and reflection strength on the basis of petrophysical data and numerical modelling.

– Considering a transversely isotropic lower crust, *Bohlen et al.* study sensitivity of an iterative linearized 3-D travel-time inversion on two types of systematic errors: erroneous velocity and/or interface topography.

– *Pohl et al.* present a modelling study on a laminated anisotropic lower crust using the reflectivity method. Laboratory measurements of rocks from three different sections of exposed lower crust served as input models.

(b) New Concepts of Seismic Migration and Interpretation

– *Buske* applied a Kirchhoff pre-stack depth migration to 3-D seismic data collected at the site of the German continental deep drill hole. Depth sections and time slices provide a 3-D image of a steeply dipping major fault zone, the SE1-reflector, and a bright mid-crustal reflection, the so-called Erbendorf body.

– *Bleibinhaus et al.* used pre-stack migration of wide-angle data for imaging the boundary between the Saxothuringian and Moldaunubian terranes.

– *Tillmanns and Gebrande* present a slowness-driven isochrone migration for 2-D seismic data. The combination of slowness information with migration leads to an improved depth image.

– *Gajewski et al.* present an approach to integrate structural geology and reflection seismic data based on seismic modelling and geophysical borehole data.

– *Toutou* incorporates the effects of absorption into Kirchhoff migration. A complex velocity is introduced into the Helmholtz equation leading to an anti-dissipation operator which is convolved with the wave field.

(c) Seismic Signal Processing and Velocity Determination

– *Buttkus and Bönnemann* show that adaptive filtering is an attractive alternative to conventional Wiener filtering. In order to increase the signal-to-noise ratio three different adaptive filters involving multichannel seismic data are presented and applied to DEKORP data.

- *Bönnemann and Buttkus* review the potential of the τ-p transform for the determination of velocities from deep seismic reflection data. Special properties of the transform are exploited to enhance signal-to-noise ratio.
- *Müller et al.* use the wave-field directivity to obtain velocity information from a walk away VSP-experiment at the KTB-site. Local slant-stacks serve as a tool to extract slowness information.
- *Polom* developed a new method to detect and eliminate impulse type noise from vibroseis data before stack.
- *Fertig et al.* review key elements of digital seismic processing, such as wavelet processing and coherency filtering, emphasizing their role as a precondition to reliable interpretation.

In part, the articles open the perspective to new and future research. In part, they are intended as a documentation of research activity triggered or inspired by technical and interpretational questions raised by DEKORP field work and profiling results. Many of the presented methods can find immediate application in industrial seismic prospecting, just as for many years the work for and with DEKORP became an excellent starting point for students and their careers.

Future Perspectives of Deep Seismic Research

The DEKORP project started as a national initiative to study the structure of the deep crust of the Variscan mountain belt in central Europe, and to develop appropriate methods of data processing and interpretation. The final phase was characterized by international cooperation in science and funding, by the diversification of geological targets, representing different tectonic processes, and by including the marine component to the terrestrial acquisition technique.

After DEKORP concluded and based on the results of deep seismic projects worldwide, we can summarize the status and future perspectives of deep seismic research by the following statements:

- Seismic reflection imaging, combined with refraction and wide-angle measurements, is established as a reliable standard technology of deep crustal research. The current standard implies the use of compressional waves and a 2-D profiling approach. The aim of future methodical efforts may be to identify new wave-field attributes which improve the definition of seismic subsurface models.
- The complex three-dimensional character of many geological key targets and the need to develop efficient 3-D technology for exploring deep and superdeep targets has become evident. Such technology will have to combine both high resolution reflection seismics and passive seismological methods cost-effectively.

- Seismic response of tectonic features and the origin of reflectivity are still under discussion. Progress can only be achieved by increased efforts in petrophysics, shear-wave technology and seismic anisotropy, by modelling, and, of course, by drilling and borehole investigations such as the International Continental scientific Drilling Program (ICDP).

- The dynamic aspects of crustal development, differences in the evolution of upper and lower crust, the influence of rheology, stress and temperatures of deformation have shown the need for comprehensive interdisciplinary studies, preferably in regions which are tectonically active at present. Deep crustal research in seismically active areas will represent a close link between scientific and social needs.

- An increasing number of superdeep reflections from the earth's upper mantle is being reported. Understanding mantle dynamics as the driving force of tectonic processes will require the enhanced application of both active reflection and passive seismic methods focussing on reflectivity, anisotropy and the scale length of seismic heterogeneity.

- Studying crust-mantle interaction requires large-scale research programs which combine transects of both oceanic and continental lithosphere with a focus on active and passive continental margins.

- Even the need for classical seismic studies of the crust is far from being saturated. The repeated occurrence of unexpected crustal features, such as the bright spots in Tibet, illustrates that the catalogue of deep crustal conditions is incomplete and too small to allow predictions of deep structure and processes without seismic imaging.

In the framework of international reflection programs, DEKORP has broadened our insight into the evolution of the continental crust. It eventually led to new frontiers to be opened by the next generation of research initiatives dedicated to the dynamics of the earth's interior.

Acknowledgements

The following persons substantially contributed to this volume by serving as reviewers or as partners in discussing future perspectives in a spontaneous opinion poll. We are grateful to: G. Bokelmann, Ch. Bönnemann, M. Bopp, H.-J. Brink, S. Buske, B. Buttkus, N. Ettrich, J. Fertig, E. Flüh, W. Franke, K. Fuchs, H. Gebrande, J.-H. Götze, A. Henk, R. Kind, D. Kläschen, M. Korn, Ch. Krawczyk, P. Krajewski, G. Lambaré, E. Lüschen, H. Lykke-Andersen, R. Marschal, J.

Mechie, R. Meissner, B. Milkereit, W. Mooney, G. Müller, O. Oncken, I. Pšenčík, D. Ristow, V. Schenk, J. Schleicher, S. Siegesmund, M. Simon, M. Stiller, F. Theilen, Ph. Thierry, H. Thybo, Ch. Vasudevan, M. Weber, R. Wenk, and F. Wenzel.

W. Rabbel
Institute of Geosciences
University of Kiel
Otto-Hahn-Platz 1
D-24118 Kiel
Germany

D. Gajewski
Institute for Geophysics
University of Hamburg
Bundesstr. 55
D-20146 Hamburg
Germany

 To access this journal online:
http://www.birkhauser.ch

Pure appl. geophys. 156 (1999) 7–28
0033–4553/99/010007–22 $ 1.50 + 0.20/0

Pure and Applied Geophysics

Nature of Crustal Reflectivity along the DEKORP Profiles in Germany in Comparison with Reflection Patterns from Different Tectonic Units Worldwide: A Review

R. MEISSNER[1] and W. RABBEL[1]

Abstract—Reflectivity of the continental crust displays many different patterns. The DEKORP lines are used as a basis for comparing and reviewing reflectivity in different tectonic units. The (brittle) upper crust generally exhibits only two types of reflectivity. It is either rather "transparent," preferably in some extensional provinces, or/and it shows traces of thrust and shear zones of former or present ruptures. As these zones have a low impedance interior (with few exceptions), their first reflection onsets have a negative polarity and evince strong, but short signals, which sometimes can be correlated over several kilometers. The (generally ductile) lower crust displays a completely different reflectivity. In warm, extensional and thin crusts the lower part is full of reflecting lamellae. It is suggested that this type of reflectivity has a thermo-rheological origin. The creation of lamellae must take place in a ductile material with contrasting impedance under extensional stresses. It can be associated with mineral alignment and corresponding seismic anisotropy. Destruction of lamellae may take place by a cooling process, transforming parts of the lower crust into a brittle regime. Small stresses might deform or break the lamellae and leave a certain dispersed reflectivity like that in some old (and cold) shields. There are no observations of reflecting lamellae in the upper crust or in the upper mantle. In all areas the Moho is the last reflecting band (reflection Moho), which most often is identical with the classical refraction Moho. There are isolated, mostly dipping, reflections in the uppermost mantle in zones where the last tectonic event, a delamination or subduction, was not succeeded by a heating process. The uppermost mantle is brittle again in most areas and may keep the memory of a (cold) collision over billions of years.

Key words: Continental crust, seismic reflectivity, lamellae, crustal viscosity.

Introduction

Except for the DEKORP lines in Northern Germany and in the Baltic Sea, the bulk of DEKORP profiles is located in the Variscan mountain belts (MEISSNER and BORTFELD, 1990; DEKORP—BASIN RESEARCH GROUP, 1998). There are remarkable differences between the rather uniform crustal reflectivity in the Variscan internides, that of the northern externides, and that of the North German Basin and the Baltic Sea (SADOWIAK et al., 1991). Even stronger differences exist between our data and those of ancient or very recent tectonic units. Surveys in the

[1] Institut für Geowissenschaften, Universität Kiel, Otto-Hahn-Platz 1, D-24118 Kiel, Germany.

Proterozoic Baltic or North-American shields or those in the Tertiary Alpine or Himalayan belts indicate very specific reflectivity patterns (MUSACCHIO et al., 1997; GODWIN and THOMPSON, 1988; PFIFFNER et al., 1988). On the other hand, strong similarities exist between the data of the Variscan internides and those of the marine BIRPS—profiles, crossing various basins around Great Britain, the ECORS line through the Paris Basin, and to the COCORP lines when crossing extensional units in the U.S. (RESTON, 1990; FUIS and CLOWES, 1993; BROWN, 1991). Discussion will concern how the various reflectivity patterns might be generated, how they might survive extremely long periods, and how they might be destroyed. Considerable evidence has been accumulated for the dominant role of thermal evolution in a specific tectonic area, and it will be suggested that a temperature-dependent rheology and its variation with time and depth are the main ingredients for the observed reflectivity patterns. This has been indicated in some of our previous studies (DEKORP RESEARCH GROUP, 1990, 1991; MEISSNER and SADOWIAK, 1992), concentrating on DEKORP lines, and by REY (1993) for Phanerozoic areas in general, concentrating on French and British profiles. The final goal of this study is to understand the various reflectivity patterns and to relate them to specific tectonic units and their thermal evolution.

In general, several conditions are necessary to make a reflection visible and correlatable

– The reflection coefficients must be at least 0.01 in order to exceed the noise level (WARNER, 1990).
– Reflectors must be rather plane, in order to allow imaging and correlation; some reflected rays from strongly curved reflectors might not reach a normal geophone spread (MEISSNER, 1996).
– Also the dip of reflectors should be moderate and not exceed about 60°; otherwise imaging requires special seismic field configurations and processing efforts (GEBRANDE et al., 1991).
– Reflectors must be rather continuous, their length exceeding the Fresnel Zone radius (SHERIFF and GELDART, 1982) which is a few kilometers at lower crustal levels.
– A display of true amplitudes often can give convincing evidence for the arrival of reflecting energy. Sophisticated migration and stacking support modern imaging procedures.

There are general observations which seem to apply to crustal reflection studies in every tectonic unit. These will be summarized in the following, before DEKORP's observations in Germany are described.

1. In the (brittle) upper crust most reflections are from (mostly, but not only) compressional shear zones. Inside the shear zones the impedance $V \cdot \rho$ is smaller than outside, resulting in a negative reflection polarity for the first onset (MEISSNER, 1996). As shear zones are limited in width, preferably short reflecting signals are generated, sometimes over long distances.

2. Often, upper crust is reported to be "transparent" (MATTHEWS and the BIRPS GROUP, 1990). This transparency is caused by low impedance contrasts within the mainly felsic rocks or, in some cases, by steep inclination of folded or faulted layers (DUYSTER et al., 1993). Such steeply dipping structures, as well as diffractions from small inhomogeneities, are often suppressed in standard seismic data processing, causing a kind of "apparent transparency" (STEENTOFT and RABBEL, 1995).

3. Most often the Moho represents the deepest crustal reflection (= reflection Moho) (MEISSNER and BROWN, 1991), either as the deepest reflection of a thick package of reflecting lamellae or of otherwise reflecting lower crust, sometimes as a rather isolated sequence of reflections. The "Reflection Moho" most frequently coincides with the (classical) refraction and wide angle Moho (MCGEARY and WARNER, 1985).

4. In some areas reflections from the uppermost mantle are observed. They appear generally as isolated and dipping reflectors. Their polarity is difficult to determine, and their appearance seems to be connected with subduction or delamination zones (KRISHNA et al., 1996).

Specific DEKORP Observations in Germany and in the SW Baltic Sea

Figure 1 shows a location map of DEKORP lines in Germany with some tectonic boundaries (FRANKE, 1992). It will be shown that reflectivity in the Variscan internides differs from that in the externides and from that of the North German Basin and its northern rim. Actually, the term "lamellae" was coined during a special seismic investigation of the Urach heatflow anomaly in the Variscan interior (BARTELSEN et al., 1982). Lamellae are defined as a "densely packed, preferably subhorizontal, sequence of reflecting bands and elements, correlatable over at least a few kilometers;" by British authors (BIRPS) sometimes called "anastomozing" or "laminated" (shear) zones (RESTON, 1987, 1988). It will be shown that such laminated lower crusts are characteristic for extensional structures worldwide.

The *Variscan internides* include the area of the Saxothuringian (ST) and the Moldanubian (MN), SW of the Bohemian Massif (Fig. 1). Three examples of this area are shown in Figure 2 (after BARTELSEN et al., 1982; MEISSNER and BORTFELD, 1990). We summarize the main observations as follows:

1. Compared to the global average of continental crust, crustal thickness is small (30 km) and rather uniform.

2. The (brittle) upper crust is rather free of strong continuous reflections except for some shear zones, mostly compressional, marking tectonic boundaries. In some cases an "apparent transparency" was observed, although some structural

details could be revealed by refined data processing (STEENTOFT and RABBEL, 1995; SIMON, 1998).

3. There are strong lower crustal lamellae with a thickness of 4–5 s two-way travel time (TWT), corresponding to a depth range up to 15 km.

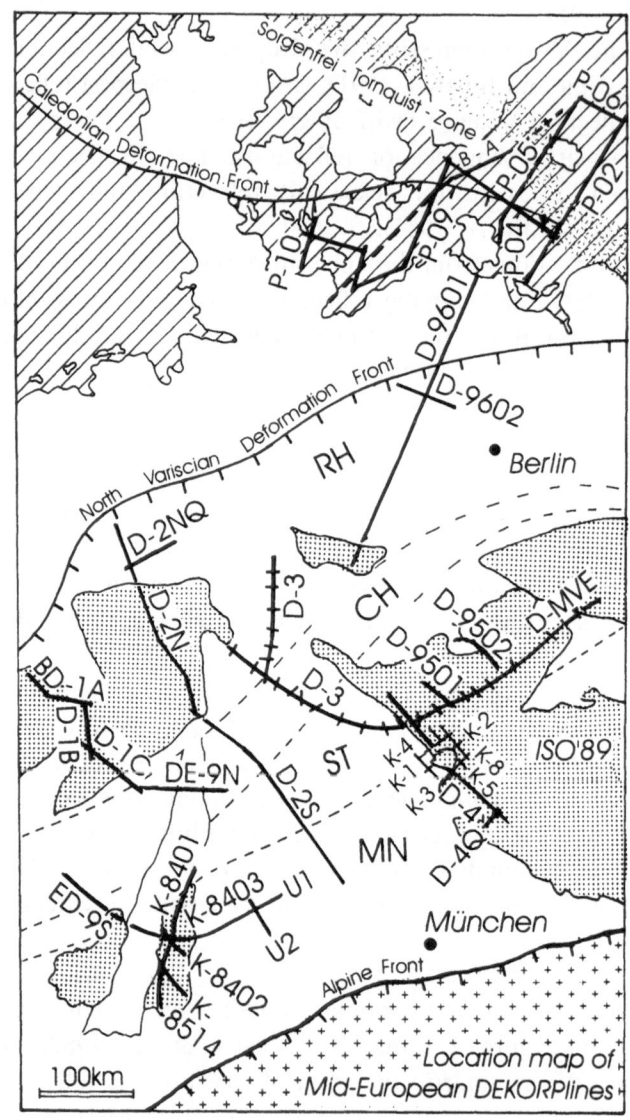

Figure 1

Location of DEKORP lines with simplified tectonic units. CH = Crystalline High, RH = Rhenohercynian, ST = Saxothuringian, MN = Moldanubian, dotted areas = Geologic units older than Variscan, Seismic Lines: D = DEKORP; P = DEKORP/BGR, K = KTB, E = ECORS, B = BELCORP, U = URACH, B-A = BABEL.

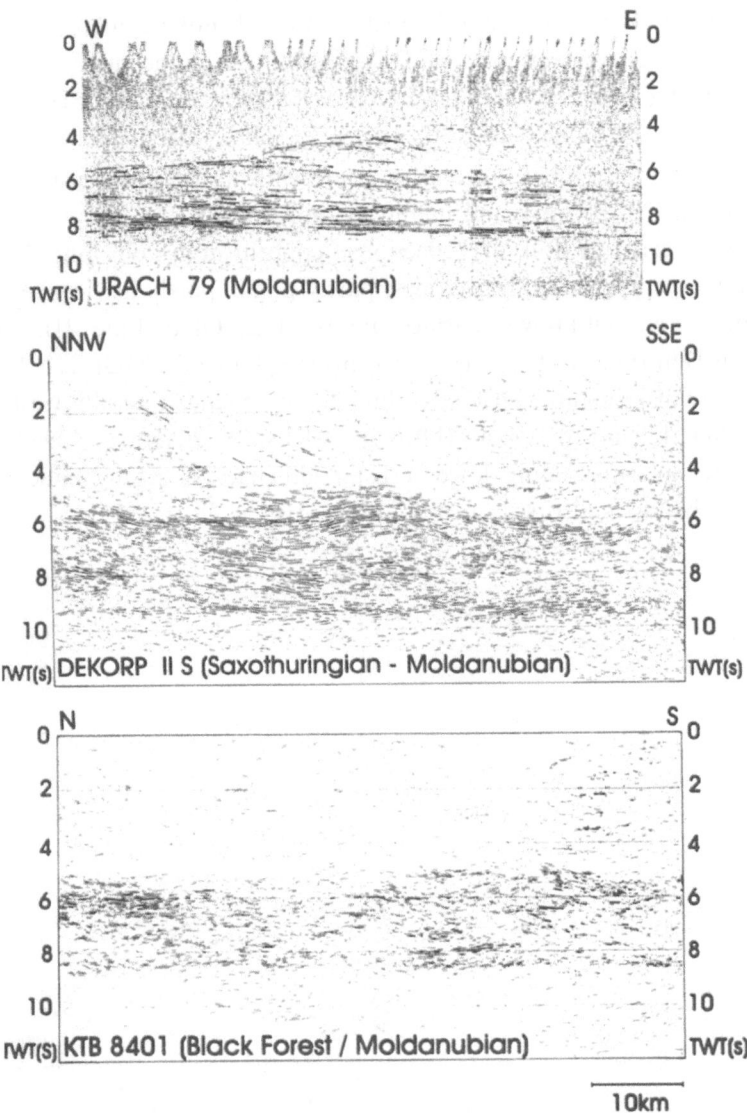

Figure 2
Three seismic sections through the Variscan Internides. URACH79 is the original pre-DEKORP seismogram section with interpretation. DEKORP II S and BLACK FOREST lines are automatic line drawings (from BARTELSEN et al., 1982 and MEISSNER and BORTFELD, 1990).

4. The dense subhorizontal lamellae with their contrasting impedance suggest seismic anisotropy. Such an anisotropy was detected by a special experiment in the Urach area (RABBEL and LÜSCHEN, 1996).

5. No reflections from the upper mantle have been observed, in spite of specific field surveys (BARTELSEN et al., 1982).

The northern *Variscan externides* include the Rhenohercynian (RH in Fig. 1). Figure 3 provides two examples of DEKORP lines in the northern part of the Rhenohercynian (TAIT *et al.*, 1996; DEKORP RESEARCH GROUP, 1991; MEISSNER and BORTFELD, 1990). We summarize the main observations as follows:

1. Compared to the global average of continental crust, the crustal thickness is rather small (30–35 km), and manifests undulations.
2. The *whole* crust is full of reflections, concentrating on long bands, most probably compressional shear zones, partly plane, partly listric, the upper ones merging with well-known surface faults (DEKORP Line II). The North Variscan Deformation Front (Aachen thrust = Faille du Midi) is a 100-km-long ramp and flat structure (DEKORP line I) and extends laterally westwards over nearly 2000 km up to Ireland (DEKORP RESEARCH GROUP, 1990).
3. There are no outstanding lower crustal lamellae. A strong concentration of reflecting bands is also seen in the lower crust.

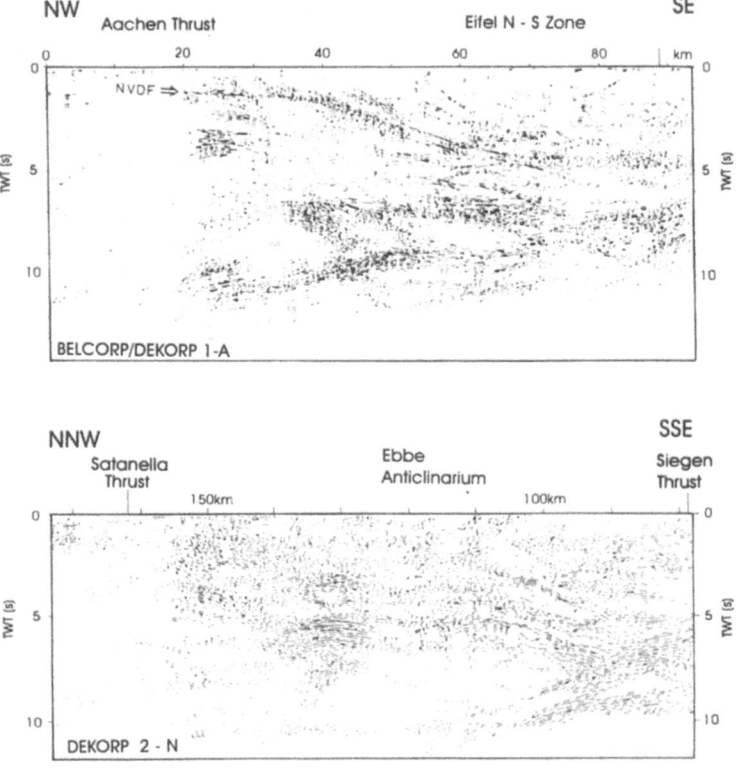

Figure 3
Two seismic sections through the northern Variscan externides. The sections are manual line drawings across the northern rim of the Rhenohercynian. Line DEKORP 1 A shows the North Variscan Deformation Front (NVDF); note the divergence of reflection shear zones toward the NW.

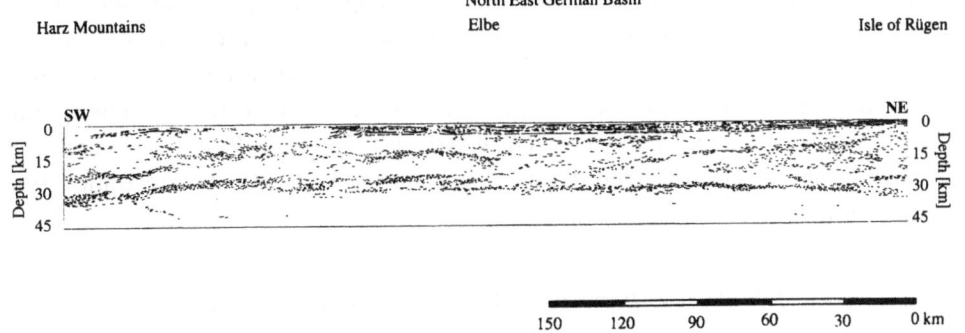

Figure 4
Automatic line drawing of DEKORP line 9601 through the North German Basin. Note small updoming, very thin lower crustal lamellae, poorly reflecting Moho, broad bands of reflectivity in the middle crust (mafic intrusions?), tectonic impacts on both ends (indentations in the NE).

4. There are no significant reflections in the mantle, although in the north reflection elements attributed to the Moho dip into mantle depth below the Sub-Variscan foredeep.

The *North German Basin* is covered by a large number of industry profiles and recently by DEKORP—Basin lines (DEKORP—BASIN RESEARCH GROUP, 1999; Fig. 4). The main observations are summarized in the following:

1. Crustal thickness varies between 25 km in the NW and 32 km in the E.
2. No large crustal shear zones are observed, except for the northern rim.
3. There are very thin lower crustal lamellae above Moho.
4. Some strange mid-crustal reflectors may be interpreted as indications of intrusions.
5. No reflections are observed in the upper mantle.

In the *Southwest Baltic Sea* (BABEL WORKING GROUP, 1993; DEKORP—BASIN RESEARCH GROUP, 1999) there are many new marine DEKORP profiles (Fig. 1) and only the northern part of line P-9 is shown as an example (Fig. 5). We summarize the most important observations in the following:

1. There is a small to intermediate crustal thickness (30–40 km).
2. The well-known Caledonian Deformation Front (CDF), as mapped by earlier seismic studies (BABEL WORKING GROUP, 1991, 1993), shows reflections from Avalonian thrusts onto Baltica, dipping S to SW in the upper crust, and also reflections dipping N to NE in the uppermost mantle. These observations could be confirmed and extended by the three new marine profiles (P-02, P-05, P-09). The N to NE dipping mantle reflectors probably represent traces of the subducted Tornquist Ocean. To the NE of the CDF, the Baltic Shield begins (next section), but crustal thickness does not increase significantly up to the Bornholm area.

3. The appearance of Moho is in the form of a single or slightly laminated band of reflectors, deepening significantly toward the NE eastward of Bornholm.
4. E of Bornholm there is a zone of extremely dense, diverging, and widening reflectivity in the whole crust. It represents a Proterozoic terrane accumulation, seen also on the BABEL line (ABRAMOVITZ *et al.*, 1997). This zone of convergence, similar to the CDF, marks a typical indenter tectonics (see discussion).

Observations of Reflection Patterns in Various Tectonic Units Outside Germany

Three types of tectonic units are selected for a comparison, namely, Proterozoic shields and platforms (not present in Germany), Phanerozoic extensional zones (comparable to similar structures along DEKORP profiles), and young, mainly Tertiary, compressional zones, e.g., recent continent-continent collisions (not present in Germany except for the northern Alps).

Many seismic surveys were performed in *Proterozoic shields or platforms* (by COCORP in the USA; by BABEL in Fennoscandia; by LITHOPROBE in Canada). Reflectivity varies strongly, but in general crustal fine structure, faults, and shear zones are well imaged. This high quality seems to be an effect of the old and cold crustal units in which the brittle regime extends to greater depths, and seismic absorption is smaller than in younger (and warmer) crustal units. We summarize the observations:

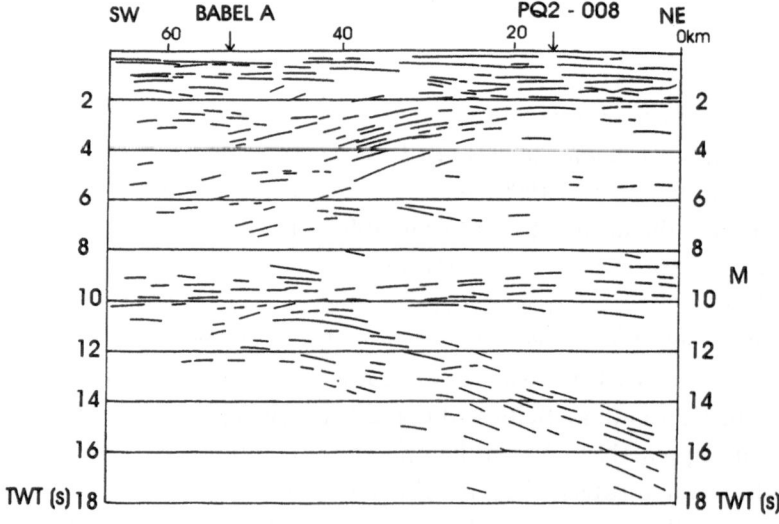

Figure 5

Manual line drawing of the NE part of the marine line PQ2–009 (P-09 in Fig. 1). Bivergent collision (indentations) along the Caledonian Deformation Front (CDF) clearly observed; different dips of reflective bands in the upper crust and uppermost mantle.

1. Crustal thickness (40–55 km) is generally larger than in the Variscides.
2. Often the whole crust is full of reflections, sometimes concentrating on long shear zones, similar to the reflectivity in the Variscan externides (BABEL WORKING GROUP, 1991). Only in a few areas crustal reflectivity is poor. The Moho is always visible. Although no pronounced lamellae are seen in the lower crust, it is often full of short reflectors and terminated by a reflecting Moho.
3. In the upper crust many dolerite sills are observed around the Siljan borehole (JUHLIN, 1990). Their sequence and reflection patterns match exactly widespread observations in the Bay of Bothnia (VASEDUVAN et al., 1996; KLÄSCHEN and FLÜH, 1996) and in parts of American shields (HAMMER and CLOWES, 1997; BEZDAN and HAJNAL, 1998).
4. There are signs of Proterozoic compressional processes such as terrane accretion or subduction (ABRAMOVITZ et al., 1997); BABEL WORKING GROUP, 1990). The Canadian Lithoprobe lines particularly show the most spectacular imaging of ancient tectonic processes (CALVERT et al., 1995).
5. There are mantle reflectors in various ancient provinces, probably from frozen-in subducting slabs (ALLMENDINGER et al., 1987; BABEL WORKING GROUP, 1990). Also north of Caledonian Scotland very prominent reflections are seen in the uppermost mantle (WARNER and McGEARY, 1987).

For *Phanerozoic extensional zones* (BIRPS, around British Isles, ECORS, France, COCORP, U.S) the following observations can be noted:
1. Crustal thickness is as small (25–35 km), as in the Variscides and the North German Basin.
2. Only very few reflections are observed in the upper crust (BIRPS "transparent upper crust," CHEADLE et al., 1987), but flat-lying extensional faults are seen in COCORP's profiles through the Basin and Range Province (SMITHSON, 1986).
3. The North Variscan Deformation Front is observed in the Irish Sea and even west of Ireland in the middle and upper crusts, similar to its image along DEKORP I (BIRPS and ECORS, 1986; DEKORP RESEARCH GROUP, 1990). This compressional feature has survived later extensions, at least in the upper crust, matching similar observations in the German Variscides.
4. There are ample lower crustal lamellae, as observed along the BIRPS profiles in all young basins around the British Isles (CHEADLE et al., 1987) and by the ECORS profile through the Paris Basin (BOIS et al., 1986), very similar to the reflective images in the Variscan internides.
5. The Iapetus Suture around the British Isles is observed along several BIRPS profiles by various N-dipping reflections in the lower crust and the uppermost mantle (KLEMPERER and MATTHEWS, 1987).
6. No reflections in the uppermost mantle are observed. An exception is a part of the North Sea (and the Scagerak), where during the MONA LISA project some deep reflections were observed (ABRAMOVITZ et al., 1998; MONA LISA WORKING GROUP, 1997), probably signs of ancient subductions.

In *young compressional belts* (Swiss Working Group, NFP 20; ECORS, CROP, INDEPTH) one observes the following:

1. In the *Alps* there is considerable crustal thickening (Moho depth > 55 km), and highly asymmetric crustal root (HEITZMANN *et al.*, 1991).
2. Many compressional shear zones can be observed in the upper and middle crusts.
3. Large indentations of slivers from the Adriatic upper mantle and lower crusts into the (weak) middle and lower European crusts (crustal shortening) have taken place (PFIFFNER *et al.*, 1988).
4. Some thin lamellae mark the S and SE-dipping European lower crust.
5. The European lower crust and uppermost mantle seem to indicate the onset of delamination (MEISSNER and MOONEY, 1998).
6. Small indentations are also observed in the *Pyrenees* (ECORS PYRENEES TEAM, 1988).
7. In the *Himalaya* (INDEPTH I) reflections from Moho peter out at depths > 70 km (ZHAO *et al.*, 1993).
8. Crustal thickness reaches more than 70 km in the northern Himalaya and southern Tibet (INDEPTH II; NELSON *et al.*, 1996).

Possible Explanations of Crustal Reflectivity by Rheological and Tectonic Models

It is well known that heatflow density values q_0 and temperature T show an anticorrelation with crustal age: the older the crust, the lower are q_0 and T (CHAPMAN, 1985; ČERMAK, 1993). Processes such as crustal growth, metamorphism or intrusions are associated with heat transfer from the mantle to the crust. Such a thermal peak can be identified, for example, with the ascent of a mantle plume or a delamination. We define delamination (often wrongly called "continental subduction") as a decoupling process in the ductile lower crust by which units of the lowermost crust plus upper mantle become unstable and descend (ENGLAND and HOUSEMAN, 1989; MEISSNER and MOONEY, 1998, and references therein). This onset of delamination, mainly during compression, is later followed by the ascend of magma from the mobilization of the asthenosphere (mature delamination).

The crustal age of stable old and cold shields and cratons correlates with crustal thickness. Except for recent zones of convergence, young crusts are thinner than old crusts, a relationship which is not well understood. It might be speculated that possibly the percentage of outpouring partial melts in old cratons was higher than today, or that processes of compressional accretion and convergence were stronger and post-orogenic extension was smaller during ancient crust forming processes.

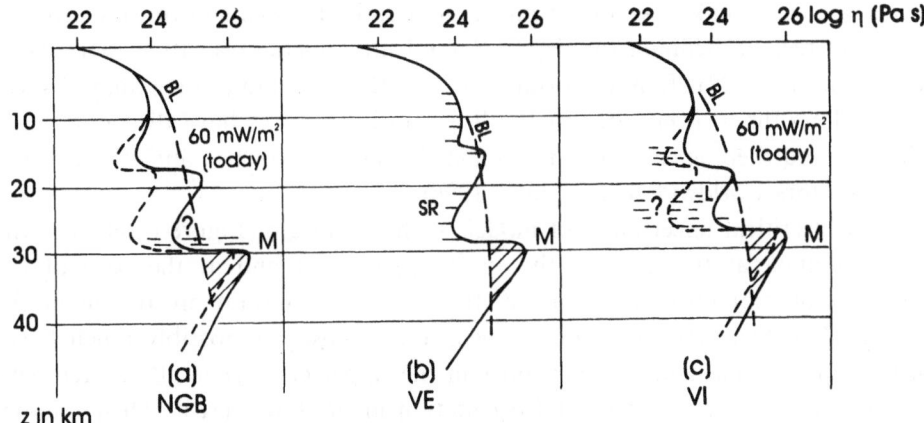

Figure 6

Viscosity depth curves for: (a) the North German Basin with its mafic lower crust; heatflow of 90 mW/m² assumed for time of origin, 60 mW/m² assumed for today. (b) the Variscan Externides (60 mW/m²). (c) the Variscan Internides with their lamellae; heatflow of 90 mW/m² assumed for time of origin, 60 mW/m² assumed for today. Crustal structure from the various DEKORP studies; M = Moho, L = lamellae, SR = single reflectors, BL = Byerlee's "Law" (BYERLEE, 1978). Smooth curves from Weertman's "creep law" (WEERTMAN, 1970) with correction from OHNAKA (1995). Parameters in the equation were adapted to crustal velocities according to Meissner (1996).

$$\tau = \tau_0 + \mu^* \sigma \tag{1}$$

where τ = shear strength, τ_0 = strength at zero-pressure, μ^* frictional coefficient, σ = confining pressure related to depth.

$$\varepsilon = C_n \tau^n \exp[-(E_c^*/RT)], \quad \text{or} \quad \ln \eta = (1/\eta)[E_c^*/RT) + (1-n) \ln C_n] \tag{2}$$

where: C_n = a material constant: E activation energy; R = gas constant; T = temperature in °K; n = exponent $(1, \ldots, 4)$; τ = shear stress; ε = creep rate.

From the varied geotherms different models of crustal and lithospheric rheology have been derived (SIBSON, 1982; MEISSNER and STREHLAU, 1982). These models consist of pressure-dependent brittle sections in the upper crust and sometimes in the uppermost mantle, according to Byerlee's relationship (BYERLEE, 1978), and of temperature- and strain rate-dependent ductile sections in the middle and lower crusts and in the deeper mantle, according to Weertman's power-law creep equations (WEERTMAN, 1970).

Figure 6 presents some simple models for different q_0-values, corresponding to different crustal ages. Various modifications of these simple models have been discussed (HANDY, 1989; RUTTER, 1999; MEISSNER and MOONEY, 1998). The transition zone between the upper brittle and the lower ductile regime seems to be rather smooth (OHNAKA, 1995; KARATO et al., 1986). A pore pressure larger than hydrostatic will modify and deepen the brittle part of the curves (MEISSNER and WEVER, 1992). Rupture processes and the formation of shear and fault zones in the upper brittle crust is a general phenomenon. Moderate to strong tectonic stresses will generate earthquakes, and their maximum depth has been shown to depend on the thermal regime, defining the depth of the brittle-ductile transition (MEISSNER and STREHLAU, 1982). It is important to note that old fault zones may survive extremely long times, even after the fault-generating stresses have died out. Their mechanical and rheological structure remains largely undisturbed and still provides strong reflectors worldwide (MEISSNER, 1996).

The most critical uncertainty of models such as those of Figure 6 refers to the viscosity minima at the base of the sialic upper crust and at the base of the (assumed gabbroic) lower crust. Here, the use of a constant creep rate in the equations (Fig. 6) is certainly not justified, and viscosity is possibly much lower than indicated in the figure. Shear zones in the upper crust generally flatten with depth and assume a near-horizontal orientation in the lower crust. Their internal weakness adds to that of the warm lower crust, resulting in a bulk viscosity considerably lower than indicated in the simple models such as Figure 6. Creep rate will increase under any nonvanishing tectonic stress regime, enhancing an ordering process creating mineral alignment and/or (metamorphic) layering. Especially during or shortly after the time of a heat peak, ample subhorizontal flow or creep will take place. As we must consider the reflecting lamellae as being composed of thin layers with strongly contrasting impedance, e.g., sialic or even pelitic material versus mafic or even ultramafic splinters, the different rheological behavior (creeping or flowing sialic material, ordering the mafic splinters) seems to be a valid explanation for generating the reflecting lamellae of the lower crust in young and warm areas.

This suggestion for the generation of reflecting lamellae is not new (MOONEY and MEISSNER, 1992). It might involve magmatic additions from the mantle (SINGH and McKENZIE, 1993) instead of substantial sialic contribution, however in this case cooling and solidification of magma will soon reduce impedance contrasts. It is more likely that mafic magmas will mobilize sialic, granulitic, dioritic or even metasedimentary material. For an ordering process of the inhomogeneous material some stress and strain are required, and extensional processes with the advection of magma or heat are the preferred candidates because they support the mechanical weakening of the crust. The observation of lower crustal lamellae, preferably in areas where the last tectonic process was associated with a strong heat pulse (Variscan internides) or at least extensional (the various Mesozoic basins around Great Britain, BIRPS), seems to confirm the thermo-rheological explanation of the seismic lamellae (MOONEY and MEISSNER, 1992).

As mentioned earlier, laminated reflecting patterns could also be caused by metasediments which were emplaced into the deep crust during a previous orogeny. In this case the underlying mafic material of the former (orogenic) crustal root could have been correlated with a delamination process, the necessary instability being partly caused by a transition of gabbroic into eclogitic facies in an overthickened crust (ENGLAND and HOUSEMAN, 1989). The new (rejuvenated) Moho might be flat and shallow, as in the Variscan internides, and might represent a type of decoupling unit (KAY and MAHLBURG-KAY, 1993). Also, mylonitic shear zones and the involvement of pore fluid are suggested, although most granulites from the lower crust seem to be dry (DOWNES, 1993; YARDLEY and VALLEY, 1997).

Regardless of the reason for the appearance of laminated reflection patterns and its subhorizontal layering, reflection imaging alone might not discriminate between the various genetic possibilities. However, any kind of layering should cause seismic anisotropy, and hence special anisotropy experiments could provide further information. Seismic anisotropy is connected with geological ordering processes causing structural order of rocks, such as sedimentation or ductile flow. From the petrological point of view, anisotropy may be connected with finely layered rocks, aligned fractures, and—most important for deep crustal and mantle rocks—rock fabrics, namely the lattice-preferred orientation of highly anisotropic minerals, such as biotite, hornblende, and olivine (BACKUS, 1962; CRAMPIN, 1989; WERNER and SHAPIRO, 1998; WEISS et al., 1999). In order to verify seismic anisotropy of the lower crust, near-vertical seismic reflection profiling must be complemented by special wide-angle experiments. Preferably the compressional and shear-wave reflections from the crust-mantle boundary ($P_M P$ and $S_M S$, respectably) penetrate the lowermost crust. Seismic anisotropy experiments must take into account velocity variations with both azimuth and dip of the ray paths. In anisotropic rocks, shear waves split into two orthogonally polarized wave trains travelling with different velocities (S-wave birefringence or splitting). Since this wave splitting is a reliable indicator of anisotropy, appropriate experiments should focus on recording $S_M S$ arrivals with three-component receivers.

A clear example of split $S_M S$, arrivals, indicating lower crustal anisotropy, is shown in Figure 7 (RABBEL and LÜSCHEN, 1996). It was recorded at the Urach geothermal anomaly where the lamellae are especially strong, plane, and long (see Fig. 2). Both reflectivity pattern and the degree of anisotropy strongly resemble those of metapelite layers found in the exposed lower crust of Ivrea and Calabria, Italy. The suggestion that the Urach lamellae represent metasediments is further supported by local xenoliths. Also the finding that magmatic intrusions, even if they are layered, are considerably less anisotropic than found at the Urach site, supports this thesis (RABBEL et al., 1998). A differentiated overview of the anisotropy of lower crustal rock candidates and as regards the problem of identifying them are provided by WEISS et al. (1999) and BOHLEN et al. (1999).

The observation that the mantle is generally free of any lamellae or similar reflectivity, and that the base of crustal reflectors correlates with the Moho, might also be explained by our rheological concept. The strong viscosity jump from crust to mantle, which is a consequence of the large difference in activation energy between crustal and mantle rocks, makes the uppermost mantle substantially stronger than the lower crust. Creep and related ordering processes are not possible in a brittle material. Moreover, mineralogy of mantle rocks is rather uniform; the mineral olivine plays a dominant role compared to about ten different minerals and rock units in the crust. The appearance of single, mostly dipping, reflectors in the uppermost mantle seems to be restricted to areas which never suffered from a strong heat pulse (SNYDER and FLACK, 1990; KRISHNA et al., 1996). Remnants of orogenic compressional processes, such as subduction or the onset of delamination, might still be observed. The associated mantle reflectors seem to be faults and shear zones in a rigid material, similar to those in the upper crust, or they might represent an eglogitization of oceanic mafic crust from a subduction process. As faults and shear zones have a low-velocity, low-viscosity interior, the first onset of the reflection should be negative. In case of an eclogitic splinter, a positive reflection coefficient should result. Further studies must clarify this.

It should also be stressed that the widespread observation of mantle anisotropy (e.g., CHRISTENSEN, 1984) is not a contradiction to the presence or absence of mantle faults or the absence of seismic lamellae. Strong seismic anisotropy was detected rather early in the oceanic upper mantle (HESS, 1964), and also continental upper mantle often shows some degree of anisotropy (BAMFORD et al., 1979; FUCHS, 1983). In the oceanic environment the fastest velocities are always parallel

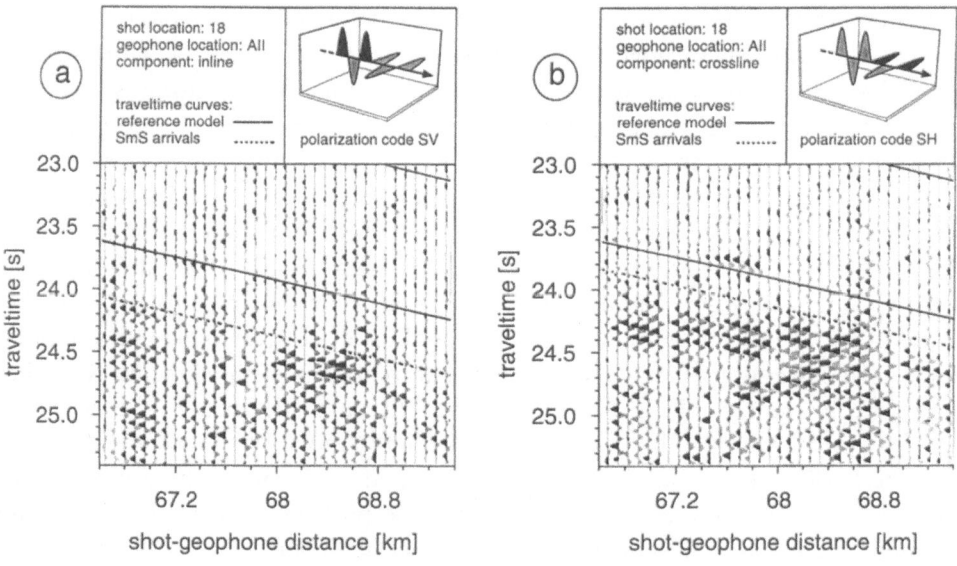

Figure 7

Shear-wave splitting of reflections from the Moho ($S_M S$) at the area of the Urach geothermal anomaly (SW Germany). Figures 7 (a) and (b) show readings of horizontal geophones oriented parallel ("inline") (a) and perpendicular ("crossline") (b) to the profile. The arrivals in (a) and (b) are reflected SV- and SH-type waves, polarized in the vertical and horizontal planes, respectively. Note the different arrival times of SV- and SH-type waves, indicated by dashed lines in both diagrams. The time difference (= shear-wave splitting) is approximately 250 ms. The solid lines correspond to $S_M S$—arrivals expected for an isotropic lower crust. For details see RABBEL and LÜSCHEN (1996) and RABBEL et al. (1998).

to the paleoplate spreading direction, and crystallographic axes of minerals (here olivine) are aligned or realigned due to tectonic stress in the warm and deformable environment near the spreading axes. A similar ordering process for the olivines during a thermal peak must be responsible for the weak anisotropy of continental mantle. Any faulting must be restricted to time intervals with cold and brittle thermo-rheological conditions, and the survival of the faults depends on the subsequent thermal history.

The thermo-rheological concept, applied to mantle anisotropy and reflectors, also explains another observation for the Variscan interior, which suffered from a very strong heat pulse during the collapse of the mountain ranges. No mantle reflectors are observed anywhere, in spite of some special experiments with strong explosive charges and long recording times (BARTELSEN et al., 1982). Certainly during the Variscan orogeny deep crustal roots and subducting slabs were formed, as in any convergence zone (BEAUMONT et al., 1994). However, the strong heat pulse toward the end of the collision eliminated all structures in the mantle and later changed the former crustal root into a new and rejuvenated, flat and shallow Moho. There are many arguments for a Variscan delamination toward the end of the orogeny, mainly the peak of granitic magmatism, the HT/LP metamorphism at about 330 Ma and the complete disappearance of the mountain root (MEISSNER and MOONEY, 1998). A delamination involves a mobilization of the asthenosphere with rising magmas, carrying, considerable heat. It could explain the missing mantle faults, the mobilization of so many granites in the Variscan internides, and is possibly associated with the generation of mafic-granulitic-metapellitic lamellae. The final re-equilibration and the formation of a flat Moho might have occurred during a later extension by a combination of far-field stresses and gravitational instabilities (ZIEGLER, 1992; HENK, 1997).

In the various *Paleozoic and Mesozoic basins around the British Isles* many extensional structures, including an upcoming of the Moho, are observed (MATTHEWS and the BIRPS GROUP, 1990). Their generation was possibly accompanied by heat pulses, which supported the generation of extensive lower crustal lamellae. The upper crust atop the lamellae appears "transparent" except for former compressional features like the North Variscan Deformation Front in the Irish Sea (BIRPS and ECORS, 1986). Extensional faults are mostly near-vertical and cannot be observed easily by conventional seismic methods.

In the *North German Basin* the condition is more complex. There is no strong upcoming of the Moho, and the zone of lamellae above it is not pronounced, being only about 2 to 3 km thick (DEKORP—BASIN RESEARCH GROUP, 1999). From gravity studies and some mid-crustal reflections one must conclude that massive (Permian) intrusions have occurred, not only in the sediments (where they are well known), but also in the middle crust (SCHECK, 1997). As there are about 10 km of sediments and a rather shallow Moho, the crystalline crust is only 15–20 km thick on average, consequently 1/3 of a normal crustal thickness (about 30 km) has disappeared. This means that there must have been a massive extension with an outward "flow" of crustal material, at least in the W to SW part of the basin, while heavy mafic intrusions in the Permo-Carboniferous preferred the eastern parts of the basin. Compared to other extensional areas such as the Basin and Range province, the Paris Basin or the many basins in the Irish Sea and around the British Isles, the North German Basin lacks the presence of strong lower crust lamellae. Presently we can only speculate that the massive intrusions of mafic magmas and the displacement of former material of the lower crust are responsible for this observation. Whether the magmas were mobilized by an early stretching event, by a plume in the Permo-Carboniferous, or by a previous delamination (from the northern Variscides), or by some combination is an open question.

In order to understand the reflectivity of *old and cold shields and platforms* (ALLMENDINGER *et al.*, 1987; BABEL WORKING GROUP, 1991), we have to consider the long cooling time which has finally resulted in the observed low heat-flow values, typical for old and cold crusts. Any previous lower crustal lamellae were transferred into the brittle regime, and tectonic stresses might have deformed or destroyed them. Today one observes a reflective crust, sometimes with enormous details and spectacular resolution (COOK *et al.*, 1997; NEMETH and HAJNAL, 1998), but without the long and strong lamellae of young extensional regimes. It seems that the absence of any heat pulse in the post-orogenic history is a guarantee for high resolution images of old compressional events. Even an Early-Proterozoic subduction of 2.2 Ga could be seismically imaged in the Gulf of Bothnia (BABEL WORKING GROUP, 1990), and an accumulation of three Late-Proterozoic terranes is observed northwest of Bornholm by its strong, long and diverging reflectors, representing prominent compressional shear zones in the whole crust (ABRAMOVITZ *et al.*, 1997).

As mentioned earlier, crustal thickening and shortening is observed in *young collisional belts*. In addition to compressional uplift and the formation of nappes, strong indentations are observed in the middle and lower crusts, and the formation of asymmetric crustal roots is evident, especially in the Alpine belt (PFIFFNER *et al.*, 1988). These indentations and interfingering of hard Adriatic mantle rocks are marked by rather strong reflections from the shear zones around the indenter, which intrudes the weak European middle and lower crusts. The European lower crust, exhibits some lamellae before it is partly guided down together with the underlying mantle, partly serving as a decoupling zone for the onset of delamination. In the upper and middle crust the complex nature of collision is represented by many bands of reflections.

Moho reflections in the southern Himalaya are rather weak and display only a few lamellae; however all Moho reflections disappear below the Central Himalaya and southern Tibet (ZHAO *et al.*, 1993). Refraction work and seismological receiver functions indicate that the Moho here is deeper than 60 km. At this depth the phase transition from gabbroic to eclogitic facies will take place. It may be a rather smooth transition (MENGEL and KERN, 1992) expressed by a velocity-gradient zone with no sharp impedance contrasts. This could explain the missing near-vertical reflections from the Moho. In southern Tibet the situation is even more complex because the middle and lower crusts are extremely hot and ductile. There are at least four magma chambers around 20 km depth, marked by strong clusters of reflections (NELSON et al., 1996). In contrast, the intruding Indian mantle penetrating the Asian crust is cold and rigid. It has reached Tibet, as seen by high P_n velocities, by a strong anisotropy and by earthquakes between 80 and 100 km depth (BEGHOUL et al., 1993). However, no near-vertical reflections are observed from the deep crust-mantle boundary.

Discussion and Conclusion

It has been shown that the thermo-rheological concept for the generation of reflecting interfaces may explain various types of reflectivity at different depths. The rigid upper crust, also called the seismogenic zone, contains fault and shear zones which often provide excellent reflectors because of their low-velocity, low-viscosity interior, as generated by various weakening processes during rupture (MEISSNER, 1996). These zones might survive the active seismic period and retain some of their mechanical properties, especially their mineral orientation and diminution, although some recrystallization may take place. They generally provide rather short reflecting signals, sometimes over several kilometers distance. It seems that the upper crust preserves the memory of ancient tectonic processes, at least that of the latest tectonic event. There is never a lamellae-type reflectivity in the rigid upper crust or in the upper mantle.

Changes of the stress system will have interesting consequences. Small changes of direction or changes around 180° also use the same former rupture zone for the new rupture, as seen by many inversion and flower structures in sediments (DEEKS and THOMAS, 1995). This concentration of a large range of stress directions into one perhaps prominent fault zone, might explain the limited number of faults in the upper crust, an important consequence for their detectability.

The rheological behavior cf the lower crust differs strongly from that of the upper crust. The lower, crust is not seismogenic, except for some very large rupture processes penetrating the lower crust, and its ductility has been mentioned several times in this study. It has a weak matrix; shear zones from the upper crust often turn subhorizontal when approaching the lower crust, and during heat pulses the matrix becomes very weak. The change from the Variscan mountain roots to a flat and shallow, rejuvenated lower crust is the most dramatic example of the mobility of the lower crust. The change of stress from compression to extension, which accompanies this process, seems to play a major role also in the generation of the lower crustal lamellae. At the Urach site their anisotropy is best explained by alternating layers of metasediments, with mineral alignment in a low-viscosity matrix. Only a directed flow or creep can create the densely packed sequence of layers with strongly contrasting impedance and contrasting viscosities. No contrasting rock units and no lamellae are found in the mantle which however indicates anisotropy resulting from aligned olivine crystals, and some reflections from fault zones.

The lower crustal lamellae, the occurrence of which is correlated with previous heat pulses during extension, can also be destroyed. This might take place preferably in a new thermal process, possibly involving intrusions, however also by a long cooling process, transforming the ductile regime into a brittle environment, a process which may have taken place in old shields and platforms. Here, reflectivity generally covers the entire brittle crust, and collision of terranes and old subduction zones is observed with surprising clarity. One condition for the long survival of tectonic structures, as mentioned repeatedly, is that no new thermal event has taken place since their formation. The Moho is always the last reflecting band, also in these ancient tectonic units, although some isolated traces of a former collision in the upper mantle might also be observed.

Young collisional processes with crustal thickening and shortening provide the most spectacular reflection images, often closely resembling pictures of ancient compression except for a pronounced crustal root. Architecture and evolution can be seen by following the reflecting shear zones. Especially features at mid-crustal depth (the indentations of hard into weak layers), or in the lower crust (the asymmetric root zones), some containing lamellae can be followed and reveal the internal structure of particular mountain belts.

Certainly the thermo-rheologic concept of reflectivity is not yet complete. However, based on a wide range of observations in various geological units, the general concept seems to be uniform, widely applicable, and not contradictory. Future observations will confirm or negate the concept.

Acknowledgements

The authors gratefully acknowledge funding by the German Federal Ministry of Research BMBF grant 03GT9410. Two anonymous reviewers helped to improve the article significantly.

REFERENCES

ABRAMOVITZ, T., BERTHELSEN, A., and THYBO, H. (1997), *Proterozoic Sutures and Terranes in the Southwestern Baltic Shield, Interpreted from BABEL Deep Seismic Data*, Tectonophysics 270, 259–278.
ABRAMOVITZ, T., THYBO, H., MONA LISA WORKING GROUP (1998), *Seismic Structure across the Caledonian Deformation Front along the MONA LISA Profile 1 in the Northeastern North Sea*, Tectonophysics 288, 153–176.

ALLMENDINGER, R. W., NELSON, K. D., POTTER, C. J., BARANZANGI, M., BROWN, L. D., and OLIVER, J. E. (1987), *Deep seismic reflection Characteristic of the Continental Crust*, Geology *15*, 304–310.

BABEL WORKING GROUP (1990), *Early Proterozoic Plate Tectonics: Evidence from Seismic Reflection Profiles in the Baltic Shield*, Nature *384* (6296), 34–38.

BABEL WORKING GROUP (1991), *Deep Seismic Images from the Tornquist Zone Beneath the Southern Baltic Sea*, Geophys. Res. Lett. *18*, 1091–1094.

BABEL WORKING GROUP (1993), *Deep Seismic Reflection/Refraction Interpretation of Crustal Structure along BABAl Profiles A and B in the Southern Baltic Sea*, Geophys. J. Int. *112*, 325–343.

BACKUS, G. E. (1962), *Long-wave Elastic Anisotropy Produced by Horizontal Layering*, J. Geophys. Res. *67*, 753–770.

BAMFORD, D., JENTSCH, M., and PRODEHL, C. (1979), P_n *Anisotropy Studies in Northern Britain and the Eastern US*, Royal Astron. Soc. Geophys. J. *57*, 387–429.

BARTELSEN, H., LUESCHEN, E., KREY, T., MEISSNER, R., CHSMOLL, H., and WALTER, C., *The combined seismic reflection—refraction investigation of the Urach geothermal anomaly*. In *The Urach Geothermal Project* (ed. Haenel, R.) (Schweizerbart, Stuttgart 1982) *36*, pp. 501–518.

BEGHOUL, N., BARANZANGI, M., and ISACKS, B. L. (1993), *Lithospheric Structure of Tibet and Western North America: Mechanism of Uplift and a Comparative Study*, J. Geophys. Res. *98 B2*, 1997–2016.

BEAUMONT, C., FULLSACK, P., and HAMILTON, J. (1994), *Styles of Crustal Deformation in Compressional Orogens Caused by Subduction of the Underlying Lithosphere*, Tectonophysics *232*, 119–132.

BEZDAN, S., and HAJNAL, Z. (1998), *Expanding Spread Profiles across the Trans-Hudson Orogen*, Tectonophysics *288*, 83–91.

BIRPS and ECORS (1986), *Deep Seismic Reflection Profiling between England, Ireland, and France (SWAT)*, J. Geol. Soc., London *143*, 45–52.

BOHLEN, T., RABBEL, W., WEISS, T., SIEGESMUND, S., and POHL, M. (1999), *Recovering Shear-wave Anisotropy of the Lower crust: The Influence of Systematic Errors on Traveltime Inversion*, Pure appl. geophys. *156*, 123–138.

BOIS, C., and 8 others, *Deep seismic profiling of the crust in northern France: The ECORS Project*. In *Reflection Seismology: A Global Perspective* (eds. Barazangi, M., and Brown, L. D.), (Am. Geophys. Union, Geodyn. Ser. *13*, 1986) pp. 21–30.

BROWN, L. D. (1991), *A New Map of Crustal "Terranes" in the United States from COCORP Deep Seismic Reflection Profiling*, Geophys. J. Inst. *105*.

BYERLEE, J. D. (1978), *Friction of Rocks*, Pure appl. geophys. *116*, 615–626.

ČERMAK, V. (1993), *Lithospheric Thermal Regimes in Europe*, Phys. Earth Planet. Int. *79*, 79–193.

CHAPMAN, D. S., *Continental heat flow data*. In *Physics of the Solid Earth, the Moon and the Planets* (eds. Fuchs, K., and Soffel, H.) Landolt Börnstein, Newseries, 2, Subvo.b (Springer, Berlin 1985) pp. 1–19.

CHEADLE, M., MCGEARY, S., WARNER, M. R., and MATTHEWS, D. H. (1987), *Extensional structures on the UK continental shelf: A review*. In *Continental Extension Tectonics* (eds. Coward, M., Dewey, J. F., and Hancock, P. L.). Geol. Soc. London, Spec. Publ. *28*, 445–465.

CHRISTENSEN, N. I. (1984), *The Magnitude, Symmetry and Origin of Upper Mantle Anisotropy Based on Fabric Analyses of Ultramafic Tectonites*, Royal Astron. Soc. Geophys. J. *76*, 89–111.

COOK, F. A., VAN DER VELDEN, A., HALI, K., and ROBERTS, B. R. (1997), *Lithoprobe SNORCLE-ing Beneath the Northwestern Canadian Shield*, Lithoprobe Res. Notes *10* (1), 59–64.

CRAMPIN, S. (1989), *Suggestions for a Consistent Terminology for Seismic Anisotropy*, Geophys. Prosp. *37*, 753–770.

DEEKS, N. R., and THOMAS, S. A. (1995), *Basin inversion in a strike slip regime: The Tornquist Zone, southern Baltic Sea*. In *Basin Inversion* (eds. Buchanan, B., and Buchanan, C.), Geol. Soc. Spec. Publ., *88*, 319–338.

DEKORP RESEARCH GROUP (1990), *Reflectivity Patterns in the Variscan Mountain Belts and Adjacent Areas*, Tectonophysics *173*, 361–378.

DEKORP RESEARCH GROUP (1991), *Results from DEKORP 1 Deep Seismic Reflection Studies in the Western Part of the Rhenish Massif*, Geophys. J. Int. *106*, 203–227.

DEKORP BASIN RESEARCH GROUP (1999), *The North German Basin and its Development*, Pure ppl. geophys. *27*, 55–58.

DOWNES, H. (1993), *The Nature of the Lower Continental Crust of Europe: Petrological and Geochemical Evidence from Xenoliths*, Phys. Earth Planet. Int. *79*, 195–218.

DUYSTER, J., and 15 co-authors, *The lithological profile of the KTB-Hauptbohrung, 6000–7200 m*. In (eds. Emmermann, R., Lauterjung, J., and Umsonst, T.), KTB-report *93-2*; (Schweizerbart, Stuttgart 1993) pp. 15–75.

ENGLAND, P., and HOUSEMAN, G. (1989), *Extension during Continental Convergence with Application to the Tibet Plateau*, J. Geophys. Res. *94*, 17,561–17,579.

ECORS PYRENEES TEAM (1988), *The ECORS Deep Reflection Seismic Survey across the Pyrenees*, Nature *331*, 508–511.

FRANKE, W., *Phanerozoic structures and events in Central Europe*. In *The European Geotraverse* (eds. Blundell, D., Freeman, R., and Mueller, S.) (Cambridge University Press, Cambridge 1992) pp. 164–179.

FUCHS, K. (1983), *Recently Formed Elastic Anisotropy and Petrological Models for the Continental Subcrustal Lithosphere in Southern Germany*, Phys. Earth Planet. Int. *31*, 93–118.

FUIS, G. S., and CLOWES, R. M. (1993), *Comparison of Deep Structure along Three Transects of the Western North American Continental Margin*, Tectonics *12* (6), 1420–1435.

GEBRANDE, H., BOPP, M., MEICHELBÖCK, M., and NEUROEDER, P., *3-D wide-angle investigation in the KTB surroundings, first results*. In *Continental Lithosphere: Deep Seismic Reflections* (eds. Meissner, R., Brown, L., Dürbaum, H. J., Franke, W., Fuchs, K., and Seifert, F.) (AGU Washington, D.C. 1991) pp. 147–160.

GODWIN, E. B., and THOMPSON, G. A. (1988), *The Seismically Reflective Crust beneath Highly Extended Terraines: Evidence for its Origin in Extension*, Geol. Soc. Am. Bull. *100*, 1616–1626.

HANDY, M. R. (1989), *Deformation Regimes and the Rheological Evolution of Fault Zones in the Lithosphere*, Tectonophysics *163*, 119–152.

HEITZMANN, P., FREI, W., LEHNER, P., and VALASEK, P., *Crustal indentation in the Alps—An overview of reflection seismic profiling in Switzerland*. In *Continental Lithosphere* (Meissner, R., Brown, L., Dürbaum, H. J., Franke, W., Fuchs, K., and Seifert, F.) (AGU Geodynamic Series, *22*, AGU, Washington 1991).

HENK, A. (1997), *Gravitational Orogenic Collapse vs. Plate Boundary Stresses: A Numerical Modelling Approach to Permo-Carboniferous Evolution of Central Europe*, Geol. Rundschau *86*, 39–55.

HESS, H. (1964), *Seismic Anisotropy in the Uppermost Mantle under Oceans*, Nature *203*, 629–631.

JUHLIN, C. (1990), *Interpretation of the Reflections in the Siljan Ring Area Based on Results from Gravberg-1 Borehole*, Tectonophysics *173*, 345–360.

KARATO, S., PATERSON, M., and FITZGERALD, J. (1986), *Rheology of Synthetic Olivine Aggregates: Influence of Grainsize and Water*, J. Geophys. Res. *91*, 8151–8176.

KAY, W., and MAHLBURG-KAY, S. (1993), *Delamination and Delamination Magmatism*, Tectonophysics *219*, 177–189.

KLÄSCHEN, D., and FLÜH, E., *Reprocessing of BABEL Line 1 normal incidence data*. In *The BABEL Project* (eds. Meissner, R., Blundell, D., Snyder, D., and McBride, J.) (European Commission, Brussels 1996) pp. 61–76.

KLEMPERER, S. L., and MATTHEWS, D. H. (1987), *Iapetus Suture Located beneath the North Sea by BIRPS Deep Seismic Reflection Profiling*, Geology *15*, 195–198.

KRISHNA, V. G., MEISSNER, R., and THOMAS, S. A. (1996), *Unusual sub-Moho Events near Bornholm, Baltic Sea: Modelling of BABEL Seismic Wide-angle Data and Tectonic Interpretation*, Geophys. J. Int. *125*, 193–198.

MATTHEWS, D. H., and the BIRPS GROUP (1990), *Progress in BIRPS Deep Seismic Profiling around the British Isles*, Tectonophysics *173*, 387–396.

MCGEARY, S., and WARNER, M. (1985), *Seismic Profiling of the Continental Lithosphere*, Nature *317*, 795–797.

MEISSNER, R., and STREHLAU, J. (1982), *Limits of Stresses in Continental Crusts and their Relationship to the Depth-frequency Distribution of Shallow Earthquakes*, Tectonics *1*, 73–89.

MEISSNER, R., and BORTFELD, R. K. (eds.), *The DEKORP Atlas* (Springer Verlag, Berlin 1990) (80 seismic sections).

MEISSNER, R., and BROWN, L. D. (1991), *Seismic Reflections from the Earth's Crust. Comparative Studies of Tectonic Patterns*, Geophys. J. Int. *105*, 1–2.

MEISSNER, R., and SADOWIAK, P. (1992), *The Terrane Concept and its Manifestation by Deep Reflection Studies in the Variscides*, Terra Nova 4, 598–607.

MEISSNER, R. (1996), *Faults and Folds, Fact and Fiction*, Tectonophysics 264, 279–293.

MEISSNER, R., and WEVER, T. (1992), *The Possible Role of Fluids for the Structuring of the Continental Crust*, Earth Sci. Rev. 32, 19–32.

MEISSNER, R., and MOONEY, W. D. (1998), *Weakness of the Lower Continental Crust: A Condition for Delamination, Uplift, and Escape*, Tectonophysics 296, 47–60.

MENGEL, K., and KERN, H. (1992), *Evolution of the Petrological and Seismic Moho—Implications for the Continental Crust-mantle Boundary*, Terra Nova 4, 109–116.

MONA LISA WORKING GROUP (1997), *MONA LISA—Deep Seismic Investigation of the Lithosphere in the Southeastern North Sea*, Tectonophysics 269, 1–19.

MOONEY, W. D., and MEISSNER, R., *Multi-genetic origin of crustal reflectivity: A review of seismic reflection profiling of the continental lower crust and Moho*. In *Continental Lower Crust* (eds. Fountain, D. M., Arculus, R., and Kay, R. W.) (Elsevier, Amsterdam 1992) pp. 45–79.

MUSACCHIO, G., MOONEY, W. D., LUETGERT, J. H., and CHRISTENSEN, N. I. (1997), *Composition of the Crust in the Grenville and Appalachian Provinces of North America Inferred from V_P and V_S Ratios*, J. Geophys. Res. 102 (B7), 15,225–15,241.

NEMETH, B., and HAJNAL, Z. (1998), *Structure of the Lithospheric Mantle beneath the Trans-Hudson Orogen, Canada*, Tectonophysics 288, 93–104.

NELSON, K. D., and 24 coauthors, (1996), *Partially Molten Middle Crust beneath Southern Tibet: Synthesis of Project INDEPTH Results*, Science 274, 1684–1696.

OHNAKA, M. (1995), *A Shear Failure Strength Law of rocks in the Brittle-plastic Regime*, Geophys. Res. Lett. 22 (1), 25–28.

PFIFFNER, O. A., FREI, W., FINCK, P., and VALASEK, P. (1988), *Deep Seismic Reflection Profiling in the Swiss Alps: Explosion Seismology Results for Line NFP 20-East*, Geology 16, 987–990.

RABBEL, W. and LÜSCHEN, E. (1996), *Shear-wave Anisotropy of Laminated Lower Crust at the Urach Geothermal Anomaly*, Tectonophysics 264, 219–233.

RABBEL, W., SIEGESMUND, S., BOHLEN, TH., POHL, M., and WEISS, T. H. (1998), *Shear-wave Anisotropy of Laminated Lower Crust beneath Urach (SW Germany)—A Comparison with Exposed Lower Crustal Sections*, Tectonophysics 298, 337–356.

RESTON, T. J. (1987), *Spatial Interference, Reflection Character and the Structure of the Lower Crust under Extension, Results from 2-D Seismic Modelling*, Ann. Geophys. Ser. B. xx: 339–348.

RESTON, T. J. (1988), *Evidence for Shear Zones in the Lower Crust Offshore Britain*, Tectonics 7, 929–945.

RESTON, T. J. (1990), *Shear in the Lower Crust During Extension: Not so Pure and Simple*, Tectonophysics 173, 175–183.

REY, P. (1993), *Seismic and Tectono-metamorphic Characters of the Lower Continental Crust in Phanerozoic Areas: A Consequence of Post-thickening Extension*, Tectonics 12 (2), 580–590.

RUTTER, E. H. (1999), *On the Relationship between the Formation of Shear Zones and the Form of the Flow Law for Rocks*, Tectonophysics, submitted.

SADOWIAK, P., MEISSNER, R., and BROWN, L. D. (1991), *Seismic reflectivity patterns: comparative investigations of Europe and North-America*. In *Continental Lithosphere: Deep Seismic Reflections* (eds. R. Meissner et al.) (Am. Geophys. Geodyn Ser. 33, 1991) pp. 363–369.

SIBSON, R. M. (1982), *Fault Zone Models, Heat Flow, and the Depth Distribution of Earthquakes in the Continental Crust of the United States*, Bull. Seismol. Soc. Am. 72, 151–163.

Scheck, M. (1997), *Dreidimensionale Strukturmodellierung des Nordostdeutschen Beckens unter Einbeziehung von Krustenmodellen*, GFZ Potsdam, Scient. Techn. Report STR97/10, 1–125.

SIMON, M. (1998), *AVO Analysis by Offset-limited Prestack Migrations of Crustal Data*, Tectonophysics 286, 143–153.

SINGH, S. C., and McKENZIE, D. (1993), *Layering in the Lower Crust*, Geophys. J. Int. 113, 622–628.

SHERIFF, R. E., and GELDART, L. P., *Explosion Seismology, Vol 1* (Cambridge Univ. Press, Cambridge, England 1982) pp. 1–249.

SMITHSON, D. B. (1986), *A physical model of the lower continental crust from North America based on seismic data*. In *The Nature of the Continental Crust* (Dawson, J. B., Carswell, D. A., Hall, J., and Wedepohl, K. H.) Geol. Soc. London Spec. Publ. 24, 23–34.

SNYDER, D. B., and FLACK, C. A. (1990), *A Caledonian Age for Reflectors within the Mantle Lithosphere North and West of Scotland*, Tectonics *9*, 903–922.

STEENTOFT, H., and RABBEL, W. (1995), *Conflicting Dips and the Transparency of the Upper Crust*, Terra Nostra *9518*, 127–130.

TAIT, J. A., BACHTADSE, V., SOFFEL, H. C., and FRANKE, W. (1996), *Paleomagnetic Constraints on the Evolution of the European Variscan Foldbelt*, Geologia Bavaria *101*, 221–232.

VASEDUVAN, K., MAIER, R., GEIGER, H., and COOK, F. A. (1996), *Reprocessing of deep crustal reflections acquired in the Gulf of Bothnia*. In *The Babel Project* (eds. Meissner, R., Blundell, D., Snyder, D., and McBride, J.) (European Commission Brussels 1996) pp. 87–94.

WARNER, N. (1990), *Basalt, water, or shear zones in the lower continental crust?*, In *Seismic Probing of Continents and their Margin* (eds. Leven, J. M., Finlayson, D. M., Wright, C., Dooley, J. C., and Kennett, B. L. N.) Tectonophysics *173*, 163–174.

WARNER, M., and MCGEARY, S. (1987), *Seismic Reflection Coefficients from Mantle Fault Zones*, Geophys. J. R. Astron. Soc. *89*, 223–230.

WEERTMAN, J. (1970), *The Creep Strength of the Earth Mantle*, Rev. Geophys. Space Phys. *6*, 145–168.

WERNER, U., and SHAPIRO, S. A. (1998), *Intrinsic Anisotropy and Thin Multilayering—Two Anisotropy Effects Combined*, Geophys. J. Int. *132*, 363–373.

WEISS, TH., SIEGESMUND, S., BOHLEN, TH., POHL, M., and RABBEL, W. (1999), *Seismic Velocities in the Lower Continental Crust: An Anisotropic Perspective*, Pure appl. geophys. *156*, 97–122.

YARDLEY, B., and VALLEY, J. (1997), *The Petrological Case for a Dry Lower Crust*, J. Geophys. Res. *102*, 12,173–12,185.

ZHAO, W. J., NELSON, K. D., CHE, J., GUO, D., WU, C., LIU, X., BROWN, L. D., HAUCK, M. L., KUO, J. T., KLEMPERER, S., and MAKOWSKY, Y. (1993), *Deep Seismic Reflection Evidence for Continental Underthrusting beneath Southern Tibet*, Nature *366*, 557–559.

ZIEGLER, P. A., *Geological Atlas of Western and Central Europe*, 2nd ed. (Shell, The Hague, Netherlands 1992).

(Received February 18, 1998, revised January 21, 1999, accepted January 25, 1999)

 To access this journal online:
http://www.birkhauser.ch

Pure appl. geophys. 156 (1999) 29–52
0033–4553/99/010029–24 $ 1.50 + 0.20/0

Pure and Applied Geophysics

Scales of Heterogeneities in the Continental Crust and Upper Mantle

MARC TITTGEMEYER,[1] FRIEDEMANN WENZEL,[1] TROND RYBERG[2] and KARL FUCHS[3]

Abstract—A seismological characterization of crust and upper mantle can refer to large-scale averages of seismic velocities or to fluctuations of elastic parameters. Large is understood here relative to the wavelength used to probe the earth.

In this paper we try to characterize crust and upper mantle by the fluctuations in media properties rather than by their average velocities. As such it becomes evident that different scales of heterogeneities prevail in different layers of crust and mantle. Although we cannot provide final models and an explanation of why these different scales exist, we believe that scales of inhomogeneities carry significant information regarding the tectonic processes that have affected the lower crust, the lithospheric and the sublithospheric upper mantle.

We focus on four different types of small-scale inhomogeneities: (1) the characteristics of the lower crust, (2) velocity fluctuations in the uppermost mantle, (3) scattering in the lowermost lithosphere and on (4) heterogeneities in the mantle transition zone.

Key words: Inhomogeneous media, scattering, lower crust, upper mantle, transition zone.

Introduction

Seismologists characterize crust and upper mantle in two ways. They either specify the velocities of elastic waves, or their fluctuations.

In the first case the earth is described by a velocity-depth function, which typically shows P-wave velocities between 6.4 and 7.2 km s^{-1} in the lower crust, a jump to velocities between 8.0 and 8.3 km s^{-1} at the crust-mantle boundary, a

[1] Geophysical Institute, Karlsruhe University, Hertzstr. 16, 76187 Karlsruhe, Germany. E-mail: marc.tittgemeyer@phys.uni-karlsruhe.de
[2] GeoForschungZentrum, Telegrafenberg, 14473 Potsdam, Germany.
[3] Geophysical Institute, Karlsruhe University, Hertzstr. 16, 76187 Karlsruhe, Germany.

small gradient of increasing velocities in the uppermost mantle, a low-velocity zone between 180 and 220 km depth and another positive gradient zone above and between the discontinuities at 410 and 660 km, respectively. This velocity-depth function may change laterally, and will show modification in values for different tectonic regimes.

Velocity models for the lithosphere are usually derived from refraction seismic data. The elastic waves generated by controlled seismic sources are received at the surface along seismic lines, which extend from the source to distances of hundred, or several hundreds to thousands of km, depending upon the target. The waves that propagate through the crust and upper mantle along different paths appear as seismic phases in the seismogram sections. The travel times of those phases can be inverted for the velocity-depth structure.

Inversion of travel times from correlated phases results in a model in which velocities between interfaces are averaged over many wavelengths. In fact, the physical notion of a phase relies on ray theory which is only applicable for media with large blocks of velocities which do not change strongly on the scale of a wavelegth. The individual blocks (or layers) can be separated by sharp interfaces (AKI and RICHARDS, 1980).

Fluctuations of the medium cause wave scattering and coda effects on transmitted waves (AKI and CHOUET, 1975). Scattering analysis is widely conducted for crustal earthquakes and array records of teleseismic events, but rarely for controlled source data other than near-vertical reflection data.

To date, controlled source seismology (seismic reflection and refraction data) and long-period passive seismology provide the impression that the upper mantle is rather homogeneous. For near-vertical reflections the mantle appears transparent as compared to the reflective lower crust (e.g., BOIS and ECORS, 1991; HAMMER and CLOWES, 1997; MATTHEWS, 1986; MEISSNER and DEKORP, 1991; THOMPSON and HILL, 1986). Mantle reflections are sometimes observed (e.g., BREWER *et al.*, 1983; THOUVENOT *et al.*, 1995; WARNER and MCGEARY, 1987), but as discrete reflections rather than spatially distributed reflectivity. Refraction data allow the observation of a P_n phase which serves to define a sub-Moho velocity or velocity gradient (cf., SHERIFF and GELDHARD, 1982). Long-range observations allow for the definition of discrete velocity contrasts in the mantle, but again the models display a mantle which is mostly homogeneous on the scale of several wavelengths.

Descriptions of the earth's mantle by its velocity fluctuations rather than by its average velocity are mostly restricted to interpretation of studies of wave fluctuation beneath seismological arrays (WU and FLATTÉ, 1990), to modelling wave forms for propagation in oceanic lithosphere (e.g., GETTRUST and FRAZER, 1981; MALLICK and FRAZER, 1990) and more recently to the understanding of a high-frequency phase observed in Russian long-range data with nuclear explosions as the source (RYBERG *et al.*, 1995).

Those data from the Russian Peaceful Explosion Program (PNE) offer a unique possibility to systematically assess the scattering properties of the upper mantle from the Mohorovičić discontinuity (Moho) to the bottom of the transition zone at 670 km depth.

In this paper we endeavor to characterize crust and upper mantle not primarily by their well-known average velocities but rather by the fluctuations in media properties. In doing this it follows that different scales of heterogeneities prevail in different layers of crust and mantle. Although we cannot provide final models and explanations of why these different scales exist, we believe that scales of inhomogeneities carry significant information on the tectonic processes when they affect the lower crust, the lithospheric and the sublithospheric upper mantle.

We focus on four different types of small-scale inhomogeneities: (1) the characteristics of the lower crust, (2) velocity fluctuations in the uppermost mantle, (3) scattering in the lowermost lithosphere and on (4) heterogeneities in the mantle transition zone.

Characteristics of the Lower Crust

Important information relative to the continental lithosphere is provided by near-vertical reflections backscattered from the lithosphere along reflection seismic profiles. These data carry little information concerning the crustal velocities, but permit the generation of seismic record sections—either stacked sections or migrated sections—which provide an image of the velocity and/or density fluctuations. If the earth were as homogeneous as most models inverted from refraction seismic experiments suggest, reflection seismology would have never developed for lithospheric applications.

However, the earth's crust does contain fluctuations which can be imaged by reflection seismology (cf., MOONEY and MEISSNER, 1992). In the upper crust these reflections can be correlated with features of surface geology and represent deep sedimentary layers, volcanic intrusions, thrust and normal faults. In general faults become listric at midcrustal levels and rarely cut through the lower crust (e.g., ALLMENDINGER and COCORP, 1983; COOK et al., 1981; MATTHEWS, 1986).

The lower crust is very often characterized by numerous subhorizontal or shallow dipping reflections which seem to be randomly distributed. They terminate abruptly at a certain depth (MEISSNER, 1986; OLIVER, 1986), which usually coincides with the Moho as found from wide-angle refraction seismic data (cf., MOONEY and BROCHER, 1987). This widespread observation gave rise to the notion of a laminated (GLOCKE and MEISSNER, 1976) or layered (FOUNTAIN, 1987) lower crust.

A variety of explanations were initially discussed as the physical cause of these reflection patterns. The most widely accepted is that the layered lower crust is the result of intensive stretching (see e.g., REY, 1993) under thermodynamic conditions in which rocks deform in plastic flow (KIRBY and KRONENBERG, 1987; KUSZNIR, 1991). This flow affects material of different impedance and generates anisotropic orientation of minerals by localized deformation (e.g., LÜSCHEN *et al.*, 1990; SMITHSON, 1989, cf. POHL and WENZEL, *Realistic Models of Anisotropic Laminated Lower Crust*, this volume). This effect would produce the observed predominantly horizontal reflection pattern.

Figure 1 shows a seismic reflection record section from the central Black Forest (LÜSCHEN *et al.*, 1987). Its reflectivity is concentrated within the lower crust, the upper mantle appears as transparent, only a few reflections are discernible in the upper crust.

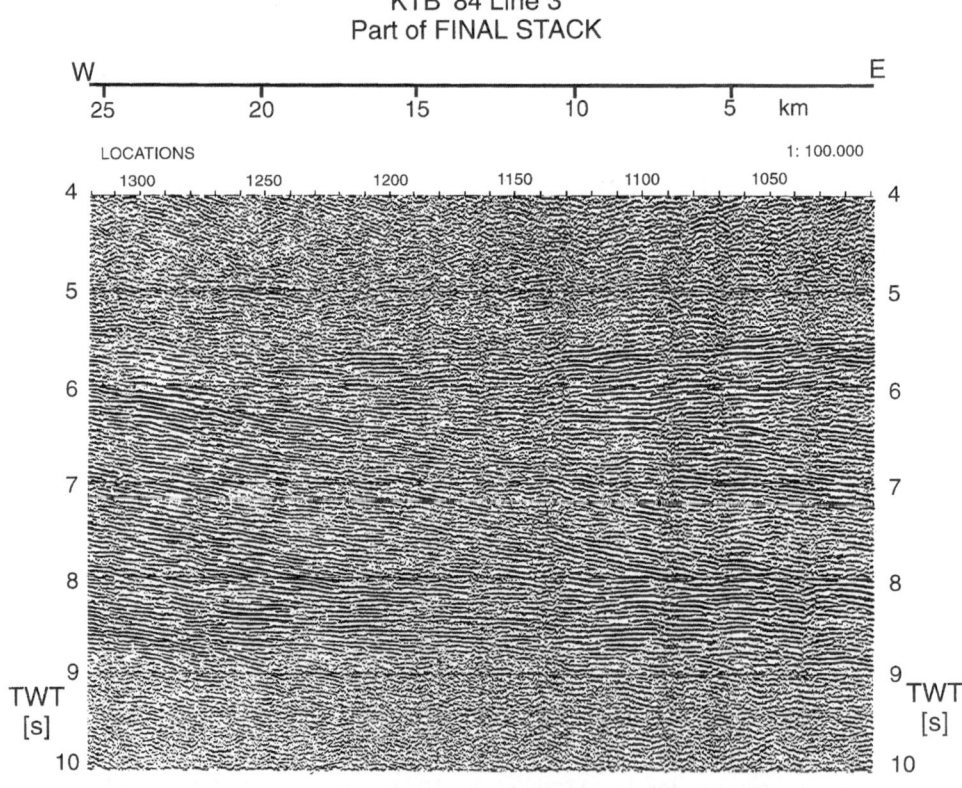

Figure 1

Lower crustal seismic lamellae as observed on a seismic reflection profile recorded in the Black Forest, SW-Germany (LÜSCHEN *et al.*, 1987). Crustal reflectivity increases abruptly at about 5.5 s two-way time (TWT) and terminates at 9 s at the Moho. The high amplitudes of the reflections require the existence of significant velocity contrasts (± 5% or larger) in the lower crust. Lower crustal seismic lamellae have been observed throughout the world with varying amplitudes, lateral continuity, and dips depending on the tectonic setting (adopted from MOONEY and MEISSNER, 1992).

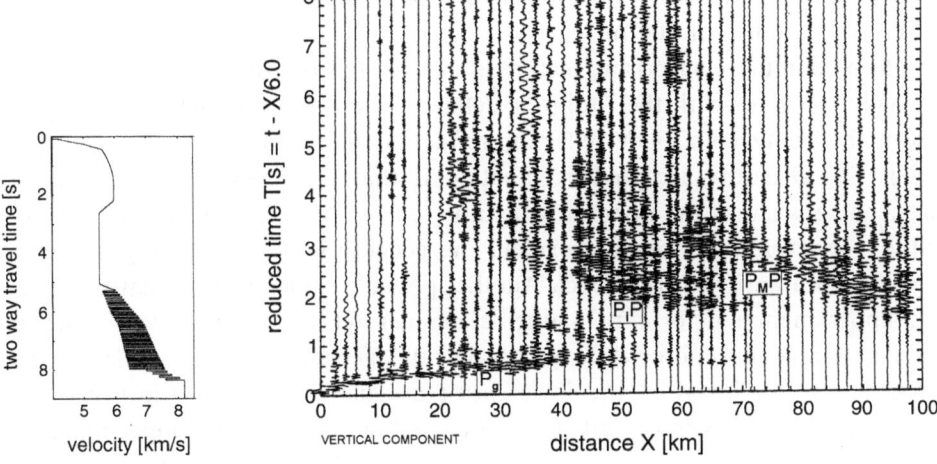

Figure 2
Record section of the refraction profile in the Black Forest. The right panel shows the vertical component. Reduction velocity is 6.0 km s^{-1}. Upper crustal phase (P_g) and mantle reflections ($P_M P$) are recognized. Data are scaled to maximum trace amplitude. The left panel shows the preferred 1-D model with strong fluctuation in the lower crust, causing the $P_i P$ coda (adopted from SANDMEIER and WENZEL, 1986).

Velocity depth distribution and reflectivity pattern carry different information. The first reflects the state of chemical differentiation of the earth and the metamorphic conditions at depth. If models derived from reflection data are grouped according to the tectonic setting, one finds characteristic differences between old shields, rifted areas, crust thickened by ongoing orogeny, volcanic arcs a.s.o. (cf., MOONEY and MEISSNER, 1992; REY, 1993).

Reflectivity of the lower crust is usually attributed to extension and/or magmatic intrusions. Extension can be a result of late to post-orogenic collapse or active or passive rifting of the lithosphere. In both scenarios lower crustal temperatures are high enough to promote plastic flow which tends to align minerals and material of different composition horizontally. The dynamic processes that affect the lower crust generate a particular scale (vertical and horizontal) that can be observed with reflection seismic experiments. Those typically invoke frequencies between 10 Hz and 50 Hz and near-vertical wave propagation.

The lower crustal reflectivity and wide-angle seismic observations are related in two ways (BRAILLE and CHIANG, 1986). The velocity (or impedance) fluctuations within the lower crust are also visible in high-resolution seismic refraction data. They appear as coda wavetrain that follows the midcrustal reflection ($P_i P$) from the Conrad discontinuity. This has been observed from and modelled for the data set shown in Figure 2 (SANDMEIER and WENZEL, 1986). It served as proof that impedance fluctuations are actually causing the reflections and allowed to quantify the size of fluctuations. For a velocity model based on this data set LEVANDER and

GIBSON (1991) introduced isotropic two-dimensional random fluctuation with a correlation distance of 200 m.

The second more surprising feature is that the bottom of the reflectivity pattern normally coincides with the Moho as defined from refraction data. This has been shown for many coincident reflection/refraction lines (MOONEY and BROCHER, 1987). The generally accepted explanation is that the lower crust deforms in a plastic state whereas the uppermost mantle, due to its different lithology, may be brittle. However, this does not explain the generation of the heterogeneities and their disappearance at Moho level. Another possible explanation is that lower crust and upper mantle deform in a ductile manner (cf., RANALLI, 1995), although different lithologies and/or different strain rates generate different scales of structures. Those developed in the lower crust are detectable with the near-vertical reflection seismic geometry and frequency window; those in the upper mantle are not.

From reflection seismic records, the notion of reflective lower crust and transparent upper mantle arose. It must be emphasized, however, that the mantle is only transparent for the frequency range and the geometry of this particular method, namely for near-vertical incidence. There is no reason to believe that the mantle is homogeneous on a scale of tens of kilometers, that it has not been affected by tectonic processes and that those processes have not left their fingerprints as impedance fluctuations on a particular scale. In fact, inspection of wide-angle data and their coda properties reveals that a significant change in scale of fluctuations occurs especially between lower crust and upper mantle (ENDERLE *et al.*, 1997). The particular scale at Moho level is imaged in the coda properties of the wide-angle Moho reflection ($P_M P$). The scale of mantle fluctuations is contained in long-range refraction data which allow observation of the so-called teleseismic P_n and S_n phases.

Fluctuations in the Uppermost Mantle

Long-range refraction seismic data analyzed for the teleseismic P_n and S_n phases belong to the system of deep seismic sounding profiles carried out by Russian scientists from 1971 to 1990 (Fig. 3) (cf., EGORKIN and MIKHALTSEV, 1990). Peaceful Nuclear Explosions (PNE) and numerous chemical explosions were used as powerful sources. In particular, the seismic data from the QUARTZ profile, located in northern Eurasia, were used to investigate the teleseismic P_n (RYBERG *et al.*, 1995). About 400 short-period (1–2 Hz) three-component, analogue recording systems were deployed to record the ground motion which resulted in maximum observation distances of about 3200 km. The average station spacing of about 10 km along the profiles provides a unique data density for studies of the upper mantle's reflection response in an unprecedented resolution.

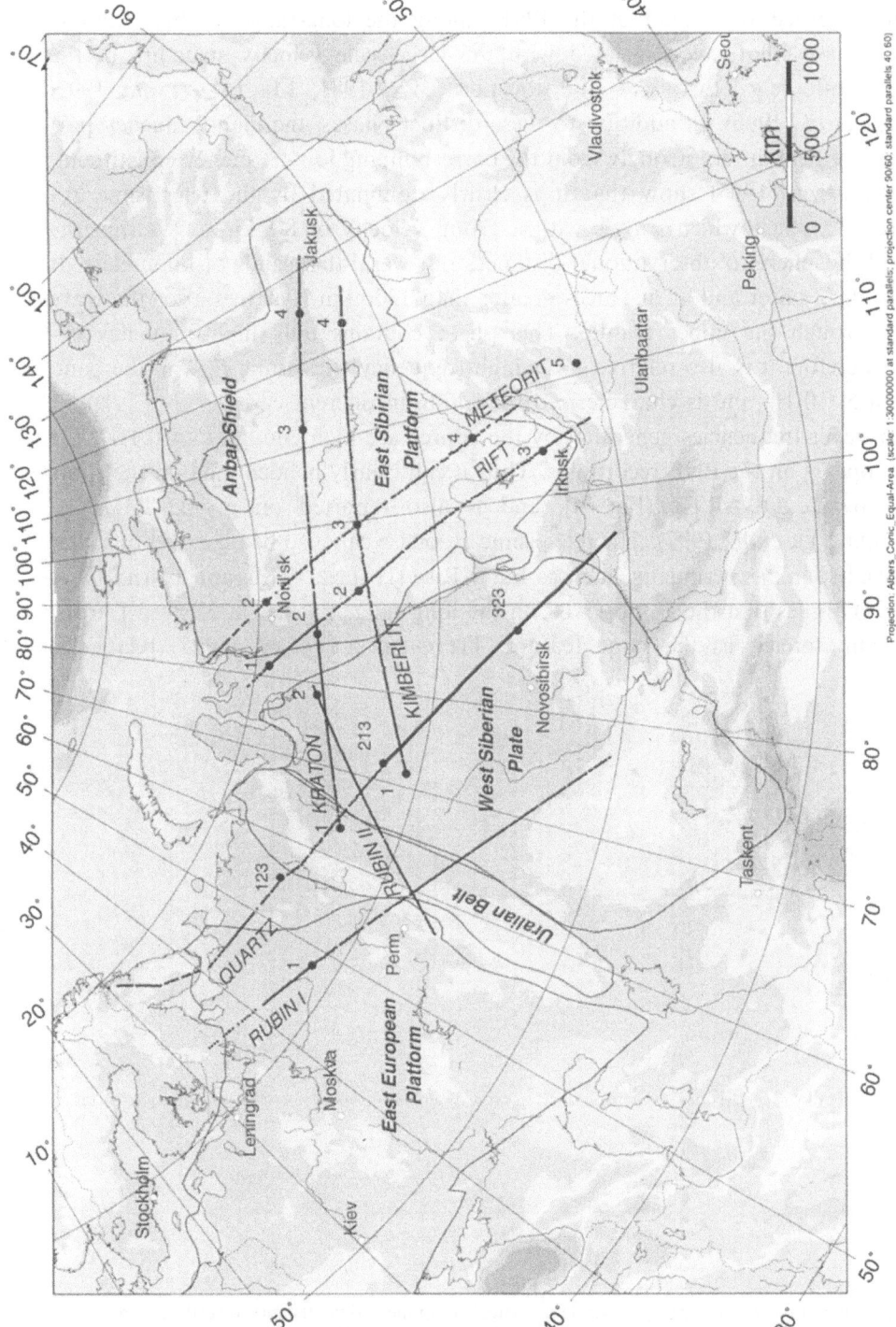

Figure 3

Map of northern Eurasia with the seven major DSS profiles (heavy dotted lines) and the locations of the Peaceful Nuclear Explosions (filled circles). The major tectonic provinces are indicated. The projects were managed by GEON, the geophysical exploration unit of the Ministry of Geology of the former USSR.

The recorded wave field of the PNEs along the long-range profiles exhibits distinct phases that are associated with the large-scale velocity structure of the upper mantle (e.g., EGORKIN and PAVLENKOVA, 1981; MECHIE *et al.*, 1993; RYBERG *et al.*, 1996). In addition to these distinct phases, the high-frequency part of the wave field differs strongly from the corresponding low-frequency constituent. RYBERG *et al.* (1995) show that it is clearly dominated by the teleseismic P_n, defined as a phase which travels with a group velocity of 8.1 km s^{-1} within the mantle lithosphere to observational distances of several thousands of kilometers. It has no sharp onset and at distances greater than 1300 km it arrives after the wave diving through the upper mantle. This phase contains only high-frequency energy—therefore it is also referred to as high-frequency teleseismic P_n—in the band between 5–10 Hz and is characterized by a long incoherent coda.

Whenever frequencies generated by the source are high enough (5–10 Hz), the phase appears on the PNE recordings. Thus it can clearly be identified on the shots of the profile QUARTZ (Fig. 4), and is also reported on profile RUBY I (TITTGEMEYER *et al.*, 1997). The teleseismic P_n phase can also be observed on other controlled source experiments, such as Early Rise (HALES, 1972), and ENDERLE *et al.* (1997) argues that the P_n observed on the long-range profiles in western Europe can be interpreted as the same feature. TITTGEMEYER *et al.* (1997) relate the

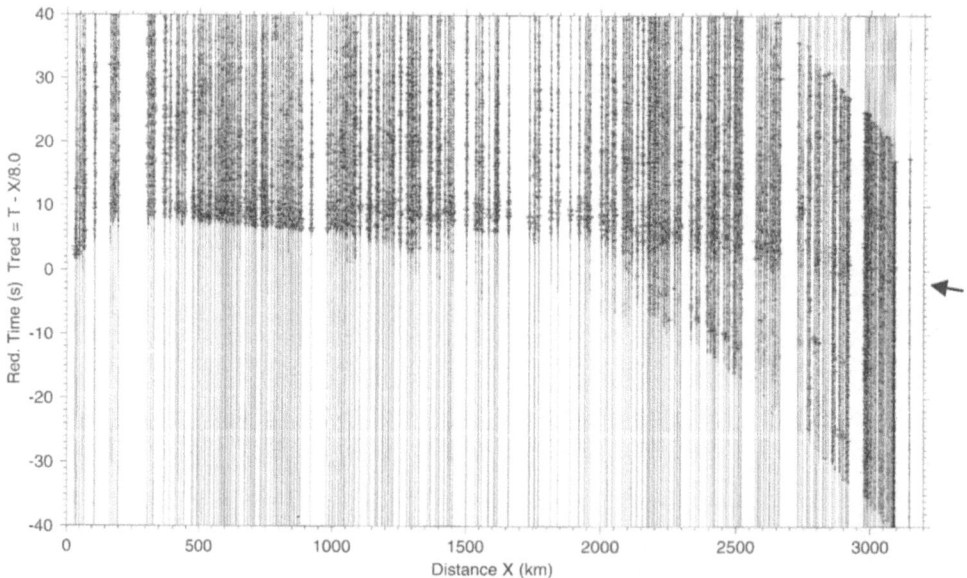

Figure 4

High-pass filtered (corner frequency $f_c = 5$ Hz) vertical-component time-distance record section (reduction velocity $v_{red} = 8.0$ km s^{-1}) on profile QUARTZ for shotpoint 323 recorded in the Northwest. It is trace normalized individually to maximum trace amplitude. The marked strong wave band traveling with a group velocity of ~ 8.1 km s^{-1} is the high-frequency teleseismic P_n.

appearance of the teleseismic P_n on the PNE profiles to its correspondent in the S wave field, the teleseismic S_n.

The notion teleseismic P_n/S_n has been used previously by MOLNAR and OLIVER (1969), who investigated efficient long-range propagation of mostly S_n waves generated by earthquakes. Those waves could be observed to travel with upper mantle velocities for several thousands of kilometers and are reported as high frequency signals both for continental and oceanic paths. Unfortunately the spareness of the net of earthquake stations did not allow plotting seismic record sections and thus visualization of the evolution of the teleseismic P_n/S_n with distance.

As a likely explanation for the efficient propagation of the P_n, several authors consider the existence of a sub-Moho waveguide. For this waveguide several theories have been proposed: transmission of energy in a low-velocity zone beneath the Moho (SUTTON and WALKER, 1972); tunneling of low-frequency waves through thin high-velocity layers (FUCHS and SCHULZ, 1976); transmission of energy in a high-velocity layer beneath the Moho as 'lid' and normal modes (MANTOVANI et al., 1977; STEPHENS and ISACKS, 1977); transmission of energy as a whispering-gallery wave multiple reflected at the crust-mantle boundary (MENKE and RICHARDS, 1980); and propagation in models containing random velocity fluctuations in the crust and/or upper mantle (MENKE and CHENG, 1984; RICHARDS and MENKE, 1983). By simulations of the wave propagation in analog seismic models of scattering earth structure, MENKE and RICHARDS (1983) demonstrated that layers of scatterers within the mantle could explain, in certain respects, the coda phenomenon of the teleseismic P_n. TITTGEMEYER et al. (1996) showed that random successions of high- and low-velocity layers acting as a waveguide can quantitatively explain the main features of the teleseismic P_n as observed on profile QUARTZ, such as coda length, enhancement of high frequencies, and a group velocity of 8.0 km s^{-1} to 8.1 km s^{-1}.

Our favorite upper mantle model is shown in Figure 5. It includes a standard crustal structure with homogeneous upper and lower crust with P-wave velocities of 6.0 km s^{-1} for the upper and 6.5 km s^{-1} for the lower crust between 16 and 35 km depth. Moho is marked by a first-order discontinuity at 35 km depth. The upper 75 km of the mantle are characterized by a slightly positive gradient zone. For this zone we assume a constant Q_P of 1400 and a constant Q_S of 600. The density-depth distribution is calculated according to Birch's relation (BIRCH, 1961).

The observations on long-range seismic profiles call for a class of upper mantle models which are characterized by random fluctuations in elastic parameters. According to the results of TITTGEMEYER et al. (1996), we focus on P-wave velocity fluctuations whilst the S-wave velocity is kept constant in our model. Thus, the P-wave velocity variations are accompanied by variations in Poisson's ratio. If the velocity fluctuations are in the range of $\pm 3\%$, a new feature in the wave field, the teleseismic P_n, arises. If they are smooth, low-frequency, in comparison to the frequency content of the incident source signal, the notion of a heterogeneous

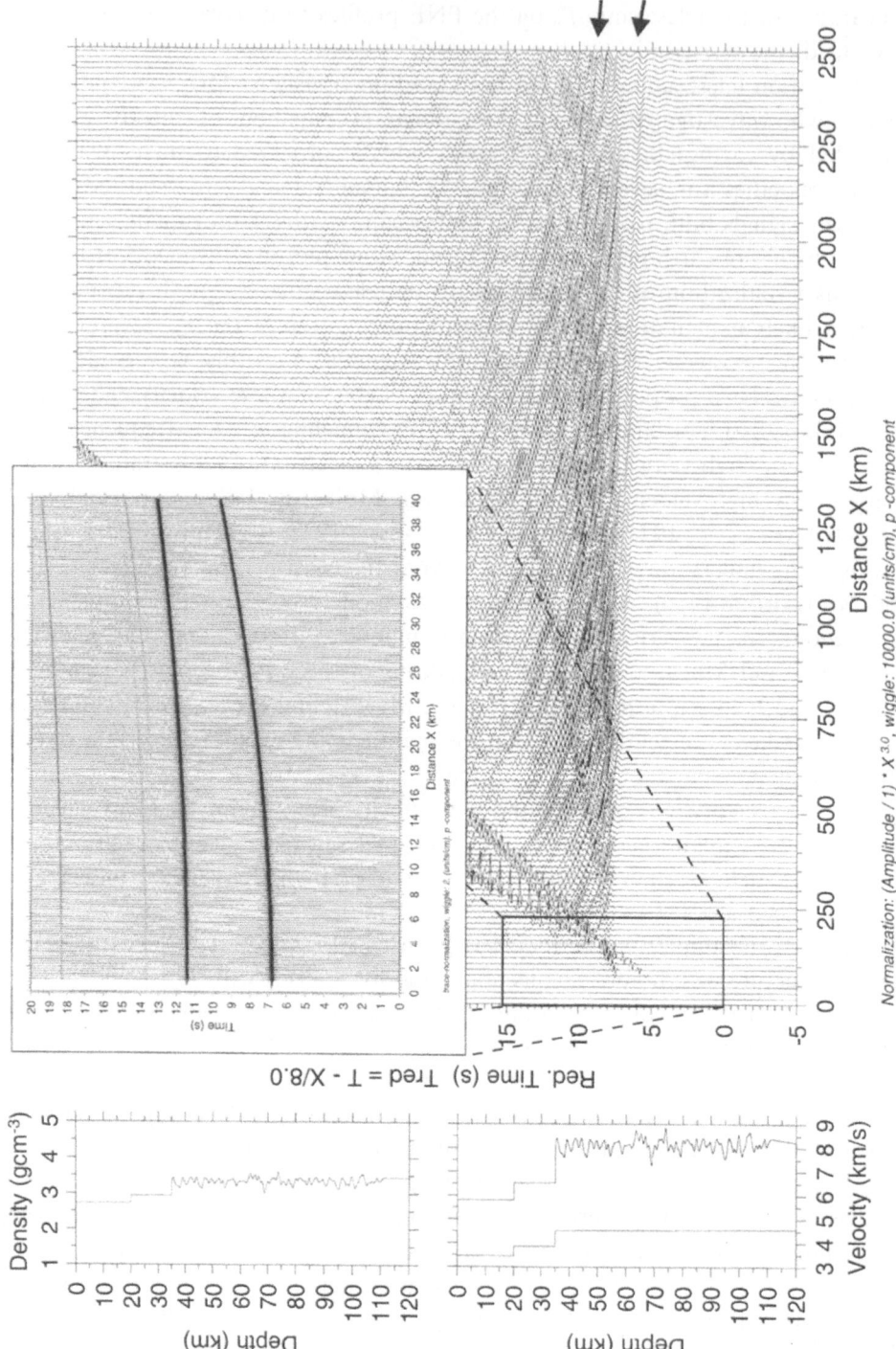

Normalization: (Amplitude / 1) * X^3.0, wiggle: 10000.0 (units/cm), p -component

mantle is still consistent with the initially discussed observation under the reflection seismic aperture and frequency content, that the upper mantle appears to be transparent.[4]

In Figure 6 the theoretical seismic response of a model similar to that shown in Figure 5 is displayed. The appearance of the fluctuations in this model is more rough, high-frequency, owing to the different correlation function.

Random models of heterogeneity have the advantage of being concisely described by a spatial correlation function and the magnitude of variation in physical properties (FRANKEL and CLAYTON, 1986).

The power spectrum of the medium corresponds to the Fourier transform of its autocorrelation function (Theorem of WIENER-CHINTSCHIN, cf. REIF, 1987). Therefore a simple way to characterize a random medium is through a monotonically decaying autocorrelation function (cf., HOLLIGER and LEVANDER, 1992). The rate of decay of the autocorrelation function is a measure for the degree of randomness of the medium. The most prominent analytical correlation functions are the Gaussian and the exponential correlation function (cf., FRANKEL, 1989). The spectrum of a Gaussian correlated medium indicates a slowly decaying spatial correlation at short to intermediate lags and dominance of low to intermediate wave numbers. Its rapid exponential decay results in smooth medium with little contribution from high wave-number heterogeneities. The spectrum of an exponential correlated medium behaves differently. The spectrum displays a much broader range of wave numbers which contribute to the heterogeneity spectrum. The medium is rougher and more complex than the Gaussian medium due to its wider range of contributing wave numbers.

The model depicted in Figure 6 is based on an exponential correlation function, whereas the fluctuations on the other (Fig. 5) are Gaussian correlated.

To calculate the wide-angle response of both models, a P-wave point source was located in 100 m depth, and a source time function with a dominant frequency of 4 Hz was used. Representing the typical frequency content of reflection seismograms,

---

Figure 5

P-wave theoretical record sections calculated with the reflectivity method (FUCHS and MÜLLER, 1971). The velocity-depth functions that the calculation of the seismograms is based on are displayed to the left. IASPEI-91 (KENNETT and ENGDAHL, 1991) serves as a reference background model to which P-wave velocity fluctuations are superimposed. The fluctuations are characterized by a spatial correlation length of 700 m and a standard deviation of 3%. They are distributed according to a Gaussian correlation function. Note that a teleseismic P_n is generated with its significant characteristics, high-frequency content (compared to the lower frequency precursor), extensive coda and a constant group velocity of 8.0 km s^{-1}. The near-vertical response (see inset) of that model remains the same as if the model would be not heterogeneous.

---

[4] The vertical change in velocity gradient ($\Delta v / \Delta z$) could act as a filter on near-vertically incident signals (WOLF, 1937).

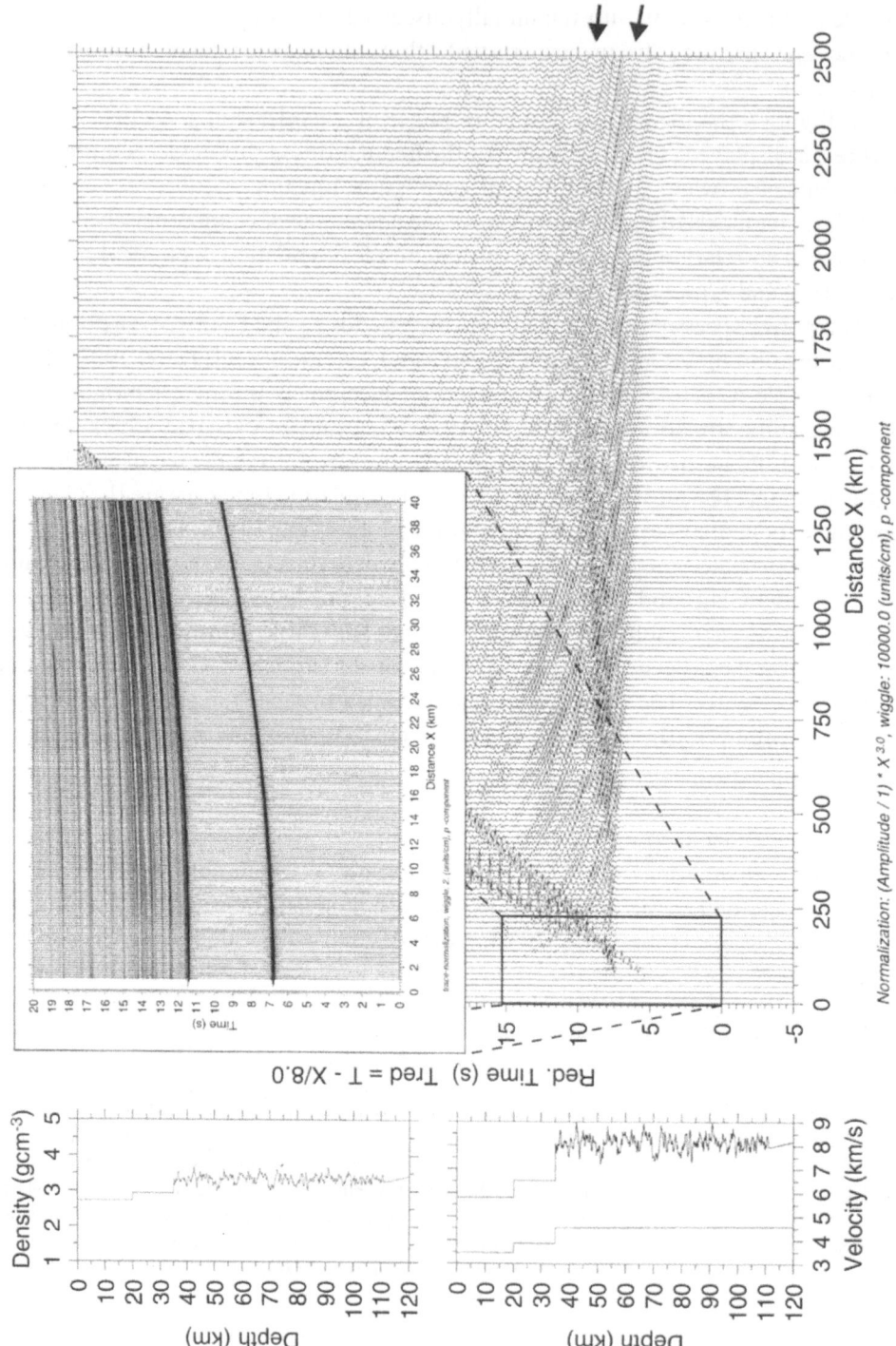

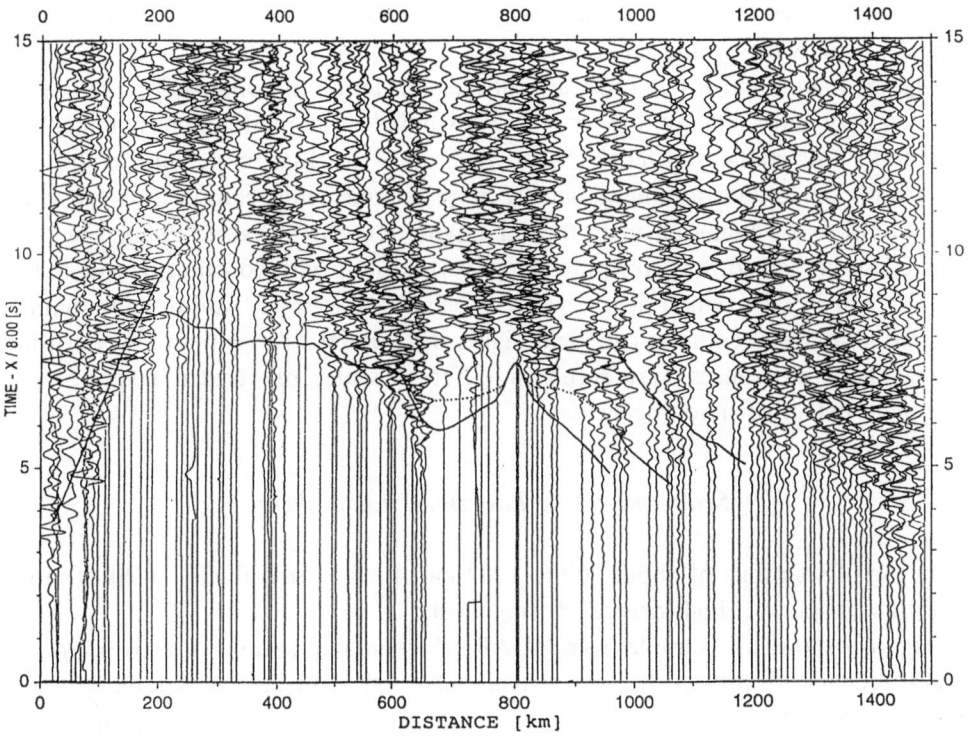

Figure 7
Vertical-component time-distance record section (reduction velocity $v_{red} = 8.0$ km s^{-1}) on profile
QUARTZ (for the location see Fig. 3) for shotpoint 323 recorded in the Northwest. The offset range
between 0 and 1500 km is shown. At distances between 700 and 1500 km the wave field is hardly
described by a phase. The wave field becomes very irregular. This is taken as evidence that strong
scattering occurs at the respective depth of about 100–180 km.

the theoretical record sections depicted in the inset of Figures 5 and 6 were
calculated with a dominant frequency of the source signal of 30 Hz.

Whereas the wide-angle response of both models is quite similar to that which
is observed, the near-vertical response of the rough model is in strong contrast to
that which is widely seen.

Because the actual size of fluctuations will probably change with the dimension-
ality of the models, we do not want to discuss petrological implications in a

Figure 6
P-wave theoretical record sections calculated with the reflectivity method (FUCHS and MÜLLER, 1971).
The velocity-depth functions that the calculation of the seismograms is based on are displayed to the left.
IASPEI-91 (KENNETT and ENGDAHL, 1991) serves as a reference background model to which P-wave
velocity fluctuations are superimposed. The fluctuations are distributed according to an exponential
correlation function. Statistically, they are characterized by a spatial correlation length of 700 m and a
standard deviation of 3%. Note that even though the wide-angle response is quite similar to that which
is observed, the near-vertical response (see inset) is in strong contrast to that which is widely seen.

quantitative sense. It is conceivable that the fluctuations represent preferred orientation of olivine with a random component.

Upper mantle and lower crust are different in terms of composition, compositional variety and rheology. Lower crust is assumed to be predominantly mafic but to contain both mafic and felsic rocks which allow for high impedance contrasts. The mantle could have a mineralogically more homogeneous composition but with highly anisotropic material, in particular olivine (cf., ANDERSON, 1989). Fluctuations caused by anisotropy are pure velocity variations with constant density. This would result in fairly small near-vertical impedance contrasts. Additionally mantle material at a given geotherm can yield higher differential stresses than lower crustal rocks. These differences may result in different scales of vertical layering and horizontal extent of inhomogeneities.

Scattering in the Lowermost Lithosphere

In the offset range of about 700 to 1200 km the first-arrival wave field of most PNE data displays a high degree of irregularity.

Figure 7 shows an example from QUARTZ where fluctuations of arrival time of several seconds are observed. The arrivals have apparent velocities ranging from about 8.0 to 8.7 km s^{-1}. For an explanation MECHIE *et al.* (1993) proposed a model consisting of alternating high- and low-velocity layers. THYBO and PERCHUĆ (1997a,b) interpret this property as wave-field fluctuation caused by velocity variatons between the 8° discontinuity at ~ 100 km depth and the Lehmann discontinuity (LEHMANN, 1964) at a depth of ~ 220 km (its depth being variable, depending on the thermal regime). Similar features were also observed on the FENNOLORA (HAUSER *et al.*, 1990) profiles (cf., PERCHUĆ and THYBO, 1996) which are, like QUARTZ, located in an old stable continent. No real phase correlation is possible in this particular offset range. The associated depth range is about 100 to 180 km, which corresponds to the lowermost part of the lithosphere.

The waves beyond 1200 km are more regular as regards their first arrivals. It can thus be concluded that the lower lithosphere is characterized by intensive scattering. The underlaying medium is clearly more homogeneous. Although we have no quantitative model of lower lithosphere scattering, it is obvious that the spatial scale of scatterers in this part of the lithosphere must be larger than in the uppermost mantle.

The Mantle Transition Zone

The upper mantle transition zone has been mapped by passive seismology using various data sets. Accordingly, *P* to *S* conversion of long-period data (PETERSEN *et al.*, 1993), *P* wide-angle underside reflections and conversions (ESTABROOK and

KIND, 1996; KENNETT and ENGDAHL, 1991), long-period shear waves (*SS* and *ScS*) (REVENAUGH and JORDAN, 1991; SHEARER, 1991, 1996; SHEARER and MASTERS, 1992) tomographic methods (CREAGER and JORDAN, 1984, 1986; FUKAO *et al.*, 1992; VAN DER HILST *et al.*, 1991; WIDIYANTORO and HILST, 1996) and near-vertical *P'P'* underside reflections (BENZ and VIDALE, 1993) were used to constrain the elastic properties of the mantle transition zone.

Most of these methods are based on relatively long-period signals (> 5 s) and require substantial data processing (deconvolution stacking, etc.) of wave forms and thus inherently average over large spatial regions. Consequently, they can hardly serve as an indication for small-scale variabilities of the transition zone. Short-period data are more sensitive to small velocity fluctuations than long-period data. This is usually viewed as disadvantageous, but can be taken as a measure of the spatial scale of these fluctuations and can possibly answer questions concerning the degree of chemical homogenization of the mantle.

Controlled source seismic data sets (such as the PNEs recorded in northern Eurasia) have several potential advantages in contrast to earthquake data. Well-known shot times and locations and a relatively simple source function in conjunction with a relatively dense spatial sampling of the wave field provide high-quality data sets.

The amplitudes of seismic phases from the upper mantle discontinuities can be influenced by several factors. The energy distribution along the wide-angle phases, mainly the top-side reflection at the boundary and the refraction below it, depends substantially on the velocity contrast and the sharpness of the discontinuity. The depth of a reflector and its possible topography will produce changes in the energy distribution along the phases due to caustics and focussing/defocussing effects. Scatterers in the overlaying upper mantle and/or in the crust will also have an impact on the wave field by introducing an incoherent scattering coda and other amplitude fluctuations. Local site effects are well-known from strong motion analysis to easily change the amplitudes by a factor of 5 to 10. Tuning by resonance effects can enhance frequency range and thus considerably change the wave forms.

On the other hand short-period data are more sensitive to small-scale lateral fluctuations in the elastic properties of the medium. The wave form is influenced by the medium properties along the seismic ray path and region around it that is described by the Fresnel zone. Short-period *P'P'* underside reflections (BENZ and VIDALE, 1993) inferred sharp discontinuities at 410 and 660 km depth. These reflections are very sensitive to the width of the transition at both respective depths. The fact that they appear and disappear from place to place is indicative of substantial lateral variations of at least this property of the transition zone. Previous work by PAULSSEN (1988) with short-period data recorded by the NARS array used *P* to *S* conversions. The author notes substantial lateral variations in impedance contrast across the discontinuities.

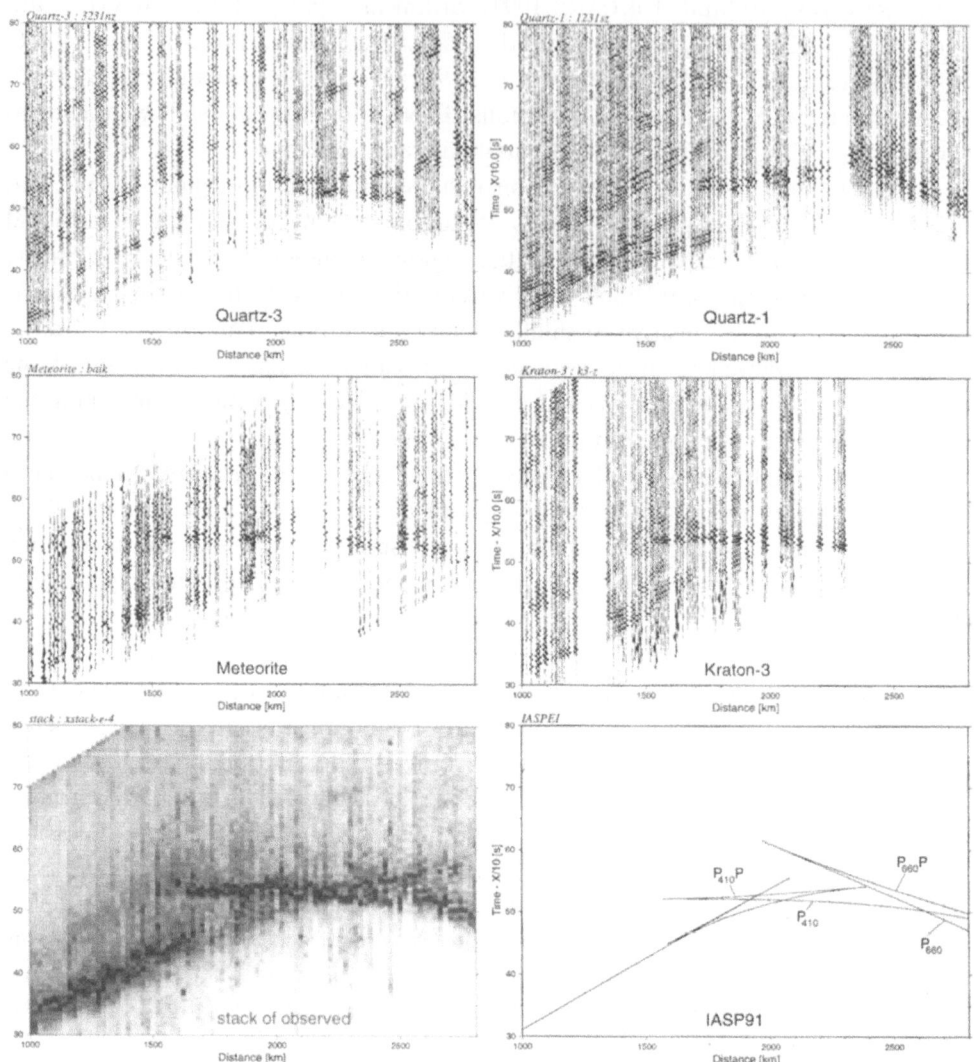

Figure 8

Examples of vertical component record sections recorded along PNE profiles in northern Eurasia. Depicted are the recordings of two shots along QUARTZ (top right and left), of the southernmost shot along METEORITE (middle left), and the recordings shot 3 recorded along KRATON (middle right) from east to west. All data are shown trace normalized and time reduced (reduction velocity $v_{red} = 10.0$ km s^{-1}), and to further enhance the mantle phases a low-pass filter (corner frequency $f_c = 1.25$ Hz) was applied. The sketch at the bottom right provides an overview of the arrival time for the reflected ($P_{410}P$, $P_{660}P$) and refracted (P_{410}, P_{660}) seismic phases associated with the mantle transition zone. This is based on the model IASPEI-91 (KENNETT and ENGDAHL, 1991). The panel to the bottom left shows a stack of all available PNE data (18 shots recorded along 7 profiles, cf.. RYBERG *et al.*, 1998).

Figure 8 shows the examples of the seismic record sections with phases of the transition zone for 4 PNE shotpoints recorded along the profiles QUARTZ, KRATON and METEORITE. A reduction velocity has been chosen as 10 km s^{-1} so that the 410 km response is aligned horizontally. The travel-time curves predicted by the model of the International Association of Seismology and Physics of the Earth's Interior 1991 (IASPEI-91) (KENNETT and ENGDAHL, 1991) are displayed in the right lower corner. In addition one panel shows the stack of all available PNE data (18 shots recorded along 7 profiles, cf., RYBERG et al., 1998). This panel represents the average structure of the 410 km discontinuity under northern Eurasia and is fully compatible with the IASPEI-91 model. While the travel times of the reflected and refracted phases from the mantle transition zone are in good agreement with the times predicted by IASPEI-91, there is significant variation of their (relative) amplitudes. The four data sets (Fig. 8) show characteristic differences of the 410 km response with regard to the amplitude of the wave refracted beneath the discontinuity and the wide-angle reflection from the discontinuity. KRATON and METEORITE indicate a critical point at about 1500 km distance; in contrast, the critical point of the QUARTZ profiles seems to be shifted to larger offsets. This, however, may be an effect of the relative amplitude scaling of the seismograms. Nonetheless the main difference can be seen in the amplitude ratio of wide-angle reflection and refraction. All profiles, apart from QUARTZ-1 display a high-amplitude refraction. This is indicative of a relatively large velocity gradient beneath the discontinuity and/or for variable velocity contrasts across or for topographic effects along the discontinuity. THYBO et al. (1997) even supplied an interpretation concluding a second reflector in the depth range of 400 km.

The refraction in QUARTZ-1 is barely visible, which must be related to a relatively small velocity gradient of the refractor. The sites of the mantle sampled by QUARTZ-1 and -3 are within a few hundred kilometers distance (cf., RYBERG et al., 1996). Thus the properties of the transition zone can obviously change on this lateral scale. The wide-angle reflection is very well established in QUARTZ-1 and -3, but much less pronounced in both other profiles. This calls for different reflection properties of the 410 km discontinuity beneath northeast Eurasia. Similar observations of the relative amplitudes for the 660 km phases hold true for the 670 km discontinuity (RYBERG et al., 1998), except for the reduced velocity step across that discontinuity. While the stacking method applied by RYBERG et al. (1998) tends to average the variability of the 410 and 660 km discontinuity, the individual record sections show a substantial and systematic scatter of amplitudes for the phases of the mantle transition zone. The relatively consistent behavior of the amplitudes along the seismic phases allows us to exclude strong site effects (crustal scatterers, tuning due to the crust, etc.). Therefore, most of the observed amplitude features should be caused by structures in the upper mantle.

Concluding from the PNE data base alone, it would be hard to tell whether focussing/defocussing effects due to topography along the boundaries or a more complicated velocity structure (transitional discontinuities with varying thicknesses, small-scale velocity fluctuation acting as scatterers, or several phase transitions occurring at slightly different depths: γ spinel to perovskite and magnesiowüstite and/or a transition from garnet to perovskite for the 660 km discontinuity (DUFFY and ANDERSON, 1989; IRIFUNE, 1987; ITA and STIXRUDE, 1992; ITO and TAKAHASHI, 1989; KUSKOV and PANFEROV, 1991) are responsible for the observation.

Conclusions

The data presented to date indicate that the crust/mantle system shows conspicuous fluctuations in elastic parameters with different scales: Reflection data show a layered or reflective lower crust with velocity fluctuations as high as 10%, with variable vertical and lateral scales, the latter typically being in the 100 m range. At the crust/mantle boundary (Moho) the scale of heterogeneities changes.

PNE data from super-long refraction seismic profiles in Russia give evidence of random fluctuations in the upper mantle. 1-D modeling provides a vertical scale of the inhomogeneities in the km range. The horizontal correlation length is unknown. We estimate it to be in the 10 km range in order to trap the elastic energy and propagate it out to 3000 km offset from the source. Observations of S_n waves recorded at permanent seismic stations and data with oceanic travel path indicate that these upper mantle fluctuations may also represent a wide-spread feature. Those fluctuations of the uppermost mantle have a larger correlation length and are smoother as compared to the lower crust. This change in scale is in fact the physical reason for the coincidence of reflection and refraction Moho. Figure 9 portrays a cartoon which summarizes the described properties.

The deepest portion of the lithosphere seems to consist of heterogeneous material with correlation lengths of several wavelengths which lead to distortions of first-arrival times of P waves.

Figure 9
Long distance wave propagation of P_n and S_n is achieved by statistical fluctuations of elastic parameters in the uppermost mantle, which differ markedly in the scale of structure and in the magnitude of variation from the overlaying lower crust. Here a typical lower crustal structure which could reproduce what is widely observed, is generated after HOLLIGER *et al.* (1994). Statistically this is represented by heterogeneity with a fractal dimension of 2.7, a bimodal velocity distribution of 5% and a horizontal and vertical characteristic scale of 800 and 200 m, respectively. With the anticipation that the results from 1-D modeling still determine the 2-D case, mantle, heterogeneities are spatially Gaussian distributed and have a characteristic vertical scale length of 700 m. The aspect ratio is set arbitrarily to 4 (as in the lower crust). The standard deviation of the velocity distribution of 3% is assumed to be smaller for 2-D modeling rather than in the 1-D case. Both targets, lower crust and uppermost mantle, provide a generic description of geologic heterogeneity and represent end-member models in the description of reflecting structures.

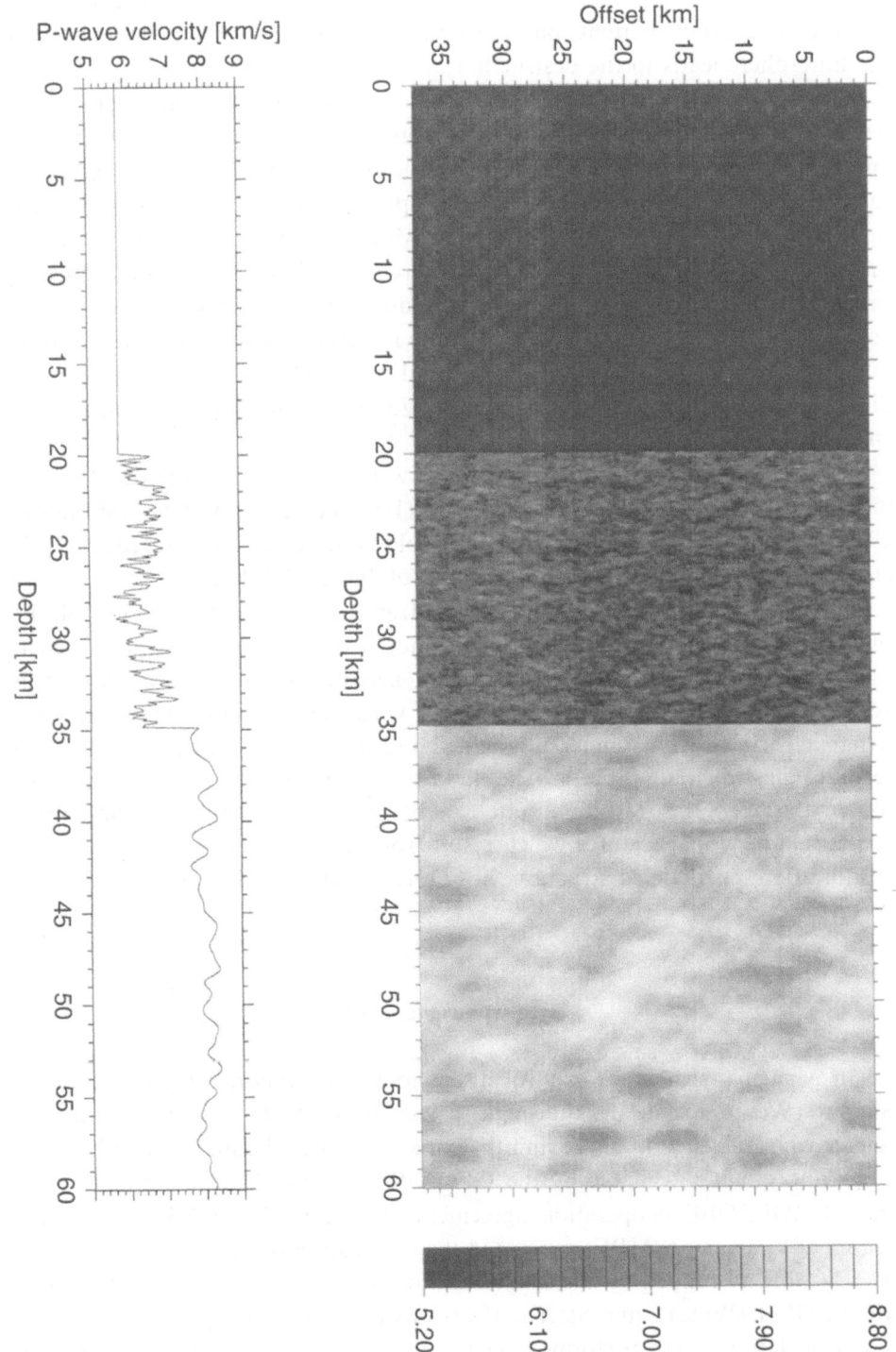

The different scales and the associated scattering have significant influence on wide-angle seismic data. The fluctuation characteristics for the uppermost mantle generate a new (non-Fermat) phase, whereas the coarse structure at the bottom of the lithosphere leads to the destruction of a clear refracted phase.

Once the scales and their variations are established, their geological and petrological significance becomes the key problem. At this point only speculation and options can be discussed. The layered lower crust may represent the evidence for a decoupling of motion between the upper crust and mantle during tectonic processes acting at high temperatures. The Moho with its conspicuous crust/mantle mix (cf., ENDERLE *et al.*, 1997) could have acted as detachment during these processes. Upper mantle and lower crust are different in terms of composition, compositional variety and rheology. Mantle material at a given geotherm can yield higher differential stresses than lower crustal rocks. These differences may result in different scales of vertical layering and horizontal extent of inhomogeneities (TITT-GEMEYER *et al.*, 1996).

The fluctuations in the deep lithosphere were interpreted as the result of partial melting processes by THYBO and PERCHUĆ (1997a) (cf., also THYBO and PERCHUĆ, 1997b), assuming that the geotherm of a cold continental area exceeds the solidus curve of upper mantle material at a depth of 100 to 200 km.

We believe that the PNE images of the transition zone reflect the complexity of this particular regime. Modern seismic tomography evidences that slabs penetrate the transition zone, however those do not necessarily do so. Also the transition zone may represent the source of hot spot volcanism. It represents a layer of the dynamic earth to which material is added and from which material is extracted. A result of these ongoing processes should be a certain amount of lateral inhomogeneity. In this view the transition zone can be compared with the Moho. Both are recognized as global features which result from the earth's evolution and chemical stratification. Its detailed structure, however, reflects and depends considerably on its history.

Acknowledgments

In particular, for many fruitful discussions, the authors wish to thank Sergei Shapiro. We appreciate the thorough reviews by Hans Thybo and an anonymous referee which assisted in improving the paper. The digital data for the PNE profiles were kindly made available by the GEON/Russian Geological Committee within the EUROPROBE cooperation agreement. The ministry of Education and Research of Germany (BMBF) supported the digitization of the PNE data at GEON. M.T.'s work is funded by the Deutsche Forschungsgemeinschaft (grant WE1394/10–1). GMT (WESSEL and SMITH, 1995) was used for plotting most of the figures. The calculations were performed on the SNI VPP300/4 of the computer center,

Karlsruhe University and on a Cray T3E of the German National Supercomputer Center in Stuttgart.

REFERENCES

AKI, K., and CHOUET, B. (1975), *Origin of Coda Waves: Source, Attenuation, and Scattering Effects*, J. Geophys. Res *80*, 3322–3342.

AKI, K., and RICHARDS, P. G., *Quantitative Seismology: Theory and Methods*, vols. I and II (W. H. Freeman and Company, San Francisco 1980).

ALLMENDINGER, R. W., and COCORP (1983), *Cenozoic and Mesozoic Structures of the Eastern Basin and Range Province, Utah, from COCORP Seismic-reflection Data*, Geology *11*, 532–536.

ANDERSON, D. L., *Theory of the Earth* (Blackwell Scientific Publications, Boston, Oxford, London, Endinburgh, Melbourne 1989).

BENZ, H. M., and VIDALE, J. E. (1993), *Sharpness of Upper-mantle Discontinuities Determined from High-frequency Reflections*, Nature *365*, 147–149.

BIRCH, F. (1961), *The Velocity of Compressional Waves in Rocks to 10 kbars, Part 2*, J. Geophys. Res. *66*, 2199–2224.

BOIS, C., and ECORS, *Late- and post-orogenic evolution of the crust studied from ECORS deep seismic profiles*. In *Continental Lithosphere: Deep Seismic Reflections* (eds. R. Meissner, L. Brown, H.-J. Dürbaum, W. Franke, K. Fuchs, and F. Seifert), vol. 22 (AGU Geodynamics Series, 1991).

BRAILLE, L., and CHIANG, C. S., *The continental Mohorovičić discontinuity; results from near-vertical and wide-angle seismic refraction studies*. In *Reflection Seismology: A Global Perspective* (eds. M. Barazangi and L. D. Brown), vol. 13, Geodyn. Series ed. (Am. Geophys. Union 1986) pp 257–272.

BREWER, J. A., MATTHEWS M. R., WARNER, M. R., HALL, J., SMYTHIE, D. K., and WHITTINGTON, R. J. (1983), *BIRPS Deep Seismic Reflection Studies of the British Caledonies—The WINCH Profile*, Nature *305*, 206–210.

COOK, F. A., BROWN, L. D., KAUFMAN, S., OLIVER, J., and PETERSEN (1981), *COCORP Seismic Profiling of the Apalachian Orogen beneath the Costal Plain of Georgia*, Geol. Sco. Am. Bull. *92*, 738–748.

CREAGER, K. C., and JORDAN, T. H. (1984), *Slab Penetration into the Lower Mantle*, J. Geophys. Res. *89*, 3031–3049.

CREAGER, K. C., and JORDAN, T. H. (1986), *Slab Penetration into the Lower Mantle beneath the Mariana and Other Island Arcs of the Northwest Pacific*, J. Geophys. Res. *91*, 3573–3589.

DUFFY, T. S., and ANDERSON, D. L. (1989), *Seismic Velocities in Mantle Minerals and the Mineralogy of the Upper Mantle*, J. Geophys. Res. *94*, 1895–1912.

EGORKIN, A. V., and PAVLENKOVA, N. I. (1981), *Studies of Mantle Structure of U.S.S.R. Territory on Long-range Seismic Profiles*, Phys. Earth Planet. Inter. *25*, 12–26.

EGORKIN, V. A., and MIKHALTSEV, A. V., *The results of seismic investigations along geotraverses*. In *Super-deep Continental Drilling and Deep Geophysical Sounding* (eds. K. Fuchs, Y. A. Kozlovsky, A. I. Krivtsov, and D. Zoback) (Springer Berlin, Heidelberg, New York 1990) pp. 111–119.

ENDERLE, U., TITTGEMEYER, M., ITZIN, M., PRODEHL, C., and FUCHS, K., *Scales of structure in the lithosphere—Image of processes*. In *Stress and Stress Release in the Lithosphere—Structure and Dynamic Processes in the Rifts of Western Europe* (eds. K. Fuchs, R. Altherr, B. Müller, and C. PRODEHL), vol. 275, Special Issue (Tectonophysics 1997) pp. 165–198.

ESTABROOK, C., and KIND, R. (1996), *The Nature of the 660 km Upper-mantle Seismic Discontinuity from Precursors to the PP Phase*, Science *274*, 1179–1182.

FOUNTAIN, D. M., *Geological and geophysical nature of the lower continental crust as revealed by exposed cross sections of the continental crust*. In *Geophysics and Petrology of the Deep Crust and Upper Mantle* (eds. S. J. Noller, S. H. Kirby, and J. E. Nelson-Pike), vol. 956 (U.S. Geol. Survey Circular 1987) pp. 25–26.

FRANKEL, A. (1989), *Review of Numerical Experiments on Seismic Wave Sattering*, Pure appl. geophys. *131*, 639–685.

FRANKEL, A., and CLAYTON, R. W. (1986), *Finite-difference Simulations of Seismic Scattering: Implications for the Propagation of Short-period Seismic Waves in the Crust and Models of Crustal Heterogeneity*, J. Geophys. Res *91*, 6465–6489.

FUCHS, K., and MÜLLER, G. (1971), *Computation of Synthetic Seismograms with the Reflectivity Method and Comparison to Observations*, Geophys. J. Royal Astron. Soc. *23*, 417–433.

FUCHS, K., and K. SCHULZ (1976), *Tunneling of Low-frequency Waves through the Subcrustal Lithosphere in Southern Germany*, J. Geophys *42*, 175–190.

FUKAO, Y., OBAYASHI, M., INOUE, H., and NENBAI, M. (1992), *Subducting Slabs Stagnant in the Mantle Transition Zone*, J. Geophys. Res. *97*, 4809–4822.

GETTRUST, J. F., and FRAZER, L. N. (1981), *A Computer Model Study of the Propagation of the Long-range P_n Phase*, Geophys. Res. Lett. *8*, 749–752.

GLOCKE, A., and MEISSNER, R., *Near-vertical reflections recorded at the wide-angle profile in the Rheinish Massif. In Explosion Seismology in Central Europe* (eds. P. Giese, C. Prodehl, and A. Stein) (Springer Verlag, Berlin 1976) pp. 252–256.

HALES, A. L. (1972), *The Travel Times of P Seismic Waves and their Relevance to the Upper Mantle Velocity Distribution*, Tectonophysics *13*, 447–482.

HAMMER, P. T. C., and CLOWES, R. M. (1997), *Moho Reflectivity Patterns—A Comparison of Canadian LITHOPROBE Transects*, Tectonophysics *269*, 179–198.

HAUSER, F., PRODEHL, C., and SCHIMMEL, M. (1990), *Fennolora, Open File Report 90–2*, Geophysical Institute.

HOLLIGER, K., and LEVANDER, A. R. (1992), *A Stochastic View of Lower Crustal Fabric Based on Evidence from the Ivrea Zone*, Geophys. Res. Lett. *19*, 1153–1156.

HOLLIGER, K., LEVANDER, A., CARBONELL, R., and HOBBS, R. (1994), *Some Attributes of Wavefields Scattered from Ivrea-type Lower Crust*, Tectonophysics *119*, 497–510.

IRIFUNE, T. (1987), *An Experimental Investigation of the Pyroxene-garnet Transformation in a Pyrolite Composition and its Bearing on the Constitution of the Mantle*, Phys. Earth Planet. Inter. *45*, 324–336.

ITA, J., and STIXRUDE, L. (1992), *Petrology, Elasticity, and Composition of the Mantle Transition Zone*, J. Geophys. Res. *97*, 6849–6866.

ITO, E., and TAKAHASHI, E. (1989), *Postspinel Transformation in the System Mg_2SiO_4-Fe_2SiO_4 and Some Geophysical Implications*, J. Geophys. Res. *94*, 10,637–10,646.

KENNETT, B. L. N., and ENGDAHL, E. R. (1991), *Travel Times for Global Earthquake Location and Phase Identification*, Geophys. J. Int. *105*, 429–465.

KIRBY, S., and KRONENBERG, A. (1987), *Rheology of the Lithosphere: Selected Topics*, Rev. Geophys. Space Phys. *25*, 1219–1244.

KUSKOV, O. L., and PANFEROV, A. B. (1991), *Constitution of the Mantle. Part 3: Density, Elastic Properties and the Mineralogy of the 400-km Discontinuity*, Phys. Earth Planet Inter. *69*, 85–100.

KUSZNIR, N. J., *The distribution of stress with depth in the lithosphere: Thermorheological and geodynamic constraints*, vol. 194. In *Tectonic Stress in Lithosphere* (The Royal Society, London 1991).

LEHMANN, I. (1964), *On the Velocity of P in Upper Mantle*, Bull. Seismol. Soc. Am. *54*, 1097–1103.

LEVANDER, A., and GIBSON, B. S. (1991), *Wide-angle Seismic Reflections from Two-dimensional Random Target Zones*, J. Geophys. Res. *96*, 10251–10260.

LÜSCHEN, E., NOLTE, B., and FUCHS, K., *Shear-wave evidence for an anisotropic lower crust beneath the Black Forest, SW Germany. In Seismic Probing of Continents and their Margins* (eds. J. H. Leven, D. M. Finlayson, C. Wright, J. C. Dooley, and B. L. N. Kennett), vol. 173, *Special Series* (Tectonophysics 1990) pp 483–493.

LÜSCHEN, E., *et al.* (1987), *Near-vertical and Wide-angle Seismic Surveys in the Black Forest, SW Germany*, J. Geophys. *62*, 1–30.

MALLICK, S., and FRAZER, L. N. (1990), *P_0/S_0 Synthetics for a Variety of Oceanic Models and their Implications for the Structure of the Oceanic Lithosphere*, Geophys. J. Int. *100*, 235–253.

MANTOVANI, E., SCHWAB, F., LIAO, H., and KNOPOFF L. (1977), *Teleseismic S_n: A Guided Wave in the Mantle*, Geophys. J. Royal Astr. Soc. *51*, 709–726.

MATTHEWS, D. H., *Seismic reflections from the lower crust around Britain. In Nature of the Lower Continental Crust* (eds. J. B. Dawson, D. A. Carswell, J. Hall, and K. H. Wedepohl), vol. 24 (Geol. Soc. London, Spec. Publ. 1986) pp. 11–21.

MECHIE, J., EGORKIN, A. V., FUCHS, K., RYBERG, T., SOLODÌLOV, L., and WENZEL, F. (1993), *P-wave Mantle Velocity Structure beneath Northern Eurasia from Long-range Recordings along the Profile QUARTZ*, Phys. Earth Planet. Inter. *79*, 269–286.

MEISSNER, R., *The Continental Crust* (Acad. Press, Orlando 1986).

MEISSNER, R., and DEKORP, *The DEKORP surveys: Major achievements for tectonical and reflective styles*. In *Continental Lithosphere: Deep Seismic Reflections* (eds. R. Meissner, L. Brown, H.-J. Dürbaum, W. Franke, K. Fuchs, and F. Seifert), vol. 22 (AGU Geodynamics Series 1991) pp. 69–76.

MENKE, W. H., and CHENG, R. (1984), *Numerical Studies of the Coda Fall-off Rate of Multiple Scattered Waves in Randomly Layered Media*, Bull. Seismol. Soc. Am. *74*, 1605–1621.

MENKE, W. H., and RICHARDS, P. G. (1980), *Crust-mantle Whispering Gallery Phases: A Deterministic Model of Teleseismic P_n Wave Propagation*, J. Geophys. Res. *85*, 5416–5422.

MENKE, W. H., and RICHARDS, P. G. (1983), *The Horizontal Propagation of P Waves through Scattering Media: Analog Model Studies Relevant to Long-range P_n Propagation*, Bull. Seismol. Soc. Am. *73*, 125–142.

MOLNAR, P., and OLIVER, J. (1969), *Lateral Variation of Attenuation in the Upper Mantle and Discontinuities in the Lithosphere*, J. Geophys. Res. *74*, 2648–2682.

MOONEY, W. D., and BROCHER, T. M. (1987), *Coincident Seismic Reflection/Refraction Studies of the Continental Lithosphere: A Global Review*, Rev. Geophys. *25*, 723–742.

MOONEY, W. D., and MEISSNER, R., *Multi-generic origin of crustal reflectivity: A review of seismic reflection profiling of the continental lower crust and Moho*. In *Continental Lower Crust* (eds. D. M. Fountain, R. Arculus, and K. W. Kay) (Elsevier Amsterdam 1992) pp. 179–199.

OLIVER, J., *A global perspective on seismic reflection profiling of the continental crust*. In *Reflections Seismology: A Global Perspective* (eds. M. Barazangi and L. Brown), vol. 13 (AGU Geodynamic Series 1986) pp. 1–3.

PAULSSEN, H. (1988), *Evidence for a Sharp 670-km Discontinuity as Inferred from P–S Converted Waves*, J. Geophys. *93*, 10,489–10,500.

PERCHUĆ, E., and THYBO, H. (1996), *A New Model of Upper Mantle P Waves below the Baltic Shield: Indication of Partial Melt in the 95 to 160 km Depth Range*, Tectonophysics *253*, 227–245.

PETERSEN, N., VINNIK, L., KOSAREV, G., KIND, R., ORESHIN, S., and STAMMLER, K. (1993), *Sharpness of Mantle Discontinuities*, Geophys. Res. Lett. *20*, 859–862.

RANALLI, G., *Rheology of the Earth*, 2nd ed. (Chapman and Hall, London 1995).

REIF, F., *Statistical Physics and Theory of Heat (in German)*, 3rd ed. (Walter de Gruyter, Berlin, New York 1987).

REVENAUGH, J. S., and JORDAN, T. H. (1991), *Mantle Layering from ScS Reverberations, 2. The Transition Zone*, J. Geophys. Res. *96*, 19,763–19,780.

REY, P. (1993), *Seismic and Tectonometamorphic Characters of the Lower Continental Crust in Phanerozoic Areas: A Consequence of Post-thickening Extension*, Tectonics *112*, 580–590.

RICHARDS, P. G., and MENKE, W. (1983), *The Apparent Attenuation of a Scattering Medium*, Bull. Seismol. Soc. Am. *73*, 1005–1021.

RYBERG, T., FUCHS, K., EGORKIN, A. V., and SOLODILOV, L. (1995), *Observation of High-frequency Teleseismic P_n on Long-range QUARTZ Profile across Northern Eurasia*, J. Geophys. Res. *100*, 18,151–18,163.

RYBERG, T., WENZEL, F., MECHIE, J., EGORKIN, A. V., FUCHS, K., and SOLODILOV, L. (1996), *2D Velocity Structure beneath Northern Eurasia Derived from the Super Long-range Seismic Profile QUARTZ*, Bull. Seismol. Soc. Am. *86*, 857–867.

RYBERG, T., WENZEL, F., EGORKIN, A. V., and SOLODOILOV, L. (1998), *Properties of the Mantle Transition Zone in Northern Eurasia*, J. Geophys. Res. *103*, 811–822.

SANDMEIER, K.-J., and WENZEL, F. (1986), *Synthetic Seismograms for a Complex Model*, Geophys. Res. Lett. *13*, 22–25.

SHEARER, P. M. (1991), *Constrains on Upper Mantle Discontinuities from Observations of Long-period Reflected and Converted Phases*, J. Geophys. Res. *96*, 18,147–18,182.

SHEARER, P. M. (1996), *Transition Zone Velocity Gradients and the 520 km Discontinuity*, J. Geophys. Res. *101*, 3053–3066.

SHEARER, P. M., and MASTERS, T. G. (1992), *Global Mapping of Topography on the 660 km Discontinuity*, Nature *355*, 791–796.

SHERIFF, R. E., and GELDHARD, L. P., *Exploration Seismology; Vol. 1, History, Theory and Data Acquisition* (Cambridge Univ. Press 1982).

SMITHSON, S. B., *Contrasting types of lower crust.* In *Properties and Processes of Earth's Lower Crust,* (ed. R. F. Mereu) vol. 51 (AGU Geophys. Monograph, Washington D.C. 1989) pp. 53–63.

STEPHENS, C., and ISACKS, B. L. (1977), *Toward an Understanding of S_n: Normal Modes of Love Waves in Oceanic Structure,* Bull. Seismol. Soc. Am. *67,* 69–78.

SUTTON, G. H., and WALKER, D. A. (1972), *Oceanic Mantle Phases Recorded on Seismographs on the Northwestern Pacific at Distances between 7° and 40°,* Bull. Seismol. Soc. Am. *62,* 631–655.

THOMPSON, G. A., and HILL, J. L., *The deep crust in convergent and divergent terranes: Laramide uplifts and Basin and Range rifts.* In *The Continental Crust* (eds. M. Barazangi and L. D. Brown), vol. 14, In *Geodyn. Ser.* (Am. Geophys. Union 1986) pp. 243–256.

THOUVENOT, F., KASHUBIN, S. N., POUPINET, G., MAKOVSKIY, V. V., KASHUBINA, T. V., MATTE, P., and JENATTON, L. (1995), *The Root of the Urals: Evidence from Wide-angle Reflection Seismics,* Tectonophysics *250,* 1–13.

THYBO, H., and PERCHUĆ, E. (1997a), *The Seismic 8° Discontinuity and Partial Melting in Continental Mantle,* Science *275,* 1626–1629.

THYBO, H., and PERCHUĆ E., *A partially molten zone beneath the global 8° discontinuity at ~ 100 km depth with a new interpretation of the Lehmann discontinuity.* In *Upper Mantle Heterogeneities from Active and Passive Seismology* (ed K. Fuchs) vol. 17. In *NATO ASI Series* (Kluwer Academic Publishers, Dordrecht, NL 1997b) pp. 343–350.

THYBO, H., PERCHUĆ, E., and PAVLENKOVA, N., *Two reflectors in the 400 km depth range revealed from peaceful nuclear explosion seismic sections.* In *Upper Mantle Heterogeneities from Active and Passive Seismology* (ed. K. Fuchs) vol. 17. In *NATO ASI Series* (Kluwer Academic Publishers, Dordrecht, NL 1997) pp. 97–104.

TITTGEMEYER, M., WENZEL, F., FUCHS, K., and RYBERG, T. (1996), *Wave Propagation in a Multiple Scattering Upper Mantle—Observation and Modelling,* Geophys. J. Int. *127,* 492–502.

TITTGEMEYER, M., RYBERG, T., FUCHS, K., and WENZEL, F., *Observation of teleseismic P_n/S_n on super long-range seismic profiles in northern Eurasia and their implications for the structure of the lithosphere.* In *Upper Mantle Heterogeneities from Active and Passive Seismology* (ed. K. Fuchs) vol. 17. In *NATO ASI Series* (Kluwer Academic Publishers, Dordrecht, NL 1997) pp. 63–73.

VAN DER HILST, R. D., ENGDAHL, E. R., SPAKMAN, W., and NOLET, G. (1991), *Tomographic Imaging of Subducted Lithosphere below Northwest Pacific Island Arcs,* Nature *353,* 37–43.

WARNER, M., and MCGEARY, S. (1987), *Seismic Reflection Coefficients from Mantle Fault Zones,* Geophys. J. Royal Astron. Soc. *89,* 223–230.

WESSEL, P., and SMITH, W. H. F. (1995), *New Version of the Generic Mapping Tools (GMT),* version 3.0 released, Eos Trans. AGU *76,* 329.

WIDIYANTORO, S., and VAN DER HILST, R. D. (1996), *Structure and Evolution of Lithospheric Slab beneath the Sunda Arc, Indonesia,* Science *271,* 1566–1570.

WOLF, A. (1937), *The Reflection of Elastic Waves from Transition Layers of Variable Velocities,* Geophysics *2,* 357–363.

WU, R.-S., and FLATTÉ, S. M. (1990), *Transmission Fluctuation across an Array and Heterogeneities in the Crust and Upper Mantle,* Pure appl. geophys. *132,* 175–196.

(Received October 1, 1997; revised March 5, 1998)

Pure appl. geophys. 156 (1999) 53–81
0033–4553/99/010053–29 $ 1.50 + 0.20/0

│ Pure and Applied Geophysics

Seismic, Structural and Petrological Models of the Subcrustal Lithosphere in Southern Germany: A Quantitative Revaluation

T. WEISS[1], S. SIEGESMUND[1,2] and T. BOHLEN[3]

Abstract—Anisotropy in the subcontinental lithosphere becomes increasingly important, because it is observed in many seismic studies especially for P_n-waves. Typical rocks of the uppermost mantle are peridotites, which predominantly exhibit a pronounced elastic anisotropy. This anisotropy is mainly caused by the anisotropic elastic properties and the lattice preferred orientation (here referred to as texture) of olivine. To evaluate the elastic anisotropy of peridotites from the subcontinental lithosphere, specimens of the Northern Hessian Depression (Germany) and the Balmuccia Ultramafic Massif (Northern Italy) have been used. They comprise four olivine texture types, which are characteristic for olivine textures observed worldwide. The bulk rock elastic properties have been calculated using olivine and orthopyroxene textures, their single-crystal elastic constants at ambient pressure/temperature conditions and their volume fraction. Clinopyroxene and spinel are assumed to be randomly distributed. The effect of four different orientations of the foliation within the uppermost mantle has been evaluated, since this orientation is usually unknown.

Two of the olivine textures have a pronounced azimuthal dependence of compressional waves when a horizontal foliation within the uppermost mantle is presumed. These variations cause significant azimuthal variations of the P-wave reflections coefficients at the Moho. Primarily, we predict a significant azimuthal dependence of the critical points where the reflected amplitude increases from approximately 15% to 95%. Possibly, these azimuthal variations can be detected by seismic reflection measurements carried out at earth surface.

The remaining two texture types only manifest a small directional dependence. When anisotropy of compressional waves is observed in seismic studies, these latter types can only be of subordinate importance. However, all of the peridotites investigated are able to explain the seismically observed azimuthal variations of compressional waves when a vertical foliation is proposed. This ambiguity can be substantially reduced when shear waves (S-waves) are considered. The directional distribution of S-wave velocities and of the S-wave splitting exhibits characteristic patterns for the different olivine texture types. This could be used to discriminate between different texture types and orientations of the foliation within the uppermost mantle. A fundamental requirement for a more comprehensive interpretation is the availability of detailed S-wave observations. The maximum S-wave splitting in the peridotites investigated coincides with the maximum of the faster (leading) S-wave. This may be of importance to detect S-wave splitting in future seismic studies.

Key words: Seismic anisotropy, uppermost mantle, shear waves, compressional waves, peridotites, elastic properties.

[1] Institut für Geologie und Dynamik der Lithosphäre, Universität Göttingen, Goldschmidtstr. 3, 37077 Göttingen, Germany.
[2] Geologisch-Paläontologisches Institut, Bernoullistr. 23, CH-4056 Basel, Switzerland.
[3] Institut für Geowissenschaften, Universität Kiel, Otto-Hahn-Platz 1, 24118 Kiel, Germany.

1. *Introduction*

In the last several decades, many studies have been performed in order to better understand the seismic response of the earth's uppermost mantle. A fundamental observation was the directional dependence of seismic wave velocities beneath the oceanic crust (HESS, 1964). Additionally, a comparable anisotropy has been observed for the subcontinental lithosphere e.g., in Western Europe (BAMFORD *et al.*, 1979; FUCHS, 1983; BABUŠKA *et al.*, 1984, 1987), the Western and Northern USA (VETTER and MINSTER, 1981; SAVAGE and SILVER, 1993; SILVER and KANESHIMA, 1993), Southern Africa (VINNIK *et al.*, 1995) and Russia (CHESNOKOV and NEVSKIG, 1977). Anisotropy may be an important factor even at greater depths (MONTAGNER and ANDERSON, 1989; MONTAGNER and TANIMOTO, 1991). The origin of the seismic anisotropy in the subcrustal lithosphere has been essentially attributed to the anisotropic single-crystal elastic properties and the lattice preferred orientation (here referred to as texture) of olivine (e.g., NICOLAS, 1995; VINNIK *et al.*, 1995). The development of the texture may be caused by a variety of physical processes including heterogeneous crystallization, annealing, viscous and plastic deformation. Furthermore, present-day deformation (e.g., mantle-flow, absolute plate motion) or previous "frozen in" deformation events can be responsible for the development of a texture.

Numerous seismic refraction experiments in Southern Germany revealed comprehensive knowledge of the type and degree of P-wave velocity anisotropy in the uppermost mantle (e.g., EDEL *et al.*, 1975; GAJEWSKI and PRODEHL, 1985, 1987; ZEIS *et al.*, 1990). The evaluation of seismic travel times and amplitude information from Southern Germany led to a three-dimensional P_n-velocity model in the uppermost mantle, the so-called "anvil-model" (FUCHS, 1983). This model represents an approximation to the seismically observed P_n data using calculated elastic properties of hypothetical peridotites. For these peridotites Fuchs (1983) considered different modal compositions typical for mantle rocks consisting of olivine (ol), orthopyroxene (opx), clinopyroxene (cpx) and spinel (spi). With the exception of olivine, all other minerals are assumed to be randomly distributed. For olivine a specific amount of preferentially aligned crystals was considered by the simplified use of single-crystal orientations.

In contrast, the scope of this study is to compare the seismic signature of naturally deformed peridotites with the seismic observations. The observed fabric and olivine-texture types of the peridotites are characteristic for upper mantle rocks. The respective elastic properties of the peridotites are calculated considering the modal composition, frequently observed and distinct texture types of olivine and orthopyroxene, and depth and temperature corrections. Furthermore, the effect of different spatial orientations of the flow-planes on the elastic properties within the uppermost mantle will be investigated. In order to cover a wide range of olivine textures, peridotite xenoliths from the Northern Hessian Depression (NHD; Germany) and peridotites from the Balmuccia-Massif (Northern Italy) are used for the calculations. They comprise four representative types of olivine textures which are observed in peridotites worldwide. For these peridotites the elastic tensor was calculated. The first comparative parameter is the contribution of these peridotites to the observed directional dependence of compressional wave velocities (seismic P_n-waves, V_p) in the subcontinental lithosphere in Southern Germany. In the next step we investigate whether S-wave properties (i.e., the directional variation of V_{s_1} and V_{s_2} or the corresponding S-wave splitting (ΔV_s)) may provide further constraints for the interpretation of seismic investigations.

2. Seismic Anisotropy in the Uppermost Mantle of Southern Germany

The analysis of more than 700 P-wave velocities (V_p) registered in Southern Germany supplied evidence that a pronounced P_n-anisotropy exists in the uppermost mantle (e.g., BAMFORD, 1973, 1977). Bamford's P_n-velocities vary in the range of 7.8 km/s to 8.3 km/s, corresponding to a V_p-anisotropy of 7–8%. Generally, the uppermost ten kilometers beneath the Mohorovicic discontinuity (Moho) in Southern Germany can be regarded as an anisotropic gradient zone for P-wave velocities (e.g., FUCHS, 1983; ENDERLE et al., 1996). A significant increase of V_p with depth was proposed for the fast velocity direction. In his "anvil-model" FUCHS (1983) suggested an increase for $V_{p_{max}}$ from approximately 8.30 to 9.00 km/s within the first ten kilometers in the uppermost mantle (Fig. 1). In contrast, a corresponding $V_{p_{max}}$ increase from 8.03 to 8.60 km/s is estimated from a revaluation

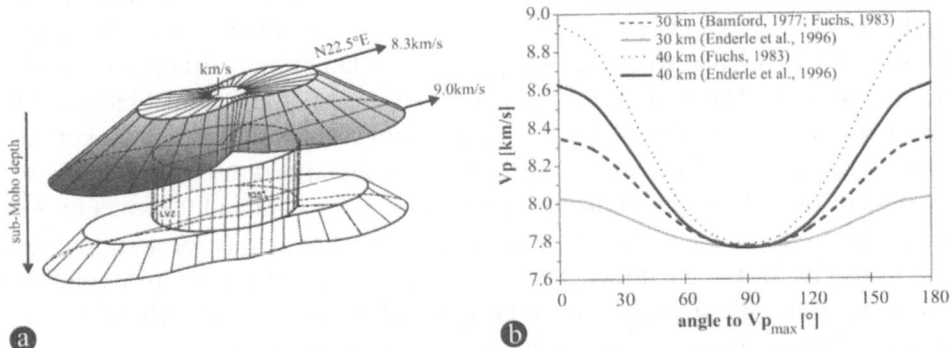

Figure 1

P_n velocities in the uppermost mantle of Southern Germany: (a) The "anvil"-model as a description of the distribution of compressional wave velocities in the uppermost mantle (FUCHS, 1983), and (b) azimuthal dependence of compressional waves at different depths. The Moho is presumed to be at 30 km. Graphs are given for depths of 30 km and 40 km, calculated from the data reported by FUCHS (1983) and ENDERLE et al. (1996). The velocity distribution of FUCHS (1983) directly below the Moho (30 km) corresponds to the velocity fit of BAMFORD (1977).

of the anvil-model, including recently observed travel times by ENDERLE *et al.* (1996). The proposed velocity gradient for the slow velocity direction $V_{p_{min}}$ is small or even zero (FUCHS, 1983; ENDERLE *et al.*, 1996).

The increase of $V_{p_{max}}$ with depth has been attributed to a varying portion (ENDERLE *et al.*, 1996) and orientation (e.g., FUCHS, 1983) of preferentially aligned olivine at the respective depth level.

3. Pressure and Temperature Dependence of the Elastic Properties

Pressure and temperature are competing parameters which effect the elastic properties of rocks. Numerous studies revealed that with increasing temperature and decreasing pressure the elastic wave velocities decrease (e.g., CHRISTENSEN, 1979). Thus the elastic constants must be corrected for ambient pressure and temperature conditions in the uppermost mantle.

For olivine, the single-crystal elastic properties and/or the pressure and temperature derivatives were taken from ISAAK (1992) and KUMAZAWA and ANDERSON (1969), for orthopyroxene from FRISILLO and BARSCH (1972), for clinopyroxene from LEVIEN *et al.* (1979) and for spinel from ANDERSON *et al.* (1968).

Investigations of surface heat flow reveal a mean value of 85 mW/m^2 (BLUNDELL *et al.*, 1992) for Southern Germany. Using a geotherm of POLLACK and CHAPMAN (1977) for the continental lithosphere gives P-/T-conditions of 0.85 GPa/700°C directly below the Moho (≈ 30 km depth) and 1.17 GPa/870°C at a sub-Moho depth of 10 km (≈ 40 km depth). The average *P*- and *S*-wave velocities of the rock-forming minerals as a function of depth can be summarized as follows:

At a depth of 30 km spinel (9.75 km/s, 5.5 km/s) exhibits the highest mean single crystal *P*-wave ($\overline{V_p}$) and *S*-wave ($\overline{V_s}$) velocities followed by olivine (8.25 km/s, 4.73 km/s), clinopyroxene (7.82 km/s, 4.45 km/s) and orthopyroxene (7.46 km/s, 4.54 km/s). The values for ($\overline{V_p}$) and ($\overline{V_s}$) decrease at around 0.2 km/s and 0.1 km/s from 30–40 km, respectively. Compared with surface conditions, the consideration of ambient pressures and temperatures for the uppermost mantle leads to a decrease of the velocities with depth up to 0.4 km/s for ($\overline{V_p}$) and 0.2 km/s for ($\overline{V_s}$). This velocity reduction is almost twice as large as when it was formerly proposed by FUCHS (1983). His anvil-model is based on pressure and temperature conditions calculated from the geotherm of WERNER and KAHLE (1980) which produces significantly lower temperatures at 30 and 40 km, respectively.

4. Compositional Variability in the Earth's Uppermost Mantle

The assumption of a reasonable modal composition is a crucial parameter for the calculation of the elastic properties of a rock. For the present calculations

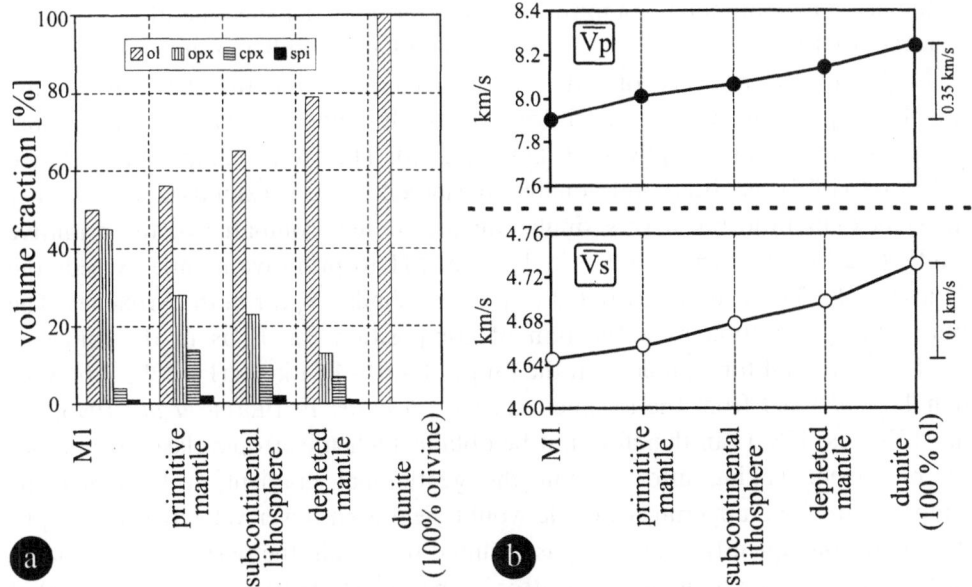

Figure 2

Compositional variability within the uppermost mantle (a), and the effect of the different modal compositions on the averaged P- and S-wave velocities; (b) at a depth of 30 km (0.85 GPa/700°C).

modal compositions have been used which represent the overall compositional variability of upper mantle rocks (e.g., HARTMANN and WEDEPOHL, 1990). The uppermost mantle in the Variscan Europe is probably in the spinel-peridotite stability field (MENZIES and BODINIER, 1993). The main rock-forming minerals are olivine, orthopyroxene, clinopyroxene and spinel. Based on these dominant minerals, three types of modal compositions are distinguished (see Fig. 2a):

(1) The average composition of fertile (primitive) mantle rocks e.g., from Central Mongolia is 58% ol, 22.5% opx, 14.6% cpx and 2.7% spi (PRESS *et al.*, 1986). This composition is similar to that of pyrolite (55% ol, 27.1% opx, 15% cpx, 1.5% spi) as defined by RINGWOOD (1973). The spinel peridotite from Balmuccia (Ivrea Zone, Italy) in particular contains an average of 56% ol, 28% opx, 14% cpx and 2%, which corresponds to a moderate depletion in basaltic components of only 4.5% (HARTMANN and WEDEPOHL, 1993).

(2) An intermediate member is represented by a worldwide compilation of mantle rocks which describes the composition of the subcontinental lithosphere. The average volume fraction of olivine is 65% (HARTMANN and WEDEPOHL, 1990).

(3) The depleted mantle types describes a modal composition of a peridotite, which is depleted in basaltic components by partial melting. The conversion from a primitive to a depleted mantle composition is expressed by a decrease in clinopyroxene and, to a lesser degree, of orthopyroxene. This decrease is partially compensated by a relative increase in olivine ($\approx 72\%$; HARTMANN and WEDEPOHL, 1990).

Similar ranges of modal compositions from fertile to depleted mantle rocks have been reported by JI *et al.* (1994) from the Canadian Cordilleran and Alaska.

Neglecting the presence of lattice preferred orientation, the effect of varying modal compositions is illustrated for commonly observed mantle rocks using average velocities for a depth of 30 km (Fig. 2b). The "M1" composition relative to FUCHS (1983) has been additionally considered, since it exhibits a significantly higher opx/cpx-ratio but a very similar olivine content compared to the primitive mantle (see Fig. 2a). The increase in $\overline{V}_p$ from M1 to primitive mantle exceeds the changes in $\overline{V}_s$. This can be attributed to the very similar $\overline{V}_s$ for both pyroxenes. The total velocity variation as a function of composition for $\overline{V}_p$ is in the range of 7.9–8.25 km/s and for $\overline{V}_s$ merely in the range of 4.64–4.74 km/s (Fig. 2b). The very high $V_{p_{max}}$ derived from the seismic data ($\approx$ 8.6 km/s: ENDERLE *et al.*, 1996; 9.0 km/s; FUCHS, 1983) can therefore not be obtained when isotropic elastic properties are presumed. Taking into account the geothermal gradient, a homogeneous composition of the uppermost mantle would yield even lower velocities at a depth of 40 km. Consequently, anisotropy is required to explain the seismic observations. Furthermore, it is obvious that an isotropic uppermost mantle causes no azimuthal dependence for P_n-waves. Finally, it is well known from experimental and petrofabric studies, that ultramafic rocks are predominantly anisotropic (e.g., PESELNICK *et al.*, 1974; CHRISTENSEN and LUNDQUIST, 1982; CHRISTENSEN, 1984; MAINPRICE and SILVER, 1993).

5. *Velocity Anisotropy Controlled by Rock Fabrics*

Most rocks are commonly elastically anisotropic as a result of the rock fabrics. Due to the single-crystal anisotropy, a polycrystal consequently displays an overall anisotropy caused by the crystallographic preferred orientation of the rock-forming minerals. At shallower crustal levels, a preferred orientation of microcracks must be considered (e.g., SIEGESMUND *et al.*, 1993). Preferred orientations can be produced by a number of processes (for details see SKROTZKI, 1994). The main rock-forming minerals in peridotites (olivine, orthopyroxene and clinopyroxene) exhibit a pronounced single-crystal anisotropy. For the calculation of the anisotropic elastic properties of the peridotites from the NHD and Balmuccia, olivine and orthopyroxene textures were considered. Clinopyroxene is assumed to be randomly distributed since it often shows weak textures in specimens from the subcontinental lithosphere (JI *et al.*, 1994).

Olivine and orthopyroxene exhibit the maximum, intermediate and minimum V_p along the *a*-, *c*- and *b*-axes, respectively (Fig. 3). The maximum single-crystal V_p-anisotropy is 21.9% for olivine and 16.2% for orthopyroxene ($A[\%] = 100(V_{p_{max}} - V_{p_{min}})/V_{p_{max}}$). The *S*-wave-splitting patterns ($\Delta V_s = V_{s_1} - V_{s_2}$) are more complicated. For olivine the minimum *S*-wave splitting can be observed in the direction of the *a*-axis, whereas the maximum *S*-wave splitting (0.9 km/s) is in the direction between the *a*- and *c*-axis. The orthopyroxene single-crystal shows the minimum *S*-wave splitting in the directions of the crystallographic axes, whereas a pronounced velocity difference between the split *S*-wave (0.6 km/s) can be observed at an angle of approximately 45° between each of the crystallographic axes (Fig. 3).

The fabrics of peridotites observed worldwide exhibit remarkably constant types. A microstructural classification on a few thousand xenoliths from Western Europe, Hawaii, Ethiopia, Baja California and South Africa revealed the existence of three main types of peridotite fabrics, though there exist transitional stages (MERCIER and NICOLAS, 1975); (i) protogranular, (ii) porphyroclastic, partially recrystallized and (iii) equigranular recrystallized. These fabric variations are also represented in the samples from the NHD and the Balmuccia peridotite massif

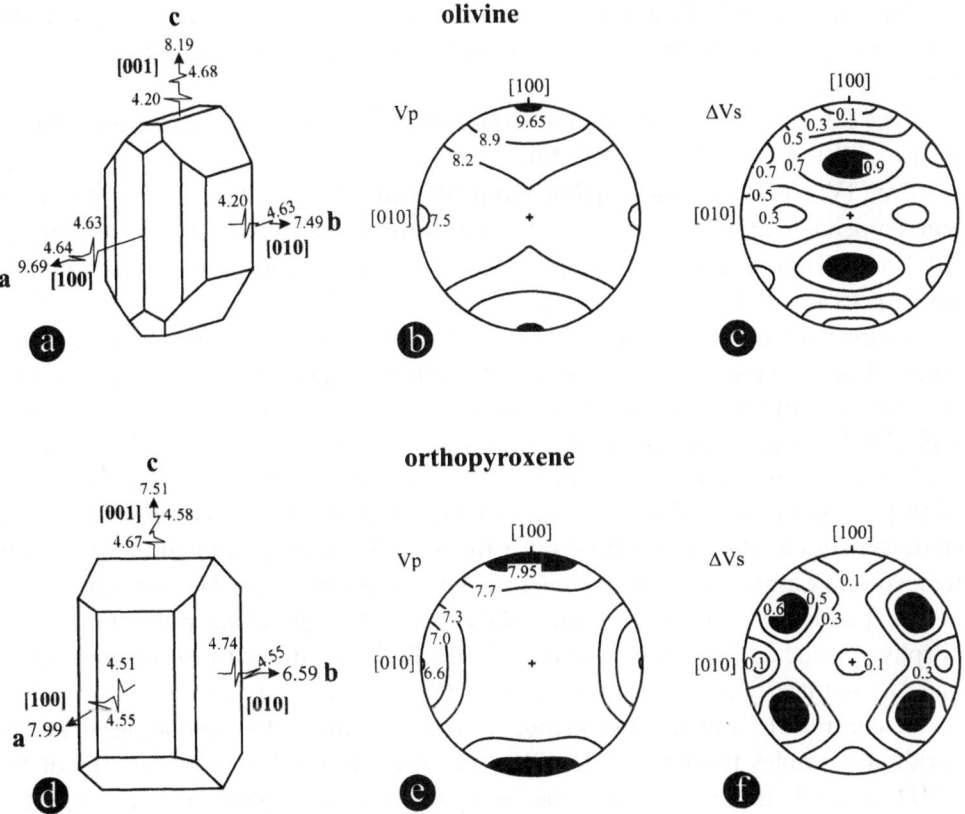

Figure 3

Single crystal elastic properties of olivine (a–c) and orthopyroxene (d–f). For the single crystals, V_p and V_s and the polarization directions of V_s are given in the directions of the crystallographic reference frame marked by the [100]-, [010]- and [001]-axis (a, d). Additionally, the spatial distribution of V_p and $\Delta V_s (\Delta V_s = V_{s_1} - V_{s_2})$ of olivine (b, c) and orthopyroxene (e, f) is shown (equal area projection, lower hemisphere, contours in km/s). The elastic properties are corrected for pressure-/temperature conditions at a depth of 30 km (0.85 GPa, 700°C).

(WEDEL, 1990; SKROTZKI *et al.*, 1990; WEDEL *et al.*, 1992; JIN *et al.*, 1998). In order to define a reference system for the description of fabric and texture data, generally a coordinate system (x, y, z) is used. The xy-plane of this coordinate system corresponds to the metamorphic foliation, frequently marked by changes in composition or by elongated grains. The x-direction is defined by a parallel alignment of grains in the form of a mineral or stretching lineation. Most of the peridotites from the NHD display a more or less developed foliation and a lineation marked by spinel and pyroxene grains (WEDEL, 1990).

Olivine textures from peridotites of the NHD and from Balmuccia can be divided into four characteristic types which are representative of worldwide observed textures (Fig. 4):

Type I: strong [010] maximum normal to the foliation, [100] and [001] girdles perpendicular to it (WEDEL *et al.*, 1992; NHD).

Type II: strong [100] maximum in the foliation or slightly inclined, [010] and [001] girdles perpendicular to it, often with subordinate maxima (WEDEL *et al.*, 1992; NHD).

Type III: strong [001] maximum, [100] and [010] girdles or maxima perpendicular to it (WEDEL *et al.*, 1992; NHD).

Type IV: single maxima of [100] and [010] parallel to the lineation (x-direction) and perpendicular to the foliation (z-direction), respectively. The corresponding [001] axes form a maximum perpendicular to the lineation within the foliation (SKROTZKI *et al.*, 1992; Balmuccia).

A direct correlation between fabric- and texture-type is lacking (WEDEL *et al.*, 1992). For example, a study of porphyroclastic (high-temperature event) and equigranular (medium-temperature event) peridotites from Balmuccia reveal a typical high-temperature creep microstructure consisting of dynamically recrystallized grains, subgrains and free dislocations in between (SKROTZKI *et al.*, 1990). Both peridotites, the prophyroclastic and the equigranular exhibit similar textural characteristics as shown in Figure 4 for the Type IV texture, though the stress and temperature estimations from the microstructures are different. The grain structures in the peridotite xenoliths of the NHD vary strongly from protogranular to porphyroclastic and to equigranular with the above-mentioned textural types (I, II and III) for nearly all individual locations (WEDEL *et al.*, 1992).

In contrast to olivine, the orthopyroxene texture is less pronounced in the peridotite samples from the NHD (WEDEL *et al.*, 1992). For the xenoliths of the NHD, a weak trend exists for the main slip direction [001] of orthopyroxene perpendicular to the main slip direction [100] of olivine. SKROTZKI *et al.* (1990) conclude that this orthopyroxene texture may result from a rigid body rotation around [001] in the ductile olivine matrix analogous to the rotation of prolate grains in a viscous fluid described by TAYLOR (1923). The specimen from the Balmuccia peridotite (type IV texture; Fig. 4) exhibits an orientation of the orthopyroxene c-axes perpendicular to the lineation, whereas the b- and a-axes are

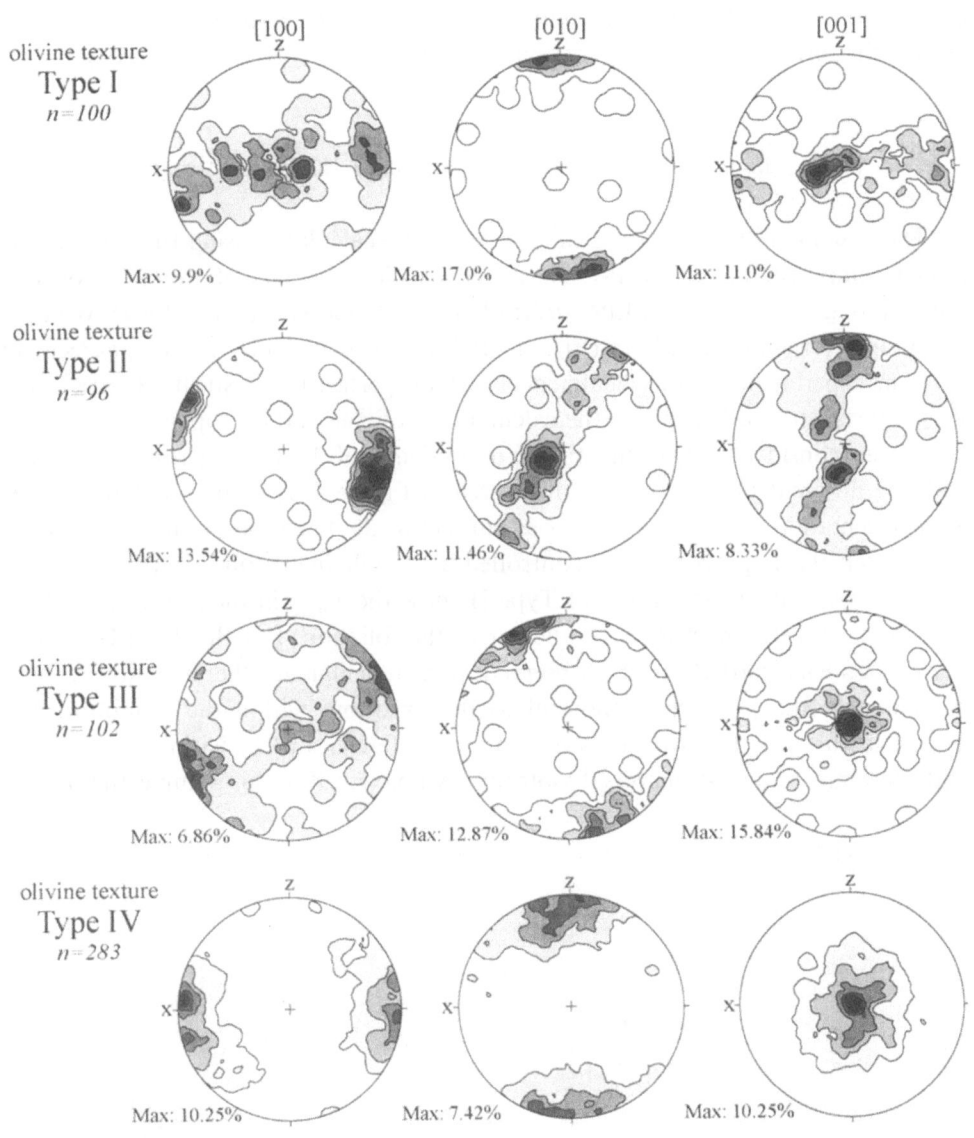

Figure 4

Preferred orientation of olivine [100], [010] and [001] in an equal area projection, lower hemisphere, for the different texture types (I–IV). The textures are shown with respect to the main fabric elements foliation (*xy*-plane) and lineation (*x*-direction). The contour intervals are 1, 3, 5, 7%, . . . of the number of measurements per 1% of the hemisphere. The number of measurements is given by the value of *n*, the maximum concentration by the value of Max (Type I–III WEDEL *et al.*, 1992; Type IV SKROTZKI *et al.*, 1992).

arranged along girdles perpendicular to the c-axes. The dominant slip system in both orthopyroxene and clinopyroxene is (100)[001] (e.g., MERCIER, 1985).

6. Directional Dependence of Compressional Waves in Peridotites

The overall elastic properties of a rock can be calculated using the texture, the modal composition and the P-/T-corrected elastic constants of the rock-forming minerals (e.g., CROSSON and LIN, 1971; MAINPRICE and HUMBERT, 1994). We used the Voigt-averaging technique (VOIGT, 1928), which provides a close agreement between experimental and calculated elastic properties (e.g., SIEGESMUND *et al.*, 1994; SIEGESMUND, 1996). For the calculations, olivine and orthopyroxene textures have been considered. The directional dependence of V_p for the respective rocks with the different texture types is illustrated in Figure 5. A common characteristic is the orientation of $V_{p_{max}}$ more or less parallel to the lineation or slightly inclined. The overall elastic properties are controlled by the olivine texture. Furthermore, all texture types with the exception of Type II show the $V_{p_{min}}$ in the z-direction of the coordinate system which is perpendicular to the foliation. For the Type II texture, a slow velocity girdle can be observed perpendicular to the lineation with a maximum in the y-direction (normal to the lineation, within the foliation). The Type I texture (see Fig. 4) causes only minor differences in V_p within the foliation and therefore a quasi-transversal isotropic symmetry. For this sample the a- and

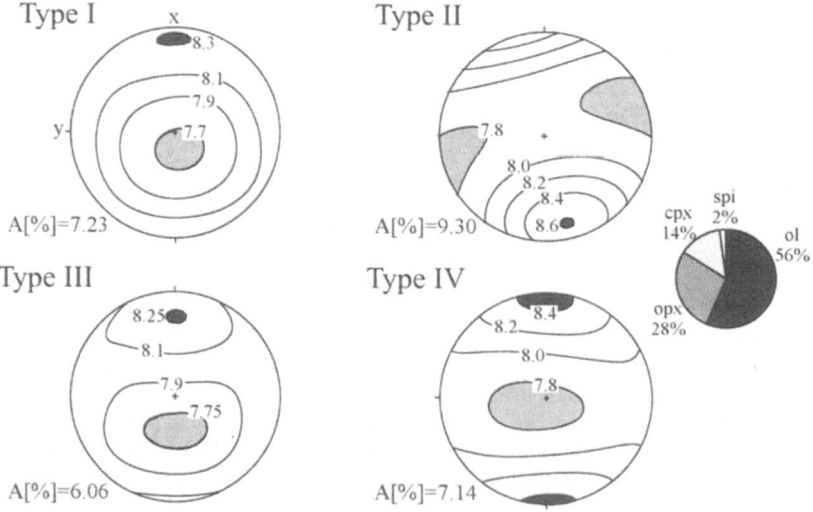

Figure 5

Directional variation of V_p calculated for the different texture types (0.85 GPa/700°C, ≈ 30 km) assuming a primitive mantle composition (as indicated). The projection plane corresponds to the foliation, the lineation (x-direction) is N–S (equal-area projection, lower hemisphere, isolines in km/s).

c-axes of olivine form girdles within the foliation. This corresponds to an averaging of intermediate and high single crystal velocities within the foliation. The texture Types III and IV are characterized by an orthorhombic symmetry with the $V_{p_{max}}$ in the direction of the lineation and the $V_{p_{min}}$ perpendicular to the foliation (Fig. 5).

At a depth of 30 km, assuming a primitive mantle composition (see Fig. 2a), $V_{p_{max}}$ varies within a range of 8.25 km/s (Type I texture) to 8.61 km/s (Type II texture). The $V_{p_{min}}$ covers a very limited range from 7.66 km/s to 7.76 km/s for the Type I and Type II textures, respectively. The V_p anisotropy varies between 6.42% (Type III texture) and 9.87% (Type II texture). At a depth of 40 km, with an assumed depleted mantle composition (see Fig. 2a), the $V_{p_{min}}$ remains nearly unchanged, whereas the $V_{p_{max}}$ increases. The result is an increasing V_p-anisotropy of about 1–2%.

If a horizontal foliation is assumed, the directional dependence of the P_n-wave velocities corresponds to the velocity distribution in the projection plane. Notice, that in this plane the maximum velocity variations do not occur, with the exception of the Type II texture (Fig. 5).

7. Directional Dependence of S-waves in Peridotites

When S-waves pass an anisotropic medium they are split into two orthogonally polarized waves which propagate with different velocities (e.g., CRAMPIN, 1981). Accordingly, the S-wave splitting (ΔV_s) is a direct measure of anisotropy in every wave propagation direction. The specimens with the texture Types II, III, and IV display a minimum of the S-wave splitting in the x-direction, which coincides with the maximum concentration of the olivine a-axes (Fig. 6). In contrast, the Type I texture shows the minimum of the S-wave splitting perpendicular to the foliation. Comparable to the V_p-distribution this can be attributed to the a- and c-axes girdles of olivine within the foliation for the Type I texture. A pronounced S-wave splitting (0.25–0.3 km/s) can be observed at an angle of approximately 45° to $V_{p_{max}}$ within the foliation for all texture types. If the $V_{p_{max}}$ direction is assumed to be at N22°E (FUCHS, 1983) or N31°E (ENDERLE et al., 1996), a maximum S-wave splitting could be expected at an angle of N67°E or N76°E, respectively. Generally, the velocity difference between the two split S-waves is on average 0.2 km/s in peridotites (CHRISTENSEN and RAMANANTOANDRO, 1971). Only a slight increase (≈ 0.08 km/s) of $\Delta V_{s_{max}}$ can be detected at a depth of 40 km as a result of a composition change from primitive to depleted mantle.

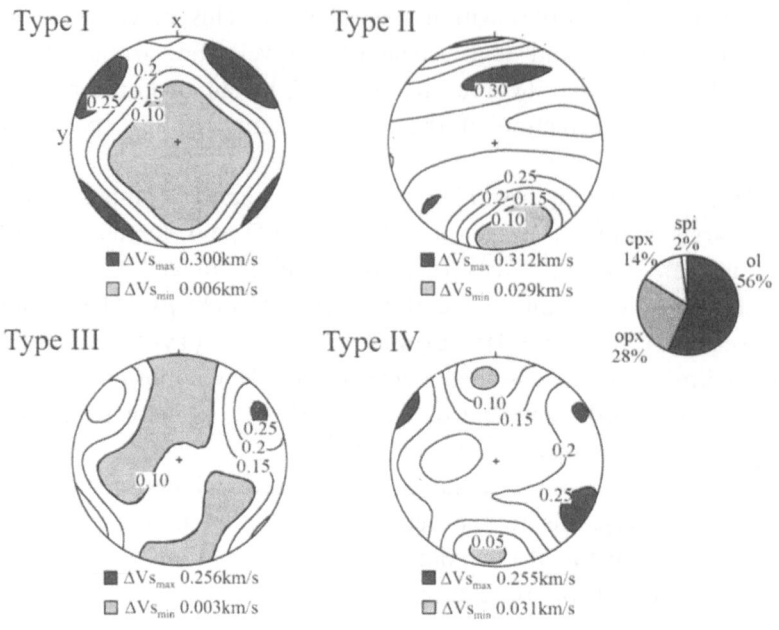

Figure 6
Directional variation of ΔV_s calculated for the different texture types and P/T conditions at 0.85 GPa/700°C (30 km). The modal composition is as indicated, the lineation (x-direction) is N–S, the foliation (xy-plane) is horizontal (equal-area projection, lower hemisphere).

8. Anisotropy of Compressional Waves at Different Depths

The velocity distributions derived from naturally deformed peridotites can be compared with velocity-fits derived from seismic data of previous studies (e.g., BAMFORD, 1973, 1977; FUCHS, 1983; ENDERLE *et al.*, 1996; see Fig. 1). Therefore, P-wave velocities of the investigated peridotites have been selected in the horizontal plane of P_n-propagation (xy-plane).

A significant azimuthal dependence can be observed for peridotites with the texture Types II and IV (Fig. 7a). At a depth of 30 km the patterns are comparable to the values proposed by BAMFORD (1977) and FUCHS (1983). Unusually high P_n-velocities of around 9.0 km/s cannot be achieved even if a depleted mantle composition is considered (Fig. 7b). The change in modal composition leads to a $V_{p_{max}}$ increase of about 0.2 km/s from 30 to 40 km only. For the texture Types I and III a directional V_p-dependence is more or less lacking and the increase of the velocities from a depth of 30 km to 40 km is less pronounced (≈ 0.05 km/s). A maximum velocity of 8.6 km/s at 40 km as proposed by ENDERLE *et al.* (1996) can easily be derived from naturally deformed peridotite xenoliths. The best fit of xenolith and seismic velocity data is given by peridotites corresponding to the Type II texture (Fig. 7c).

9. Anisotropy of Horizontally Propagating Compressional Waves as a Function of Flow Plane Geometries

The original orientation of the peridotites within the continental lithospheric mantle prior to their emplacement in the host magma is impossible to determine and may vary with the tectonic setting (BABUŠKA and CARA, 1991; MAINPRICE and SILVER, 1993; JI et al., 1994). Recent asthenospheric flow may be closely linked to absolute plate motion (APM; e.g., BORMANN et al., 1996; VINNIK, 1997). In contrast, anisotropy caused by rock fabrics in the subcrustal lithosphere is probably a remnant of ancient mantle flow (e.g., BABUŠKA et al., 1987). The resulting textures of olivine are closely related to strain and ambient temperature with respect to the flow-plane. Several models have been proposed for the orientation of the foliation within the uppermost mantle. BABUŠKA et al. (1987) proposed dipping

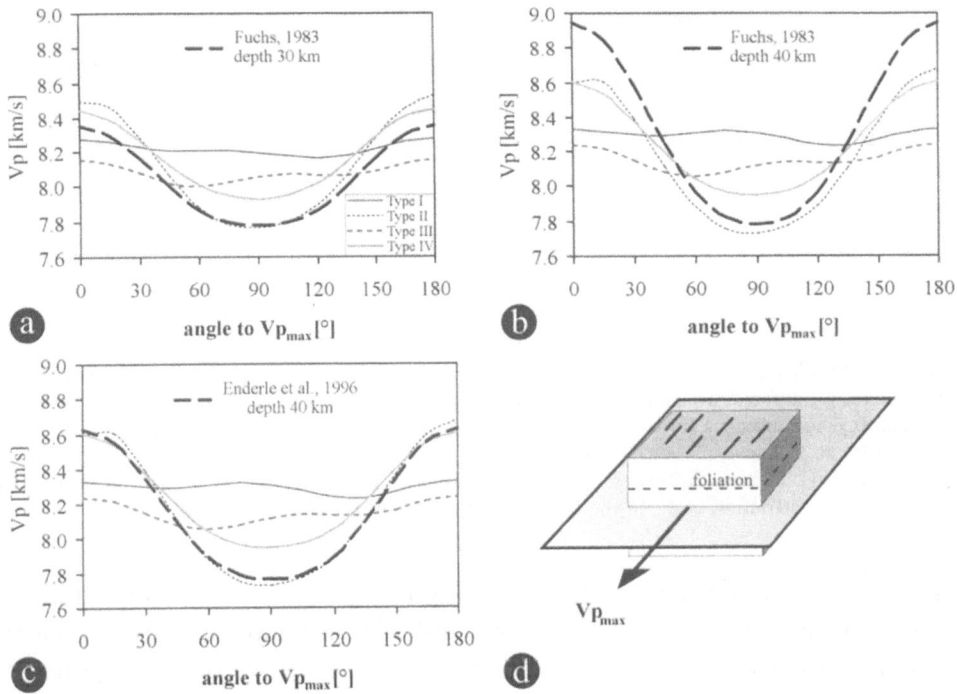

Figure 7

Comparison of the azimuthal dependence of V_p within a horizontal plane for the different olivine texture types with velocity fits to the seismic data from previous studies. The orientation of the foliation is assumed to be horizontal (d), the velocity distributions are shown as a function of angle to the $V_{p_{max}}$ direction. The V_p-distributions calculated for the peridotites are compared with (a) the P_n-distribution of BAMFORD (1977) and FUCHS (1983); (b) with the V_p-distribution of FUCHS (1983) at a depth of 40 km, and (c) with the V_p-distribution of ENDERLE et al. (1996) at 40 km. For the depth of 30 km the primitive mantle and for the depth of 40 km the depleted mantle composition has been considered for the calculations.

66 T. Weiss *et al.* Pure appl. geophys.,

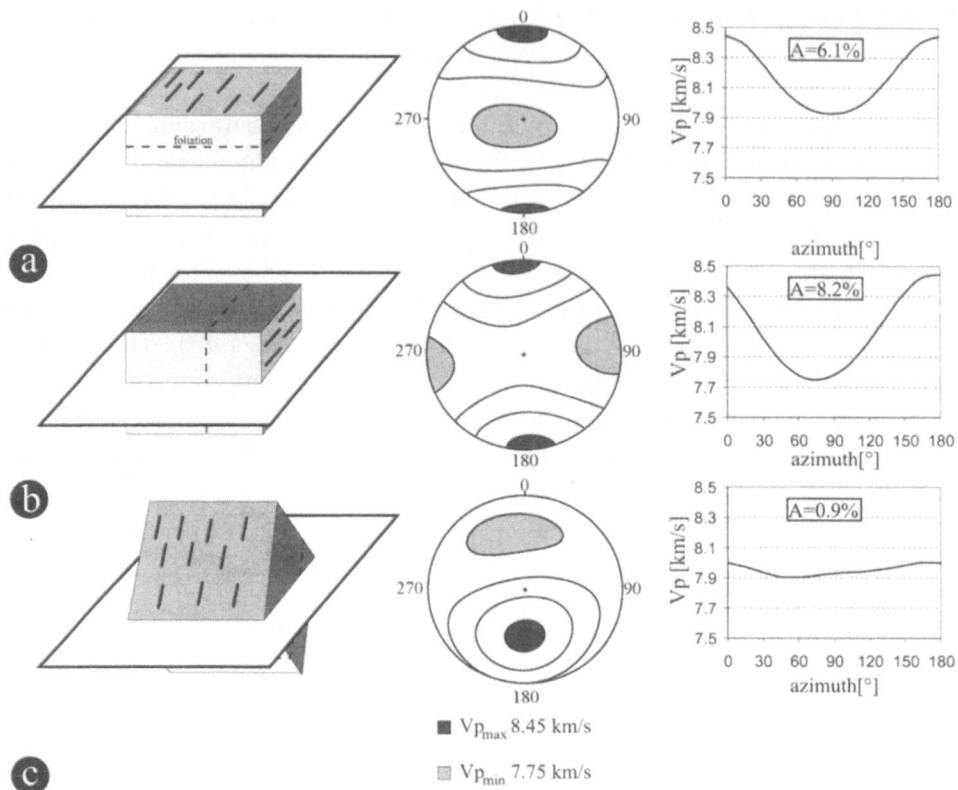

Figure 8
Azimuthal dependence of the V_p for the specimen corresponding to the Type IV texture. The velocities are calculated for different orientations of the foliation within the uppermost mantle (schematic sketches, left column). These are (a) a horizontal foliation, (b) a vertical foliation (e.g., MAINPRICE and SILVER, 1993) and (c) an inclined foliation dipping at an angle of 45° (e.g., BABUŠKA *et al.*, 1987). The spatial distribution of V_p (equal-area, lower hemisphere, middle column) and the corresponding azimuthal variation of V_p in the horizontal plane (right column) are shown.

stacks of uppermost mantle with an inclined $V_{p_{max}}$ direction for Central Europe from evaluations of teleseismic P-wave residuals. The assumption of deep-reaching strike-slip movements (e.g., MAINPRICE and SILVER, 1993; NICOLAS, 1995) results in a vertical orientation of the foliation. A horizontal orientation of the foliation is implied from investigations of ultramafic ophiolite complexes (e.g., CHRISTENSEN and LUNDQUIST, 1982).

Depending on the assumed geometry, the azimuthal dependence of V_p within the horizontal plane of P_n-propagation varies significantly. These dependencies are illustrated for a peridotite with the Type IV texture (Fig. 8).

The V_p-anisotropy is about 6%, assuming a horizontal foliation (Fig. 8a). In this case the P_n-wave propagation corresponds to the V_p-distribution in the xy-plane of the reference system. The anisotropy increases at about 8.2%, when a vertical foliation is assumed (Fig. 8b). A negligible anisotropy is observed when the $V_{p_{max}}$-direction dips at an angle of $\approx 45°$ (Fig. 8c).

When oblique textures are considered (e.g., Type II texture), the resulting velocity distribution depends on the spatial orientation of the peridotite within the uppermost mantle. Conceivable peridotite orientations within the uppermost mantle are illustrated in Figure 9a for the Type II texture. A gradual rotation of a peridotite following the "sense 1" rotation has a slight effect on the azimuthal dependence of V_p (Fig. 9b), whereas the opposite effect is observable for a "sense 2" rotation (Fig. 9c).

A vertical orientation of the foliation can be caused by deep-reaching strike-slip movements. In this case, the resulting azimuth dependence of horizontally propagating P_n-waves is distinct for all of the texture types (Fig. 10a). The Type IV texture closely fits the velocity curve of the anvil-model (FUCHS, 1983) and

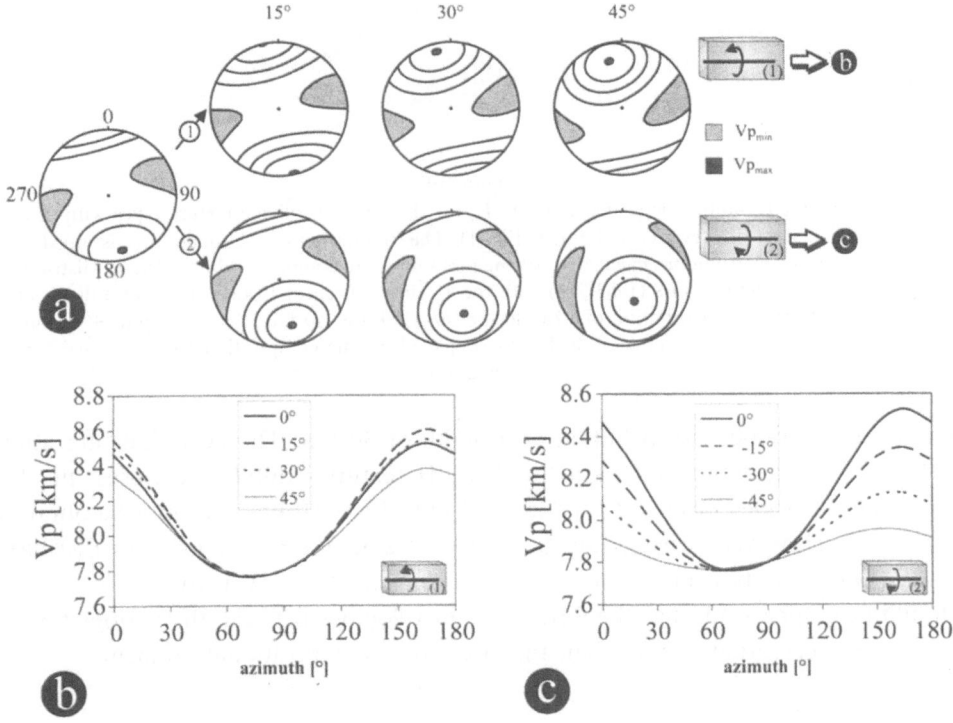

Figure 9

Velocity distribution for V_p for varying incidence angles of the lineation (x-direction) for a peridotite with an oblique olivine texture (e.g., the Type II texture) at a depth of 30 km: (a) The spatial distribution of V_p for an original horizontal foliation, which is gradually rotated applying a "sense 1" and "sense 2" rotation in steps of 15° as indicated (equal area projection, lower hemisphere). (b,c) The corresponding V_p azimuth relationships within horizontal planes for the pole figures shown in (a).

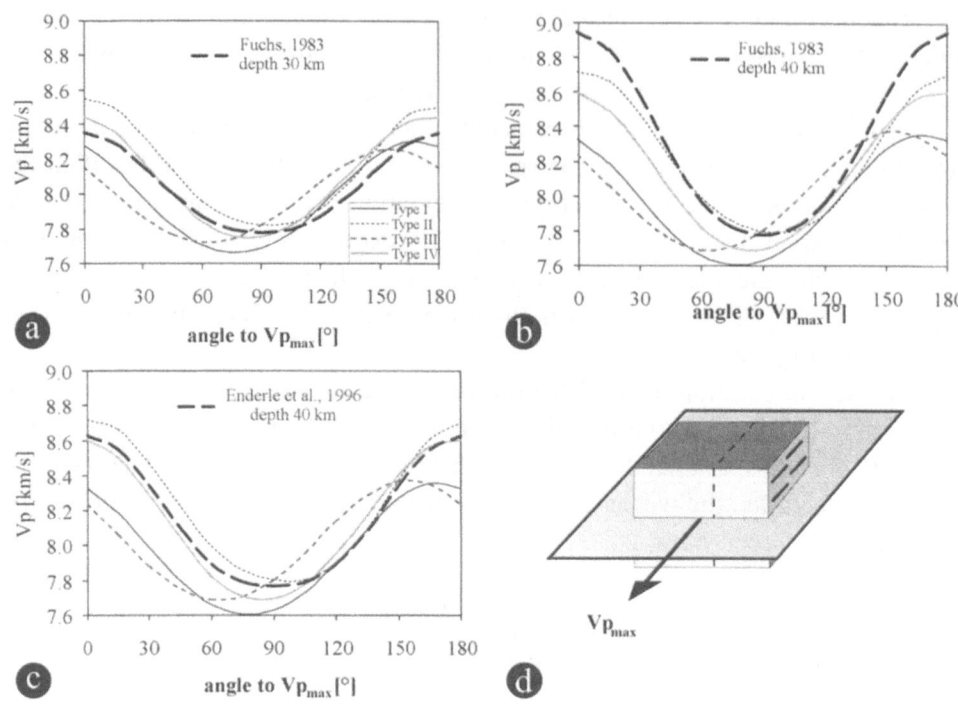

Figure 10
Comparison of the azimuthal dependence of the V_p for the different olivine texture types with velocity fits from seismic data of previous studies (see Fig. 1). The orientation of the foliation is assumed to be vertical (d). The V_p-distributions calculated for the peridotites are compared with (a) the P_n-distribution of BAMFORD (1977) and FUCHS (1983), (b) with the V_p-distribution of FUCHS (1983) at a depth of 40 km, and (c) with the V_p-distribution of ENDERLE *et al.* (1996) at 40 km. For the depth of 30 km the primitive mantle and for the depth of 40 km the depleted mantle composition has been used for the calculations.

Bamford's P_n-velocity distribution at a depth of 30 km. The deviations are about 0.05 km/s. At a depth of 40 km the Type II texture exhibits the closest fit. The maximum velocities proposed for the anvil-model (≈ 9.0 km/s) cannot be obtained (Fig. 10b). However, considering the updated velocity estimations from ENDERLE *et al.* (1996), the best fit can be achieved using the Type IV texture (Fig. 10c). It should be emphasized, that the differences in velocities between the various texture types are substantially larger than the effect of a compositional gradient.

10. S-wave Observations as a Further Constraint

Based on the evaluation of *S*-wave velocities in Southern Germany from earthquakes, no indicators were found for an *S*-wave splitting (PLENEFISCH *et al.*,

1994). The S-wave velocities from earthquakes vary in the range of 4.6–4.8 km/s (Fig. 11a) which is almost identical with the variability observed for the peridotites in the present study (Fig. 11b).

The maximum S-wave splitting for the investigated peridotites is closely linked to the $V_{s_{max}}$-direction (Figs. 12a–d). Since all the texture types produce a pronounced azimuthal variation of V_p assuming a vertical foliation, at least two texture types can be excluded when observations of an S-wave splitting were available. The peridotites corresponding to the texture Types II and IV still manifest a pronounced S-wave splitting at an angle of 45° to $V_{p_{max}}$ (Fig. 13). The Type I texture exhibits a very small S-wave splitting within the range of 45° to 135° to $V_{p_{max}}$ and the Type III texture a small S-wave splitting at an angle of 135° to $V_{p_{max}}$ (Fig. 13). Thus, for a more comprehensive interpretation, seismic profiles at both azimuths and with emphasis on S-waves are required.

Direct anisotropy indications can furthermore be derived from SKS-data. Generally, the SKS-splitting in continental regions gives an average value of 1 s (SILVER and CHAN, 1991). In Central Europe, VINNIK et al. (1994) and BORMANN et al. (1993, 1996) give an average value of 0.83 s for the SKS-splitting. Based on thermal and seismic estimates, the lithospheric thickness in central Germany ranges between 60 and 80 km (MÜLLER and PANZA, 1984; ČERMAK and BODRI, 1993).

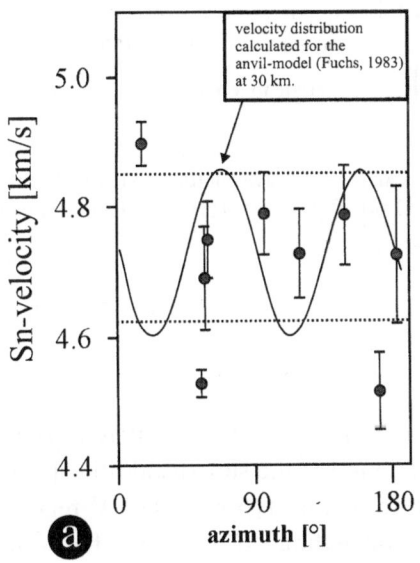
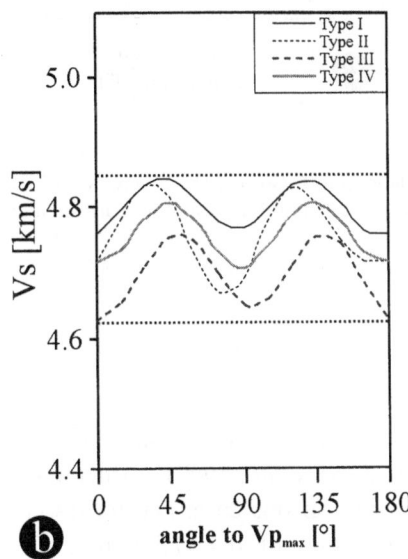

Figure 11

Azimuthal variation of S-wave velocities: (a) S_n-wave velocities derived from earthquake data (circles) and of those calculated for the anvil-model at a depth of 30 km (after PLENEFISCH et al., 1994) and (b) the directional dependence of the velocity of the faster S-wave for peridotites with the texture Types I–IV within the horizontal foliation. Notice that the fastest V_p-direction is N22.5°E.

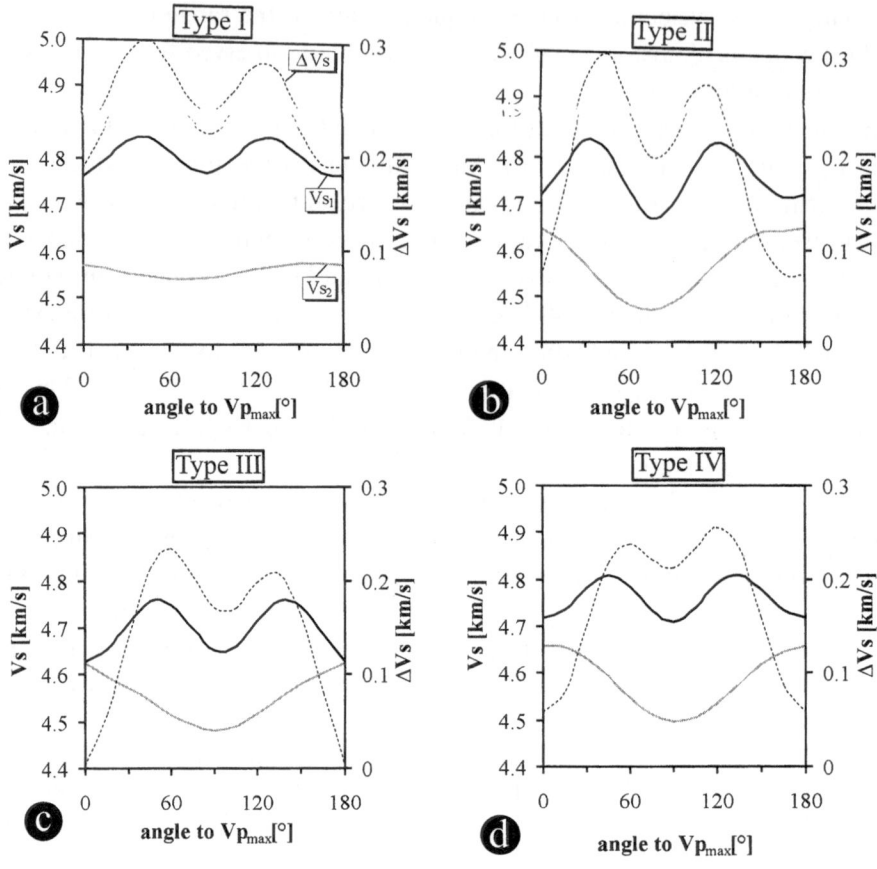

Figure 12
Relationship between both *S*-wave velocities (V_{s_1} and V_{s_2}) and the *S*-wave splitting (ΔV_s) within the plane of foliation for peridotites corresponding to the Type I (a), Type II (b), Type III (c) and Type IV (d) texture.

When the crustal thickness approaches 30 km (GIESE and PAVLENKOVA, 1991; BORMANN *et al.*, 1993), a thickness of at most 50 km for the subcrustal lithosphere remains. In this case, peridotites corresponding to texture types II and IV register delay times of 0.4–0.5 s for the vertically propagating *SKS*-waves. Thus the uppermost mantle can explain at least a part of the *SKS*-splitting as observed in Central Europe. A clear differentiation between the different texture types or peridotite orientations on the basis of SKS-observations is impossible. When a vertical foliation is assumed, all the texture types give delay times in a range of 0.4–0.6 s.

11. Directional Variations of P-wave Reflection Coefficients

An important question for reflection seismic exploration of the subcrustal (anisotropic) lithosphere is whether reflection amplitudes of compressional waves are sensitive to a directional dependence of seismic wave velocities in the uppermost mantle. It is possible, in general, only to discriminate between the four texture Types I–IV (anisotropy patterns shown in Figs. 5 and 6) using amplitude variations of reflected compressional waves with both offset (AVO) and azimuth (AVA)?

To explore the sensitivity of P-wave reflection amplitudes to the anisotropy patterns of the uppermost mantle, we computed the directional variation of PP-reflection coefficients at the Moho by means of 3-D ray tracing (GAJEWSKI and PSENČIK, 1990). In the computations the elastic tensors of the different texture Types I–IV were used for the mantle. To investigate the influence of anisotropy on the PP-reflection coefficients we also computed the coefficients for an isotropic mantle. We use the averaging procedure of VOIGT (1928) to derive effective elastic wave velocities for the texture types I–IV. In all cases the lower crust was assumed to be isotropic ($V_p = 6.4$ km/s, $V_s = 3.8$ km/s, density $= 2900$ kg/m^3).

A correlation between the directional variation of reflection coefficients (Fig. 14) and the corresponding variations of P-wave velocities (Fig. 5) and shear-wave splitting (Fig. 6) is not obvious. Generally, at the so-called critical angle (of

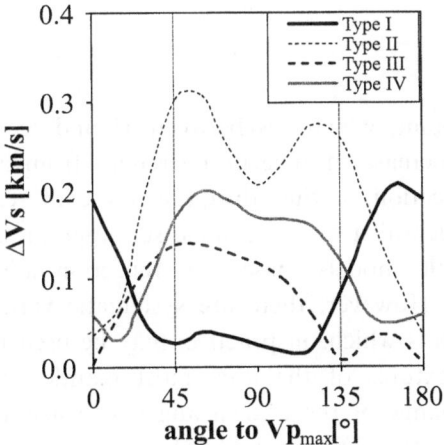

Figure 13
Azimuthal dependence of the S-wave splitting within a horizontal plane, when a vertical foliation is assumed for the different texture types (I–IV).

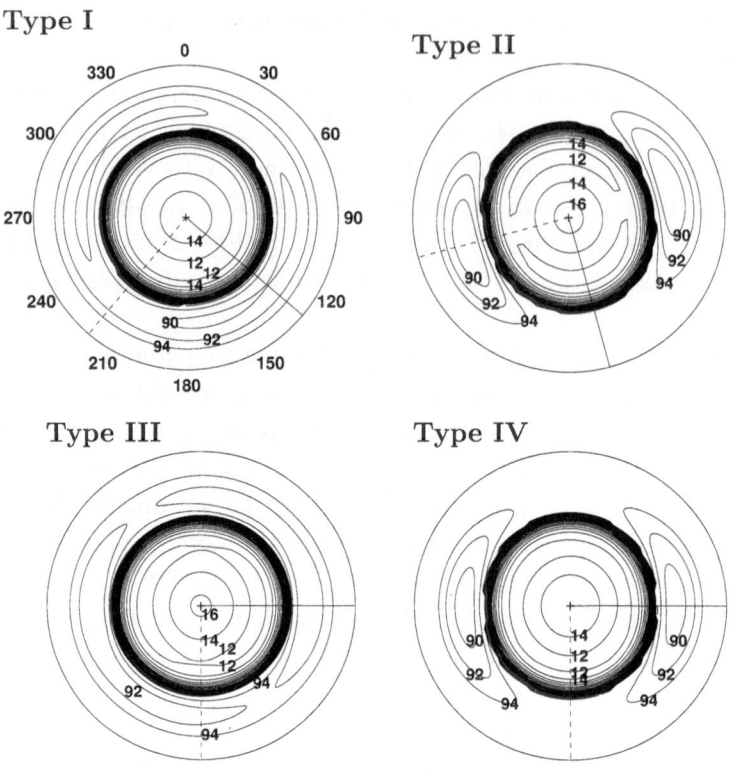

Figure 14

Directional variation of the reflection coefficient (in percent) of compressional waves at Moho (depth = 30 km) for the different texture types shown in Figures 5 and 6. The lower crust was assumed to be isotropic with V_p = 6.4 km/s, V_s = 3.8 km/s, density = 2900 kg/m^3. The contours correspond to intervals of 2%.

incidence) or critical point, which lies between 45 and 55° for all texture types, we observe a strong increase of reflection strength from approximately 15% to 95% (Fig. 15). The location of this abrupt increase of reflection amplitudes at earth surface can be identified on wide-angle seismogram sections. A discrimination of different mantle models, based on reflection amplitudes of only one azimuth, is impossible. However, there are systematic variations of the distributions of the critical points which, in practice, may be used to distinguish between different anisotropic textures of the uppermost mantle. In the case of texture Types I and III the change of the critical angle with azimuth is very small. The reflection coefficients of these textures are almost transversely isotropic. In contrast, in the case of texture Types II and IV, the location of the critical point changes significantly with azimuth. The maximum variation of the critical angle is approximately 3° and 5° for the texture Types II and IV, respectively (Fig.

15). Assuming a Moho depth of 30 km, this corresponds to azimuthal variations of approximately 10 and 15 km at the earth surface. In principle, such azimuthal variations of the critical point should be detectable by seismic reflection measurements.

12. Discussion and Conclusions

Seismic anisotropy within the subcrustal lithosphere in Southern Germany has been observed in many seismic studies. The scope of this study was to investigate,

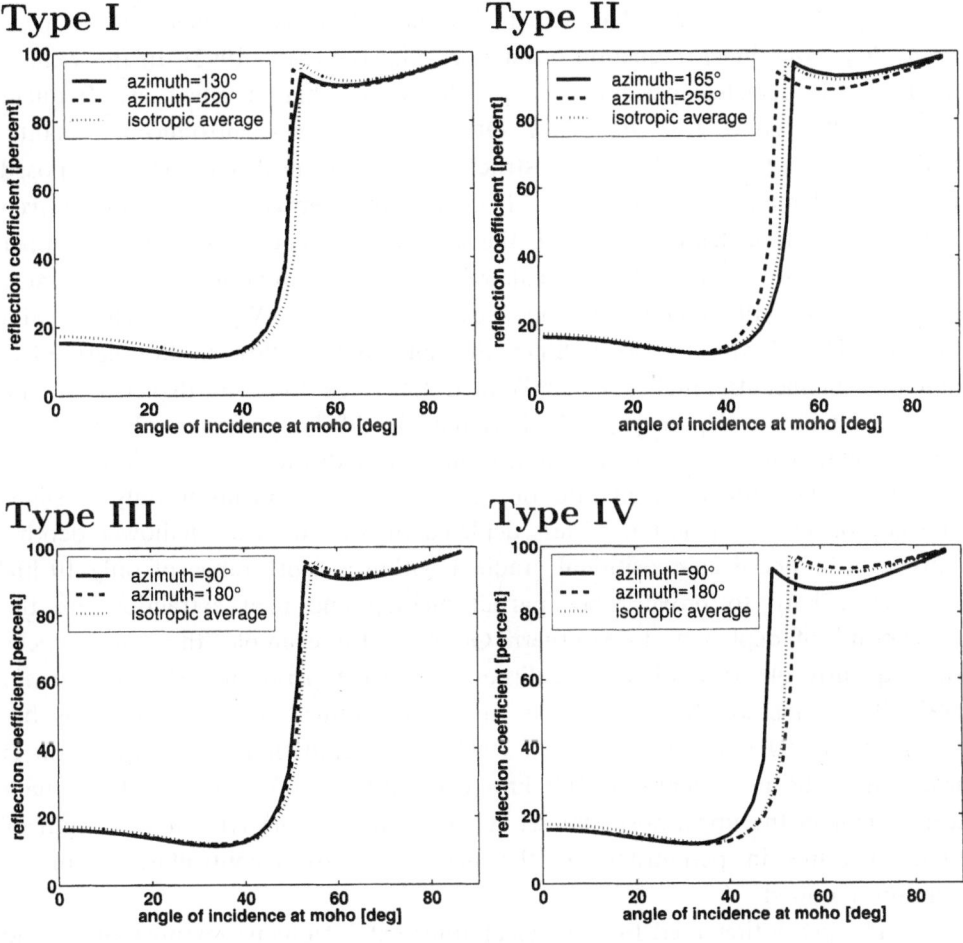

Figure 15

PP-reflection coefficient as a function of the angle of incidence at Moho. The corresponding azimuths are indicated in Figure 14. The reflection coefficients for an averaged isotropic upper mantle are shown as dotted lines.

if naturally deformed peridotites and their respective elastic properties yield further information for the interpretation of seismic data. In summary, the following conclusions can be drawn:

(1) An increase of the olivine content with depth obviously results in higher
seismic velocities. This increase is represented by a compositional variation from
primitive to depleted peridotites. However, the increase of velocities due to composition is moderate and will be diluted due to the effect of the temperature on
the elastic properties. The prevailing majority of mantle xenoliths from Southern
Germany originate from depths corresponding to the spinel stability field. The
sparse data indicate a dominance of olivine-rich (60% to 90% olivine) peridotites
(GLAHN *et al.*, 1992). A lateral rather than a vertical heterogeneity can be
proposed from mantle xenolith studies for Southern Germany. Within the area
of the Urach Volcanic Field (UVF) in Southern Germany metasomatized peridotites (e.g., wherlites) frequently occur (e.g., GLAHN *et al.*, 1992). In the vicinity
of the UVF depleted peridotites (e.g., lherzolites, harzburgites) are frequent
(SACHS, 1988; GLAHN *et al.*, 1992). OEHM *et al.* (1983) introduced a mantle
model for the NHD based on microstructural and chemical data. They proposed
that depleted mantle rocks originate from a depth between 50–80 km, derived
from equilibration temperatures of the peridotites. The corresponding upper
mantle temperatures have been calculated using two geotherms from POLLACK
and CHAPMAN (1977) and a surface heat flow of 65 mW/m^2, which is valid
for the NHD (HAENEL, 1983). Since the surface heat flow in Southern Germany is significantly higher (85 mW/m^2), the occurrence of depleted mantle
rocks at shallower depths (e.g., 40 km) may be likely. However, OEHM *et al.*
(1983) remark that there are no indications for a distinct compositional layering within the uppermost mantle of the NHD. They mention only a slight
tendency of an accumulation of less depleted mantle rocks at shallower depths.
Thus, a pronounced compositional gradient within a depth range of only 10 km
is basically hard to prove, since partial melting and melt extraction are the
mechanisms of depletion. In Southern Germany for example, the source rocks
for frequently observed olivine melilitites are garnet peridotites (GLAHN *et al.*,
1992). The transition from spinel to garnet peridotites within the UVF is between 70 and 80 km (GLAHN *et al.*, 1992). The depth of magma separation is
proposed to be at a depth of 100 km (e.g., BREY, 1977) and therefore much
deeper than in the uppermost mantle. A closer look at the type and strength of
olivine textures in peridotites, as the major anisotropy-controlling factor, is
therefore required.

(2) The peridotites used for the calculations exhibit olivine textures and fabric
types which are representative for peridotites worldwide. The type and strength
of a texture is related to strain, temperature, strain rate, strain history, deformation mechanisms, etc. (e.g., NICOLAS, 1989; NICOLAS and CHRISTENSEN, 1987;

NICOLAS et al., 1973; TAKESHITA et al., 1990; WENK et al., 1991; ZHANG and KARATO, 1995). When simple shear is the dominant mode of deformation, olique textures relative to the flow plane can develop (e.g., the Type II texture) for small strains (ZHANG and KARATO, 1995). For high strains the resulting textures may be closely linked to the flow plane geometry (e.g., the Type IV texture; ZHANG and KARATO, 1995). Peridotites which show olivine textures with the a- and c-axes forming girdles within the horizontal plane (texture Types I and III) can be explained by axial compression (e.g., CHASTEL et al., 1993). Since simple shear is probably the predominant deformation mode within the uppermost mantle (e.g., ZHANG and KARATO, 1995) the latter types may be of restricted importance. Peridotites with textures comparable to Type I are furthermore rarely reported in the literature (e.g., FRANCIS, 1978). From fabric and texture data it is impossible to assign a particular texture type to a certain depth level. So far, the different parameters controlling the texture development within the uppermost mantle beneath continents are poorly understood. For example, KARATO (1987, 1988) showed that recrystallization via subgrain rotation can form a texture similar to that resulting from deformation. The texture development can furthermore be dependent on partial melting, which leads to a change in the activated slips systems and deformation mechanisms as a function of stress and strain rate (e.g., HIRTH and KOHLSTEDT, 1995).

(3) The azimuthal dependence of the P-wave velocities within the horizontal plane (i.e., the plane of P_n-propagation) largely depends on the spatial orientation of the foliation in the uppermost mantle. Usually this orientation is unknown. If an azimuthal dependence of V_p is observed in seismic studies, only two (II and IV) of the presented texture types can be responsible when assuming a horizontal foliation. The degree and type of this V_p azimuth relation coincides with the seismic observations (e.g., BAMFORD, 1973, 1977; FUCHS, 1983) at a depth of 30 km. If a change in the modal composition within a depth range from 30–40 km is proposed, $V_{p_{max}}$ increases whereas the effect on $V_{p_{min}}$ is considerably less. However, the very high $V_{p_{max}}$ (≈ 9.0 km/s) from FUCHS (1983) cannot be obtained. A $V_{p_{max}}$ of 8.6 km/s as presented by ENDERLE et al. (1996) can be derived from the elastic properties of naturally deformed peridotites and is also reported from seismic investigations (e.g., ANSORGE et al., 1979).

A clear correlation between the texture type and the azimuth dependence of seismic waves is ambiguous, when changes in the orientation of the foliation are considered. In this case, all of the texture types presented display a pronounced azimuth dependence for horizontally propagating compressional waves. Since the S-wave patterns associated with the respective texture types are markedly different, at least two of the texture types (Types I and III) can be excluded if additional S-wave splitting observations are available.

Many peridotite xenoliths show oblique textures with respect to the reference frame (foliation and lineation). An inclined direction of $V_{p_{max}}$ may be the result of dipping stacks of uppermost mantle or oblique textures. An unequivocal interpretation of recent flow or ancient fabric is impossible.

(4) All texture types exhibit a remarkably constant maximum shear-wave splitting at an angle of $\approx 45°$ to the direction of $V_{p_{max}}$ within the (horizontal) foliation. If the $V_{p_{max}}$-direction is 22°N or 31°N a shear-wave splitting should be observable at an azimuth of approximately 67° or 76°, respectively. If a vertical foliation is assumed, the respective texture types could be clearly discriminated due to their distinct different S-wave splitting characteristics. The texture Types II and IV show a large S-wave splitting at an angle of ≈ 45–$135°$ to $V_{p_{max}}$. The Type I texture exhibits a minimum of the S-wave splitting in this particular range. The Type III texture, in contrast, only shows a minimum S-wave splitting at an angle of $\approx 135°$ to $V_{p_{max}}$.

Generally, S-wave velocities for the investigated peridotites are in the range of 4.6 to 4.8 km/s and fit to the values calculated for the anvil-model and particularly to those determined from earthquake waves despite the fact that the latter show no S-wave splitting (PLENEFISCH *et al.*, 1994). For the peridotites, the maximum S-wave splitting is closely connected to the fastest (leading) S-wave for all of the peridotites, which may be used as an indicator to identify an S-wave splitting in future seismic studies.

(5) An average S-wave splitting observed for SKS-waves of about 0.83 s in Central Europe can partially be attributed to the lithospheric mantle. For an assumed thickness of 50 km for the subcrustal lithosphere, approximately 70% of the SKS-splitting can be attributed to upper mantle peridotites corresponding to texture Types II and IV. Assuming a vertical foliation, all of the texture types exhibit a remarkable S-wave splitting and can therefore contribute to the observed SKS-splitting. However, a deviation of the SKS-polarization directions from the fast P_n-velocity direction in Southern Germany exists (e.g., BORMANN *et al.*, 1996). This contradicts the values observed for the peridotites, which exclusively show a polarization of the fast S-wave within the symmetry planes of the velocity distributions. These symmetry planes coincide roughly with the planes defined by the structural reference frame (xy-plane, xz-plane). For the Upper Rhinegraben a strike-slip deformation regime can be proposed, based on the analysis of crustal fault plane solutions of earthquakes (PLENEFISCH and BONJER, 1995). The minimum horizontal stress can be expected in the direction of N60°E which coincides roughly with the direction of absolute plane motion in Southern Germany (BORMANN *et al.*, 1996). Thus, the resulting SKS-splitting may originate from the asthenosphere and can be at least reinforced by an anisotropic uppermost mantle.

(6) Anisotropy in the uppermost mantle induces azimuthal variations of the critical angle of incidence of reflected P-waves. These variations depend on the

presumed texture and may reach 5° corresponding to variations of the critical point of up to 15 km at earth surface if a crustal thickness of 30 km is assumed. In principle, it should be possible to detect these effects by seismic reflection measurements.

Acknowledgements

We are grateful to two anonymous reviewers for their critical reviews and helpful comments. This study was financially supported by the Deutsche Forschungsgemeinschaft (Ra 496/5-1). S.S. thanks the Deutsche Forschungsgemeinschaft for a Heisenberg fellowship (Si 428/10-1).

REFERENCES

ANDERSON, O. L., SCHREIBER, E., and LIEBERMANN, R. C. (1968), *Some Elastic Constant Data on Minerals Relevant to Geophysics*, Rev. Geophys. *6*, 491–524.

ANSORGE, J., BONJER, K. P., and EMTER, D. (1979), *Structure of the Uppermost Mantle from Long-range Seismic Observations in Southern Germany and the Rhinegraben Area*, Tectonophysics *56*, 31–48.

BABUŠKA, V., and CARA, M., Seismic Anisotropy in the Earth (Kluwer Academic Publ., Dordrecht 1991).

BABUŠKA, V., PLOMEROVA, J., and SILENY, J. (1984), *Spatial Variations of P Residuals and Deep Structures of the European Lithosphere*, Geophys. J. R. Astr. Soc. *79*, 363–383.

BABUŠKA, V., PLOMEROVA, J., and SILENY, J. (1987), *Structural model of the subcrustal lithosphere in central Europe*. In *Composition, Structure and Dynamics of the Lithosphere-asthenosphere System* (eds. Fuchs, K., and Froidevaux, C.) (Am. Geophys. Un., Washington DC, *16*, 1991) pp. 239–249.

BAMFORD, D. (1973), *Refraction Data in Western Germany—A Time-term Interpretation*, J. Geophys. *39*, 907–927.

BAMFORD, D. (1977), P_n-*velocity Anisotropy in a Continental Upper Mantle*, Geophys. J. R. Astr. Soc. *49*, 29–48.

BAMFORD, D., JEUTCH, M., and PRODEHL, C. (1979), P_n-*anisotropy Studies in Northern Britain and the Eastern and Western United States*, Geophys. J. R. Astr. Soc. *57*, 397–439.

BLUNDELL, D., FREEMAN, R., and MUELLER, S., *A Continent Revealed—The European Geotraverse* (Atlas Map 13 Heat-Flow Density, Cambridge University Press 1992).

BORMANN, P., BURGHARDT, P.-T., MAKEYEVA, L. I., and VINNIK, L. P. (1993), *Teleseismic shear-wave splitting and deformations in Central Europe*, Phys. Earth Planet. Int. *78*, 157–166.

BORMANN, P., GRÜNTHAL, G., KIND, R., and MONTAG, H. (1996), *Upper Mantle Anisotropy beneath Central Europe from SKS Wave Splitting: Effects of Absolute Plate Motion and Lithosphere-asthenosphere Boundary Topography?*, J. Geodynamics *22* (1–2), 11–32.

BREY, G. (1977), *Origin of Olivine Melilitites—Chemical and Experimental Constraints*, J. Volc. Geotherm. Res. *3*, 61–88.

ČERMAK, V., and BODRI, L. (1993), *Three-dimensional Deep Temperature Modelling along the European Geotraverse*, Tectonophysics *244*, 1–11.

CHASTEL, Y. B., DAWSON, P. R., WENK, H. R., and BENNETT, K. (1993), *Anisotropic Convection with Implications for the Upper Mantle*, J. Geophys. Res. *98* (B10), 17,757–17,771.

CHESNOKOV, Y. M., and NEVSKIG, M. V. (1977), *Seismic Anisotropy Investigation in the USSR*, Geophys. J. R. Astr. Soc. *49*, 115–121.

CHRISTENSEN, N. I., and LUNDQUIST, S. M. (1982), *Pyroxene Orientation within the Upper Mantle*, Geol. Soc. Am. Bull. *93*, 279–288.

CHRISTENSEN, N. I., and RAMANANTOANDRO, R. (1971), *Elastic Moduli and Anisotropy of Dunite to 10 Kilobars*, J. Geophys. Res. *76*, 4003–4010.

CHRISTENSEN, N. I. (1979), *Compressional Wave Velocities in Rocks at High Temperatures and Pressures, Critical Thermal Gradients, and Crustal Low-velocity Zones*, J. Geophys. Res. *84* (B12), 6849–6857.

CHRISTENSEN, N. I. (1984), *The Magnitude Symmetry and Origin of Upper Mantle Anisotropy Based on Fabric Analysis of Ultramafic Tectonites*, Geophys. J. R. Astr. Soc. *76*, 89–111.

CRAMPIN, S. (1981), *A Review of Wave Motion in Anisotropic and Cracked Elastic Media*, Wave Motion *3*, 343–391.

CROSSON, R. S., and LIN, J. W. (1971), *Voigt and Reuss Prediction of Anisotropic Elasticity of Olivine*, J. Geophys. Res. *76*, 570–578.

EDEL, J. B., FUCHS, K., GELBKE, C., and PRODEHL, C. (1975), *Deep Structure of the Southern Rhinegraben Area from Seismic Refraction Investigations*, J. Geophys. *41* (4), 333–356.

ENDERLE, U., MECHIE, J., SOBOLEV, S., and FUCHS, K. (1996), *Seismic Anisotropy within the Uppermost Mantle of Southern Germany*, Geophys. J. Int. *125*, 747–767.

FRANCIS, D. M. (1978), *The Implications of the Compositional Dependence of Texture in Spinel Lherzolite Xenoliths*, J. Geol. *86*, 473–485.

FRISILLO, A. L., and BARSCH, G. R. (1972), *Measurement of Single-crystal Elastic Constants of Bronzite as a Function of Pressure and Temperature*, J. Geophys. Res. *77*, 6360–6383.

FUCHS, K. (1983), *Recently Formed Elastic Anisotropy and Petrological Models for the Continental Subcrustal Lithosphere in Southern Germany*, Phys. Earth Planet. Int. *31*, 93–118.

GAJEWSKI, D., and PRODEHL, C. (1985), *Crustal Structure beneath the Swabian Jura, SW Germany, from Seismic Refraction Investigations*, J. Geophys. *56* (2), 69–80.

GAJEWSKI, D., and PRODEHL, C. (1987), *Seismic Refraction Investigation of the Black Forest*, Tectonophysics *142* (1), 27–48.

GAJEWSKI, D., and PSENČIK, I. (1990), *Vertical Seismic Profile Synthetics by Dynamic Ray Tracing in Laterally Varying Anisotropic Structures*, J. Geophys. Res. *95*, 11,301–11,315.

GIESE, P., and PAVLENKOVA, N. J., *Map 21.1, Europe west-depth of crust-mantle boundary, 1:5000000*. In *Geothermal Atlas of Europe* (eds. Hurtig. E. Čermak, V., Haenel, R., and Zui, V. I.) (Hermann Haack-Verlagsanstalt GmbH, Gotha 1991).

GLAHN, A., SACHS, P. M., and ACHAUER, U. (1992), *A Teleseismic and Petrological Study of the Crust and Upper Mantle beneath the Geothermal Anomaly Urach/SW Germany*, Phys. Earth Planet. Int. *69* (3–4), 176–206.

HAENEL, R., *Geothermal investigations in the Rhenish Massif*. In *Plateau Uplift: The Rhenish Shield: A Case History* (eds. Fuchs, K., von-Gehlen, K., Maelzer, H., Murawski, H., and Semmel, A.) (Springer Verlag, Berlin 1983) pp. 228–246.

HARTMANN, G., and WEDEPOHL, K. H. (1990), *Metasomatically Altered Peridotite Xenoliths from the Hessian Depressian (Northwest Germany)*, Geochim. Cosmochim. Acta *54*, 71–82.

HARTMANN, G., and WEDEPOHL, K. H. (1993), *The Composition of Peridotite Tectonites from the Ivrea Complex, Northern Italy: Residues from Melt Extraction*, Geochim. Cosmochim. Acta *57* (8), 1761–1782.

HESS, H. H. (1964), *Seismic Anisotropy of the Uppermost Mantle under Oceans*, Nature *203*, 629–631.

HIRTH, G., and KOHLSTEDT, D. L. (1995), *Experimental Constraints on the Dynamics of the Partially Molten Upper Mantle 2. Deformation in the Dislocation Creep Regime*, J. Geophys. Res. *100* (B8), 15,441–15,449.

ISAAK, D. G. (1992), *High Temperature Elasticity of Iron-bearing Olivines*, J. Geophys. Res. *97* (B5), 1871–1885.

JI, S., ZHAO, X., and FRANCIS, D. (1994), *Calibration of Shear-wave Splitting in the Subcontinental Upper Mantle beneath Active Orogenic Belts Using Ultramafic Xenoliths from the Canadian Cordillera and Alaska*, Tectonophysics *239*, 1–27.

JIN, D., KARATO, S. I., and OBATA, M. (1998), *Mechanisms of Shear Localization in the Continental Lithosphere: Inference from the Deformation Microstructures of Peridotites from the Ivrea Zone, Northwestern Italy*, J. Struct. Geol. *20/2–3*, 195–209.

KARATO, S. I., *Seismic anisotropy due to lattice preferred orientation of minerals: kinematic or dynamic?* In *High Pressure Research in Mineral Physics* (eds. Manghnani, M. Y., and Syono, Y.) (Geophysical Monograph. 39, American Geophysical Union, Washington DC 1987).

KARATO, S. I. (1988), *The Role of Recrystallization in the Preferred Orientation of Olivine*, Phys. Earth Planet. Int. *51*, 107–122.

KUMAZAWA, M., and ANDERSON, O. L. (1969), *Elastic Moduli, Pressure Derivatives and Temperature Derivatives of Single-crystal Olivine and Single-crystal Forsterite*, J. Geophys. Res. *74*, 5311–5320.

LEVIEN, L., WEIDNER, D. J., and PREWITT, C. T. (1979), *Elasticity of Diopside*, Phys. Chem. Min. *4* (2), 105–113.

MAINPRICE, D., and HUMBERT, M. (1993), *Methods of calculating petrophysical properties from lattice preferred orientation data*. In *Seismic Properties of Crustal and Mantle Rocks: Laboratory Measurements and Theoretical Calculations* (ed. Burlini, L.) (Surveys in Geophysics 15(5), D. Reidel Publishing Company, Dordrecht-Boston 1994) pp. 575–592.

MAINPRICE, D., and SILVER, P. G. (1993), *Interpretation of SKS-waves Using Samples from the Subcontinental Lithosphere*, Phys. Earth Planet. Inter. *78*, 257–280.

MENZIES, M. A., and BODINIER, J. L. (1993), *Growth of the European Lithospheric Mantle: Dependence of Upper-mantle Peridotite Facies and Chemical Heterogeneity on Tectonites and Age*, Phys. Earth Planet. Int. *79*, 219–240.

MERCIER, J. C., and NICOLAS, A. (1975), *Textures and Fabrics of Upper Mantle Peridotites as Illustrated by Xenoliths from Basalts*, J. Petrol. *16*, 454–487.

MERCIER, J. C. (1985), *Olivines and pyroxenes*. In *Preferred Orientation in Deformed Minerals and Rocks: An Introduction to Modern Texture Analysis* (ed. Wenk, H. R.) (Academic Press, Orlando, United States 1985) pp. 407–430.

MONTAGNER, J. P., and ANDERSON, D. L. (1989), *Constrained Reference Mantle Model*, Phys. Earth Planet. Int. *58* (2–3), 205–227.

MONTAGNER, J.-P., and TANIMOTO, T. (1991), *Global Upper Mantle Tomography of Seismic Velocities and Anisotropy*, J. Geophys. Res. *96*, 20,337–20,351.

MÜLLER, S., and PANZA, G. G. (1984), *The lithosphere-asthenosphere system in Europe*. In *First EGT Workshop: The Northern Segment* (eds. Galson, D. A., Mueller, S., and Munch, B.) (European Science Foundation, Strasbourg 1984) pp. 23–26.

NICOLAS, A., BOUDIER, F., and BOULLIER, A. M. (1973), *Mechanisms of Flow in Naturally and Experimentally Deformed Peridotites*, Am. J. Sci. *273*, 853–876.

NICOLAS, A., and CHRISTENSEN, N. I. (1987), *Formation of anisotropy in upper mantle peridotites: a review*. In *Composition, Structure and Dynamics of the Lithosphere-asthenosphere System* (eds. Fuchs, K., and Froidevaux, C.) (Geodynamics Series. 16, American Geophysical Union, Washington DC, United States 1987) pp. 11–123.

NICOLAS, A., *Structures of ophiolites and dynamics of oceanic lithosphere* (Kluwer Academic; Petrology and Structural Geology 4, 1989).

Nicolas, A., *The Mid-oceanic Ridges: Mountains below Sea Level* (Springer Verlag, Berlin–New York–Heidelberg 1995).

OEHM, J., SCHNEIDER, A., and WEDEPOHL, K. H. (1983), *Upper Mantle Rocks from Basalts of the Northern Hessian Depression*, Tscherm. Min. Petr. Mitt. *32*, 25–48.

PESELNICK, L., NICOLAS, A., and STEVENSON, P. R. (1974), *Velocity Anisotropy in a Mantle Peridotite from the Ivrea Zone: Application to Upper Mantle Anisotropy*, J. Geophys. Res. *79*, 1175–1182.

PLENEFISCH, T., and BONJER, K. P. (1995), *The stress tensor in the Rhine Graben area derived from earthquake focal mechanisms*. In *Seismotectonics and Seismic Hazard in the Roer Valley Graben; With Emphasis on the Roermond Earthquake of April 13, 1992* (eds. van-Eck, T. and Davenport, C. A.), Geol. en Mijnbouw. *73*(2–4), 169–172.

PLENEFISCH, T., FABER, S., and BONJER, K.-P. (1994), *Investigations of S_n and P_n Phases in the Area of the Upper Rhinegraben and Northern Switzerland*, Geophys. J. Int. *119*, 402–420.

POLLACK, H. N., and CHAPMAN, D. S. (1977), *On the Regional Variation of Heat Flow, Geotherms and Lithospheric Thickness*, Tectonophysics 38, 279–296.

PRESS, S., WITT, G., SECK, H. A., EONOV, D., and KOVALENKO, V. I. (1986), *Spinel Peridotite Xenoliths from the Tariat Depression, Mongolia: I. Major Element Chemistry and Mineralogy of a Primitive Mantle Xenoliths Suite*, Geochim. Cosmochim. Acta 50, 2587–2599.

RINGWOOD, A. E. (1973), *Phase Transformations and their Bearing on the Dynamics of the Mantle*, Fortschr. Min. 50, 113–139.

SACHS, P. M. (1988), *Untersuchungen zum Stoffbestand der tieferen Lithosphäre an Xenolithen südwestdeutscher Vulkane*, Ber. Inst. Geophys. Univ. Stuttgart, 249 pp.

SAVAGE, M. K., and SILVER, P. G. (1993), *Mantle Deformation and Tectonics: Constraints from Seismic Anisotropy in the Western United States*, Phys. Earth Planet. Int. 78, 207–227.

SIEGESMUND, S., VOLLBRECHT, A., CHLUPAC, T., NOVER, G., DÜRRAST, H., MÜLLER, J., and WEBER, K. (1993), *Fabric Controlled Anisotropy of Petrophysical Properties Observed in KTB Core Samples*, Scientific Drilling 4, 31–54.

SIEGESMUND, S., HELMIG, K., and KRUSE, R. (1994), *Complete Texture Analysis of a Deformed Amphibolite: Comparison between Neutron Diffraction and U-stage Data*, J. Struct. Geol. 16, 131–142.

SIEGESMUND, S. (1996), *The Significance of Rock Fabrics for the Geological Interpretation of Geophysical Anisotropies*, Geotekt. Forsch. 85, 1–123.

SILVER, P. G., and CHAN, W. (1991), *Shear-wave Splitting and Subcontinental Mantle Deformation*, J. Geophys. Res. 96, 16,429–16,454.

SILVER, P. G., and KANESHIMA, S. (1993), *Constraints on Mantle Anisotropy beneath Pre-Cambrian North America from a Transportable Teleseismic Experiments*, Geophys. Res. Lett. 20 (12), 1127–1130.

SKROTZKI, W., *Mechanisms of texture development in rocks*. In *Textures of Geological Materials* (eds. Bunge, H. J., Siegesmund, S., Skrotzki, W. and Weber, K.) (DGM Informationsgesellschaft Verlag, Oberursel 1994) pp. 167–186.

SKROTZKI, W., WEDEL, A., and WEBER, K. (1992), *Microstructure and Texture in Peridotites from the Balmuccia Massif (NW-Italy)*, Geotekt. Forsch. 78, 55–88.

SKROTZKI, W., WEDEL, A., WEBER, K., and MÜLLER, W. F. (1990), *Microstructure and Texture in Lherzolites of the Balmuccia Massif and their Significance Regarding the Thermomechanical History*, Tectonophysics 179, 227–251.

TAKESHITA, T., WENK, H. R., CANOVOA, G. R., and MOLINARI, A., *Simulation of dislocation-assisted plastic deformation in olivine polycrystals*. In *Deformation Processes in Minerals, Ceramics and Rocks* (eds. Barber, D. J. and Meredith, P. G.) (Unwin Hyman. London, United Kingdom 1990) pp. 365–374.

TAYLOR, G. I. (1923), *The Motion of Ellipsoidal Particles in a Viscous Fluid*, Proc. R. Soc. London, Ser. A. 103, 58–61.

VETTER, E., and MINSTER, J. (1981), P_n *Velocity Anisotropy in Southern California*, Bull. Seismol. Soc. Am. 71, 1511–1530.

VINNIK, L. P., KRISHNA, V. G., KIND, R., BORMAN, P., and STAMMLER, K. (1994), *Shear-wave Splitting in the Records of the German Regional Seismic Network*, Geophys. Res. Lett. 21 (6), 457–460.

VINNIK, L. P., GREEN, R. W. E., and NICOLAYSEN, L. O. (1995), *Recent Deformations of the Deep Continental Root Beneath Southern Africa*, Nature 375, 50–52.

VINNIK, L. P. (1997), *Seismic Anisotropy and Mantle Flow*, Geowissenschaften 15, 100–104.

VOIGT, W., *Lehrbuch der Kristallphysik* (Teubner Verlag, Leipzig 1928).

WEDEL, A. (1990), *Mikrostruktur und Texturuntersuchungen an Peridotiteinschlüssen in Basalten der Hessischen Senke*, Göttinger Arb. Geo. Paläont. 45, 63 pp.

WEDEL, A., SKROTZKI, W., and WEBER, K. (1992), *Microstructure and Texture in Peridotite Xenoliths from the Hessian Depression*, Geotekt. Forsch. 78, 89–125.

WENK, H. R., BENNET, K., CANOVA, G. R., and MOLINARI, A. (1991), *Modelling Plastic Deformation of Peridotite with the Self-consistent Theory*, J. Geophys. Res. 96 (B5), 8337–8349.

WERNER, D., and KAHLE, H. G. (1980), *A Geophysical Study of the Rhinegraben. Kinematics and Geothermics*, Geophys. J. R. Astr. Soc. 62, 617–630.

ZEIS, S., GAJEWSKI, D., and PRODEHL, C. (1990), *Crustal Structure of Southern Germany from Seismic Refraction Data*, Tectonophysics *176* (1–2), 59–86.

ZHANG, S., and KARATO, S. I. (1995), *Lattice Preferred Orientation of Olivine Aggregates Deformed in Simple Shear*, Nature *375*, 774–777.

(Received October 23, 1997, revised December 3, 1998, accepted December 22, 1998)

To access this journal online:
http://www.birkhauser.ch

Pure appl. geophys. 156 (1999) 83–95
0033–4553/99/010083–13 $ 1.50 + 0.20/0

⌐Pure and Applied Geophysics

Seismic Shear Waves in Deep Seismic Reflection Surveys: Some Notes on Problems and Profits

EWALD LÜSCHEN[1]

Abstract—Shear (*S*) waves differ from compressional (*P*) waves because of their lower propagation velocities, their lower frequencies and due to the different character of their particle motion. The move-out of travel-time branches of *S*-wave reflections is different from *P* waves owing to the difference in the propagation velocities. To distinguish between *P* and *S* waves requires broadband-frequency acquisition, long receiver arrays and three-component recording. *S*-wave generation at the source and *P*-to-*S*-wave conversion at crustal interfaces can be very efficient, implying that there is a real danger of misinterpreting signals if only vertical components are used. On the other hand, integrated *P*- and *S*-wave studies promise to provide very efficient lithological discriminators in the crystalline crust, in particular concerning the quartz content, and indicators for rock anisotropy, which can be interpreted for the existence of fine layering, the direction of the recent stress regime (alignments of micro-fractures) or for the direction of palaeo-stress (alignments of minerals).

Key words: Seismic profiles, *S* waves, anisotropy, three-component seismographs.

Introduction

Compared to standard seismic reflection measurements for exploration purposes in sedimentary basins, deep crustal surveys very often suffer from a relatively low signal-to-noise ratio. Even in exploration surveys using seismic attribute maps for structural interpretation, there can be a real danger of misinterpretation because of mistaking unrecognised noise for structural signals (HESTHAMMER and FOSSEN, 1997). Among different coherent and incoherent noise classes which are normally attenuated by modern acquisition and processing methods (e.g., SHERIFF and GELDART, 1982), scattering of seismic waves at near-surface as well as at deep crustal small-scale discontinuities (e.g., LEVANDER *et al.*, 1994) and seismic *S* waves

---

[1] GeoForschungsZentrum Potsdam, Telegrafenberg, D-14473 Potsdam, Germany. Now at Institut für Allgemeine und Angewandte Geophysik, Universität München (LMU), Theresienstrasse 41, D-80333 München. Tel.: +49-89-2394-4201, Fax: +49-89-2394-4205, E-mail: lueschen@bavaria.geophysik.uni-muenchen.de

are strong candidates for misinterpretation. Examples in this paper show that vertical-component recordings, which are commonly used in deep seismic reflection surveys, contain considerable S-wave energy, although S waves are predominantly polarised in the transverse and radial directions with respect to the ray path. If unrecognised, these signals are common sources of structural misinterpretation. On the other hand, whether generated by explosive point sources, vibrator surface sources or by special S-wave sources, S waves provide a promising source for complementary lithological and micro-structural information, as already demonstrated in petroleum exploration surveys (DANBOM and DOMENICO, 1987), laboratory petrophysical experiments (KERN, 1982; CHRISTENSEN, 1989), deep crustal research (LÜSCHEN et al., 1990, 1993; WARD et al., 1991; BOPP, 1992; SANDMEIER and WENZEL, 1990; RABBEL and LÜSCHEN, 1996) and downhole measurements (CRAMPIN et al., 1986; MACBETH, 1991; RABBEL, 1996; LÜSCHEN et al., 1996). Generally, S-wave studies require three-component recording and are therefore more expensive. However, three-component recording enables discrimination of S waves and converted waves from P waves and, secondly, exploitation of additional lithological and structural information.

The Problem

Most crustal seismic reflection profiles are based on vertical-component recording. Figure 1 shows an arbitrary example of a shot gather that was obtained in the Siberian craton before input to standard processing schemes (SHAROV, pers. comm.). Direct waves of all types are visible, including P waves, S waves and surface waves (Rayleigh-), clearly distinguishable because of their apparent velocity and their frequency. Corresponding secondary arrivals, interpreted to be reflections because of their move-out, are visible after each of these direct waves. In the case of the reflections of the surface waves, this energy has been reflected very likely off-line at topographic anomalies, oriented parallel or sub-parallel to the seismic line. The bandpass-filtered versions evidence that there is still a considerable amount of reflected S-wave energy in the pass band of P waves. The travel-time ratio of these signals (Nos. 7 and 8 in Figure 1) corresponds very closely to a V_p/V_s ratio of 1.73 (or Poisson's ratio 0.25) which is valid for an elastic solid. This indicates that P- and S-wave reflections were generated at the same structure. Actually, a conventional CMP-stack of these data shows a multitude of dipping signal alignments in the middle crust (Fig. 2). Obviously, some of these signals correspond to reflected S-wave energy and could easily be misinterpreted in terms of P-wave reflecting structures, in particular as low-angle normal or thrust faults. The example presented in Figures 1 and 2 might be regarded as an extreme one, nevertheless it demonstrates the potential influence of S waves on structural interpretations that are only based on vertical receiver components.

The explosive point source (bore-hole explosives) generated a significant amount of S waves and surface waves, which are thought to be caused by mode conversion at near-source heterogeneities or at the surface (FERTIG, 1984). This phenomenon is very commonly observed in seismic experiments in crystalline environments.

Recognition by Three-component Receivers

The effect of mode-conversion at the surface, and at deep-seated structures, is probably underestimated. Figure 3 illustrates the efficiency of mode conversion using three-component recordings of the ASTRA '91 experiment in the Voronesh

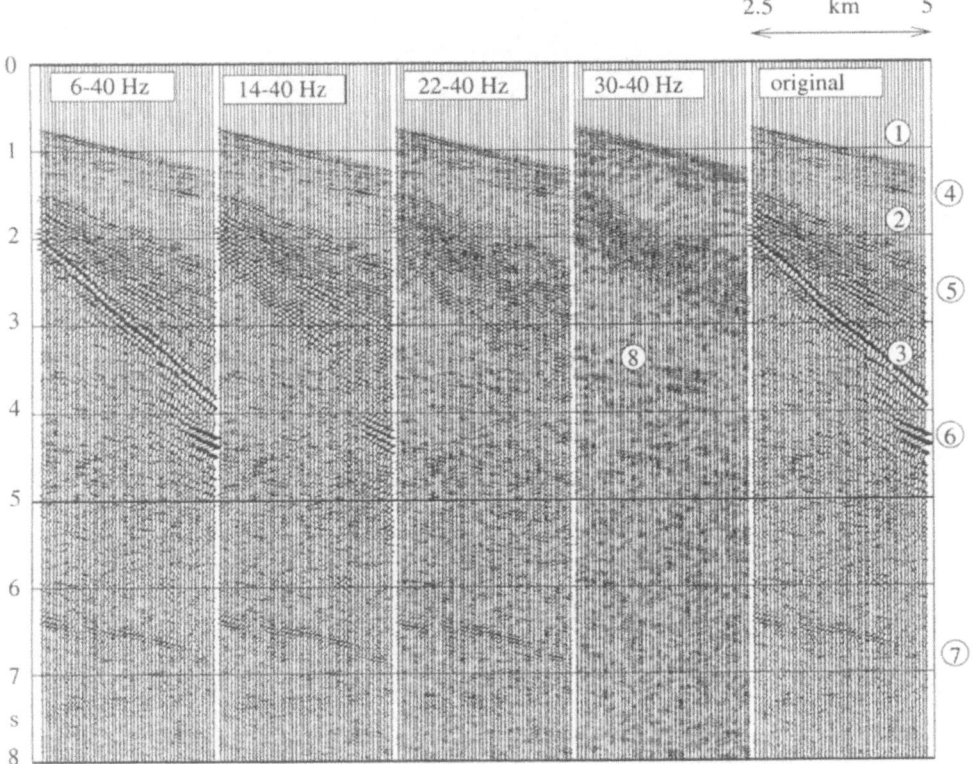

Figure 1

Dynamite shot gather from a deep crustal seismic reflection experiment in Siberia (SHAROV, pers. comm.), offset range 2.5 to 5 km (second half of the complete 96-channel gather), trace spacing 50 m, vertical component. Four different bandpass-filtered versions are shown, original version on the right, amplitude gain with time to the power of 2. (1) direct P wave, (2) direct S wave, (3) surface wave, (4) reflected P wave, (5) reflected S wave (?), (6) off-line-reflected surface wave (?), (7) reflected S wave (?), (8) reflected P wave. Note that the ratio of travel times of (7) and (8) is close to 1.73 which is the V_p / V_s ratio (corresponding to the Poisson's ratio of 0.25) of an elastic solid. This is indication that (7) and (8) stem from the same reflecting structure.

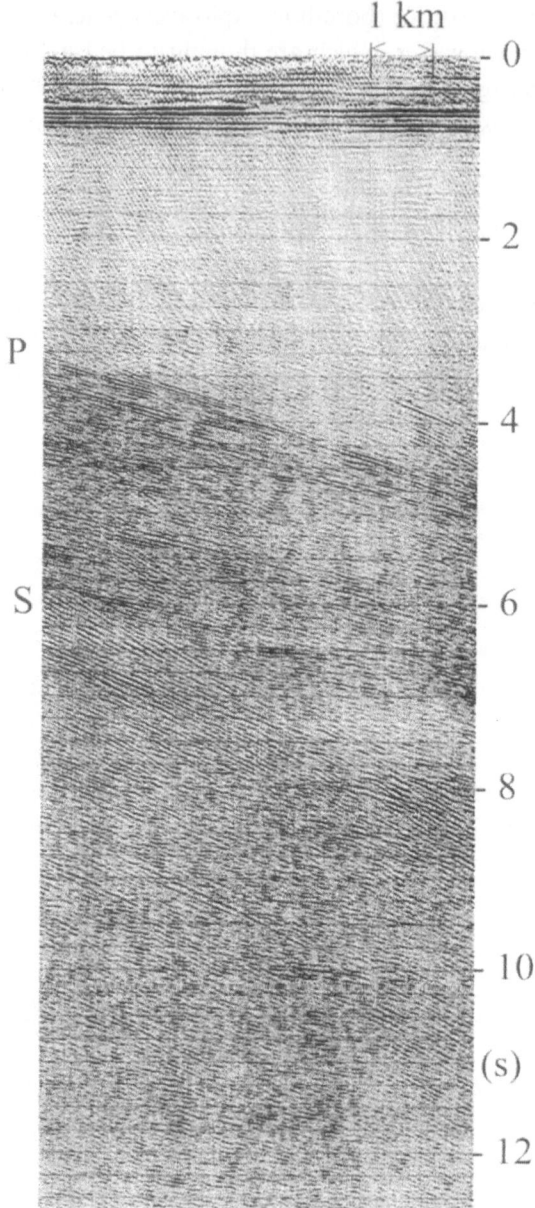

Figure 2
CMP stack along the line in Siberia (SHAROV, pers. comm.) where the shot gather of Figure 1 is from, after standard processing steps (editing, muting, bandpass-filtering, correction of geometrical spreading, CMP sorting, nmo correction), trace spacing is 25 m, horizontal exaggeration is approximately 2:1. Note two different patterns of dipping signal alignments with different dip (marked with P and S on left side). The pattern with the stronger dip (marked with S) is presumed to be of S-wave origin.

massif (KASHUBIN *et al.*, 1998). Considering the first *P*-wave arrivals, we also observe very weak amplitudes of these signals on the horizontal components, as expected because of the near-vertical incidence of the waves. However, on the radial component these arrivals are followed by much stronger signals with a constant time lag of about 200 ms. Due to their polarisation, these arrivals can be clearly identified as *S*-wave energy. Their apparent velocity which is equal to that of the *P* waves, indicates that the *S* waves were generated by the incidence of *P* waves from below and by conversion to *S* waves below the receivers. The corresponding interface is the top of the Precambrian crystalline basement, overlain by several hundred meters of Palaeozoic to Quaternary sediments. The thickness of the sedimentary layer can easily be determined with the time lag and the velocities of the direct *P* and *S* waves of the first layer (measured at short source-receiver distance). Assuming a *P*-wave velocity of 4000 m/s and a *P*-to-*S*-velocity ratio of 1.73 of the first layer, we would derive a sedimentary thickness of 1100 meters. The *P*-to-*S* converted signals also show considerable contributions on the vertical receiver component. Assuming we have only the vertical component in this case, we would certainly misinterpret the signals that follow the first *P*-wave arrivals. Here these signals could be considered as *P*-wave multiples. The conversion of *P* waves below the receivers implies that this conversion would occur in the same way below the source in the opposite direction. Conversion of *P* waves at the free surface is another possible cause of *S*-wave generation (FERTIG, 1984). Furthermore, it is obvious from the example in Figure 3 that deeper-seated interfaces, particularly if they are dipping and therefore having an oblique incidence angle, may have a similar effect. The same effect as described in Figure 3 has been exploited in seismological research which uses the 'receiver function method' to study *P*-to-*S*-converted teleseismic waves, recorded by broadband seismograph stations, to derive the crustal thickness (e.g., KIND *et al.*, 1995).

Lower crustal and crust-mantle boundary conversions and *S*-wave reflections are shown in Figure 4, using three-component recordings from the Black Forest in Southwest Germany (LÜSCHEN *et al.*, 1990). This area is characterised by a relatively transparent upper crystalline crust and a highly reflective, lamella-like lower crust between 14 and 27 km depth. The three components clearly exhibit three energy decays at nearly 10 s, at 13 s and at a little below 16 s. The first one corresponds to the crust-mantle boundary; the latter two correspond to *P*-to-*S*-wave conversions and to *S*-wave reflections, respectively, from the same boundary. This is obvious because of the enhanced energy on the horizontal components and due to the travel-time ratio of approximately 1.73 which is the well-known velocity ratio for an elastic solid. A narrower band-pass filter which uses lower frequencies enhances the *S* waves, whereas the use of higher frequencies enhances the *P* waves. Unfortunately, there is a relatively wide overlapping frequency range of approximately 10 to 20 Hz where both, P and S waves, contain significant energy. During many DEKORP surveys in the 1980s a recording time of 12 s was used. Therefore, the contribution of S waves and converted waves from the lower crust was not recognised. If longer recording times and only the vertical component were used, the energy between 10 and 16 s recording time would have been interpreted as reflected P waves from upper-mantle structures and, consequently, misinterpreted.

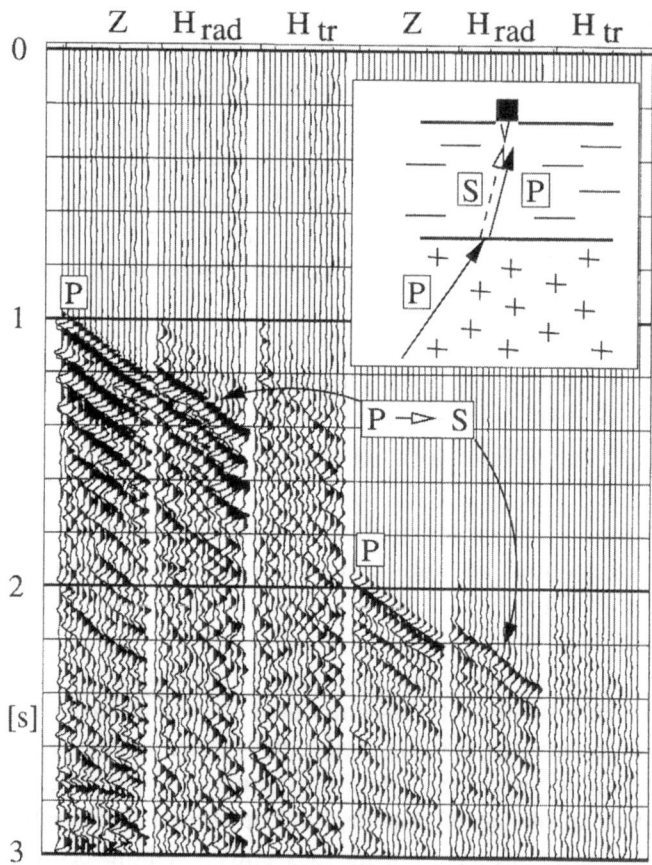

Figure 3

Two dynamite 48-channel three-component shot gathers from the ASTRA experiment in the Voronesh massif 1991. Sixteen stationary three-component stations with 100 m spacing, true amplitudes, non-filtered. Source offset to the nearest seismometer 20 km and 40 km, respectively, reduced time scale ($V_{red} = 8$ km/s) Z = vertical component, H_{rad} = horizontal radial component, H_{tr} = horizontal transverse component. Note the P-to-S conversions 200 ms behind the direct P-wave first arrivals. The inlet shows the generation of the converted phases below the receivers.

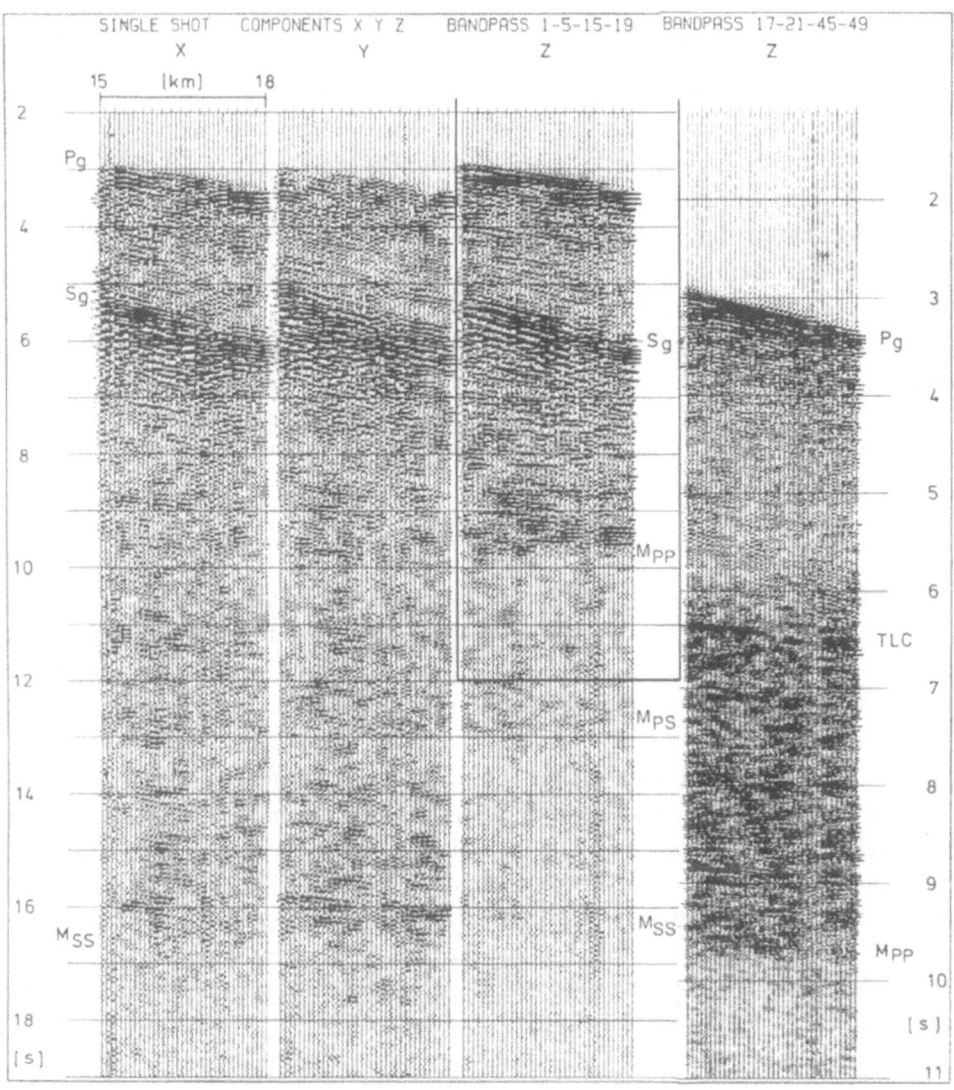

Figure 4

Three-component dynamite shot gather from the central Black Forest (LÜSCHEN *et al.*, 1990) with horizontal components (*X* = radial, *Y* = transversal) and vertical component. Offset range between 15 and 18 km, trace spacing 100 m. Amplitudes between 3 and 17 s were multiplied with a factor increasing linearly from 1 to 2 (*Z*-component between 3 and 10 s). The frame on the vertical component marks the data field that was available in previous standard reflection studies. The time scale for the vertical component (*Z*) in the right panel is stretched by a factor of 1.73 according to the mean crustal V_p/V_s ratio. Note that *P*- (*Z* panel on the right) and *S*-wave reflections (*X* and *Y* panels on the left) from the same structure can be compared directly now. Pg = direct *P* wave, Sg = direct *S* wave, TLC = top lower crust, Mpp = reflected *P* wave from crust-mantle boundary, Mps = corresponding *P*-to-*S* converted phases, Mss = reflected *S* waves from crust-mantle boundary, trapezoidal bandpass filter settings on top of the figure.

Lithological Discriminator and Rock Anisotropy

Full recognition and usage of S waves in seismic reflection studies require, first of all, three-component recordings for applying appropriate analysis tools like polarisation filters and particle motion diagrams (hodograms). Dense receiver spacing as normally used in reflection experiments, is required because of possible ambiguities due to the scattering of seismic waves at near-surface heterogeneities. Broadband recording and long receiver arrays are also advantageous, the first for including the full frequency range of both, P and S waves, and for applying appropriate bandpass filters, the latter for observing long move-out times of P- and S-wave reflections. The extent to which S waves from deep seismic reflection experiments can really contribute to crustal research under *in situ* conditions may be indicated by selected cases presented below.

Poisson's Ratio

The interval V_p/V_s ratio or Poisson's ratio can be derived from the P- and S-wave first arrivals and the average ratios from the travel-time ratio of reflections. The V_p/V_s ratio has been proven by laboratory (e.g., KERN, 1982), seismic *in situ* measurements (e.g., LÜSCHEN *et al.*, 1996), seismic reflection surveys (WARD *et al.*, 1991; LÜSCHEN *et al.*, 1990, 1993) and seismic refraction surveys (HOLBROOK *et al.*, 1988) to be a more useful lithological indicator in combination with V_p than V_p alone. In particular, felsic rocks exhibit a significantly lower V_p/V_s ratio because of their high quartz content than mafic rocks. A relatively low V_p/V_s ratio was observed for the upper crust, however a higher one was observed for the lower crust, consistent with a granitic upper crust and a mafic lower crust (e.g., WARD *et al.*, 1991; HOLBROOK *et al.*, 1988). However, the V_p/V_s ratio depends not only on the lithology and the physical state of the rocks, but also on the direction of the seismic ray path, as discussed below.

Reflection Strength

Differing reflection strengths of P and S waves are another source of information. A special S-wave reflection experiment was performed at the KTB deep drilling site employing horizontal vibrators as seismic sources with orientations cross-line and at oblique angles with respect to the seismic line and compared to a former standard P-wave reflection profile (LÜSCHEN *et al.*, 1993). Figure 5 shows a range of models which were used to explain the appearance of P-wave reflections at about 8.5-km depth where corresponding S-wave reflections were missed. All four models are consistent with a higher contrast in P-wave velocity than in S-wave velocity. Independent information, mainly from logging data and a VSP experiment (LÜSCHEN *et al.*, 1996) favoured the model of a fracture zone, where the aspect ratio of fracs and the porosity is enhanced. *In situ* velocities measured by logging and by the VSP experiment were reduced at all depths with respect to laboratory-derived velocities of KERN *et al.* (1991). This indicates that crystalline rocks *in situ* generally contain a significant volume of cracks, microcracks and pores and that seismic reflections may be caused by variations of this property. A varying strength of *S*-wave reflections obtained from the lower crust in the Rhinegraben area along a sub-horizontal structure that also was seen by pronounced *P*-wave reflections (MAYER *et al.*, 1997) indicates that the physical state (porosity, crack density, fluids) changes along this structure. The strength of the *S*-wave reflections in Figure 4 was not compatible with wide-angle reflections in the same area which were interpreted by SANDMEIER and WENZEL (1990) in terms of the V_p/V_s ratio using a petrological inversion procedure. This contrast was modelled by LÜSCHEN *et al.* (1990) by introducing alternating isotropic and anisotropic layers in the lower crust. The elastic tensor determined by SIEGESMUND *et al.* (1989) in laboratory experiments on the mineral Hornblende served as input to anisotropic reflectivity modelling.

S-wave Splitting

The fourth model in Figure 5 emphasises the effect of another important rock property, the seismic anisotropy. Seismic velocities, in particular the *S*-wave velocity, depend on propagation direction if the rock is characterised 1) by thin-layering with the thickness of single layers (even isotropic) considerably smaller than the wavelength), 2) by mineral alignments due to ductile deformation (meta-

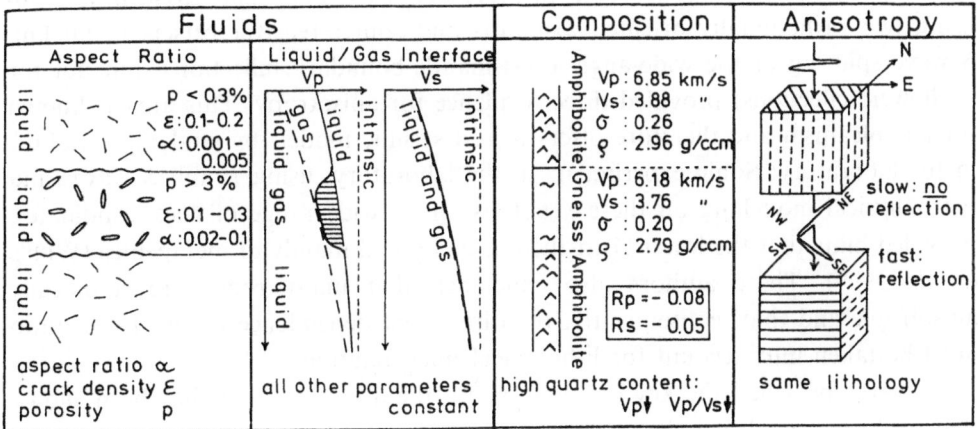

Figure 5

Model space for explaining a bright *P*-wave reflection (or diffraction) and a lacking *S*-wave response (LÜSCHEN *et al.*, 1993). From left to right: fracture zone with increased aspect ratio and porosity, liquid-gas-fluid interfaces in fracture zone, pure compositional effect, *S*-wave splitting and varying anisotropy symmetry system.

morphism) or 3) by alignments of fracs and cracks due to the recent stress regime (see RABBEL and MOONEY, 1996, for review and further references). S waves exhibit a unique property, namely birefringence or S-wave splitting in anisotropic media (CRAMPIN, 1985). The incident S wave splits into a fast and a slow phase which are orthogonally polarised to each other and which propagate with different velocities. Once recognised, the S waves provide the only direct evidence and measure of seismic anisotropy because of the splitting. The detection of this property requires three-component recording and polarisation studies. Analysis of S-wave splitting can be first used for diagnosis, and for quantifying the rock anisotropy in terms of its magnitude and symmetry system. This has been done by RABBEL (1994) in the case of the KTB data, with the result of about 10% azimuthal anisotropy over a rock volume of several kilometres in diameter, caused mainly by mineral alignment in amphibolites. In this data set, interpretation benefited from the fact that the rock volume of interest (several km in diameter) was probed by many different ray-path directions, including vertical ray paths in VSPs (LÜSCHEN et al., 1996), oblique ray paths by offset VSPs and multiple-azimuth moving-source experiments (RABBEL, 1994) and wide-angle seismic measurements (BOPP, 1992). A full description of the symmetry system of the rock anisotropy requires probing of the rock volume in all directions. The deep KTB hole enabled us to directly determine the existence and magnitude of rock anisotropy by observing not only S-wave splitting, but also continuously diverging travel-time branches of the two split S waves for the first time in crystalline rocks. The indication of seismic anisotropy in the lower crust in SW-Germany presented by LÜSCHEN et al. (1990) was of indirect character. This was motivation to study the S-wave response from the lower crust in the Urach area in more detail (RABBEL and LÜSCHEN, 1996) using two orthogonally oriented azimuths and source-receiver offsets to 90 km. S-wave splitting of the wide-angle crust-mantle-boundary reflections (but not for shallower reflections) provided direct evidence for anisotropy. This area is known for its volcanism and therefore offers to test seismic models by studying xenoliths in the laboratory. Sonic experiments in the laboratory, using P and S waves and petrophysical modelling on these xenoliths and a wide range of rock candidates, provided hints that the lower crust there is composed mainly of metapelites (WEISS et al., 1999). These authors also concluded that many rocks are seismically anisotropic and that, consequently, the directional dependence of the V_p/V_s ratio must be taken into account for lithological interpretations.

S waves provided the key observations in the above cases. It has been shown that explosives in drill holes generated a significant amount of S-wave energy in normal reflection surveys. Using sources that generated horizontal forces with defined orientation, e.g., horizontal vibrators (EDELMANN, 1985), and experiments with varying source orientations (e.g., GUT et al., 1994), helped to avoid ambiguities in detecting both split phases in case of seismic anisotropy.

Conclusions

S waves differ from *P* waves because of their lower propagation velocities, their lower frequencies and because of their different polarisation. In common shot-, receiver- and CMP-gathers the moveout of travel-time branches of *S*-wave reflections is different from *P* waves due to the difference in the propagation velocities. To distinguish between *P* and *S* waves requires broadband-frequency acquisition, receiver arrays as long as possible and three-component recording. Examples shown above demonstrate that *S*-wave generation at the source and *P*-to-*S*-wave conversion at crustal interfaces can be very efficient. There is a real danger of misinterpreting signals if only vertical components are used. Application of three-component recording in crustal reflection surveys, on the other hand, requires the number of channels to be three times larger than in conventional surveys and is hence a significant cost factor. Therefore, its application is recommended only at pre-defined critical locations or targets where crustal models exist. *S* waves can then contribute to support or to withdraw existing models or to reduce the model space. When properly applied, integrated *P*- and *S*-wave studies provide very effective lithological discriminators, in particular concerning the quartz content, and indicators for rock anisotropy, which can be interpreted for the direction of the recent stress regime (alignments of microfractures) or for the direction of palaeo-stress (alignments of minerals).

Acknowledgements

Several grants of the Deutsche Forschungsgemeinschaft (DFG, Bonn) and the Bundesministerium für Bildung, Wissenschaft, Forschung und Technik (BMBF, Bonn) within the DEKORP and KTB programmes are gratefully acknowledged. I thank Karl Fuchs, Peter Hubral and colleagues at Karlsruhe University and Hans-Jürgen Dürbaum and his colleagues of the former DEKORP management at the Niedersächsische Landesamt für Bodenforschung, as well as the KTB management and staff for creating an exciting research environment. Sharov (GEON, Moscow) kindly supplied the data for Figures 1 and 2. I also thank the editor Wolfgang Rabbel, and Michael Bopp and an anonymous reviewer for helpful comments.

REFERENCES

BOPP, M. (1992), *Shear-wave Splitting Observed by Wide-angle Measurement*, KTB Report 92-5 DEKORP Report, Niedersächsisches Landesamt für Bodenforschung, Hannover, 297–308.

CHRISTENSEN, N. I., *Pore pressure, seismic velocities, and crustal structure*. In *Geophysical Framework of the Continental United States* (eds. Parkiser, L. C., and Mooney, W. D.) (Geol. Soc. Am. Mem. *172* 1989) pp. 783–798.

CRAMPIN, S. (1985), *Evaluation of Anisotropy by Shear-wave Splitting*, Geophysics *50*, 142–152.

CRAMPIN, S., BUSH, I., NAVILLE, C., and TAYLOR, D. B. (1986), *Estimating the Internal Structure of Reservoirs with Shear-wave VSP*, Leading Edge *5* (11), 35–39.

DANBOM, S. H., and DOMENICO, S. N., *Shear-wave Exploration*, Geophysical Development Series, vol. 1 (Society of Exploration Geophysicists, Tulsa 1987).

EDELMANN, H. A. K., *Shear-wave energy sources*. In *Seismic Shear Waves, Part B: Applications* (ed. Dohr, G.) (Geophysical Press, London 1985) pp. 134–177.

FERTIG, J. (1984), *Shear Waves by an Explosive Point-Source: The Earth Surface as a Generator of Converted P-S Wave*, Geophys. Prospect. *32*, 1–17.

GUT, T., SÖLLNER, W., LÜSCHEN, E., and EDELMANN, H. A. K. (1994), *More Reliable Shear-wave Data from VSPs Using the CIPHER Technique*, First Break *12*, 123–129.

HESTHAMMER, J., and FOSSEN, H. (1997), *The Influence of Seismic Noise in Structural Interpretation of Seismic Attribute Maps*, First Break *15*, 209–219.

HOLBROOK, W. S., GAJEWSKI, D., KRAMMER, A., and PRODEHL, C. (1988), *An Interpretation of Wide-angle Compressional and Shear-wave Data in SW Germany: Poisson's Ratio and Petrological Implications*, J. Geophys. Res. *93*, 12081–12106.

KASHUBIN, S. N., DUBYANSKI, A., NADEZHKA, L., and LÜSCHEN, E. (1998), *Deep Structure and Seismicity of the Voronezh Massif*, Tectonophysics, submitted.

KERN, H., *P- and S-wave velocities in crustal and mantle rocks under simultaneous action of high confining pressure and high temperature and the effect of the rock microstructure*. In *High-Pressure Researches in Geoscience* (ed. Schreyer, W.) (Schweizerbart, Stuttgart 1982) pp. 15–45.

KERN, H., SCHMIDT, R., and POPP, T. (1991), *The Velocity and Density Structure of the 4000 m Crustal Segment at the KTB Drilling Site and their Relationship to Lithological and Microstructural Characteristics of the Rocks: An Experimental Approach*, Sci. Drill. *2*, 130–145.

KIND, R., KOSAREV, G. L., and PETERSEN, N. V. (1995), *Receiver Functions at the Stations of the German Regional Seismic Network (GRSN)*, Geophys. J. Int. *121*, 191–202.

LEVANDER, A., HOBBS, R. W., SMITH, S. K., ENGLAND, R. W., SNYDER, D. B., and HOLLIGER, K. (1994), *The Crust as a Heterogeneous "Optical" Medium, or "Crocodiles in the Mist"*, Tectonophysics *232*, 281–297.

LÜSCHEN, E., NOLTE, B., and FUCHS, K. (1990), *Shear-wave Evidence for an Anisotropic Lower Crust Beneath the Black Forest, Southwest Germany*, Tectonophysics *173*, 483–493.

LÜSCHEN, E., SOBOLEV, S., WERNER, U., SÖLLNER, W., FUCHS, K., GUREVICH, B., and HUBRAL, P. (1993), *Fluid Reservoir (?) Beneath the KTB Drillbit Indicated by Seismic Shear-wave Observations*, Geophys. Res. Lett. *20*, 923–926.

LÜSCHEN, E., BRAM, K., SÖLLNER, W., and SOBOLEV, S. (1996), *Nature of Seismic Reflections and Velocities from VSP Experiments and Borehole Measurements at the KTB Deep Drilling Site in Southeast Germany*, Tectonophysics *264*, 309–326.

MACBETH, C. (1991), *Inversion for Subsurface Anisotropy Using Estimates of Shear-wave Splitting*, Geophys. J. Int. *107*, 585–595.

MAYER, G., MAI, P. M., PLENEFISCH, T., ECHTLER, H., LÜSCHEN, E., WEHRLE, V., MÜLLER, B., BONJER, K.-P., PRODEHL, C., and FUCHS, K. (1997), *The Deep Crust of the Southern Rhine Graben: Reflectivity and Seismicity as Images of Dynamic Processes*, Tectonophysics *275*, 15–40.

RABBEL, W. (1994), *Seismic Anisotropy at the Continental Deep Drilling Site (Germany)*, Tectonophysics *232*, 329–341.

RABBEL, W., and LÜSCHEN, E. (1996), *Shear Wave Anisotropy of Laminated Lower Crust at the Urach Geothermal Anomaly*, Tectonophysics *264*, 219–233.

RABBEL, W., and MOONEY, W. D. (1996), *Seismic Anisotropy of the Crystalline Crust: What Does It Tell Us?*, Terra Nova *8*, 16–21.

SANDMEIER, K.-J., and WENZEL, F. (1990), *Lower Crustal Petrology from Wide-angle P- and S-wave Measurements in the Black Forest*, Tectonophysics *173*, 495–505.

SHERIFF, R. E., and GELDART, L. P., *Exploration Seismology, Vol. 1, History, Theory and Data Acquisition* (Cambridge University Press, Cambridge 1982).

SIEGESMUND, S., TAKESHITA, T., and KERN, H. (1989), *Anisotropy of V_p and V_s in an Amphibolite of the Deeper Crust and its Relationship to the Mineralogical, Microstructural and Textural Characteristics of the Rock*, Tectonophysics *157*, 25–38.

WARD, G., WARNER, M., and the BIRP SYNDICATE, *Lower crustal lithology from shear wave seismic reflection data*. In *Continental Lithosphere: Deep Seismic Reflections* (eds. Meissner, R., Brown, L., Dürbaum, H.-J., Franke, W., Fuchs, K., and Seifert, F.) (American Geophysical Union, Washington, D.C. 1991) pp. 343–349.

WEISS, T., SIEGESMUND, S., RABBEL, W., BOHLEN, T., and POHL, M. (1998), *Seismic Velocities and Anisotropy in the Lower Continental Crust: A Review*, Pure appl. geophys. *156*, 97–122.

(Received August 19, 1997, revised May 20, 1998, accepted May 25, 1998)

 To access this journal online:
http://www.birkhauser.ch

Pure appl. geophys. 156 (1999) 97–122
0033–4553/99/010097–26 $ 1.50 + 0.20/0

❙ Pure and Applied Geophysics

Seismic Velocities and Anisotropy of the Lower Continental Crust: A Review

T. WEISS,[1] S. SIEGESMUND,[1,2] W. RABBEL,[3] T. BOHLEN[3] and M. POHL[4]

Abstract—Seismic anisotropy is often neglected in seismic studies of the earth's crust. Since anisotropy is a common property of many typically deep crustal rocks, its potential contribution to solving questions of the deep crust is evaluated. The anisotropic seismic velocities obtained from laboratory measurements can be verified by computations based on the elastic constants and on numerical data pertaining to the texture of rock-forming minerals. For typical lower crustal rocks the influence of layering is significantly less important than the influence of rock texture. Surprisingly, most natural lower crustal rocks show a hexagonal type of anisotropy. Maximum anisotropy is observed for rocks with a high content of aligned mica. It seems possible to distinguish between layered intrusives and metasediments on the basis of *in situ* measurements of anisotropy, which can thus be used to validate different scenarios of crustal evolution.

Key words: Seismic anisotropy, lower crust, shear-waves, Poisson's ratio.

Introduction

Most of our present knowledge on the deep continental crust was obtained by seismic methods: by near-vertical reflection profiling for structural imaging and by seismic refraction and wide-angle reflection profiling for velocity determination. The ultimate goal of seismic studies is to determine both the structure and composition of the crust.

To interpret seismic field data in terms of lithology, laboratory data of rocks typical for the deep crust must be considered. Typical rocks to be expected at mid to lower crustal levels can be identified at exposed high-grade metamorphic terranes, from drill core samples and from xenoliths.

There are number of physical and lithological parameters controlling the seismic properties of polyphase rocks under *in situ* conditions (Fig. 1, for further discussion

[1] Institut für Geologie und Dynamik der Lithosphäre, Goldschmidtstr. 3, D-37077 Göttingen, Germany.

[2] Geologisch-Paläontologisches Institut, Bernoullistr. 23, CH-4056 Basel, Switzerland.

[3] Institut für Geowissenschaften, Abteilung Geophysik, Olshausenstr. 40–60, D-24098 Kiel, Germany.

[4] Geophysikalisches Institut, Hertzstr. 16, D-76187 Karlsruhe, Germany.

see for example SCHÖN, 1996; SIEGESMUND, 1996). Beginning with the pioneering work of BIRCH (1960, 1961), the general relationships between velocity and pressure, velocity and temperature, velocity and density or mean atomic weight are now well established (e.g., MANGHNANI *et al.*, 1974; RUDNICK and FOUNTAIN, 1995).

Most results on deep crustal composition, based on seismic refraction data and laboratory measurements, rely on the assumption that the deeper crust is isotropic (see the compilation by HOLBROOK *et al.*, 1992). The velocities of *P*- and *S*-waves (V_p and V_s, respectively) are considered as a measure for a distinct mineralogical composition. Poisson's ratio varies between 0.20 and 0.35 for lower crustal rocks, where quartz-rich material represents the end-member at low Poisson's numbers.

The assumption of isotropy is valid only for rocks of crustal segments with a random distribution of layered rocks, minerals, microcracks, pores, etc. In some cases, seismic anisotropy has been used as a possible explanation of the reflectivity pattern observed from deeper crustal levels (e.g., JONES and NUR, 1984; RESTON, 1987; LÜSCHEN *et al.*, 1990; CARBONELL and SMITHSON, 1991; SIEGESMUND *et al.*, 1991, 1996). The most reliable indication of anisotropy is the observation of *S*-wave splitting. To date, there are only a few such data sets available for the deep crust. One of them derives from the laminated lower crust of the Urach Geothermal Area (Southern Germany), where a pronounced *S*-wave splitting (6–13%) was observed (RABBEL and LÜSCHEN, 1996; RABBEL *et al.*, 1998). In contrast, it is well known that many rocks exhibit seismic anisotropy at laboratory scales. However, laboratory seismic data which describe the complete directional dependence of sonic velocities for a general anisotropic body are extremely difficult to obtain and are rarely reported in the literature.

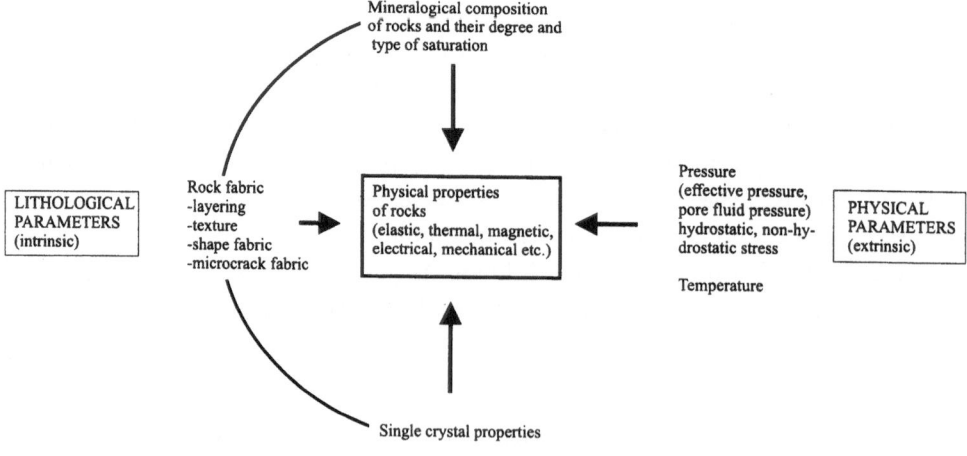

Figure 1
Factors controlling the seismic properties of crystalline rocks (from SIEGESMUND, 1996).

A comprehensive review of seismic constraints on lower crustal compositions by HOLBROOK *et al.* (1992) concluded that there are still two fundamental open questions for the interpretation of seismic studies of the deeper continental crust: the influence of anisotropy and of fluids. In the present study the role of seismic anisotropy in crustal studies is discussed, with emphasis on the so-called intrinsic seismic anisotropy caused by a pronounced rock fabric of highly anisotropic minerals at lower crustal levels (Fig. 1).

The approach used in this study to derive anisotropic seismic rock properties comprises two complementary methods: (1) calculations based on the measurements of rock fabrics and, (2) ultrasonic laboratory measurements. Both procedures will be briefly described and compared for an amphibolite.

The effect of anisotropy will be shown at different scales: (1) single-crystal anisotropy, (2) anisotropy of crustal and upper mantle rocks and (3) average anisotropy of entire crustal segments.

The classification of possible deep crustal and upper mantle rocks is based mainly on a recent compilation by RUDNICK and FOUNTAIN (1995). The exposed lower crustal sections of Calabria and the Ivrea Zone (Italy) serve as models for the calculation of the seismic velocities for a whole depth section.

Relation between Mineralogical Composition, Density and Sonic Velocities

A compilation of densities and of *P*- and *S*-wave velocities at a confining pressure of 600 MPa for selected rock groups is shown in Figure 2. It includes only laboratory investigations where sonic measurements were performed in at least three (orthogonal) directions. In general, velocities increase with increasing densities. RUDNICK and FOUNTAIN (1995) have shown that the V_p values of the different rock samples cluster with overlap between the groups. The *P*- and *S*-wave velocities and the density (ρ) are closely related to the single-crystal velocities and densities and the modal composition of a rock. Abundant garnet, sillimanite, olivine, ortho and clinopyroxene and hornblende are responsible for the relatively high velocities and densities. Low velocity and density minerals such as quartz, biotite, K-feldspar and plagioclase lead to a decrease in velocity and density.

A decrease of V_p with increasing SiO_2 content is a well-known relationship (SCHÖN, 1996). Rock samples with higher densities have a lower SiO_2 content. However, for mafic rocks this relation holds only qualitatively because of the strong influence of the TiO_2, Fe_2O_3, MgO, and Al_2O_3 content. In contrast to felsic rocks which are composed mainly of quartz, feldspar and mica, mafic rocks exhibit a wider variability in mineralogy and chemistry. The chemical composition (not shown here) is only weakly reflected by V_p, V_s and the density. Statistical investigations by BÜTTGENBACH (1990) have shown that a more refined classification of rock types within the $V_p - \rho$ relation is possible when the mean atomic weight (*M*-value after BIRCH, 1961) is considered. Higher FeO, CaO, and TiO_2 contents

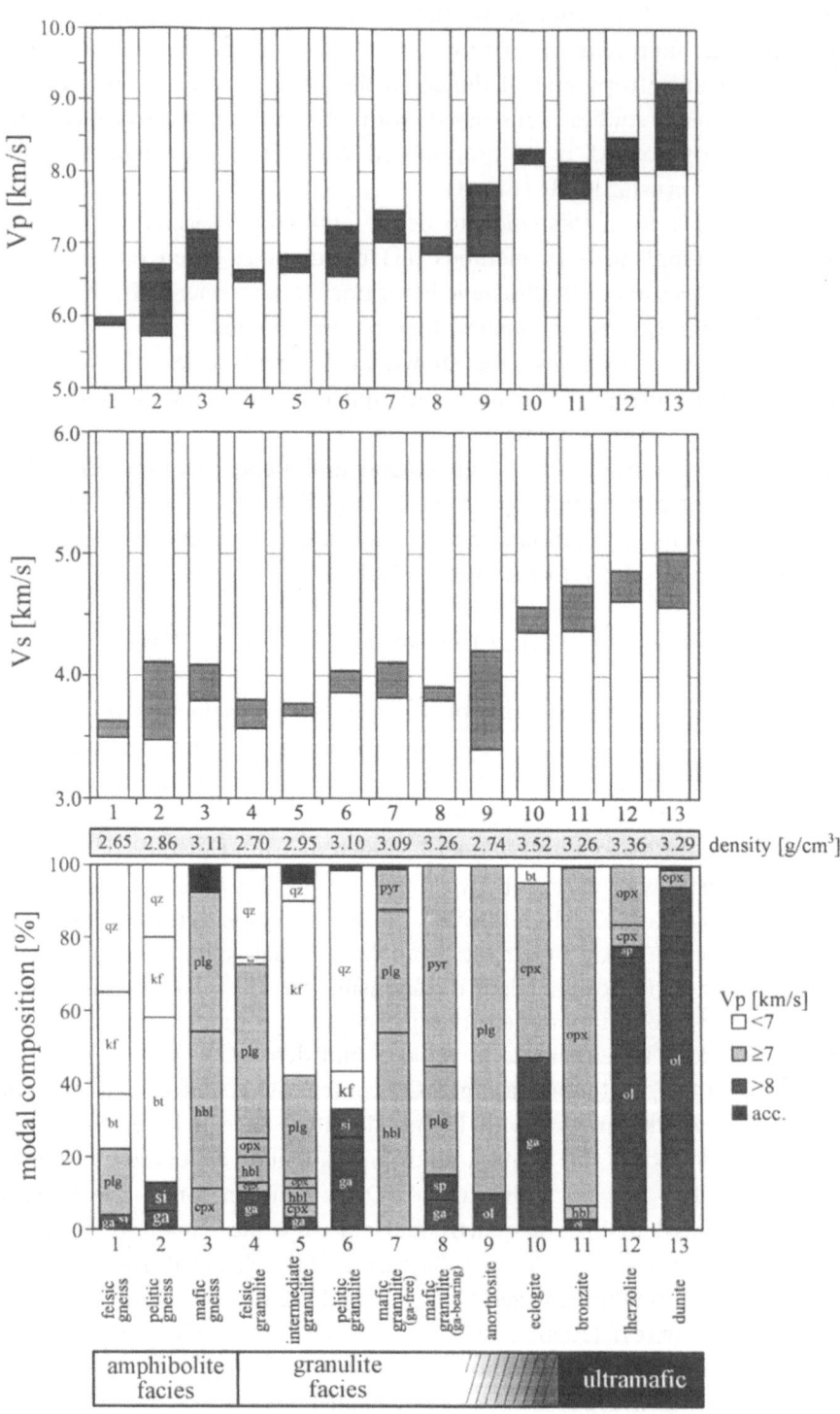

correspond to M-values between 35.5 to 26.6, whereas Al_2O_3, MgO and SiO_2 have M-values of 20.0 to 20.4. CHRISTENSEN and CROSSON (1968) found for upper mantle rocks, that with an increasing fayalite content of 10–25%, in olivine the velocity decreases about 0.25 km/s, whereas density increases about 0.2 g/cm^3. M varies between 20.1 for forsterite and 29.7 for fayalite, respectively, due to the very small atomic weight of Mg compared to Fe.

In terms of P- and S-wave velocity and anisotropy only (felsic) gneisses, eclogites and ultramafic rocks can clearly be distinguished from the other amphibolite and granulite facies rock groups (see Fig. 2). Seismic anisotropy causes an apparent scatter of the velocities preventing a simple correlation of velocity, density and V_p/V_s ratio with lithology. Therefore, a closer look at the magnitude and symmetry of possible anisotropies of typical deeper crustal rocks is required.

Determination of Anisotropic Seismic Velocities

A rock is elastically anisotropic if it consists of anisotropic minerals with a lattice (texture) or shape-preferred orientation. Moreover, at shallow crustal levels preferred orientations of microcracks can be of importance. Usually, the texture of a particular mineral phase is determined by measurements of single-crystal orientations or by statistical methods, describing the orientation of the crystal lattice planes in terms of frequency distributions (see compilation in BUNGE et al., 1994). The velocity of each mineral phase can be calculated by averaging the elastic constants of the respective mineral over all observed orientations (KLIMA and KLUHANEK, 1968; SIEGESMUND et al., 1989). Frequently used averaging methods are the so-called Voigt, Reuss and Voigt-Reuss-Hill averaging methods (e.g., MAINPRICE and HUMBERT, 1993). In the present study the Voigt average was used, supplying a sufficient agreement between calculated and experimentally determined velocities (e.g., SIEGESMUND and DAHMS, 1994). The elastic tensor computed for a single mineral phase represents a monomineralic rock with the respective texture. In

Figure 2

Mineral composition (bt = biotite, cpx = clinopyroxene, ga = garnet, hbl = hornblende, kf = K-feldspar, ol = olivine, opx = orthopyroxene, plg = plagioclase, pyr = pyroxene, qz = quartz, si = sillimanite, sp = spinel), densities, compressional- (V_p) and shear-wave (V_s) velocities for typical lower crustal rocks measured at room temperature and 600 MPa confining pressure (i.e., corresponding to the pressure at a depth of 20 km). The data were compiled from BABUŠKA (1972), MANGHNANI et al. (1974), SIEGESMUND et al. (1989), BARRUOL et al. (1992), SERONT et al. (1993), KERN and TUBIA (1993), BARRUOL and KERN (1996). The variations of the S-wave velocities, represented by the length of the grey bars, correspond to the maximum and minimum values measured in orthogonal directions. The lithological classification follows a suggestion of RUDNICK and FOUNTAIN (1995). Because of the limited number of published measurements on anisotropy of seismic waves our compilation represents a first rough estimate.

the next step, bulk rock properties are obtained by averaging the elastic tensor of the respective mineral phases according to their volume fraction. Finally, the seismic phase velocities can be calculated by solving the Christoffel equation (e.g., KLIMA and KLUHANEK, 1968).

A comparable approach is required for the calculation of the representative elastic properties of a rock unit or even an entire crustal segment. A supplementary averaging method is proposed for the following reasons:

Voigt and Reuss averages assume that either stress or strain are constant within the rock unit under consideration. In contrast, the Backus average technique (BACKUS, 1962) is based on a proper formulation of the boundary conditions of a fine-layered media. It results in an effectively anisotropic homogeneous layer even if the alternating layers are isotropic. In this specific case, the resulting effective anisotropy is transversely isotropic. The Backus average was extended to general anisotropic layered media (SCHÖNBERG and MUIR, 1989) and can thus be applied to the problems considered here, namely, for computing the effective anisotropy caused by a stack of anisotropic layers in the lower crust.

Elastic Anisotropy of Single-crystals

The relationship between seismic anisotropy and the texture of the constituent minerals can be best explained by considering the single-crystal properties. Most of the rock-forming minerals manifest a pronounced single-crystal anisotropy. The directional dependence of P- and S-wave velocities and of the S-wave splitting can be correlated with the orientation of the crystallographic axes of the respective minerals. Minerals of higher symmetry (e.g., orthorhombic, tetragonal) generally exhibit a coincidence of crystallographic planes (i.e., their normals) and the respective crystallographic axes. The a-, b- and c-axes of the single crystal correspond to the direction normal to the (100)-, (010)- and (001)-plane, respectively. For minerals of lower symmetry (monoclinic, triclinic) this relation does not hold. For hornblende, for example, the angle between the crystallographic a- and c-axes is about 105°. In this case only the b-axis corresponds to the [010]-direction and to the normal to (010)-plane (Fig. 3). The directions perpendicular to the (100)- and (001)-planes are oblique to the a- and c-axes and are termed a^*- and c^*-axes, respectively.

The anisotropic seismic velocities of typical lower crustal minerals are compiled in Figure 4. Biotite can be treated as hexagonal, since the velocity differences in the directions of the a-and b-axes are vanishingly small. A maximum of V_p and S-wave splitting occurs perpendicular to the c-axis, and a minimum parallel to the c-axis. For orthorhombic olivine, a maximum S-wave splitting can be observed between the a- and c-axes. Parallel to the a-axis, which is the fastest V_p direction, the S-wave splitting vanishes. In all other directions moderate S-wave splitting occurs. Other orthorhombic minerals are bronzite (orthopyroxene), sillimanite and cordierite. Bronzite, a typical mineral in ultramafic and granulite facies rocks, shows a similar V_p distribution as olivine although remarkably different S-wave patterns. The faster S-wave shows a small directional dependence, whereas the slower S-wave exhibits minima at an angle of 45° between the a- and b-axes and the b- and c-axes, respectively. In contrast, the maximum S-wave splitting for olivine is caused by a strong directional dependence of the faster S-wave. Sillimanite manifests a large S-wave splitting parallel to the b- and c-axes. The c-axis corresponds to the maximum V_p direction. However, the maximum S-wave splitting occurs at an angle of 45° between each of the crystallographic axes. For cordierite the S-wave splitting is significant in the planes given by the a- and b-axes and b- and c-axes, respectively. It is relatively small parallel to the c-axis. In contrast to other minerals, the c-axis of cordierite corresponds to the minimum V_p direction. Monoclinic minerals exhibit velocity distributions with only one symmetry plane. For hornblende the symmetry plane is defined by the a- and c^*-axes (see Fig. 3). A pronounced S-wave splitting can be observed in the plane marked by the b- and c^*-axes with a maximum parallel to the c^*-axis. The maximum of V_p can be observed parallel to

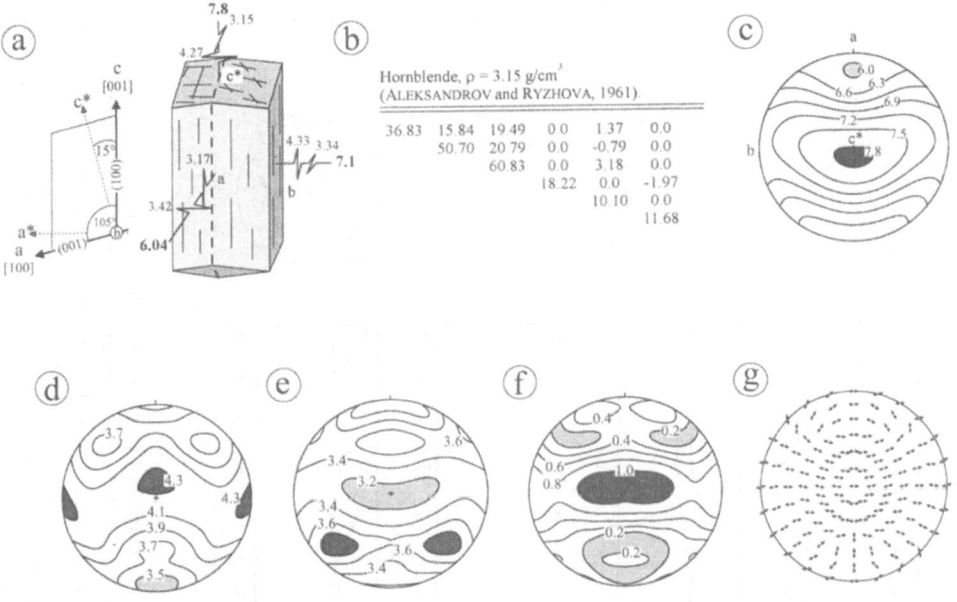

Figure 3

Relationship between the orientation of the crystallographic axes and the seismic velocities of hornblende: a) hornblende single crystal and a schematic sketch of the relationship between different crystallographic axes or planes, b) density normalized elastic tensor, and the directional dependence of V_p (c), V_{s_1} (d), V_{s_2} (e), $\Delta V_s = V_{s_1} - V_{s_2}$ in km/s (f) and the plane wave polarization of the faster S-wave (g) in equal area projections, lower hemisphere. The indices S_1 and S_2 denote faster and slower S-waves, respectively. The orientation of the crystallographic axes is indicated in c).

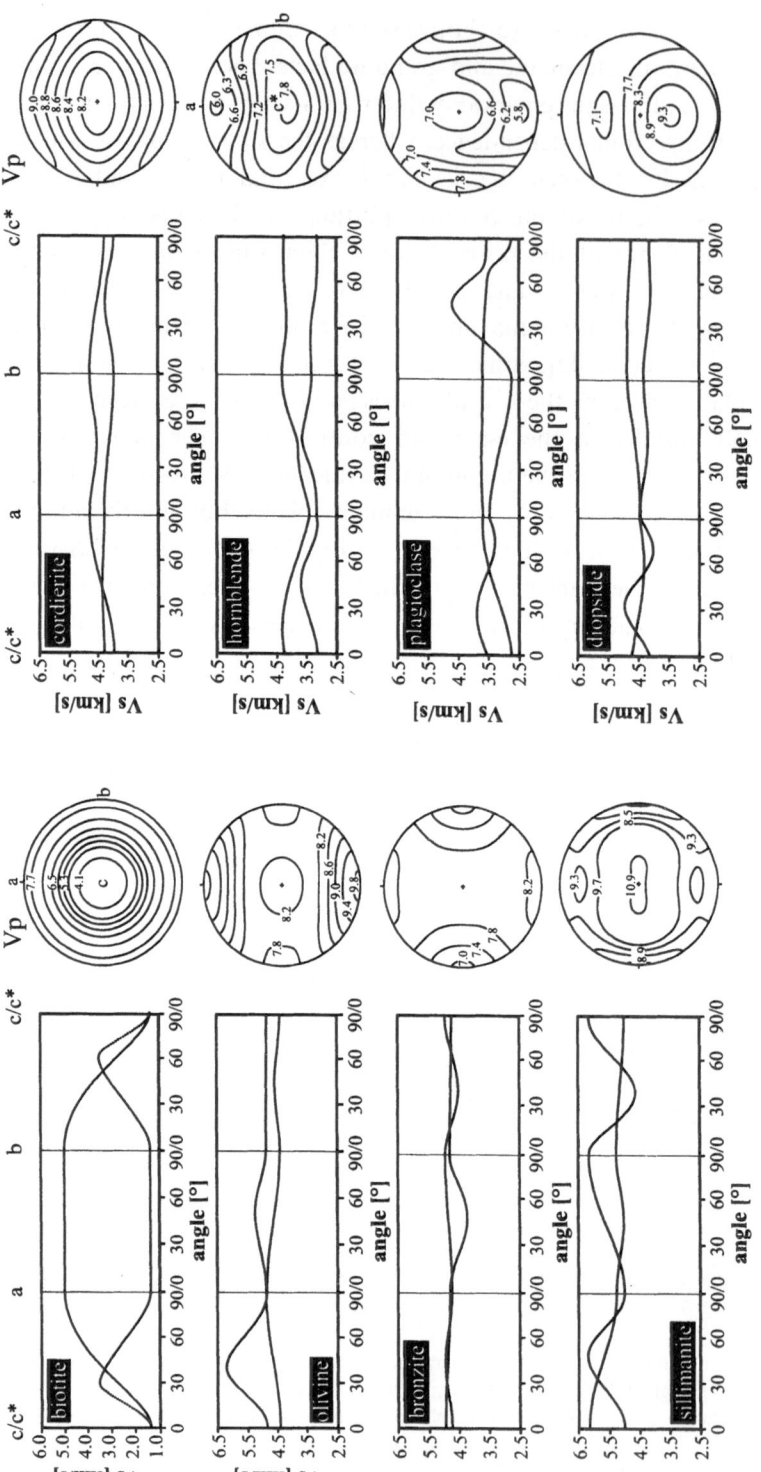

Figure 4

Single-crystal velocities of important minerals of the lower crust. The respective elastic tensors for biotite are from BELIKOV *et al.* (1970), for olivine from KUMAZAWA and ANDERSON (1969), for bronzite (orthopyroxene) from FRISILLO and BARSCH (1972), for sillimanite from VAUGHAN and WEIDNER (1978), for cordierite from TOBHILL *et al.* (1999), for hornblende (amphibole) from ALEKSANDROV and RYZHOVA (1961), for plagiocase from ALEKSANDROV *et al.* (1974) and for diopside (clinopyroxene) from LEVIEN *et al.* (1979). *S*-wave velocities are shown in planes defined by two of the crystallographic axes (*a*-, *b*-, *c*-axes for orthorhombic and *a*-, *b*-, *c**-axes for monoclinic and triclinic minerals). To illustrate the relationship between crystallographic symmetry and elasticity, the pole figure of the directional dependence of V_p is given additionally (equal area projection, lower hemisphere, isolines in km/s).

the c-axis (see Fig. 3). The single-crystal symmetry of plagioclase is triclinic, i.e., no symmetry elements can be observed. However, due to the lack of adequate single-crystal elastic constants, it is treated as monoclinic. The maximum S-wave splitting occurs parallel to the b-axis which is also the maximum V_p direction. Diopside shows a pronounced S-wave splitting in the plane defined by the b- and c^*-axes, and no simple correlation with the V_p distribution (i.e., V_p maximum or minimum) can be established. In conclusion, these typical lower crustal minerals manifest a significant S-wave splitting with different velocity patterns. The minerals show similar single-crystal symmetries and V_p distributions.

Elastic Anisotropy of a Polycrystal

In this section we show a quantitative comparison between measured sonic velocities and computed velocities based on textural data and the averaging procedures described above. To gain an estimate of the contributions of each mineral to the measured velocity anisotropy, the calculated velocity functions of each rock-forming mineral are compared with the frequency distribution of the respective crystallographic axes. All data are related to the coordinates (x, y, z) defined by the fabric elements (foliation and lineation). The sample we used for this purpose is an amphibolite from the Spessart Mountains (Germany). It comprises two mineral phases, namely hornblende (58%) and plagioclase (40%) plus 2% accessory minerals (SIEGESMUND et al., 1994). A macroscopically visible foliation is defined by the metamorphic layering of hornblende and plagioclase rich layers. Hornblende is aligned with the c-axis, which is the morphologically long axis, parallel to the lineation (x-direction; Fig. 5a). The a-, b- and c-axes of hornblende form maxima parallel to the z-, y-, and x-directions of the structural reference frame, respectively. In particular, the b-axes show a weak girdle perpendicular to the lineation. In contrast, the texture of plagioclase is less significant (Fig. 5b). The poles of the (001)-planes form a maximum perpendicular to the foliation, whereas [100] and (001) are distributed along girdles within the foliation.

The elastic tensor can be calculated from the complete distribution of V_p and the velocities of split S-waves in three directions (KLIMA, 1973; JECH, 1991). Therefore, the transition times of a spherical amphibolite sample were measured in 132 independent directions, applying the pulse transmission method at a confining pressure of 400 MPa (Fig. 5c). The velocities of the split S-waves were determined separately for three cylindrical samples each of which was drilled parallel to each axis of the reference frame (x, y, z, respectively). At low pressures, the elastic anisotropy is caused by a combination of intrinsic (crack-free) and crack-induced anisotropy. At higher pressures, above 100 MPa, cracks are mainly closed and the remaining anisotropy corresponds to the "intrinsic" rock properties. The difference between low- and high-pressure conditions can be used to quantify the symmetry and magnitude of crack-induced anisotropy (ARTS et al., 1996).

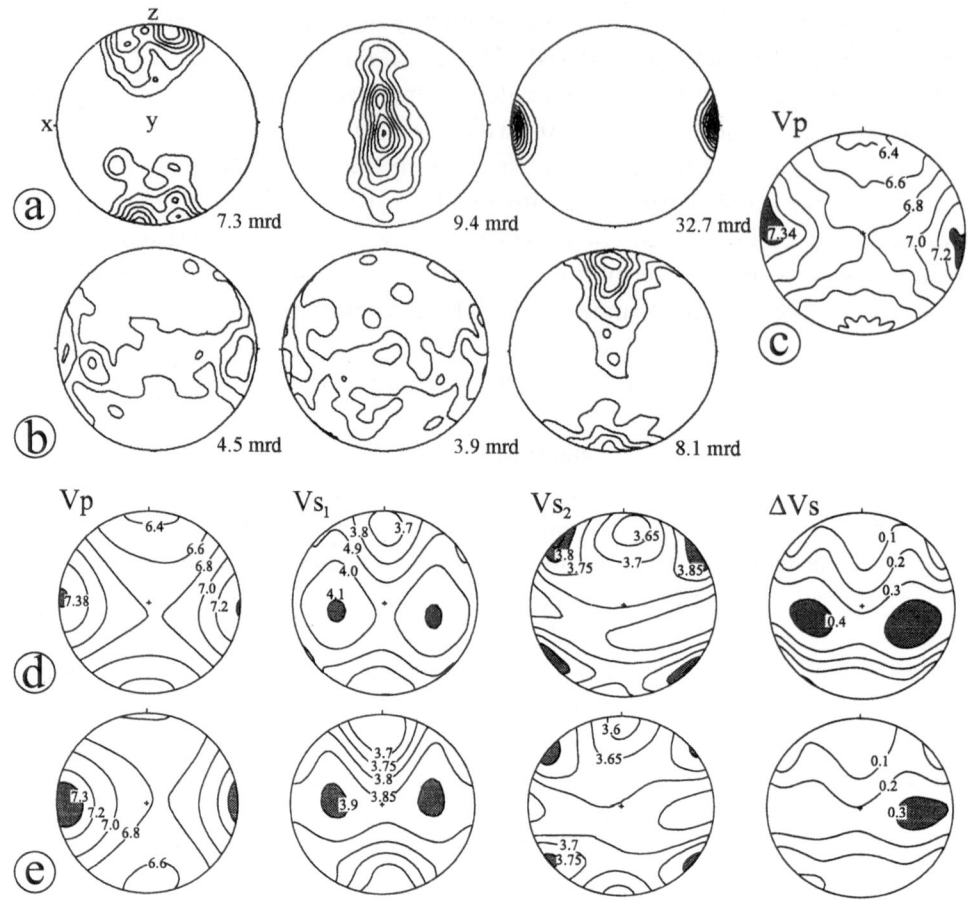

Figure 5

Stereographic projections (equal area, lower hemisphere) of seismic velocities of an amphibolite from the Spessart Mountain (Germany). Texture is represented by frequency distributions of the orientation of crystallographic axes (maximum frequency given in multiples of random distribution (mrd)). (a) Hornblende texture ($[\approx 100]$, [010], [001]). (b) Plagioclase texture ([100], (010), (001)). (c) V_p distribution at a confining pressure of 400 MPa based on the interpolation of laboratory measurements in 132 directions. (d) Velocity distributions calculated from the laboratory measurements of V_p in 132 (Fig. 5c) and from V_s in three directions. The velocity determination from the laboratory data is based on a linearized multivariate regression approach (KLIMA and KLUHANEK 1968). (e) Seismic velocities (in km/s) calculated from the texture data.

The intrinsic rock properties can also be calculated from textural data. The measured seismic wave velocities (Fig. 5d) agree well with the computed velocities based on texture data (Fig. 5e). Both complementary methods show an orthorhombic V_p distribution with a maximum V_p parallel to the lineation and a minimum V_p perpendicular to the foliation. The minimum of V_{s_1} and V_{s_2} is, likewise, observed perpendicular to the foliation, which coincides with the minimum S-wave splitting.

The maxima of V_{s_1} and of the S-wave splitting (ΔV_s) are aligned at an angle of approximately 45° between the x and y direction. In contrast, the maxima of V_{s_2} are at an angle of 45° between the lineation and the foliation normal. S-wave measurements only in the directions of the reference coordinates (x, y, z) would not be able to detect the maximum S-wave velocities and S-wave splitting in this case. Even some asymmetries, which are already visible in the experimental V_p distribution (Fig. 5c), are verified in the computed V_p distribution (Fig. 5e). Therefore, the elastic constants computed on the basis of fabric data are accurate enough to evaluate the particular anisotropic properties of a compact (crack- and pore-free) rock. The directional dependence of the seismic velocities of some typical middle to lower crustal rocks is shown in Figure 6.

Felsic gneisses frequently show a well developed foliation marked by preferentially aligned biotite (Fig. 7a). The dominant effect of the highly anisotropic biotite on the overall seismic velocities is clearly visible in the maximum S-wave splitting within the foliation (see Fig. 6). Despite the complicated S-wave patterns of monoclinic hornblende (see Figs. 3 and 4), mafic gneisses (i.e., amphibolites) likewise show a significant S-wave splitting parallel to the foliation.

The transition from amphibolite to granulite facies metamorphic conditions in metapelitic rocks correlates with the biotite/garnet-ratio. The directional dependence of seismic velocities of amphibolite facies metapelites (kinzigites) are controlled by the volume content and fabric strength of mica and sillimanite (see also BURLINI, 1994). Their metamorphic layering is defined by the preferred orientation of mica parallel to the foliation (Fig. 7b) and occasionally of sillimanite fibres parallel to the lineation. Parallel to the c-axis, the sillimanite single-crystal values exhibit maximum V_p and large S-wave splitting (see Fig. 4), which can cause a more pronounced anisotropy. Laboratory measurements reveal that the pattern for compressional waves of metapelites at a confining pressure of 600 MPa is almost transversely isotropic (BURLINI, 1994). High velocities of compressional waves can be observed within, and low velocities normal to the foliation plane. Particular examples are the metapelites from the Ivrea-Verbano Zone and from the Serie dei Laghi (Northern, Italy), where the V_p anisotropy varies between 10 and 26% (BURLINI, 1994). Correspondingly, the maximum S-wave splitting covers a range of 0.3–1.0 km/s with maximum values parallel to the foliation (cf., BURLINI and FOUNTAIN, 1993). Typical granulite facies metapelites are stronalites. The increase in garnet and the decrease in biotite is associated with increasing V_p and V_s but with a decreasing S-wave splitting. On the other hand, sillimanite can also be a frequent mineral phase in stronalites. A strong alignment of the c-axis of sillimanite can, produce a more pronounced intrinsic anisotropy in stronalites as well (Fig. 7c). Mafic rocks frequently show a small anisotropy because of the "destructive interference" of a variety of constituent minerals. The directional dependence of the seismic velocities of a magmatic gabbro from the so-called "Main Gabbro Intrusive Complex" in the Val Sesia sequence (Italy, e.g., RIVALENTI et al., 1981) is

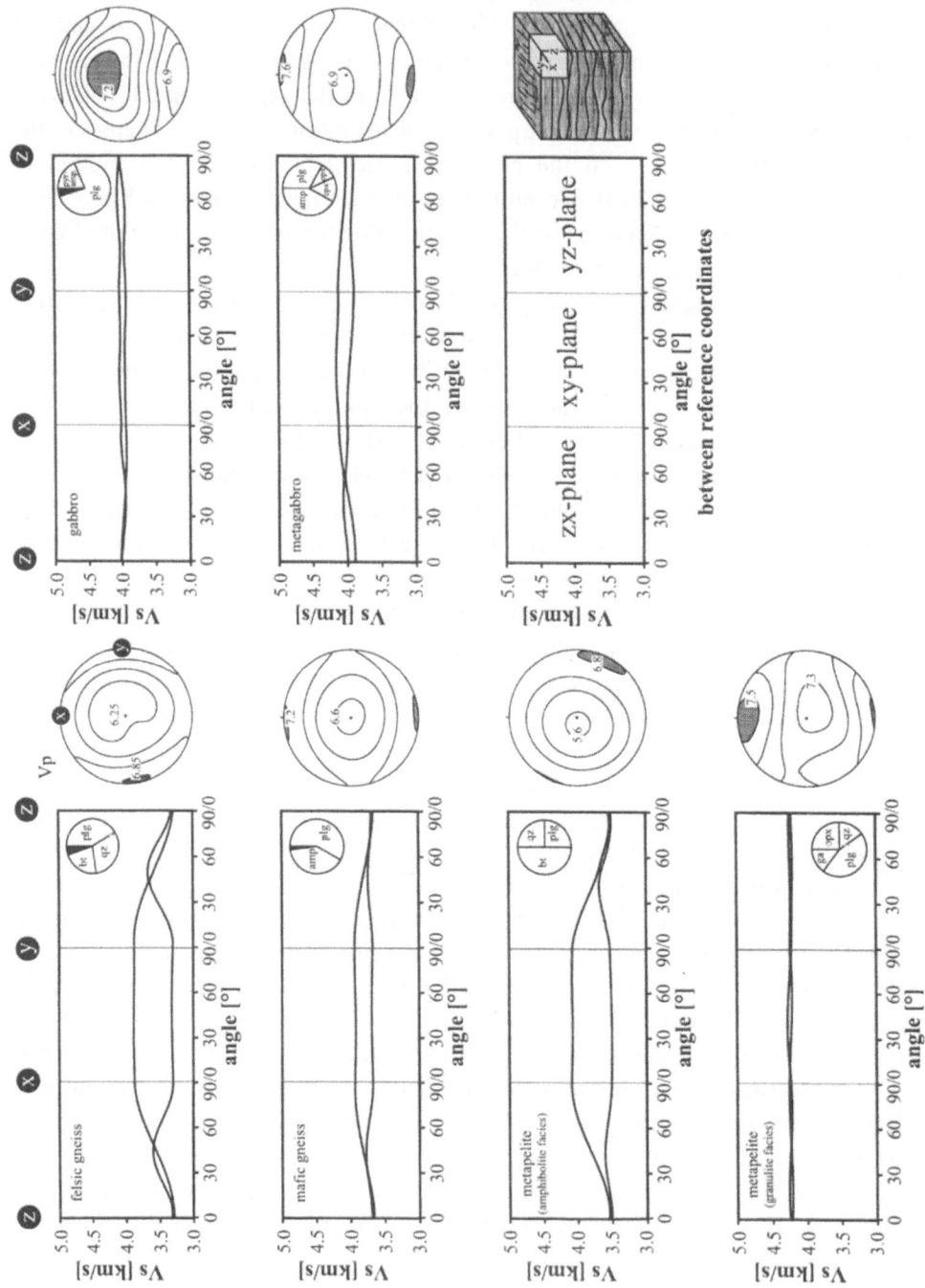

dominated by the fabric of volumetrically important plagioclase (BARRUOL and MAINPRICE, 1993). The maximum concentration of the b-axes of plagioclase, which is the fast velocity direction in the single crystal, is aligned perpendicular to the foliation. Metamorphic gabbroic rocks can contain a significant amount of garnet which leads to a general increase of the velocities (Fig. 7d). In contrast, the overall anisotropic seismic velocities of a metagabbro from Anzola (Italy) are clearly controlled by a hornblende texture which is similar to the texture of the amphibolite from the Spessart (see Fig. 5; DORNBUSCH, 1995; WEISS, 1998). Consequently, the seismic velocities of the entire rock are significantly anisotropic and show an orthorhombic symmetry. Since both types of mafic lower crustal rocks are possible at lower crustal levels, a wide variability of seismic anisotropy must be expected for this rock group.

Average Seismic Anisotropy of Exposed Lower Crustal Sections

From field measurements seismic velocities can only be determined for layers which are significantly larger than a wavelength. Often the whole lower crust is treated as consisting of one or two weakly inhomogeneous layers for which average velocities can be measured reliably. Therefore, estimates of lower crustal anisotropy must include one further step of averaging. The average has to take into account possible ensembles of different rock types which compose lower crustal units at the kilometer scale. Seismic anisotropy will be particularly strong for layered structures such as the so-called seismic lamellae; typical reflection pattern of the lower crust of young tectonic areas (MEISSNER, 1967; see also MEISSNER and RABBEL, this volume). An analogue may be found in the exposed lower crustal sections of Ivrea and Calabria (Italy). Since detailed lithological and structural information is available for both sites, reliable estimates of lower crustal anisotropy can be obtained.

The Ivrea Zone (Northern Italy) comprises both mafic and metapelitic sequences. The lower crustal section of Calabria (Southern Italy) is considered to be an almost complete profile through a Variscan lower crust (Fig. 8a) (e.g., SCHENK,

---

Figure 6

Directional dependence of seismic velocities of lower crustal rocks calculated from textures of rock-forming minerals and their volume fraction (for abbreviations see Fig. 2; black sectors correspond to accessory mineral phases). The specific rock types have been selected according to Figure 2 from the following references: felsic gneiss (SIEGESMUND et al., 1996; see Fig. 9e), mafic gneiss (SIEGESMUND, 1996; Weiss, 1998), amphibolite facies metapelite (BARRUOL, 1993), granulite facies metapelite, gabbro (BARRUOL and MAINPRICE, 1993) and metagabbro (DORNBUSCH, 1995; WEISS, 1998). S-wave velocities are shown for the planes zx, xy, and yz defined by the axes of the coordinate system. V_p is shown in an equal area projection (lower hemisphere, isolines in km/s).

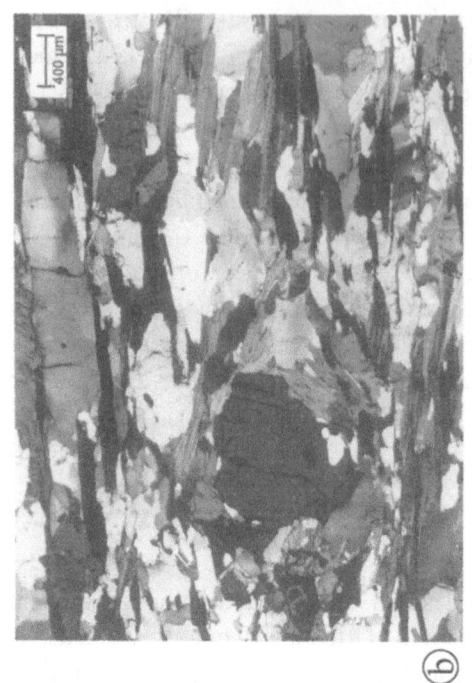

(b)

(d)

(a)

(c)

1980, 1984 or LÜSCHEN et al., 1992). Two lithological subunits can be distinguished; a granulite-pyriclasite unit at the base and a metapelite unit at the top of the lower crustal slice. The Ivrea Zone (Northern Italy) consists of two sections, namely those in the Val Strona (Fig. 8b) and the Val Sesia (Fig. 8c), respectively, which consist of different lithologies. The Val Strona is dominated by metapelites which comprise kinzigites and amphibolites in the uppermost section and stronalites and amphibolites in the lowermost part, with subordinate volumes of silicate marbles, pegmatites and microgranites (BERTOLANI, 1968). In contrast, the Val Sesia section consists predominantly of mafic to ultramafic rocks, while metapelites have only a very subordinate volume (e.g., RIVALENTI et al., 1984; VOSHAGE et al., 1990). The sequence in the Sesia Valley has been interpreted as a "multiphase layered intrusion" and is regarded as a typical example for magmatic underplating (RIVALENTI et al., 1975; GARUTI et al., 1979; RUTTER et al., 1993; QUICK et al., 1994). Detailed lithology/depth-sections have been compiled for characteristic lower crustal sections, using data available in the literature (BERTOLANI, 1968; SCHENK, 1980, 1984; VOSHAGE et al., 1990).

The elastic tensors, calculated from texture data of the different rock types, were selected from complementary data sources: BARRUOL (1993) and BARRUOL and MAINPRICE (1993) published elastic tensors of deeper crustal rocks from the Ivrea Zone (Italy). SIEGESMUND et al. (1996) characterized the modal compositions and the textures of the rock-forming minerals for the lithologies of the lower crust of Calabria. The corresponding tensors of elasticity, calculated from texture data, for the Calabrian rocks are shown in Table 1. The velocity-depth functions for the different sections are shown in Figure 8. Each layer is characterized by an elastic tensor. The metapelitic sections (Calabria and Val Strona, Figs. 8a,b) show distinct variations of density and seismic velocity with depth, and the main difference between both sections is the wavelength of the lithology variations. The Val Strona section exhibits a finer layering than, that of Calabria. In contrast, the Val Sesia profile shows a fine layering with alternating density and seismic velocity only in the lowermost part of the sequence (Fig. 8c).

Figure 7

Photomicrographs showing the fabric of selected lower crustal rocks (thin sections perpendicular to the foliation, the x-direction corresponds to the long side of the photograph, the scale is indicated). (a) Felsic gneiss: metamorphic layering is defined by parallel alignment of mica flakes between quartz- and feldspar-rich layers. Many of the grains in the quartz/feldspar matrix show a shape-preferred orientation with the long axis parallel to the lineation. (b) Amphibolite facies metapelite (kinzigite): Biotite is a frequent mineral and exhibits an alignment parallel to the foliation. Larger garnet porphyroblasts are surrounded by biotite which leads to an undulose foliation. (c) Granulite facies metapelite (stronalite): A well-defined metamorphic layering is marked by an alignment of garnet porphyroblasts, prismatic and fibrolitic sillimanite is oriented parallel to the lineation (x-direction). (d) Metagabbro: Frequent mineral phases are clinopyroxene, plagioclase and subordinately garnet. No shape-preferred orientation of the rock-forming minerals is visible, and the compositional layering is weak. When hornblende is a rock-forming mineral, it frequently presents an alignment of the c-axis (morphologic long axis) parallel to the lineation, and the fabric-induced anisotropy can be larger.

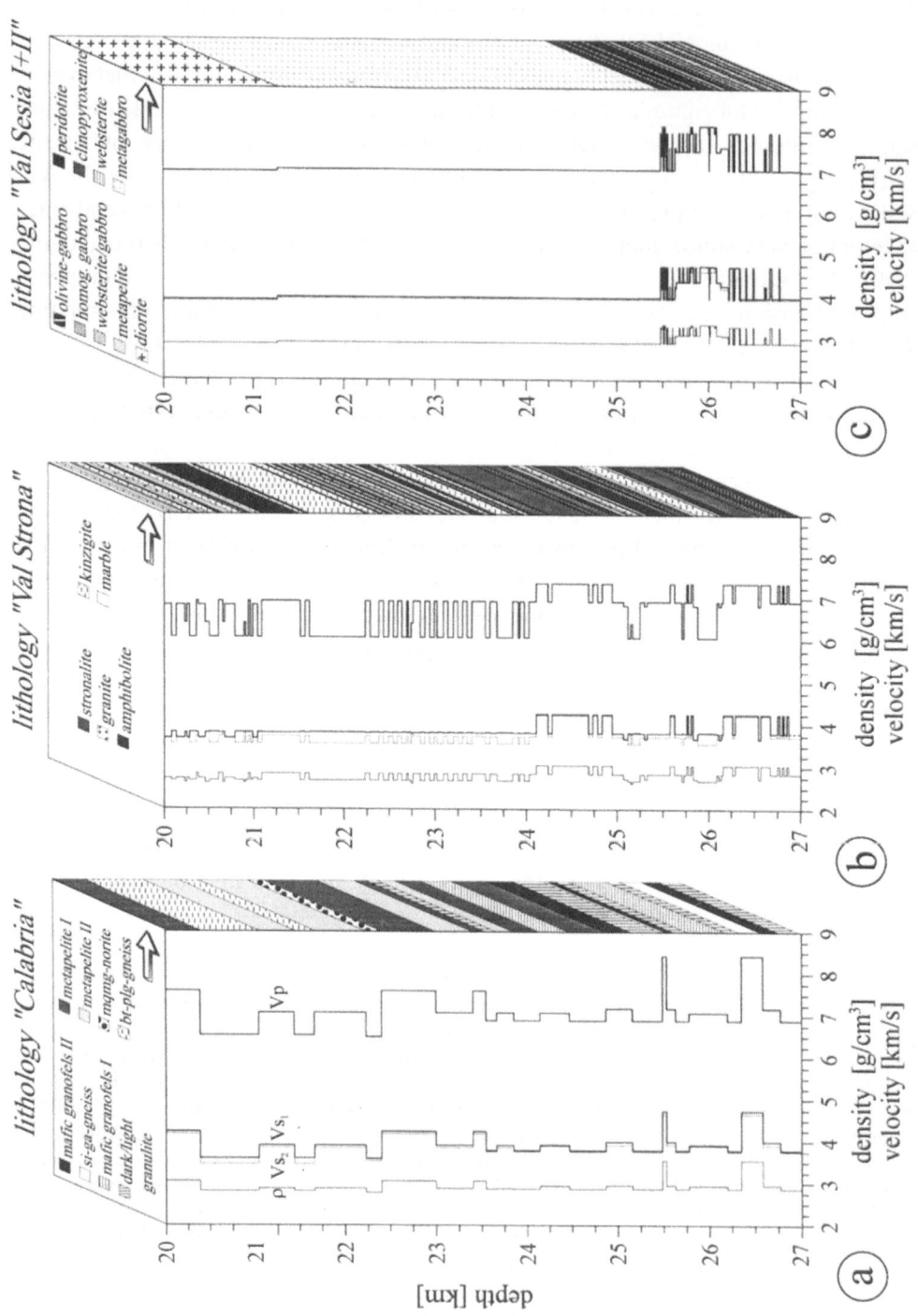

Seismic field measurements would determine only the average velocity and anisotropy of large segments of the lower crust or even of the lower crust in total. Therefore the respective average values have been computed for the total stack of layers in the Ivrea and Calabria sections. In order to consider the influence of layering on the resulting effective elastic constants, two different averaging methods, the Voigt (VOIGT, 1928) and the Backus (BACKUS, 1962; SCHÖNBERG and MUIR, 1989) averaging were used, respectively. The difference between results of both methods ultimately becomes vanishingly small. This surprising result demonstrates that the layering effect on anisotropy is less important than the mineral alignment in the case of lower crustal rocks.

The averaging procedure and the contribution of each rock type using the Calabria section are illustrated in Figure 9. Most of the lower crustal rocks show a maximum of V_p parallel to the lineation and a maximum S-wave splitting within the foliation (Figs. 9a–h). Although the individual rock types are variable in the symmetry patterns of the seismic velocities, the result is an overall quasi-transversal isotropic behavior with an S-wave splitting of 0.2 km/s observed along the direction of foliation (Fig. 9i).

The average seismic anisotropy obtained for the different models shown in Figure 8 is transversely isotropic for all the metasedimentary-composed sections (Calabria, Val Strona; Figs. 10a,b). Average V_p ranges within 6.8–7.0 km/s. The maximum V_p is parallel to the lineation (x direction) whereas the minimum V_p is oriented normal to the foliation (z direction). The transverse isotropy is also clearly visible in the V_{s_1}, V_{s_2}, ΔV_s distributions and in the polarization directions of the faster S-wave. A maximum S-wave splitting of approximately 0.2 km/s (approximately 5%) can be observed parallel to the foliation, a minimum normal to the foliation. In contrast, the mafic Val Sesia I section (Fig. 10c) shows a maximum of V_p perpendicular to the foliation and low P-wave velocities within the foliation. This effect is due to the volumetrically dominant gabbros (see Fig. 10e). There is no significant S-wave anisotropy. In order to investigate the possible range of anisotropy, we substituted all gabroic rocks of Val Sesia I numerically by the highly anisotropic Anzola metagabbro (Val Sesia II in Fig. 10d). The result is a significant effect on the seismic velocities of the entire crustal segment (Fig. 10d). The V_p distribution displays an orthorhombic symmetry with a P-wave anisotropy of approximately 7.3%. A maximum S-wave splitting of ≈ 0.2 km/s ($\approx 5\%$) occurs in the y-direction and of ≈ 0.1 km/s ($\approx 2.5\%$) in the x-direction.

Figure 8

Lithological composition, seismic velocities (V_p, V_{s_1}, V_{s_2}), for horizontally propagating waves and density as a function of depth for three exposed lower crustal sections in Calabria, Val Strona and Val Sesia, respectively. Val Strona and Val Sesia are part of the Ivrea section. The profiles, *in situ* oriented horizontally, are displayed vertically in order to simulate vertical sections through the lower crust. (For abbreviations see Fig. 2, mqmg = meta-quartz-monzo-gabbro-norite.)

Table 1

Tensor elements (in GPa) for different rock types from Calabria (Italy). The numbers refer to SIEGES-
MUND *et al. (1996), where the fabric and textures are described in detail. The mineralogical composition,
including the main constituent minerals, is shown (for abbreviations see, Figure 2, co = cordierite,
amp = amphibole). Accessory phases (i.e., ore) are not shown. The corresponding directional dependence of*
V_p *and of the S-wave splitting is shown in Figure 9*

Sample	metapelite I	metapelite II	mafic granofels I	mafic granofels II	bt-plg gneiss	mqmg-norite	granulite	si-ga gneiss
Number	3279	3351	3238	3366	3270	3265	3261	3244
density [g/cm³]	3.06	2.88	2.95	2.94	2.81	2.76	2.87	3.55
mineralogical composition [%]	8.4 plg	20.8 plg	23.6 amp	38.0 amp	41.6 plg	3.4 amp	11.3 opx	23.8 plg
	16.2 qz	33.6 qz	7.2 opx	3.3 opx	31.2 qz	2.8 opx	44.9 plg	4.2 kf
	15.6 kf	1.2 kf	16.4 cpx	8.6 cpz	1.2 kf	35.0 plg	25.7 qz	51.1 ga
	16.2 ga	6.4 ga	52.0 plg	47.0 plg	1.0 ga	42.2 qz	8.3 kf	18.4 si
	14.4 si	6.6 si		2.3 bt	21.8 bt	4.8 kf	6.5 ga	
	6.0 co	9.8 co				10.2 bt	2.1 bt	
	21.8 bt	21.0 bt						
C_{11}	166.95	141.09	146.48	141.64	111.27	116.94	142.05	253.01
C_{22}	188.77	150.57	153.59	157.94	124.69	117.63	131.25	256.61
C_{33}	177.96	149.48	146.08	140.44	131.80	118.13	135.65	252.81
C_{44}	58.55	46.03	46.76	46.98	42.16	36.82	40.88	77.70
C_{55}	51.45	41.24	43.45	41.78	30.85	33.63	41.08	75.50
C_{66}	52.65	41.74	43.96	42.68	30.74	34.12	40.28	76.80
C_{12}	61.26	55.42	59.57	57.48	44.77	46.79	53.88	90.40
C_{13}	62.76	57.32	59.07	53.88	45.27	46.50	56.28	91.80
C_{14}	−0.50	−0.40	−0.20	−0.70	−0.40	−0.40	−1.20	0.30
C_{15}	−1.60	0.70	−1.70	−0.80	−0.40	−0.90	−0.90	0.00
C_{16}	0.60	1.40	0.20	−0.10	0.50	−0.80	−0.50	0.09
C_{23}	66.36	57.21	54.47	53.58	43.47	44.00	50.98	95.80
C_{24}	0.30	0.00	−0.70	0.50	−0.50	0.80	−0.80	−0.30
C_{25}	−0.10	0.50	−1.40	−0.20	−0.50	−0.20	−0.50	−0.10
C_{26}	0.30	1.90	0.40	0.60	0.60	−1.20	0.60	−0.30
C_{34}	0.20	0.20	0.20	−0.50	−0.60	0.90	−0.10	−0.10
C_{35}	−0.30	1.80	−1.10	−1.60	−1.60	0.30	−1.10	−0.10
C_{36}	0.30	0.60	0.50	−0.20	0.10	−0.40	−0.10	−0.20
C_{45}	−0.30	0.70	−0.20	0.40	0.80	−0.50	0.20	−0.10
C_{46}	−0.30	0.50	0.00	−1.00	−0.60	0.00	0.00	−0.10
C_{56}	0.20	−0.10	−0.30	−0.20	−0.30	0.10	−0.10	−0.10

Discussion and Conclusions

The data basis concerning seismic anisotropy of the lower continental crust, in
particular relevant to S-wave splitting, is still small. Some of the questions relating
to anisotropy are the following:

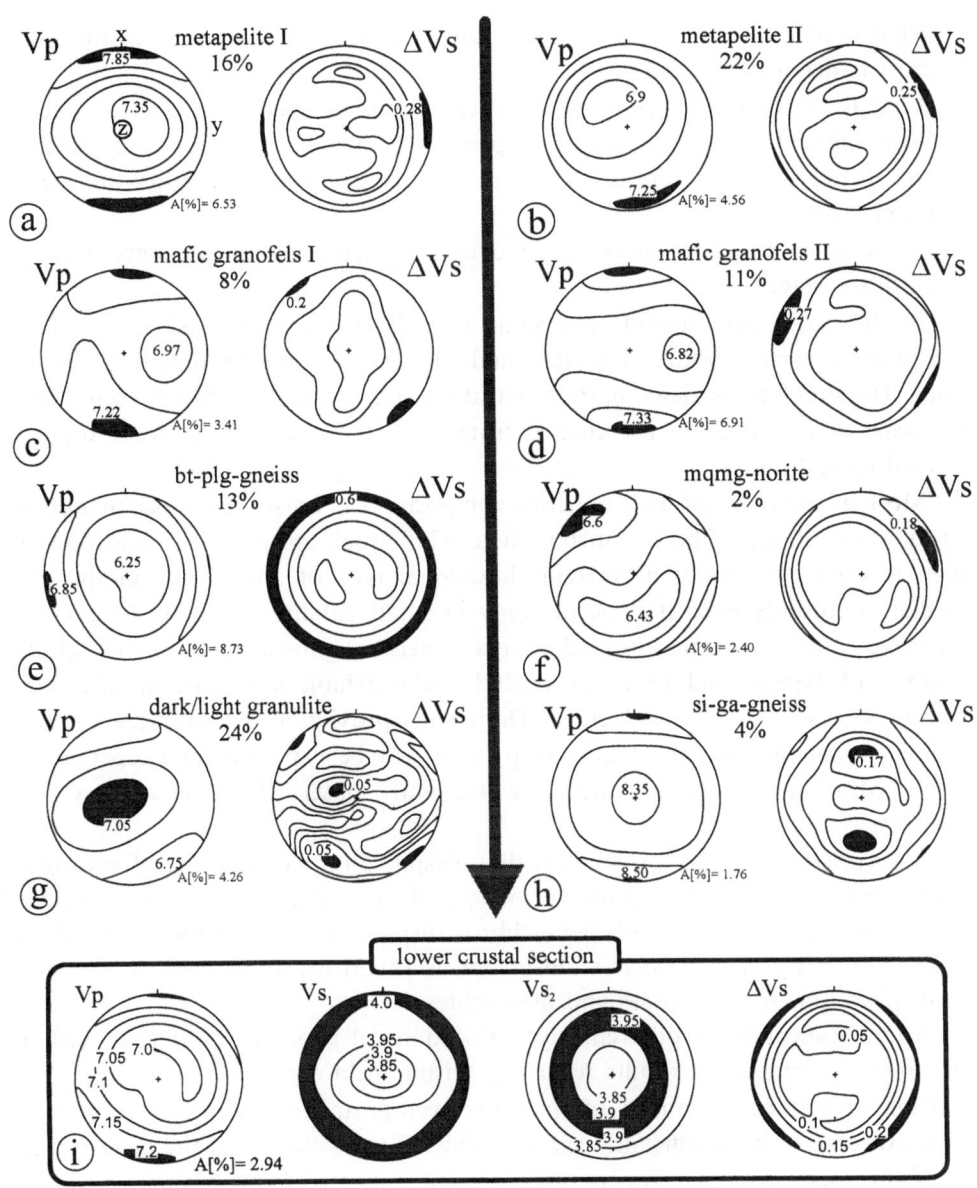

Figure 9

Velocities of compressional waves (V_p) and S-wave splitting (ΔV_s), in km/s computed from the elasticity tensor (see Table 1) for the rock types of the exposed lower crustal section in Southern Calabria (a–h). The volume fraction of the respective lithological unit is indicated in percent. Directions of maximum V_p or maximum S-wave splitting are shaded. V_p anisotropy (A) ranges from 1.76–8.73% ($A = (V_{p_{max}} - V_{p_{min}})/V_{p_{max}} * 100$). The average seismic velocities of the whole lower crust (i) display a smaller V_p anisotropy and S-wave splitting because of the superposition of different rock types. (For the abbreviations see Figs. 2 and 8.)

- What magnitude and type of seismic anisotropy can be expected in the lower crust under realistic conditions?
- What is the relative importance of rock fabrics compared to layering effects?
- What minerals, if any, leave their imprint in the whole rock anisotropy pattern?
- Are there typical differences in the average anisotropy between different types of lower crust?
- If yes, can these differences be discovered by seismic measurements from the earth's surface?

In this study an attempt was made to answer some of these questions by compiling laboratory data on crystals and whole rocks, and by numerical simulations. The simulations aimed particularly at the averaging effect which occurs when the scale of investigation is switched from crystals to rocks, and from rocks to layered crustal units.

Most of the rock-forming minerals composing the crust and upper mantle are strongly anisotropic. Their contribution to whole-rock anisotropy is controlled by their texture. Because of its extreme hexagonal type of anisotropy, biotite is of special importance even if it occurs only in small volume fractions. As a consequence of ductile deformation and related mineral alignment, most of the selected samples of typical mid to lower crustal rocks exhibit a significant directional variation of seismic wave velocities. The results provide a first overview on type, symmetry and magnitude of anisotropy that can be expected for deeper crustal levels. More than 10% anisotropy is frequently observed for laboratory scale samples.

A surprising result of this study is that, despite all the complexity of anisotropy on the crystal scale, the average anisotropy of most deep crustal rocks is quasi-hexagonal. Only gabbros and metagabbros display an orthorhombic but relative small anisotropy. The strongest anisotropy is found for rocks with a high mica content such as felsic gneisses and metapelites.

The investigation of different layered sections of exposed lower crust yielding the following important results regarding seismic field experiments.

- Compared to cm-scale rock samples, the average anisotropy of a stack of layers is expected to be significantly smaller. This effect is caused be the heterogeneity of the rock package. Five to seven percent anisotropy is a realistic maximum estimate for km-scale average values of the lower crust.
- These maximum values are expected to be found preferably in mica-bearing layered rocks of felsic composition. In contrast, mafic intrusive rocks seem to be quasi-isotropic even if they are layered.
- The primary cause of lower crustal anisotropy is rock texture (mineral alignment) rather than layering anisotropy (cf., RABBEL and LÜSCHEN, 1996).

Of crucial importance is, of course, the question, whether laboratory results may be applied to the interpretation of seismic field data. Three arguments are considered which assume that this is the case: (1) rock anisotropy can be verified quantitatively from crystal anisotropy by use of a simple averaging scheme, (2) the average anisotropy computed for a stack of lower crustal layers by the same algorithm agrees with traveltimes of synthetic wavefields computed with exact methods (cf., RABBEL et al., 1998; POHL et al., this volume). SHAPIRO and HUBRAL (1996) demonstrated that the anisotropy of thin alternating layers depends in

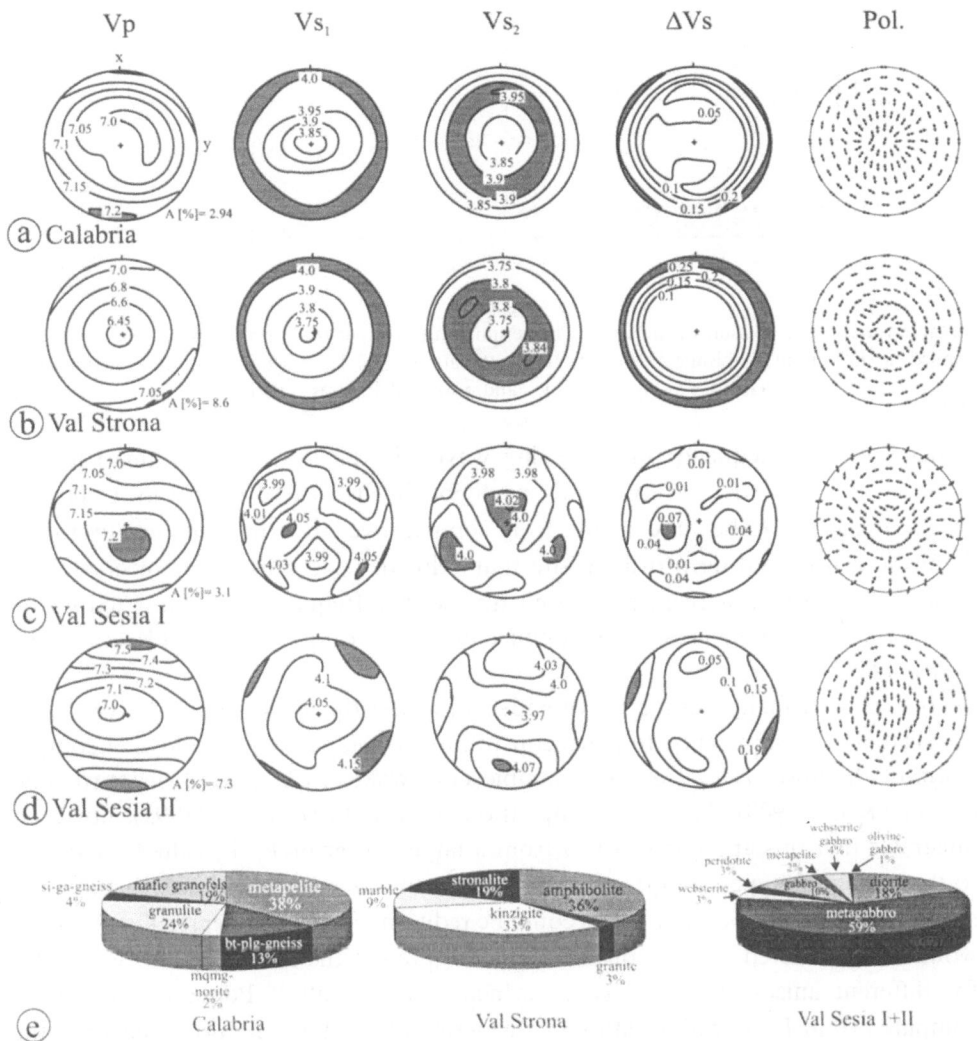

Figure 10
Average seismic velocities of the lower crustal profiles: (a) Calabria, (b) Val Strona, (c) Val Sesia I, (d) Val Sesia II (cf., Fig. 8). Val Sesia I and II differ in the elastic tensors assumed for the metagabbros. The directional dependencies of V_p, V_{s_1}, V_{s_2}, ΔV_s and of the plane wave polarization of the faster S-wave (Pol.) are shown (equal area projection, lower hemisphere, isolines in km/s). The directions of maximum velocity or S-wave splitting are shaded. Additionally, the volume fractions of the respective lithologies composing the lower crustal sections are shown (e).

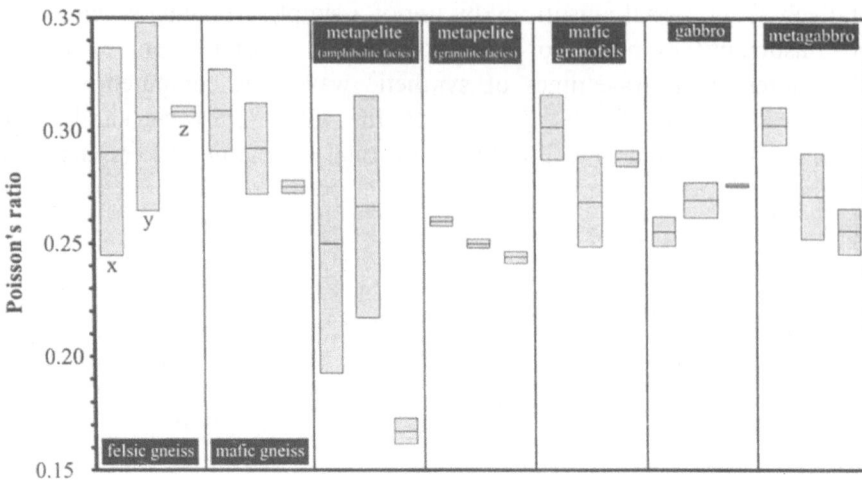

Figure 11

Possible variation of Poisson's ratio caused by seismic anisotropy if the formula for isotropic media is applied (as it is usual in seismic field studies). Input data are the V_p and minimum and maximum V_s velocities along the *x*-, *y*-, and *z*-axes of Figures 6 and 9d.

principle on the frequency of traversing waves. POHL *et al.* (this volume) found, however, that this effect is apparently negligible in the case of typical lower crustal rock sequences. (3) There is evidence from comparative laboratory and *in situ* studies from the continental deep drilling site that velocities and anisotropy of crystalline rocks register a high agreement at sonic and seismic frequencies if simulated *in situ* conditions of confining pressure are applied in the laboratory (POPP and KERN, 1994; JAHNS *et al.*, 1996).

In the past, many authors have endeavored use the V_p/V_s ratio, derived from seismic field measurements, to place constraints on the possible lithological composition of the lower crust (see, for example GAJEWSKI *et al.*, 1987; FOUNTAIN and CHRISTENSEN, 1989). The data compilation of HOLBROOK *et al.* (1992), however, illustrates that this approach suffers from a large scatter of V_p/V_s values attributed to the same rock types, and from overlap between different rock types. Possibly the scatter in the laboratory data bases could be reduced to a certain degree if anisotropy would be considered. To demonstrate this, we present the scatter in Poisson's ratio for different anisotropic rock types, which would result if Poisson's ratio were computed from V_p/V_s ratios without considering the rock orientation relative to the ray path and the polarization of the *S*-waves (Fig. 11).

A concept of how to apply the gathered laboratory knowledge regarding seismic anisotropy to the interpretation of seismic field data could be the following approach:
1. Perform seismic field measurements of deep crustal anisotropy at locations where the geological structure is known from near-vertical reflection seismology (BOHLEN *et al.*, this volume, discuss this procedure in detail).

2. Compile a spectrum of different lithological situations which are compatible with the reflection image, xenoliths, if available, and other geological information, and which stand for different scenarios of crustal evolution.
3. Use a data base of elastic tensors of the relevant rock types and of rock fabrics data to define anisotropic velocity-depth-functions corresponding to (2).
4. Compute synthetic seismograms and average anisotropic velocity functions in order to verify or disprove the hypothetical situations.

An example to which this concept has been successfully applied is the reflective lower crust at Urach Southern Germany (see RABBEL et al., 1998; WEISS, 1998). Both the Urach example and the more general data compilation of this study show that seismic anisotropy has the potential to distinguish realistic end-members of rock composition in the lower crust.

Acknowledgements

S.S. thanks the "Deutsche Forschungsgemeinschaft" for a Heisenberg Fellowship (Si 438/10-1) and D. Fountain as well as St. Schmid for their hospitality during stays in Laramie and Basel. The project was financially supported by the DFG grant Ra 496/5-1. The critical comments of E. Lüschen and one unknown reviewer helped to improve the paper significantly.

REFERENCES

ALEKSANDROV, K. S., and RYZHOVA, T. V. (1961), *The Elastic Properties of Rock-forming Minerals II: Layered Silicates*, Izv. Acad. Sci. USSR, Geophys. Phys. Solid Earth, 1165–1168.

ALEKSANDROV, K. S., ALCHIKOV, U. V., BELIKOV, B. P., ZASLAVSKII, B. I., and KRUPNYI, A. I. (1974), *Velocities of Elastic Waves in Minerals at Atmospheric Pressure and Increasing Precision of Elastic Constants by Means of EVM* (in Russian), Izv. Acad. Sci. USSR. Geol. Ser. *10*, 15–24.

ARTS, R. J., RASOLOFOSAON, P. N. J., and ZINSZNER, B. (1996), *Experimental and theoretical tools for characterizing anisotropy due to mechanical defects in rocks under varying pore and confining pressures.* In *Seismic Anisotropy* (eds. S. E. Fjaer, R. M. Holt and J. S. Rathore) (Society of Exploration Geophysicists, Tulsa, OK 1996) pp. 384–432.

BABUŠKA, V. (1972), *Elasticity and Anisotropy of Dunite and Bronzitite*, J. Geophys. Res. 77, 35, 6955–6965.

BACKUS, G. E. (1962), *Long-wave Elastic Anisotropy Produced by Horizontal Layering*, J. Geophys. Res. 67, 4427–4440.

BARRUOL, G. (1993), *Pétrophysique de la croûte inférieure. Rôle de l'anisotropie sismique sur la réflectivité et le déphasage des ondes S*, Ph.D. Thesis, Univ. des Sciences et Techniques du Languedoc, Montpellier, 271 pp.

BARRUOL, G. MAINPRICE, D., KERN, H., DE St. BLANQUAT, M., and COMPTE, P. (1992), *3-D Seismic Study of a Ductile Shear Zone from Laboratory and Petrofabric Data (Saint Barthelemy Massif, Northern Pyrenees, France)*, Terra Nova 4 (1), 63–76.

BARRUOL, G., and MAINPRICE, D. (1993), *3-D Seismic Velocities Calculated from Lattice-preferred Orientation and Reflectivity of a Lower Crustal Section: Examples of the Val Sesia Section (Ivrea Zone, Northern Italy)*, Geophys. J. Int. *115* (3), 1169–1188.

BARRUOL, G., and KERN, H. (1996), *Seismic anisotropy and shear-wave splitting in lower-crustal and upper-mantle rocks from the Ivrea Zone; experimental and calculated data.* In *Dynamics of the Subcontinental Mantle, from Seismic Anisotropy to Mountain Building* (eds. Mainprice, D. and Vauchez, A.), Phys. Earth Planet. Int. *95*, 3–4, 175–194.

BELIKOV, B. P., ALEKSANDROV, K. S., and RYZHOVA, T. V. (1970), *Elastic Properties of Rock-forming Minerals and Rocks; with Appended Tables of the Elastic Constants of the Principal Types of Rocks* (in Russian), Izd. Nauka (Akad. Nauk SSSR, Inst. Geol. Rud. Mestorozhd. Petrogr. Mineral. Geokhim.-Sib. Otd., Inst. Fiz.), 276.

BERTOLANI, M. (1968), *La petrografia della Valle Strona (Alpi occidentali italiane)*, Schweiz. Min. Petrogr. Mitt. *48*, 695–732.

BIRCH, F. (1960), *The Velocity of Compressional Waves in Rocks to 10 kilobars; Part 1*, J. Geophys. Res. *65*, 1083–1102.

BIRCH, F. (1961), *The Velocity of Compressional Waves in Rocks to 10 kilobars; Part 2*, J. Geophys. Res. *66*, 2199–2224.

BOHLEN, T., RABBEL, W., WEISS, T., SIEGESMUND, S., and POHL, M. (1999), *Recovering Shear Wave Anisotropy of the Lower Crust: The Influence of Systematic Errors of Traveltime Inversion*, Pure appl. geophys. *156*, 123–138.

BUNGE, H. J., SIEGESMUND, S., SKROTZKI, W., and WEBER, K., *Textures of Geological Materials* (DGM Informationsgesellschaft Verlag 1994) 399 pp.

BURLINI, L. (1994), *A Model for the Calculation of the Seismic Properties of Geologic Units*, Surv. in Geoph. *15*, 593–617.

BURLINI, L., and FOUNTAIN, D. M. (1993), *Seismic Anisotropy of Metapelites from the Ivrea-Verbano Zone and Serie dei Laghi (Northern Italy)*, Phys. Earth Planet. Int. *78* (3–4), 301–317.

BÜTTGENBACH, B. (1990), *Über die Schärfe von Fehlerabschätzungen bei der numerischen Lösung von Randwertproblemen durch Differenzenverfahren*, Ph.D. Thesis, Tech. Univ. Aachen.

CARBONELL, R., and SMITHSON, S. B. (1991), *Large-scale Anisotropy within the Crust in the Basin and Range Province*, Geology *19*, 698–701.

CHRISTENSEN, N. I., and CROSSON, R. S. (1968), *Seismic Anisotropy in the Upper Mantle*, Tectonophysics *6* (2), 93–107.

DORNBUSCH, J. (1995), *Gefüge-, Mikrostruktur- und Texturuntersuchungen an Hochtemperatur-Scherzonen in granulitfaziellen Metabasiten der Ivrea-Zone*, Geotekt. Forsch. *83*, 94 pp.

FOUNTAIN, D. M., and CHRISTENSEN, N. I. (1989), *Composition of the continental crust and upper mantle; a review.* In *Geophysical Framework of the Continental United States* (eds. Pakiser, L. C. and Mooney, W. D.) (Memoir, Geological Society of America, 172) pp. 711–742.

FRISILLO, A. L., and BARSCH, G. R. (1972), *Measurement of Single-crystal Elastic Constants of Bronzite as a Function of Pressure and Temperature*, J. Geophys. Res. *77*, 6360–6368.

GAJEWSKI, D., HOLBROOK, W. S., and PRODEHL, C. (1987), *A Three-dimensional Crustal Model of Southwest Germany Derived from Seismic Refraction Data*, Tectonophysics *142* (1), 49–70.

GARUTI, G., RIVALENTI, G., ROSSI, A., and SINIGOI, S. (1979), *Mineral Equilibria as Geotectonic Indicators in the Ultramafics and Related Rocks of the Ivrea-Verbano Basic Complex (Italian Western Alps): Pyroxenes and Olivine*, Proc. 2nd Symp. Ivrea-Verbano Mem. Soc. Geol. Ital. *33*, 147–160.

HOLBROOK, W. S., MOONEY, W. D., and CHRISTENSEN, N. I., *The seismic velocity structure of the deep continental crust.* In *The Continental Lower Crust* (eds. Fountain, D. M., Arculus, R. and Kay, R. W.), Developments in Geotectonics *23* (Elsevier, Amsterdam 1992) pp. 1–34.

JAHNS, E., RABBEL, W., and SIEGESMUND, S. (1996), *Quantified Seismic Anisotropy at Different Scales: A Case Study from the KTB Crustal Segment*, Zeitschrift für Geologische Wissenschaften *24*, 729–740.

JECH, J. (1991), *Computation of Elastic Parameters of Anisotropic Medium from Traveltimes of Quasi-compressional Waves*, Phys. Earth Planet. Int. *66*, 153–159.

JONES, T., and NUR, A. (1984), *The Nature of Seismic Reflections from Deep Crustal Fault Zones*, J. Geophys. Res. *89*, 3153–3171.

KERN, H., and TUBIA, J. M. (1993), *Pressure and Temperature Dependence of P- and S-wave Velocities, Seismic Anisotropy and Density of Sheared Rocks from the Sierra Alpujata Massif (Ronda Peridotites, Southern Spain)*, Earth Planet. Sci. Lett. *119* (1–2), 191–205.

KLIMA, K. (1973), *The Computation of the Elastic Constants of an Anisotropic Medium from the Velocities of Body Waves*, Stud. Geoph. Geodet. *17*, 115–132.

KLIMA, K., and KLUHANEK, O. (1968), *Quantitative Correlation between Preferred Orientation of Grains and Elastic Anisotropy of Marbel*, IEEE Geosci. Electronics, *GE-6*, 139 pp.

KUMAZAWA, M., and ANDERSON, O.L. (1969), *Elastic Moduli, Pressure Derivatives and Temperature Derivatives of Single-crystal Olivine and Single-crystal Forsterite*, J. Geophys. Res. *74*, 5311–5320.

LEVIEN, L., WEIDNER, D. J., and PREWITT, C. T. (1979), *Elasticity of Diopside*, Phys. and Chem. of Min. *4* (2), 105–113.

LÜSCHEN, E., NOLTE, B., and FUCHS, K. (1990), *Shear-wave Evidence for an Anisotropic Lower Crust beneath the Black Forest, Southwest Germany*, Tectonophysics *173*, 483–493.

LÜSCHEN, E., NICOLICH, R., CERNOBORI, L., FUCHS, K., KERN, H., KRUHL, J., PERSOGLIA, S., ROMANELLI, M., SCHENK, V., SIEGESMUND, S., and TORTORICI, L. (1992), *A Seismic Reflection-refraction Experiment across the Exposed Lower crust in Calabria (Southern Italy): First Results*, Terra Nova *4*, 77–86.

MAINPRICE, D., and HUMBERT, M. (1993), *Methods of calculating petrophysical properties from lattice-preferred orientation data*. In *Seismic Properties of Crustal and Mantle Rocks; Laboratory Measurements and Theotetical Calculations* (ed. Burlini, L.), Surveys in Geophysics *15* (5), 575–592.

MANGHNANI, M. H., RAMANANANTOANDRO, R., and CLARK, S. P. (1974), *Compressional and Shear-wave Velocities in Granulite Facies Rocks and Eclogites to 10 kb*, J. Geophys. Res. *79* (35), 5427–5446.

MEISSNER, R. (1967), *Zum Aufbau der Erdkruste—Ergebnisse der Weitwinkelmessungen im bayrischen Molassebecken*, Gerl. Beitr. Geophys. *76*, 241–254.

POHL, M., WENZEL, F., WEISS, T. SIEGESMUND, S., BOHLEN, T., and RABBEL, W. (1999), *Realistic Models of Anisotropic Laminated Lower Crust*, Pure appl. geophys. *156*, 139–155.

POPP, T., and KERN, H. (1994), *The Influence of Dry and Water-saturated Cracks on Seismic Velocities of Crustal Rocks—A Comparison of Experimental Data with Theoretical Model*, Surveys in Geophysics *15*, 443–465.

QUICK, J. E., SINIGOI, S., and MAYER, A. (1994), *Emplacement Dynamics of a Large Mafic Intrusion in the Lower Crust, Ivrea-Verbano Zone, Northern Italy*, J. Geophys. Res. 11, 21,559–21,573.

RABBEL, W., and LÜSCHEN, E. (1996), *Shear-wave Anisotropy of Laminated Lower Crust at the Urach Geothermal Anomaly*, Tectonophysics *264*, 219–233.

RABBEL, W., SIEGESMUND, S., WEISS, T., POHL, M. and BOHLEN, T. (1998), *Shear-wave Anisotropy of Laminated Lower Crust beneath Urach (SW Germany)—A Comparison with Exposed Lower Crustal Sections*, Tectonophysics *298*, 337–356.

RESTON, T. J. (1987), *Spatial interference, reflection character and the structure of the lower crust under extension; results 2-D seismic modelling*. In *The Lower Continental Crust* (Annales Geophysicae, Series B: Terrestrial and Planetary Physics *5*(4)) pp. 339–347.

RIVALENTI, G., GARUTI, G., and ROSSI, A. (1975), *The Origin of the Ivrea-Verbano Basic Formation (Western Italian Alps)—Whole Rock Geochemistry*, Boll. Soc. Geol. Ital. *94*, 1149–1186.

RIVALENTI, G., GARUTI, G., ROSSI, A., SIENA, F., and SINIGOI, S. (1981), *Existence of Different Peridotite Types and of a Layered Igneous Complex in the Ivrea Zone of the Western Alps*, J. Petrology *22* (1), 127–153.

RIVALENTI, G., ROSSI, A., SIENA, F., and SINIGOI, S. (1984), *The Layered Series of the Ivrea Verbano Igneous Complex, Western Alps, Italy*, Tscherm. Mineral. Petrogr. Mitt. *33*, 77–99.

RUDNICK, R. L., and FOUNTAIN, D. M. (1995), *Nature and Composition of the Continental Crust: A Lower Crustal Perspective*, Rev. Geophys. *33* (3), 267–309.

RUTTER, E. H., BRODIE, K. H., and EVANS, P. J. (1993), *Structural Geometry, Lower Crustal Magmatic Underplating and Lithospheric Stretching in the Ivrea-Verbano Zone, Northern Italy*, J. Struct. Geol. *15* (3–5), 647–662.

SCHENK, V. (1980), *U-Pb and Rb-Sr Radiometric Dates and their Correlation with Metamorphic Events in the Granulite-facies Basement of the Serre, Southern Calabria (Italy)*, Contrib. Mineral. Petrol. *73*, 23–38.

SCHENK, V. (1984), *Petrology of Felsic Granulites, Metapelites, Metabasics, Ultramafics and Metacarbonates from Southern Calabria (Italy): Prograde Metamorphism, Uplift and Cooling of a Former Lower Crust*, J. Petrol. *25*, 255–298.

SCHÖN, J. H., *Physical Properties of Rocks: Fundamentals and Principles of Petrophysics* (Pergamon Press 1996).

SCHÖNBERG, M. E., and MUIR, F. (1989), *A Calculus for Finely Layered Anisotropic Media*, Geophys. *54* (5), 581–589.

SERONT, B., MAINPRICE, D. M., and CHRISTENSEN, N. I. (1993), *A Determination of the Three-dimensional Seismic Properties of Anorthosite; Comparison between Values Calculated from the Petrofabric and Direct Laboratory Measurements*, J. Geophys. Res. *98* (B), 2209–2221.

SHAPIRO, S. A., and HUBRAL, P. (1996), *Elastic Waves in Finely Layered Sediments: The Equivalent Medium and Generalized O'Doherty-Anstey Formulas*, Geophys. *61* (5), 1282–1300.

SIEGESMUND, S. (1996), *The Significance of Rock Fabrics for the Geological Interpretation of Geophysical Anisotropies*, Geotekt. Forsch. *85*, 1–123.

SIEGESMUND, S., TAKESHITA, T., and KERN, H. (1989), *Anisotropy of V_p and V_s in an Amphibolite of the Deeper Crust and its Relationship to the Mineralogical Microstructural and Textural Characteristics of the Rock*, Tectonophysics *157*, 25–38.

SIEGESMUND, S., FRITSCHE, M., and BRAUN, G. (1991), *Reflectivity caused by texture-induced anisotropy in mylonites.* In *Continental Lithosphere; Deep Seismic Reflections* (eds. Meissner, R. O., Brown, L. D., Duerbaum, H. J., Franke, W., Fuchs, K. and Seifert, F.), Geodynamics Series *22*, 291–298.

SIEGESMUND, S., and DAHMS, M., Fabric-controlled anisotropy of elastic, magnetic and thermal properties. In *Textures of Geological Materials* (eds. Bunge, H. J., Siegesmund, S., Skrotzki, W. and Weber, K.) (DGM Informationsgesellschaft Verlag 1994) pp. 353–379.

SIEGESMUND, S., HELMIG, K., and KRUSE, R. (1994), *Complete Texture Analysis of a Deformed Amphibolite: Comparison between Neutron Diffraction and U-stage Data*, J. Struct. Geol. *16*, 131–142.

SIEGESMUND, S., KRUHL, J. H., and LÜSCHEN, E. (1996), *Petrophysical and Seismic Features of the Exposed Lower Continental Crust in Calabria (Italy): Field Observation versus Modelling*, Geotekt. Forsch. ·*85*, 125–163.

TOBHILL, SIEGESMUND, S., and BASS, J. D. (1999), *Elasticity of Cordierite*, Phys. Chem. Min. *26*, 333–343.

VAUGHAN, M. T., and WEIDNER, D. J. (1978), *The Relationship of Elasticity and Crystal Structure in Andalusite and Sillimanite*, Phys. Chem. Min. *3*, 133–144.

VOIGT, W., *Lehrbuch der Kristallphysik* (Teubner, Leipzig 1928).

VOSHAGE, H., HOFMANN, A. W., MAZZUCCHELLI, M., RIVALENTI, G., SINIGOI, S., RACZEK, I., and DEMARCHI, G. (1990), *Isotopic Evidence from the Ivrea Zone for a Hybrid Lower Crust Formed by Magmatic Underplating*, Nature *347*, 731–736.

WEISS, T. (1998), *Gefügeanisotropie and ihre Auswirkung auf das seismische Erscheinungsbild: Fallbeispiele aus der Lithosphäre Süddeutschlands*, Geot. Forschungen *91*, 1–156.

(Received January 20, 1998, revised/accepted November 11, 1998)

To access this journal online:
http://www.birkhauser.ch

Pure appl. geophys. 156 (1999) 123–138
0033–4553/99/010123–16 $ 1.50 + 0.20/0

⌐ Pure and Applied Geophysics

Recovering Shear-wave Anisotropy of the Lower Crust: The Influence of Systematic Errors on Travel-time Inversion

T. Bohlen,[1] W. Rabbel,[1] T. Weiss,[2] S. Siegesmund[2] and M. Pohl[3]

Abstract—Studies of seismic anisotropy *in situ* can help to discriminate between different rock types for the lower crust. In this context we investigate the sensitivity of an iterative linearized 3-D travel-time inversion scheme for transversely isotropic media with respect to two types of systematic errors: wrong velocities and interface topography of the hanging wall of the lower crust. The computations simulate realistic field conditions such as found for the Variscan crust at the Urach geothermal anomaly. The study focusses on the possible information content of split $S_M S$ arrivals observed along two orthogonal expanding spread profiles. It ensues that an imperfect knowledge of the layer geometry is of minor importance compared to errors in the velocities of the hanging wall. In particular, upper crust anisotropy has to be considered carefully. Generally, the anisotropy of transversely polarized shear waves (SH waves) was recovered with higher accuracy than the anisotropy of vertically polarized shear waves (SV waves).

Key words: Travel-time inversion, seismic anisotropy, lower crust, shear waves, Urach.

1. Introduction

Since the 1960s it is known from laboratory investigations that some rock types show a directional dependence of compressional and shear-wave velocities. This seismic anisotropy is caused by structural order within a rock package. It is expressed as thin alternating layering, aligned fractures or, most importantly, the lattice preferred orientation (fabric) of highly anisotropic minerals such as biotite and hornblende. Mineral alignment can be expected in the ductile parts of the crust, and studies of seismic anisotropy can help to discriminate between different rock types *in situ* (e.g., Rabbel and Mooney, 1996).

---

[1] Institute of Geosciences, Otto-Hahn-Platz 1, D-24118 Kiel, Germany.
[2] Institute for Geology and Dynamics of the Lithosphere, Goldschmidtstr. 3, D-37077 Göttingen, Germany.
[3] Geophysical Institute, Hertzstr. 16, D-76187 Karlsruhe, Germany.

Probing the continental lower crust for seismic anisotropy is connected with many practical problems, one of which is the subject of the present paper. We presume that we succeeded in recording reflected shear waves from the lower crust or from the Moho ($S_M S$), and that we detected shear-wave splitting (S arrivals with time delay between each other), the most reliable indicator of seismic anisotropy. How accurately could we recover the seismic anisotropy of the lower crust? How would the results of an inversion procedure be influenced by systematic errors? The aim of this paper is to investigate the relative importance of two such errors by numerical kinematic simulations, namely the influence of insufficient knowledge of the layer topography and the (anisotropic) shear-wave velocities of the hanging wall of the lower crust.

In order to study realistic conditions we simulated the same field configuration as was realized in a seismic field experiment in the area of the geothermal anomaly at Urach (SW Germany). The dashed lines in Figure 1 show two orthogonal shear-wave profiles along which expanding spread profiling was performed. It was arranged as a quasi-expanding spread in two azimuths, N60°E and N150°E. We term it quasi-expanding spread because the reflection points scatter within an area

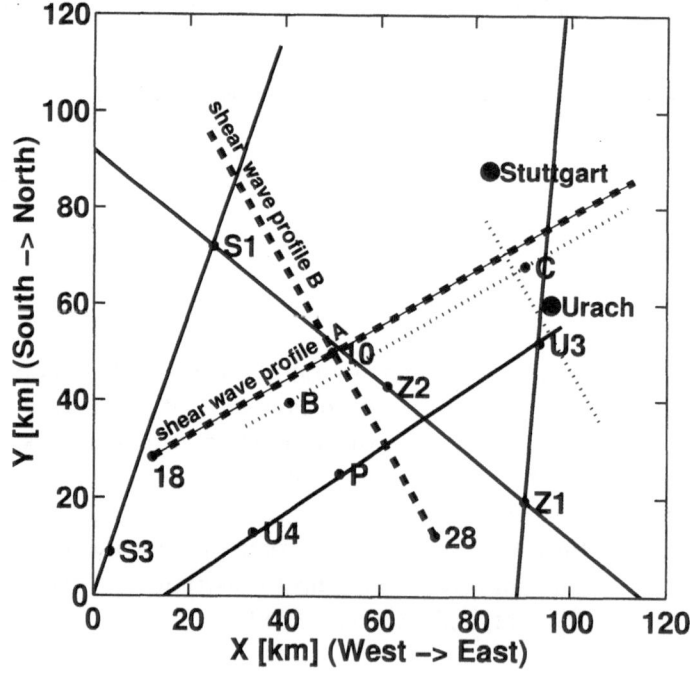

Figure 1

Location of seismic profiles in the Urach area that were used to develop the 3-D crustal model shown in Figure 2. Solid lines: refraction lines (GAJEWSKI and PRODEHL, 1987; GAJEWSKI *et al.*, 1987; HOLBROOK *et al.*, 1988) with considered shot points. Thin dashed lines: reflection lines after BARTELSEN *et al.* (1982). Thick dashed lines: orthogonal shear-wave profiles A and B (RABBEL and LÜSCHEN, 1996).

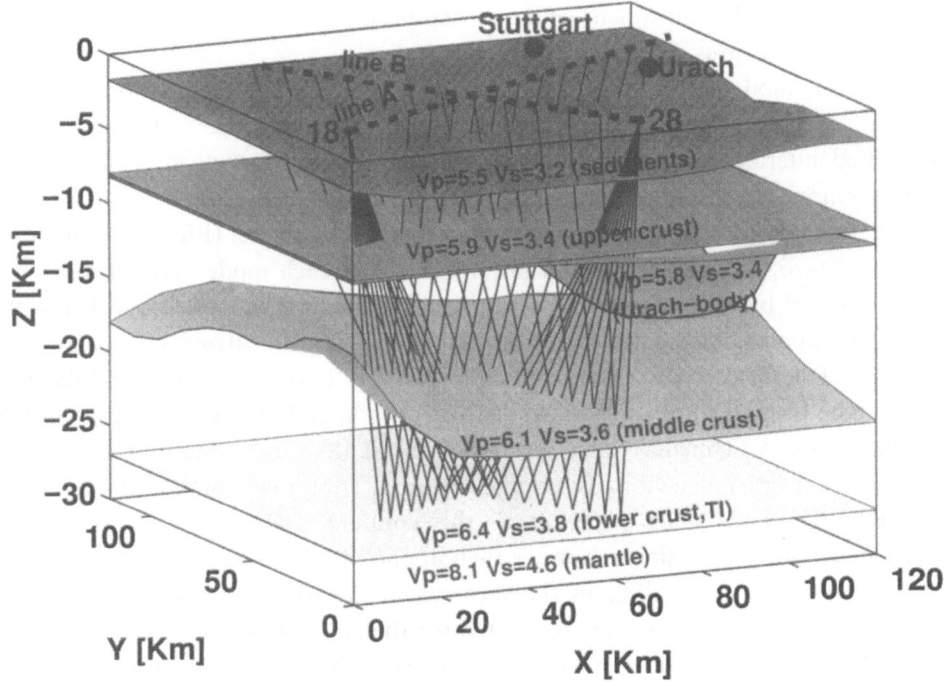

Figure 2
Crustal structure beneath the Urach area. v_s and v_p [km/s] are average isotropic phase velocities for shear and compressional waves, respectively.

of approximately 10 km radius. Details regarding the measurements and the interpretation procedure are explained in RABBEL and LÜSCHEN (1996) and RABBEL *et al.* (1998).

The most important result of the survey was the observation of up to 300 ms shear-wave splitting for $S_M S$ wide-angle reflections. The S-wave splitting could be attributed entirely to the lower crust because upper crustal S_g arrivals did not manifest any S-wave birefringence. At large, near-critical source-geophone distances (≥ 85 km) there was no problem to identify the $S_M S$ wave and to determine the time delay between the two split arrivals. These offsets correspond to approximately 70° incidence angle in the lower crust.

The Urach area had been studied beforehand by near-vertical reflection profiling in the late 1970s (MEISSNER *et al.*, 1982; BARTELSEN *et al.*, 1982). As a starting point we defined a 3-D isotropic velocity model based on the near-vertical reflection results (MEISSNER *et al.*, 1982; BARTELSEN *et al.*, 1982), and on published seismic wide-angle measurements (GAJEWSKI and PRODEHL, 1987; GAJEWSKI *et al.*, 1987; HOLBROOK *et al.*, 1988). The respective profiles are the solid lines in Figure 1. Figure 2 shows the resulting 3-D velocity model. Its compilation comprised the following steps: (1) smoothing of the originally published 2-D velocity sections, (2) averaging of the different sections into each other (removing contradictions at crossing points), (3) adjustment of layer depths in order to fit the major seismic phases of the previous refraction profiles.

The basic model of the crust beneath Urach (Fig. 2) consists of homogeneous isotropic layers separated by curved interfaces. In the following, "3-D" and "1-D" means that interfaces are curved and horizontally plane, respectively. The individual layers are always assumed to be homogeneous.

For the numerical simulations we modified the basic model (Fig. 2) by superimposing anisotropy to the upper and lower crust. For each model S_MS travel times were computed by 3-D ray tracing for anisotropic media (GAJEWSKI and PŠENČÍK, 1990). The applied kinematic travel-time inversion is a linearized approach (e.g., JECH and PŠENČÍK, 1992; OKOYE *et al.*, 1996). It is described in the following section. The formulae are valid for media with hexagonal anisotropy and a vertical symmetry axis (VTI media). This was observed at the Urach site and also at the exposed lower crustal sections Ivrea and Calabria (Italy) (RABBEL and LÜSCHEN, 1996; RABBEL *et al.*, 1998; WEISS *et al.*, this volume).

The travel-time inversion scheme is then applied to different model situations in the following section. In all cases the shear-wave anisotropy is recovered from the travel times of split S_MS arrivals computed for the crossing profiles (dashed lines in Figs. 1 and 2). The implications of this sensitivity study are discussed in the last section with respect to possible future field measurements.

2. Iterative Inversion Technique

For anisotropic materials with hexagonal symmetry and a vertical symmetry axis, the quasi shear waves are always polarized in a plane that is either vertical or horizontal. We refer to the shear wave polarized purely transverse (transversely to the shot-geophone direction) as *SH* and the shear wave polarized in the vertical plane as *qSV*. For most transversely isotropic rocks the velocity of *SH* waves increases from vertical to horizontal ray paths (WEISS *et al.*, this volume). At vertical incidence *SH* and *qSH* velocities (v_{SH} and v_{SV}, respectively) are identical, whereas $v_{SH} > v_{SV}$ for subhorizontal and horizontal ray paths. THOMSEN (1986) demonstrated that shear-wave anisotropy of such a transversely isotropic medium ('VTI-medium') can be described by only three, independent parameters: β_o, γ, and ξ. The first parameter β_o represents the shear-wave velocity in vertical direction. The second parameter γ is one of THOMSEN's parameters. It is dimensionless and specifies the anisotropy of *SH* waves (THOMSEN, 1986):

$$v_{SH}(\theta) = \beta_o[1 + \gamma \sin^2(\theta)], \tag{1}$$

where θ is the angle between the wave front normal and the vertical axis. In our simulations the crust is weakly anisotropic. In this case the wave front normal coincides with the direction of the ray.

The third parameter ξ characterizes the qSV-wave anisotropy (THOMSEN, 1986):

$$v_{SV}(\theta) = \beta_o \left[1 + \frac{\xi}{\beta_o^2} \sin^2(\theta) \cos^2(\theta) \right].$$ (2)

β_o specifies a sort of reference velocity value (minimum phase velocity at vertical incidence ($\theta = 0$) for both SH and qSV waves), whereas the parameters γ and ξ control the shape of the velocity curves:

$$\Delta v_{SH} = v_{SH}(\pi/2) - \beta_o = \beta_o \gamma,$$ (3)

$$\Delta v_{SV} = v_{SV}(\pi/4) - \beta_o = \frac{\xi}{4\beta_o}.$$ (4)

The velocity deviations Δv_{SH} and Δv_{SH} are explained in Figure 3. The shape of the velocity curves is sensitive, for example, to the texture of anisotropic minerals and density of aligned fractures in rocks.

Note that the parameter ξ can be expressed by

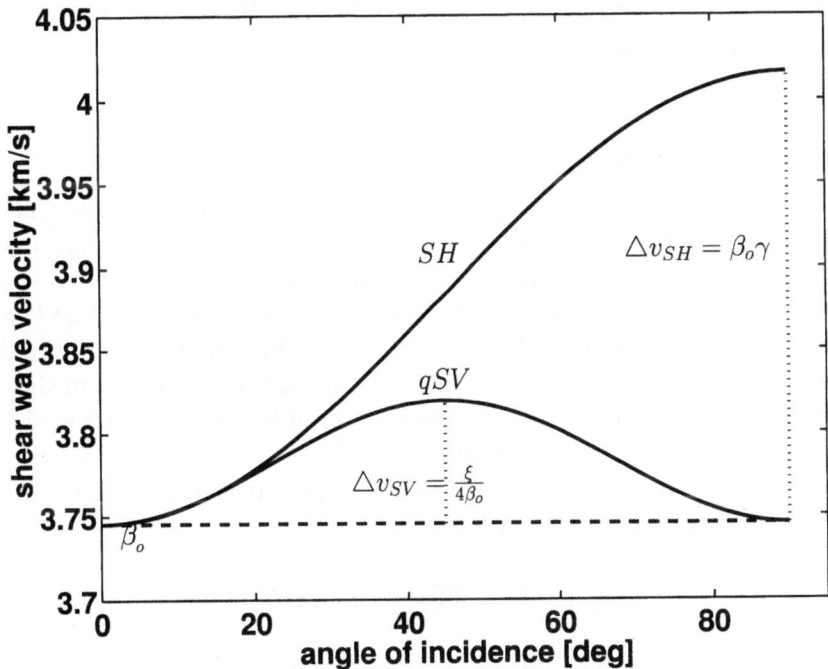

Figure 3

Directional dependence of phase velocities of SH and qSV waves travelling in a transversely isotropic lower crust. The parameters γ and ξ quantify the strength and form of SH- and qSV-wave anisotropy, respectively (see Eqs. (3) and (4)). The function shown here corresponds to the 'true' velocity function of the inversion computations.

$$\xi = \alpha_o^2(\epsilon - \delta), \tag{5}$$

where α_o is the vertical sound speed for compressional waves, and ϵ and δ are THOMSEN's (1986) dimensionless parameters, which disappear in the isotropic case.

To recover the three elastic parameters β_o, γ and ξ from travel-time observations of *SH*- and *qSV*-wave reflections at Moho ($S_M S$) we follow the linear perturbation approach of ČERVENÝ and FIRBAS (1984) for weak anisotropy:

$$\beta_o^2 = [\beta_o^2]^o + \Delta[\beta_o^2],$$

$$\gamma = \gamma^o + \Delta\gamma,$$

$$\xi = \xi^o + \Delta\xi, \tag{6}$$

where $[\beta_o^2]^o, \gamma^o, \xi^o$ are the unperturbed parameters of an initial model and $\Delta[\beta_o^2], \Delta\gamma, \Delta\xi$ are small perturbations or corrections to be determined from observed travel times. These parameters are specified in a Cartesian coordinate system x_i ($i = 1, 2, 3$), the axis of symmetry (x_3) being parallel to the vertical model axis (Z axis of Fig. 2). The axes x_1 and x_2 are situated in a horizontal plane, x_1 and x_2 point to the east and to the north, respectively (X and Y axes of Figs. 1 and 2, respectively). The first-order perturbations of the arrival times for *SH* waves (τ_i^{SH}) and *qSV* waves (τ_i^{SV}) corresponding to a weakly anisotropic homogeneous VTI-layer can be written as follows:

$$\tau_i^{SH} = -[2\beta_o^2 \, \Delta\gamma \, \sin^2(\theta_i) + \Delta[\beta_o^2]]\tau_0(\theta_i)/(2v_{SH}^2(\theta_i)), \tag{7}$$

$$\tau_i^{SV} = -[\Delta[\beta_o^2] + 2 \, \Delta\xi \, \sin^2(\theta_i) \cos^2(\theta_i)]\tau_0(\theta_i)/(2v_{SV}^2(\theta_i)), \tag{8}$$

where i is the index of observation. Similar equations were obtained by ČERVENÝ and SIMÕES-FILHO (1991) and JECH and PŠENČÍK (1992). θ_i and $\tau_0(\theta_i)$ denote angle of incidence and travel time, respectively, of the i-th ray in the background medium. $v_{SH}(\theta_i)$ and $v_{SV}(\theta_i)$ denote shear-wave velocities of *SH* and *qSV* waves, respectively, in the background. They are defined by Equations (1) and (2). θ_i and $\tau_0(\theta_i)$ must be computed by two-point 3-D ray tracing for anisotropic media. They are influenced considerably by the curved interfaces of the initial crustal model (Fig. 2) of the travel-time inversion procedure.

The goal of the inversion is the determination of the perturbations $\Delta[\beta_o^2], \Delta\gamma, \Delta\xi$ minimizing the function (JECH and PŠENČÍK, 1992):

$$I = \sum_{i=1}^{N} (\Delta t_i^M - \tau_i^M)^2 \qquad M \in \{SH, SV\}, \tag{9}$$

where $\Delta t_i^M = t_i^M - T_i^M$ are the differences between computed arrival times (t_i^M) and 'observed' arrival times (T_i^M) of *SH* ($M = SH$) and *qSV* waves ($M = SV$) on profiles A, B, respectively. N is the number of observations, i.e., rays from the shotpoints to the receivers located at the earth's surface. Within each step of iteration two-point 3-D ray tracing for anisotropic media was performed. For these

Procedure for an iterative travel-time inversion

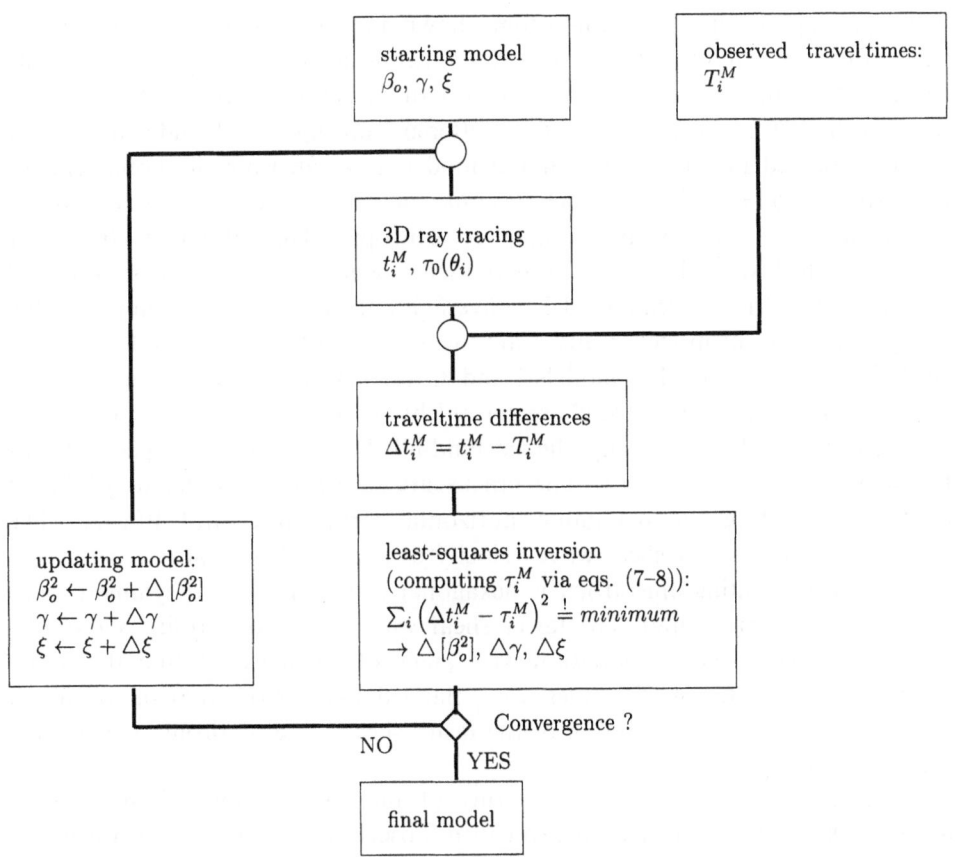

Figure 4
Flow chart for the estimation of lower crustal anisotropy from 'observed' arrival times T_i^M. M denotes the type of split shear wave (SH-type or SV-type) observed on profiles A and B. Within each step of iteration arrival times (t_i^M) 'recorded' on earth's surface and travel times within the lower crust (τ_o) are computed by 3-D ray tracing for anisotropic media. A simultaneous least-squares inversion (using Eqs. (7)–(9)) of observed arrival times leads to a new estimation of the perturbations that control the lower crustal anisotropy of shear waves.

computations we applied the FORTRAN-program 'ANRAY' by GAJEWSKI and PšENČíK (1990). The travel-time perturbations depend linearly on the perturbations of the chosen optimization parameters and nonlinearly on the incidence angles θ_i. Therefore we applied the Levenberg-Marquardt least-squares algorithm (LINES and TREITEL, 1984) iteratively in order to find the perturbations minimizing I in equation (9). As shown in Figure 4, the result of each iterative step is a new estimate of the perturbations $\Delta[\beta_o^2], \Delta\gamma, \Delta\xi$.

3. Investigation of Systematic Errors

By applying the inversion algorithm shown in Figure 4 to various starting models of the crust and to different simulated travel-time observations, it is possible to explore the range of solutions and to test the accuracy of this technique with respect to the lower crust. We used four different starting models indicated by the symbols explained in Table 1. All initial models have an isotropic homogeneous lower crust of $\beta_o = 3.8$ km/s, which becomes anisotropic during the iterations. Starting models 1 and 3 are completely isotropic. They differ in the layer topography which was adopted from the Urach model (Fig. 2) in the case of model 3. Model 1 was derived from model 3 by averaging the layer thickness horizontally. The layer velocities of models 1 and 3 are identical with the Urach model (Fig. 2). Models 2 and 4 correspond to models 1 and 3, respectively, but are modified in the upper crustal velocity. They are characterized by an anisotropic upper crust. This anisotropy is caused by vertically aligned fluid-filled cracks with an aspect ratio of 0.1 and a crack density of 0.1. The cracks are orientated parallel to profile B (N150°E), simulating the maximum horizontal stress in central Europe. The directional dependence of shear-wave velocity caused by these cracks is shown in Figure 5a. The resulting anisotropy is hexagonal with a horizontal symmetry axis (HTI, azimuthal anisotropy). The faster shear waves ($qS1$) on profile A (N60°E) show the strongest velocity variation with angle of incidence. Within the upper crustal layer the faster shear waves are polarized inline (SV-type) on profile B (N150°E), whereas the polarization of the faster shear wave on profile A is mostly transverse (SH-type).

By replacing the isotropic lower crust of models 1–4 by a homogeneous anisotropic layer (Fig. 5b) we derive so-called 'observed' or 'true' models for which the 'observed' travel times were calculated. These models are denoted by the numbers I–IV, and were assumed to be reality. In order to simulate realistic conditions we inserted elastic constants deduced from the exposed lower crustal section of Ivrea. Both reflectivity and anisotropy of this Ivrea section show high similarity to findings at the Urach area (RABBEL *et al.*, 1998). This type of lower crust is almost transversely isotropic (VTI) and can be described by the average parameters: $\beta_o = 3.743$ km/s, $\gamma = 0.0768$, $\xi = 1.032$ km²/s² (see Table 2). The diagrams of phase velocities versus angle of incidence for these parameters were computed via Equations (1) and (2). They are shown in Figure 3. The shape of the inverted velocity curves is characterized by the velocity deviations Δv_{SH} and Δv_{SV} as defined in Equations (3) and (4) (see also Fig. 3). The anisotropy of qSV waves is rather small ($\Delta v_{SV}^{\text{true}} = \xi/(4\beta_o) = 69$ m/s $\approx 1.8\%$), whereas the SH waves exhibit significant velocity variation with angle of incidence ($\Delta v_{SV}^{\text{true}} = \gamma\beta_o = 286$ m/s $\approx 7.7\%$). The fast shear wave (qS1) is polarized transversely (SH-type) on both profiles (Fig. 5b). Again, the program ANRAY was used to compute travel times of Moho reflections of SH and qSV waves for the models I to IV. On profile A (shotpoint 28) 59 and on profile B (shotpoint 18) 50 receivers were considered. The distance between neighbouring receivers was 2 km on both lines. The corresponding $S_M S$ arrival times served as 'observations' (T_i^M) in the solution of the inverse problem and were inverted simultaneously to recover the shear-wave anisotropy of the lower crustal sections of models I–IV.

The inverted values of β_o, γ and ξ after four iterations are listed in Table 2 for all plausible combinations of starting models (1–4) and 'observed' models (I–IV). They were converted to velocity variations as a function of the angle of incidence at Moho level, using Equations (1) and (2). These velocity curves are displayed in Figures 6 and 7 (dashed lines). The 'true' phase velocity function used for the computation of the 'observed' arrival times is indicated by solid lines (same velocity curves as shown in Fig. 3). We refer to them as 'true' velocity curves. To quantify the accuracy of the inversion we computed the differences of Δv_{SV} and Δv_{SH}

Table 1

Starting models. Description of the models 1–4 used as starting models within the inversion procedure. The lower crust of the starting models is isotropic. In order to simulate 'observed' $S_M S$ travel times for an anisotropic lower crust, we replaced the isotropic lower crust by the velocity function shown in Figure 5b. These models are indicated by the numbers I–IV and are also referred to as 'observed' or 'true' models

Starting models with an isotropic lower crust	Description
1	Averaged horizontally layered crust, isotropic upper crust
2	Averaged horizontally layered crust, anisotropic upper crust (cracks)
3	Layered crust with 3-D curved interfaces (Fig. 2), isotropic upper crust
4	Layered crust with 3-D curved interfaces (Fig. 2), anisotropic upper crust (cracks)

a) upper crust

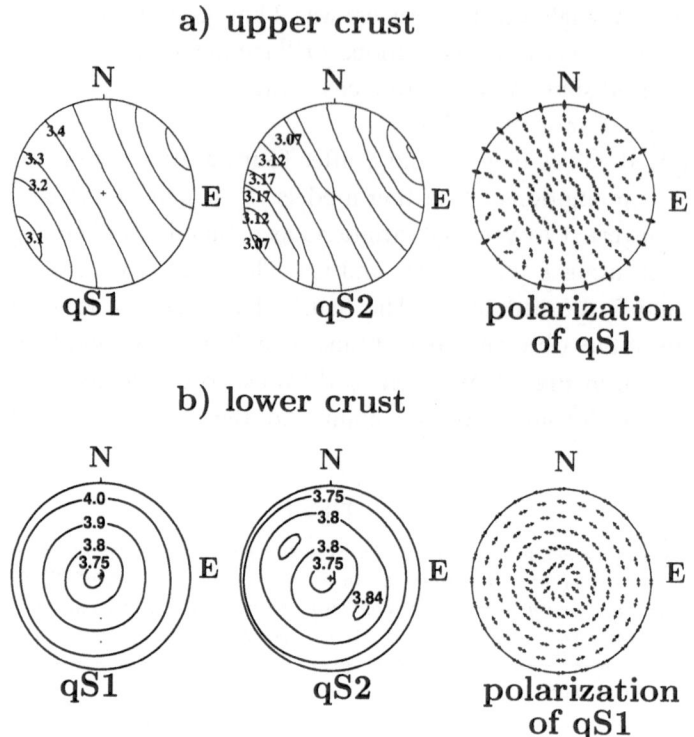

b) lower crust

Figure 5
Polar diagrams of directional dependence of shear-wave velocities within the upper (a) and lower (b) crust. (a) The anisotropy of the upper crust is caused by vertically aligned fluid-filled cracks (aspect ratio 0.1, crack density 0.1) orientated parallel to profile B (N150°E). Whereas on profile A (N60°E) the faster shear waves (qS1) are mostly polarized transverse (SH-type), they are polarized inline (SV-type) on profile B. (b) The 'observed' or 'true' lower crust is effectively transversely isotropic. On profile B the faster shear waves (qS1) are polarized inline, in contrast to the transverse polarization on profile B.

between inverted and true values (given above): $\Delta v_{SH} - \Delta v_{SH}^{\text{true}}$ and $\Delta v_{SV} - \Delta v_{SV}^{\text{true}}$, respectively. These values are displayed in Figure 8. They illustrate the corresponding deviations of the anisotropy parameters γ and ξ.

For the considered combinations of initial and 'true' models, four cases can be distinguished, which classify the sensitivity of the inversion procedure with respect to different boundary conditions. In the first case (case A in Figs. 6–8) the initial and 'true' models are identical in layer topography and upper crustal velocity. In this case the inverted velocity curves of Figures 6 and 7 are also nearly identical with the 'true' lines. The maximum deviations of the inverted shapes of velocity curves, i.e., deviations for Δv_{SV} and Δv_{SH}, are less than 15 m/s (case A of Fig. 8). The small differences arise from the small azimuthal deviations from ideal transverse isotropy within the 'true' lower crustal layer (Fig. 5b). In the second case (case B in Figs. 6–8) the initial and 'true' models are different in layer topography although the initially assumed anisotropy within the upper crust is still correct. Here, the shape of inverted velocity curves exhibits somewhat stronger deviations from the 'true' velocity curves (still less than 55 m/s, see case B of Fig. 8).

In case C (Figs. 6–8), initial and observed models differ only in upper crustal anisotropy. Now the shapes of the inverted velocity curves exhibit significant errors. The deviations for SH waves (less that 80 m/s) are smaller than the deviations for qSV waves (less than 200 m/s) (see case C of Fig. 8). Additionally the inverted velocity curves are shifted to lower velocity values (caused by a shift of β_o, see Table 2) when the 'true' upper crust is anisotropic, and to higher velocity values in the case of an initial upper crustal anisotropy. The reason for the velocity shifts is that the cracks within the upper crust generally decrease the phase velocities (see Fig. 5a) compared to intact rock ($\beta_o = 3.4$ km/s). Generally, the same applies to case D, where starting model and 'true' model are different in both the upper crustal anisotropy and the interface topography.

The iteration can be stopped when the difference of the mean travel-time residual between two iteration steps is less than 10 ms. Although the initial model of the lower crust is isotropic ($\beta_o = 3.8$ km/s), the inversion results already become stable after one iteration step (see Fig. 9). This means that the influence of the

Table 2

Initial and inverted anisotropy parameters. Values of the anisotropy parameters β_o, γ and ξ of the 'true' lower crust (Fig. 5b), of the initial (isotropic) lower crust, and of the inversion results after four iterations. The corresponding velocity curves for SH and qSV waves were computed using Equations (1) and (2). They are shown in Figures 6 and 7, respectively

	β_o [km/s]	γ	ξ [km^2/s^2]
'True' lower crust	3.743	0.0768	1.0319
Initial lower crust (isotropic)	3.800	0.0000	0.0000
Considered combinations	Results after four iterations		
I–1	3.745	0.0750	1.1275
II–2	3.745	0.0751	1.1080
III–3	3.746	0.0750	1.1260
IV–4	3.745	0.0752	1.1355
III–1	3.740	0.0918	1.5408
IV–2	3.738	0.0887	1.1972
II–1	3.527	0.1065	3.8345
IV–3	3.592	0.0902	2.2641
I–2	3.912	0.0603	−0.3355
III–4	3.883	0.0651	0.2936
IV–1	3.551	0.1111	3.1263
III–2	3.909	0.0758	0.0774

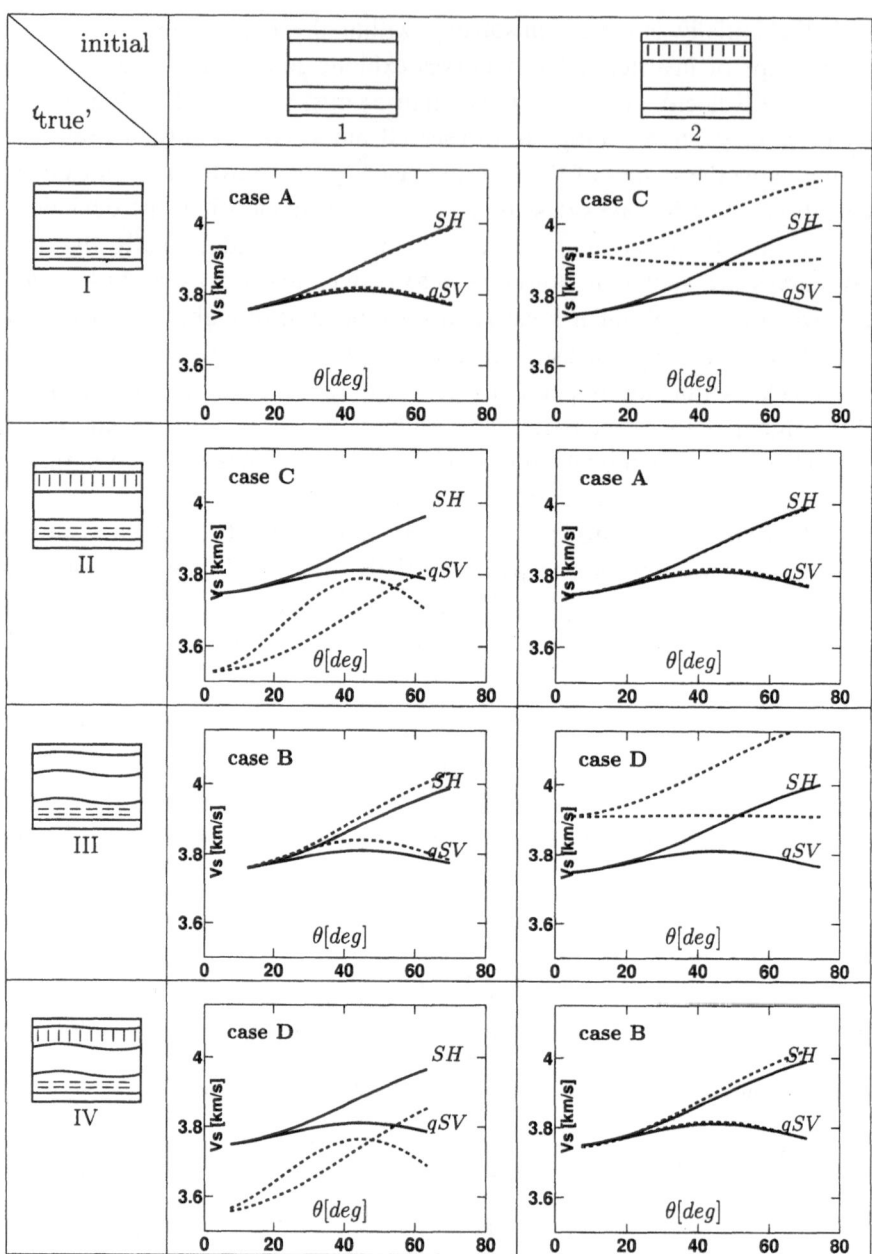

Figure 6
Inverted phase velocities of *SH* and *qSV* waves as functions of the angle of incidence after four iterations
(dashed lines). The 'observed' arrival times (T_i^M) were computed using models I–IV. They correspond
to the 'true' velocity curves (solid lines, same as in Fig. 3). Models 1–2 were used as starting models.
Case numbers A, B, C, D refer to Figure 8.

nonlinearity introduced by the angle of incidence θ (see Eqs. (7) and (8)) is very small. Clearly, this result cannot be generalized for more complex cases in which velocity and interface topography both would be considered in the inversion.

4. Discussion and Conclusions

Linearized travel-time inversion, based on reflected split shear waves, is a suitable approach to quantify shear-wave anisotropy at deep crustal levels if an appropriate starting model is known. Errors of the starting model are directly projected into free parameters of the inversion scheme. There are basically two types of systematic errors which can obscure the inversion results: wrong seismic velocities within the hanging wall, and insufficient knowledge of the layer topography. In our numerical experiments we tried to evaluate the relative importance of these errors under the assumption of realistic field conditions. The structure of the Variscan crust at Urach, chosen for our tests, shows a moderate 3-D inhomogeneity. The assumption of transverse isotropy for the lower crust was inspired by

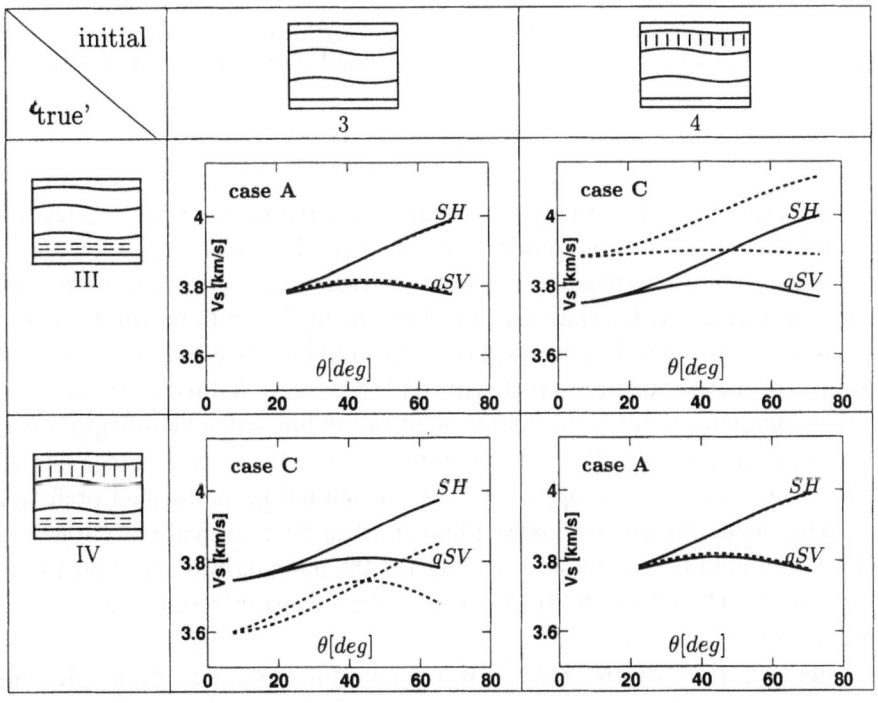

Figure 7
Same as Figure 6 but for other combinations of initial models (3–4) and observations (III–IV). Case numbers A, B, C, D refer to Figure 8.

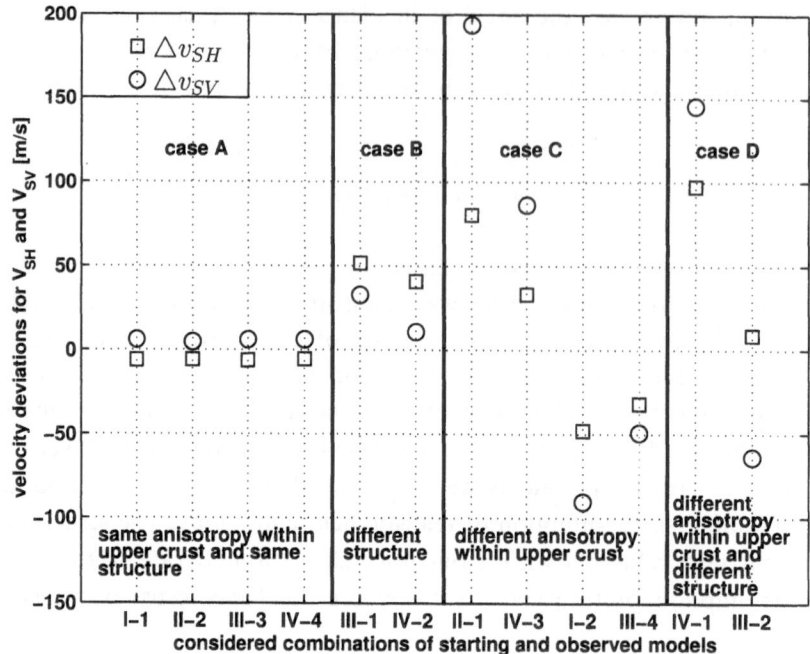

Figure 8

Accuracy of the inverted velocity curves. The differences between 'true' and inverted values for Δv_{SH} and Δv_{SV} (see Fig. 3) are shown for different groups of model combinations (cf. Figs. 6 and 7).

similar anisotropic velocity functions found at the exposed lower crust sections of Ivrea and Calabria. It eventuated that an imperfect knowledge of the layer geometry is of minor importance compared to errors in the velocities of the hanging wall (compare cases A, B versus C, D in Figs. 6–8). If the latter are set correctly, even simple starting models composed of horizontal interfaces allow recovery of the anisotropic properties of deep crust with high accuracy. A favorable precondition for this encouraging result is the choice of an expanding spread configuration along two orthogonal profiles. This observation pattern helps to average structural inhomogeneity. Also, anisotropy measurements should be performed preferably in areas where the depth structure is well known from near-vertical reflection surveys.

The assumption of inaccurate velocities for the hanging wall can lead to serious errors (cases C, D in Figs. 6–8). Regarding the inversion results we must distinguish two aspects:

1. a systematic shift of the recovered velocity function, i.e., of β_o, leading to systematically higher or lower velocities than in the 'true' case, and
2. an additional 'deformation' of the shape of the velocity function leading to erroneous estimates of shear-wave anisotropy parameters γ and ξ.

In our simulations both effects are caused by fluid-filled fractures. They are associated with a decrease of average velocity (effect 1) and with azimuthal anisotropy (effect 2) in the upper crust. Under real field conditions, effect 1 would not be a serious problem since average upper crustal velocities are usually well determined from refracted waves and shallow reflections. Effect 2 can be avoided only if the anisotropy of the upper crust can be quantified. Determining upper crustal anisotropy, however, requires special efforts in data acquisition. In particular the combination of sedimentary layers and subvertical fractures can result in complicated orthorhombic or monoclinic types of anisotropy. These cannot be recovered by two crossing, expanding spread profiles alone. If appropriate data for the upper crust are available, the method of linearized travel-time inversion can be applied downwards stepwise—depth interval after depth interval. Often the data basis is not sufficient to quantify upper crustal anisotropy. In this case its possible influence on the inversion results for the lower crust must be tested numerically to explore the possible range of solutions.

In our study we did not focus on the influence of errors in arrival picking. If they have a Gaussian probability distribution (like random noise) they will be smoothed out by the least-squares criterion of the inversion procedure.

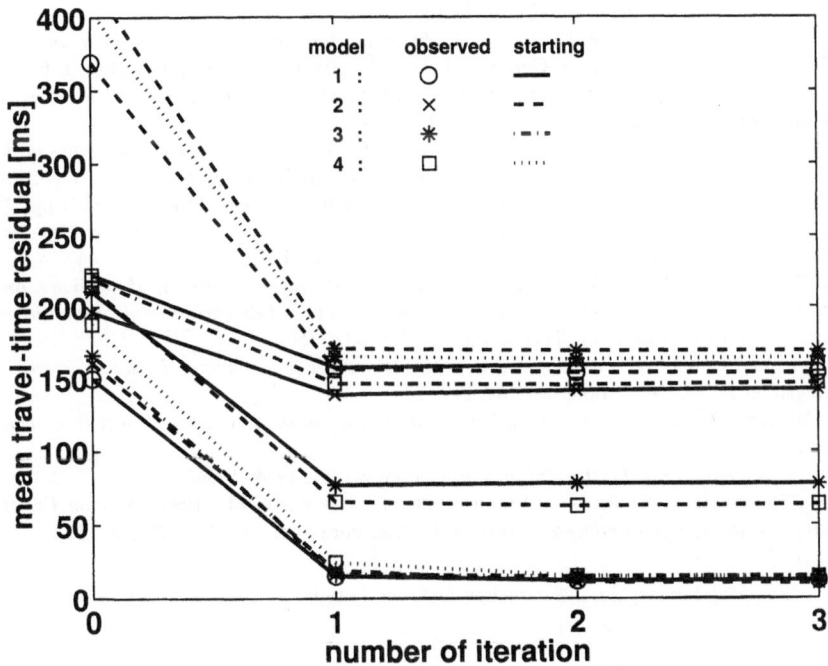

Figure 9
Mean travel-time residuals as a function of iteration number. The inversion shows convergence already after the first step.

5. Acknowledgements

The authors are grateful to I. Pšenčík and an anonymous reviewer for their helpful comments on the original manuscript. This study was financially supported by the Deutsche Forschungsgemeinschaft (Ra 496/5-1).

REFERENCES

BARTELSEN, H., LÜSCHEN, E., KREY, TH., MEISSNER, R., SCHMOLL, H., and WALTHER, C., *The combined seismic reflection-refraction investigation of the Urach Geothermal Anomaly*. In *The Urach Geothermal Project* (ed. Händel, R.) (Schweitzerbart, Stuttgart 1982) pp. 247–262.

ČERVENÝ, V., and FIRBAS, P. (1984), *Numerical Modelling and Inversion of Travel Times of Seismic Body Waves in Inhomogeneous Anisotropic Media*, Geophys. J. R. astr. Soc. *76*, 41–51.

ČERVENÝ, V., and SIMÕES-FILHO, I. A. (1991), *The Travel-time Perturbations for Seismic Body Waves in Factorized Anisotropic Inhomogeneous Media*, Geophys. J. Int. *107*, 219–229.

GAJEWSKI, D., and PRODEHL, C. (1987), *Seismic Refraction Investigation of the Black Forest*, Tectonophysics *142*, 27–48.

GAJEWSKI, D., HOLBROOK, W. S., and PRODEHL, C. (1987), *A Three-dimensional Crustal Model of SW Germany Derived from Seismic Refraction Data*, Tectonophysics *142*, 48–70.

GAJEWSKI, D., and PŠENČÍK, I. (1990), *Vertical Seismic Profile Synthetics by Dynamic Ray Tracing in Laterally Varying Anisotropic Structures*, J. Geophys. Res. *95*, 11,301–11,315.

HOLBROOK, W. S., GAJEWSKI, D., KRAMER, A., and PRODEHL, C. (1988), *An Interpretation of Wide Angle Compressional and Shear Wave Data in SW Germany: Poissons's Ratio and Petrological Implications*, J. Geophys. Res. *93*, 12,081–12,106.

JECH, J., and PŠENČÍK, I. (1992), *Kinematic Inversion for qP and qS Waves in Inhomogeneous Hexagonally Symmetric Structures*, Geophys. J. Int. *108*, 604–612. *Erratum*, Geophys. J. Int. *110*, 397.

LINES, L. R., and TREITEL, S. (1984), *A Review of Least-squares Inversion and its Application to Geophysical Problems*, Geophys. Prospect. *32*, 159–186.

MEISSNER, R., BARTELSEN, H., KREY, TH., and SCHMOLL, J., *Detecting velocity anomalies in the region of the Urach geothermal anomaly by means of seismic field arrangements*. In *Geothermics and Geothermal Energy* (eds. Čermak, V., and Hänel, R.) (Schweitzerbart, Stuttgart 1982) pp. 285–292.

OKOYE, P. N., ZHAO, P., and UREN, N. F. (1996), *Inversion Technique for Recovering the Elastic Constants of Transversely Isotropic Materials*, Geophysics *61*, 1247–1257.

RABBEL, W., SIEGESMUND, S., WEISS, T., POHL, M., and BOHLEN, T. (1998), *Shear Wave Anisotropy of Laminated Lower Crust Beneath Urach (SW Germany): A Comparison with Xenoliths and with Exposed Lower Crustal Sections*, Tectonophysics *298*, 337–356.

RABBEL, W., and LÜSCHEN, E. (1996), *Shear Wave Anisotropy of Laminated Lower Crust at the Urach Geothermal Anomaly*, Tectonophysics *264*, 219–233.

RABBEL, W., and MOONEY, W. D. (1996), *Seismic Anisotropy of the Crystalline Crust: What Does it Tell Us?* Terra Nova *8*, 16–21.

THOMSEN, L. (1986), *Weak Elastic Anisotropy*, Geophysics *51*, 1954–1966.

WEISS, T., SIEGESMUND, S., RABBEL, W., BOHLEN, T., and POHL, M. (1999), *Seismic Velocities and Anisotropy of the Lower Continental Crust: A Review*, Pure appl. geophys. *156*, 97–122.

(Received January 20, 1998, revised October 2, 1998, accepted October 29, 1998)

 To access this journal online:
http://www.birkhauser.ch

Pure appl. geophys. 156 (1999) 139–155
0033–4553/99/010139–17 $ 1.50 + 0.20/0

Pure and Applied Geophysics

Realistic Models of Anisotropic Laminated Lower Crust

MELANIE POHL,[1] FRIEDEMANN WENZEL,[1] THOMAS WEISS,[2]
SIEGFRIED SIEGESMUND,[2] THOMAS BOHLEN[3] and WOLFGANG RABBEL[3]

Abstract—The genesis of the laminated lower crust has been attributed to extensional processes leading to structural and textural ordering. This implies that the lower crust might be anisotropic. Laboratory measurements of lower crustal rock samples and xenolithes show evidence of anisotropy in these rocks due to oriented structure.

In this paper we investigate the seismic shear-wave response of realistic anisotropic lower crustal models using the anisotropic reflectivity method. Our models are based on representative petrophysical data obtained from exposed lower crustal sections in Calabria (South Italy), Val Strona and Val Sesia (Ivrea Zone, Northern Italy). The models consist of stacks of anisotropic layers characterized by quantified elastic tensors derived from representative rock samples which provide alternating high and low velocity layers.

The seismic signature of the data is comparable to seismic observations of *in situ* lower crust. For the models based on the Calabria and Val Strona sequences shear-wave splitting occurs for the Moho reflection at offsets beyond 70 km with travel-time delays up to 300 and 500 ms, respectively. The leading shear wave is predominantly horizontally polarized and followed by a predominantly vertically polarized shear wave. Contrastingly, the Val Sesia model shows no clear evidence of birefringence. Isotropic versus anisotropic modelling demonstrates that the shear-wave splitting is clearly related to the intrinsic anisotropy of the lower crustal rocks for the Val Strona sequence. No evidence of birefringence caused by thin layering is found.

Key words: Anisotropy, reflectivity.

Introduction

Seismic images of laminated lower crust found in the Variscan belts of western Europe (MATTHEWS, 1986; LÜSCHEN *et al.*, 1987; LE GALL, 1990) are characterized by a high reflectivity seen in near-vertical and wide-angle data. They are associated with reflectors between the Conrad discontinuity and the Moho. In contrast to the lower crust, upper crust and upper mantle appear as almost transparent (MOONEY and MEISSNER, 1992). The layering of the lower crust has

---

[1] Geophysical Institute, Karlsruhe University, Karlsruhe, Germany.
[2] Institute of Geology and Dynamics of the Lithosphere, Göttingen University, Göttingen, Germany.
[3] Geophysical Institute, Kiel University, Kiel, Germany.

been attributed to extensional processes (e.g., SMITHSON, 1986; MCCARTHY and THOMPSON, 1988; RABBEL and LÜSCHEN, 1996) that cause alignment of lower crustal minerals and bodies by ductile flow in the lower crust. Most rocks that form the lower crust are anisotropic. The alignment of rock-forming minerals is regarded as one major cause of seismic anisotropy (CRAMPIN, 1987; BABUŠKA and CARA, 1990). Based on laboratory measurements of typical lower crustal rocks (CHRISTENSEN, 1989; SIEGESMUND *et al.*, 1996) and modelling studies (HALE and THOMPSON, 1982; SANDMEIER and WENZEL, 1990) compositional layering is regarded as one major cause of the strong lower crustal reflectivity.

Exposed lower crustal sections can be regarded as windows to the lower crust and provide access to lower crustal parameters. The lower crustal profiles in Calabria (South Italy) and the Ivrea Zone (Northern Italy) are excellent examples of a petrologically well-defined layering. Previous studies demonstrated that these profiles provide realistic inputs to seismic modelling by means of reflectivity and petrology (HURICH and SMITHSON, 1987; SIEGESMUND *et al.*, 1996). Laboratory measurements of lower crustal rocks found in these regions provide evidence for seismic anisotropy caused by lattice-preferred orientation of minerals due to crustal deformation and metamorphism (KERN, 1982; SIEGESMUND *et al.*, 1989). Recently, SIEGESMUND *et al.* (1996) composed lower crustal models based on petrophysical data from representative lower crustal rocks of Calabria. Velocities and reflection coefficients of these models are comparable with those derived by other authors (e.g., HALE and THOMPSON, 1982; SANDMEIER and WENZEL, 1986; LÜSCHEN *et al.*, 1990).

Seismic characteristics of the laminated lower crust are well established due to many reflection and refraction surveys (HOLBROOK *et al.*, 1988; LÜSCHEN *et al.*, 1990). The records are dominated by strong reverberations in wide-angle *P*-wave data (whereas no such phases were observed for *S*-waves) and also strong *S*-wave reflections in the near vertical distance range. The Urach (Swabian Jura) wide-angle shear-wave experiment revealed evidence of shear-wave splitting in the *S*-reflections from the crust-mantle boundary (RABBEL and LÜSCHEN, 1996). The observation of shear-wave splitting—or birefringence—is a reliable indicator of anisotropy. Even for small anisotropy, incident shear waves are polarized into two orthogonal directions travelling at different velocities (CHRISTENSEN, 1971; CRAMPIN, 1987; BABUŠKA and CARA, 1990).

The objective of this paper is the investigation of the effect of intrinsic anisotropy (lattice-preferred orientation of minerals) in laminated lower crust on seismic wavefields. In particular, we focus on the observation of shear-wave splitting as a clear indicator of seismic anisotropy. Considering detailed information from exposed lower crustal sections in Italy, we composed realistic lower crustal models by means of petrophysics and petrology. Our models provide representative examples of anisotropic laminated crust. With the anisotropic reflectivity technique (BOOTH and CRAMPIN, 1983; NOLTE, 1988) we calculate shear-wave sections for these models from near vertical to wide-angle range. Firstly, we consider the general elastic properties in order to compute expectation values for the travel times and polarisation of the shear waves. Secondly, we compare the general elastic properties with the full wave response of laminated anisotropic crust. Finally, we investigate the possibility of thin layering effects for corresponding isotropic models.

Method

Our models were constructed using lithological sequences from exposed fossil lower crust in Calabria (South Italy), Val Strona and Val Sesia (Ivrea Zone, Northern Italy). Both segments are regarded as Variscan type of lower crust (SCHENK, 1981) and were uplifted and tilted at the end of Variscan time. Petrologically, the Calabria and Val Strona sections are mainly dominated by metapelitic rocks, whereas the Val Sesia profile is composed of mafic and ultramafic rocks. Composition and physical properties of the exposed rocks are known due to numerous laboratory measurements (KERN and SCHENK, 1985; SCHENK, 1990; BARRUOL and MAINPRICE, 1993; SIEGESMUND et al., 1996; BARRUOL and KERN, 1996). Those petrophysical studies revealed that many of the exposed rocks are anisotropic due to lattice-preferred orientation of crystals. Mainly felsic rocks display significant anisotropy. The intrinsic anisotropy of these rocks is closely related to birefringence and rock structure, e.g., foliation and lineation. Maximum shear-wave splitting corresponds to the foliation plane with the fast shear-wave polarized parallel to the foliation plane. On the other hand, birefringence is small at steep angles to the foliation plane. Mafic rocks such as gabbros are weakly anisotropic and provide complex relationships between rock structure and birefringence. However, olivine-rich rocks provide strong shear-wave splitting within the foliation plane. The rock structure is roughly constant throughout the exposed sections and the foliation is found to be nearly parallel to the lithological boundaries. Common to all profiles is the petrological fine structure that results in an alternating high and low velocity layering as indicated in Figure 1. High reflection coefficients can be expected. Synthetic sections with this feature were computed by HALE and THOMPSON (1982); HURICH and SMITHSON (1987); SIEGESMUND et al. (1996).

For each sequence—Calabria, Val Strona, Val Sesia—we specified a basic lower crustal model. The lithologies were rotated into their original position, e.g., before the uplift, to obtain a vertical layer stack. Layer thicknesses were defined by the petrological layering of the sequences. The final layer stacks were then scaled to obtain a uniform lower crustal thickness. Each layer is represented by an elastic tensor. The tensors result from the detailed analysis of rock fabric of the exposed rocks (SIEGESMUND et al., 1996; BARRUOL and KERN, 1996; BARRUOL and MAINPRICE, 1993) and are rotated into their original position according to the rock

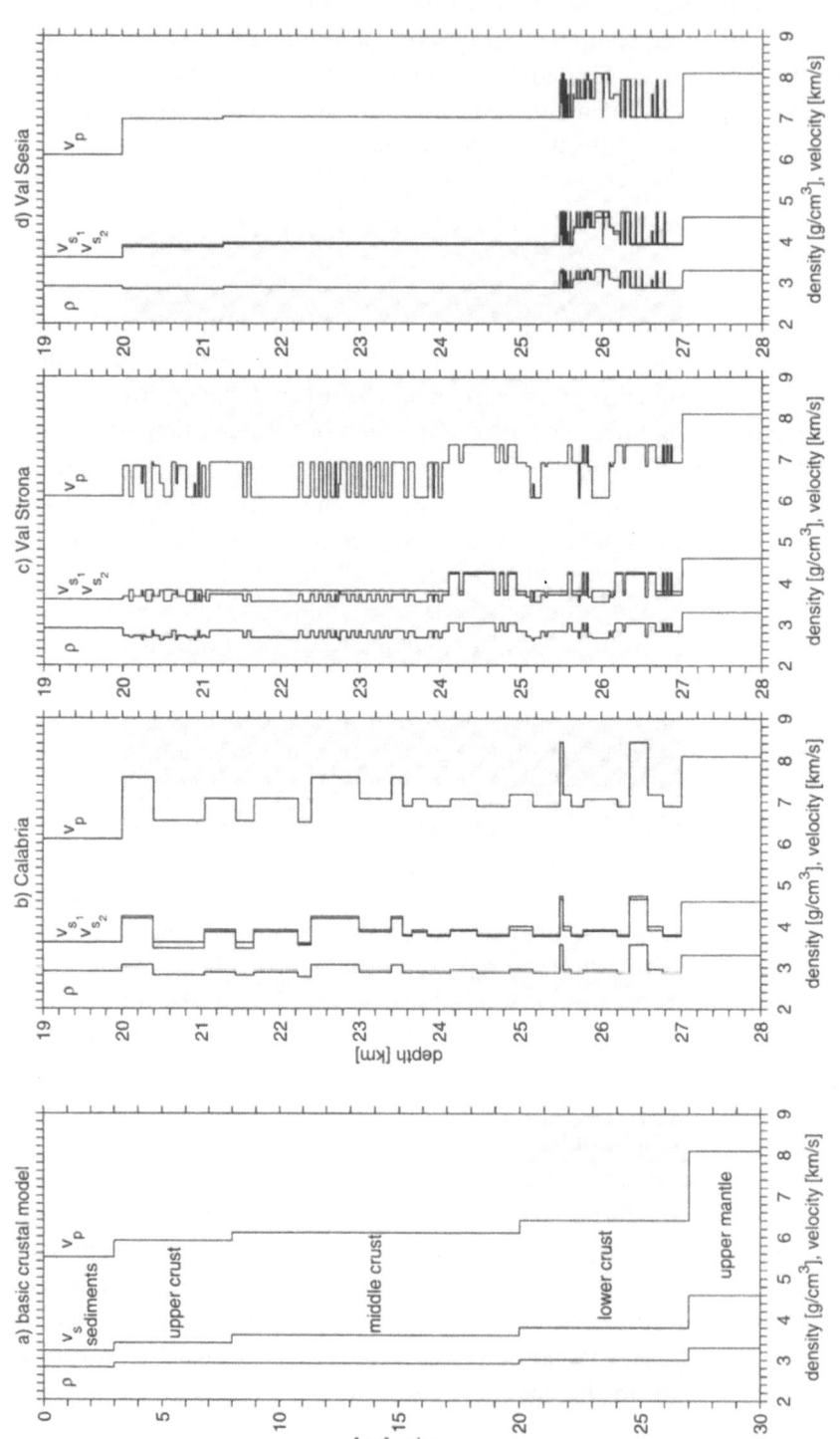

Figure 1

Velocity and density depth functions. The velocities are the averaged tensor velocities. (a) Basic crustal model of the Urach area in southwest Germany developed after HOLBROOK *et al.* (1988). (b)–(d) Lower crustal sequences of Calabria, Val Strona and Val Sesia. Lower crustal thicknesses have been modified to obtain a uniform 7-km-thick stack according to the Urach model. The stacks comprise 24 layers for Calabria, 98 layers for Val Strona and 83 layers for Val Sesia and provide an increase of seismic velocities with depth. The lowermost velocities of the mainly mafic dominated Val Sesia are higher than for Calabria and Val Strona.

structure. Since Calabria and Val Strona are composed of metapelitic rocks, we might expect high shear-wave splitting for horizontally travelling waves and weak splitting at steep angles of incidence. For Val Sesia, with its mafic composition, amounts of shear-wave splitting should be small. Table 1 summarizes the physical properties of the three sequences.

With our basic models we specified three input models (I, II, III) which differ in terms of reflectivity, anisotropy and thin layering.

For model type I the lower crust is regarded to be anisotropic as a whole. The individual tensors of the many layers of the real rock sequences are replaced by a single tensor which describes the mean elastic properties derived by a thickness weighted Voigt averaging applied to the components of all elastic tensors. Figure 2 shows the complete shear-wave velocity distribution and the corresponding shear-wave anisotropy in the horizontal plane (foliation plane). Mean shear-wave anisotropy provides almost transversely isotropic behavior for the Calabria and Val Strona sections with maximum and minimum shear-wave velocities in horizontal and vertical directions (WEISS, 1997). Shear-wave splitting is small for oblique rays, but becomes larger (5–6%) for subhorizontally travelling rays. Thus significant shear-wave splitting would be expected at offsets corresponding to incidence angles > 55° for Calabria and > 45°; for Val Strona. The S-wave velocity variations with angle of incidence are very small for the Val Sesia profile (< 1%), resulting in a negligible shear-wave splitting.

Model type II comprises the basic models. Figure 1 displays the corresponding velocity and density depth functions. P- and S-velocities are calculated by averag-

Table 1

Physical properties of the three lithological profiles

	Calabria	Val Strona	Val Sesia
Number of layers	24	98	83
Layer thickness [m]	40–405	25–430	15–4200
v_{qp} [km/s]	6.23–8.50	5.56–7.46	6.85–8.45
v_{qs_1} [km/s]	3.32–4.81	3.54–4.27	3.86–4.81
v_{qs_2} [km/s]	3.29–4.71	3.50–4.25	3.82–4.69
qp-anisotropy [%]	0.89–7.06	0.00–8.68	0.00–4.34
qs_1-anisotropy [%]	0.51–7.77	0.00–7.22	0.00–2.66
qs_2-anisotropy [%]	0.74–4.96	0.00–1.70	0.00–2.44
Δv_s [km/s]	0.02–0.56	0.00–0.50	0.00–0.22
Composition	2.8.2% metapelite	36.6% amphibolite	70.4% gabbro
	23.9% granulite	33.3% kinzigite	18.1% diorite
	12.6% bi-plag-gneiss	18.8% stronalithe	3.5% websterite-gabbro
	19.0% pyriclasite	8.7% marble	3.4% websterite
	3.9% si-ga-gneiss	2.5% granite	2.8% peridotite
	2.4% norite		1.7% metapelite
			0.2% pgm. cpxnite

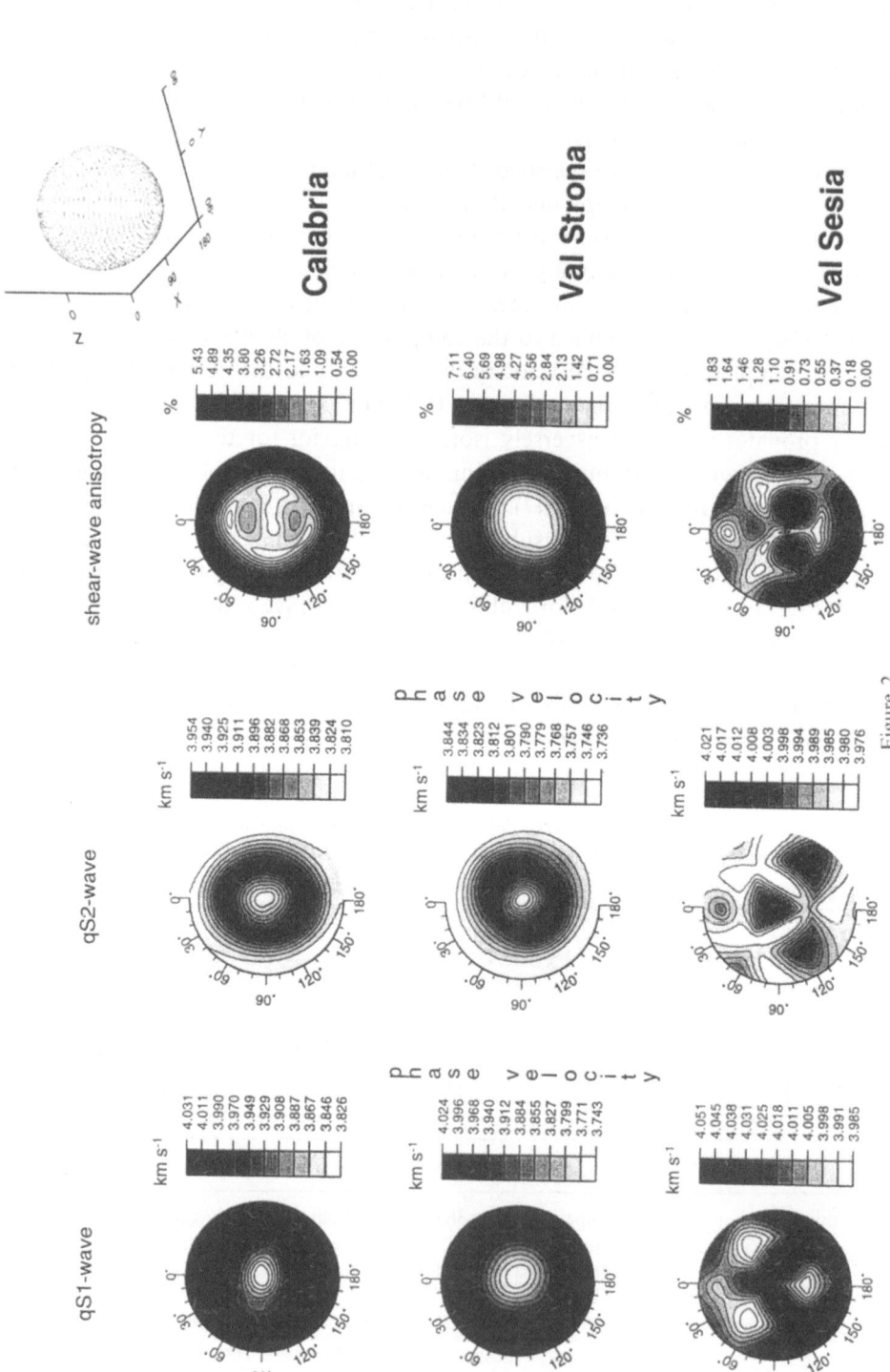

Figure 2

S-wave velocities and anisotropy of the averaged tensors (model type 1) for the different lower crustal sequences. The elastic properties are displayed in a stereographic projection. Projections are done in the *x*-*y*-plane (foliation plane). *qS*1 denotes the fast split shear wave and *qS*2 the slower one. Calabria and Val Strona provide a transversely isotropic behavior with shear-wave splitting which increases with the angle of incidence. Val Sesia, however, shows no significant difference in *qS*-wave velocities.

ing the tensor velocities over all directions within each layer. The Calabria, Val Strona and Val Sesia profiles comprise 24, 98 and 83 layers, respectively. Velocity fluctuations are on the order of 8% for all sequences. In particular, Val Sesia indicates very thin layering with average thicknesses of 20 m in the lowermost part of the sequence. There the velocities are higher than for the other profiles due to the more mafic composition. Val Strona displays increasing densities and velocities at 24 km, related to a change in petrological layering.

Finally, in model type III the elastic tensors of model II are replaced by isotropic elastic parameters, since isotropic thin-layered media of alternating velocities behave like homogeneous transversely isotropic media with the elastic parameters derived from the layered medium (BACKUS, 1962). The isotropic elastic parameters are obtained by Voigt averaging the tensor components over all directions. The number of layers in models II and III is identical.

Model type I describes the general elastic properties of the lower crustal sections. It allows the calculation of the S_MS traveltimes with no interference of lower crustal reflections. We used Model type II to evaluate the reflection signature of anisotropic laminated crust and model type III to investigate the effect of thin layering.

For our modelling procedure we chose a crustal model for the Urach area in southwest Germany, where the lower crust is strongly laminated. The crustal structure is well known due to numerous reflection and refraction experiments (BARTELSEN et al., 1982; HOLBROOK et al., 1988; RABBEL and LÜSCHEN, 1996). A shear-wave experiment in the area revealed evidence of shear-wave splitting in S_MS (RABBEL and LÜSCHEN, 1996). The model is based on the results of HOLBROOK et al. (1988) and modified to obtain a 1-D medium (Fig. 1a).

As described in the beginning of this section, the elastic parameters are obtained by the analysis of rock fabric. Thus the seismic properties correspond to crackfree rocks at room temperature. Comparison between data obtained by rock fabric analysis and by laboratory measurements at confining pressures of 600 MPa show differences up to 5% (BARRUOL and KERN, 1996). Consider that the temperature effect on seismic velocities is considerably more complicated for anisotropic than for isotropic rocks since there are no temperature derivatives available for all elastic constants. Assuming P/T conditions of 750 MPa and 650° (WEISS, 1997) we estimated P and T corrections between 0.2–0.4 km/s for $\bar{v}_p$ and 0.1–0.2 km/s for $\bar{v}_s$ for the different rock types. Velocity anisotropy and birefringence are not significantly effected by pressure and temperature (BARRUOL and KERN, 1996). Generally, velocities obtained by laboratory measurements differ from in situ velocities. Major reasons are cracks, rock alteration, fluid content, velocity dispersion to name a few. Most of these influences are effective considering upper crustal levels where cracks are opened and rocks weathered. In this study we focus on relative travel times and regard only synthetic data. Therefore we neglect those effects and correct PIT conditions.

Modelling

We used the anisotropic reflectivity code, which is an efficient method for calculating synthetic seismograms for an anisotropic laterally homogeneous earth, to calculate synthetic shear-wave sections. The reflectivity technique was designed for the calculation of synthetic seismograms from a point source in horizontally stratified isotropic media and later extended to include anisotropic layers by NOLTE (1988), based on the work of BOOTH and CRAMPIN (1983). The single slowness integration allows consideration of wave propagation in the sagittal plane (i.e., all azimuths for transversely isotropic media, or wave propagation in symmetry planes for anisotropic layers of lower symmetry). Seismic anisotropy of the layers may be arbitrarily strong in these cases. However, considering arbitrary symmetry of anisotropy along arbitrary azimuth (wave propagation outside of symmetry planes) the single slowness integration restricts the technique to weakly anisotropic media for which the wave energy does not diverge significantly from the sagittal plane. The method requires the complete elastic tensor (21 components) as input.

We used point sources radiating S_V and S_H waves with an isotropic radiation characteristic (NOLTE, 1988) and a ricker wavelet as source signal with a dominant frequency of 10 Hz. This particular point source generates only S-waves with the same amplitude. Therefore, the amplitudes of the radial and transverse component can be easily compared. The corresponding wavelengths are about 600 m for P- and 390 m for S-waves. The maximum layer thickness of the multilayered models is in the order the average S-wavelength, except the uppermost part of the Val Sesia sequence. In order to simplify the interpretation of the wavefield free surface multiples, mode conversions and internal multiples of the isotropic layers overlaying the lower crust were omitted, except for models of type III. A quality factor of 1000 was chosen for P- and S-waves for all layers. In the synthetic sections travel times are reduced by a velocity of 3.46 km/s. Amplitudes are offset-weighted and the radial sections are amplified by a factor of four compared to the transverse sections.

Figure 3 displays near vertical synthetic sections for all sequences of model type II. All three profiles provide strong near vertical reflections. Travel times of the lower crustal reflections coincide for both components.

Considering model type I (i.e., homogeneous, anisotropic lower crust), the synthetics in Figure 4a display the reflection responses in the radial and transverse components for the Val Strona profile in an offset range between 50 and 100 km. In the radial component S-to-P-converted phases (SP) appear with larger slopes due to the higher phase velocities. S-to-S-reflections from the interfaces sediment/upper crust (S_1), upper/middle crust (S_2), middle/lower crust (S_c) and lower crust/upper mantle ($S_M S$) are clearly visible. Whereas the travel times of both components coincide for the first three phases, there is a trave-time delay ($\delta\tau$) of approximately 500 ms in the $S_M S$ at 100 km. The leading $S_M S$ phase ($qS1$) is predominantly,

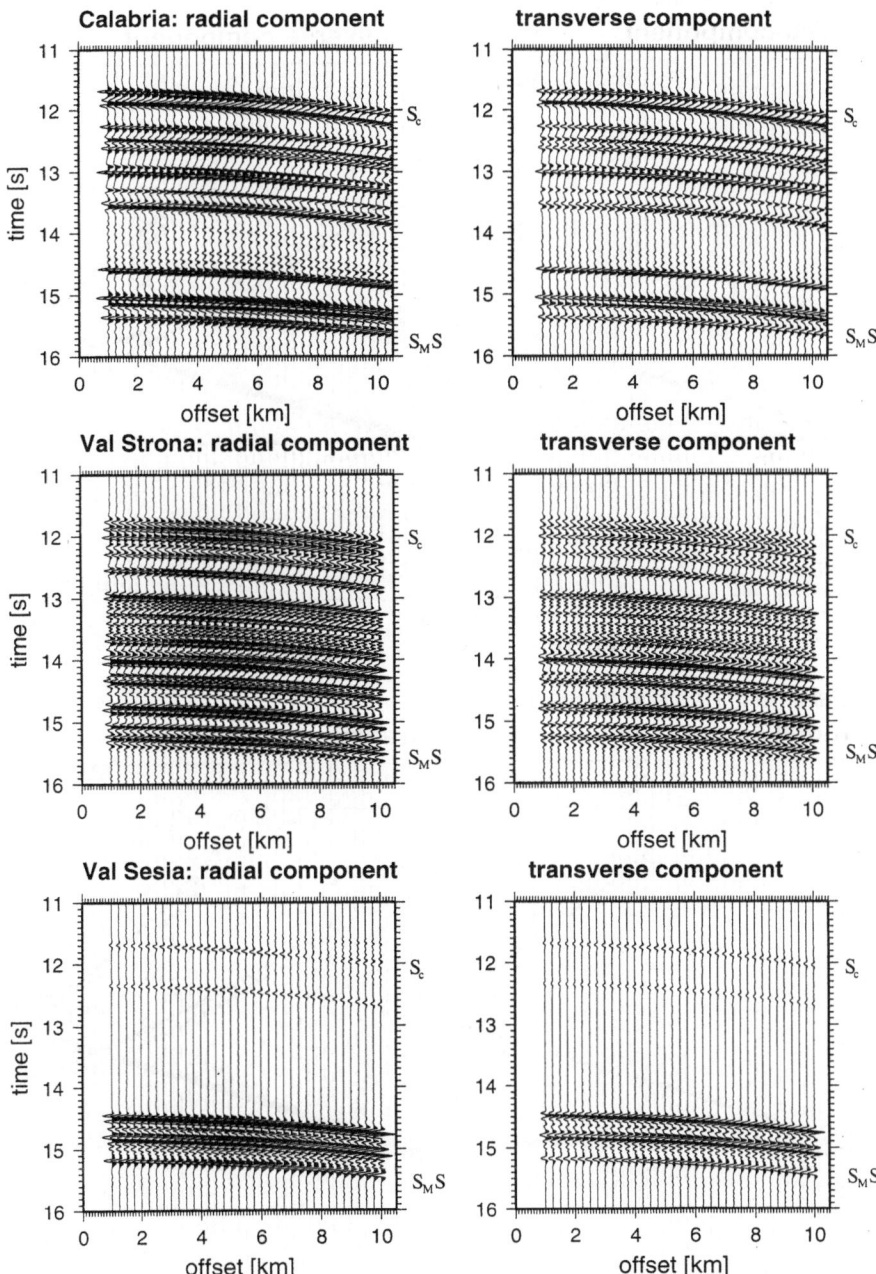

Figure 3

Near vertical sections for a laminated lower crust of anisotropic layers (model type II). The reflective signature is characterized by strong amplitudes and reflects the lithology of the corresponding profiles. There is no travel-time delay in the $S_M S$ arrivals between the radial and transverse component for these offsets.

Val Strona I: 1 anisotropic layer

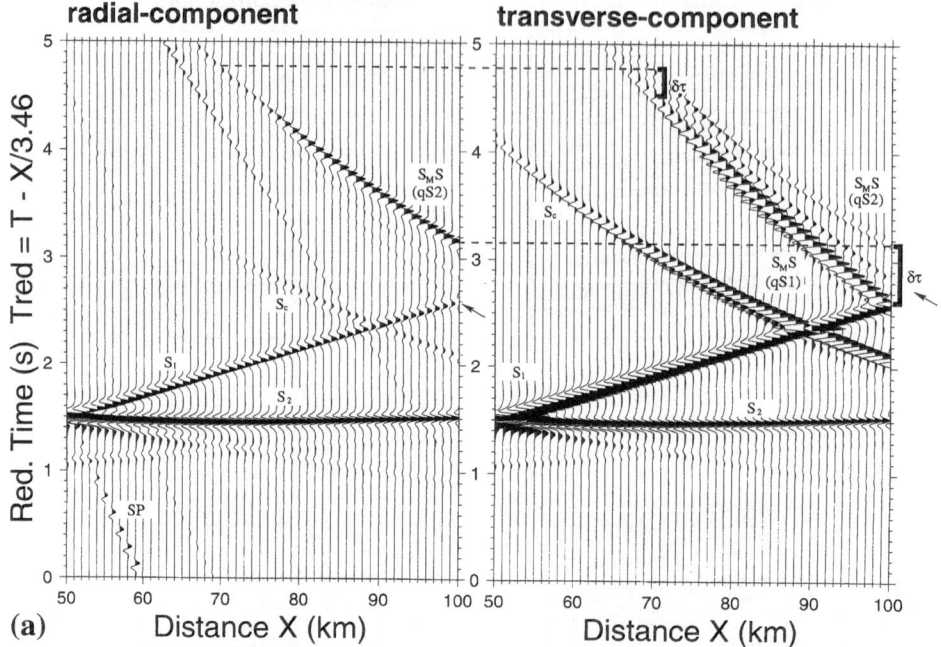

(a)

Calabria I: 1 anisotropic layer

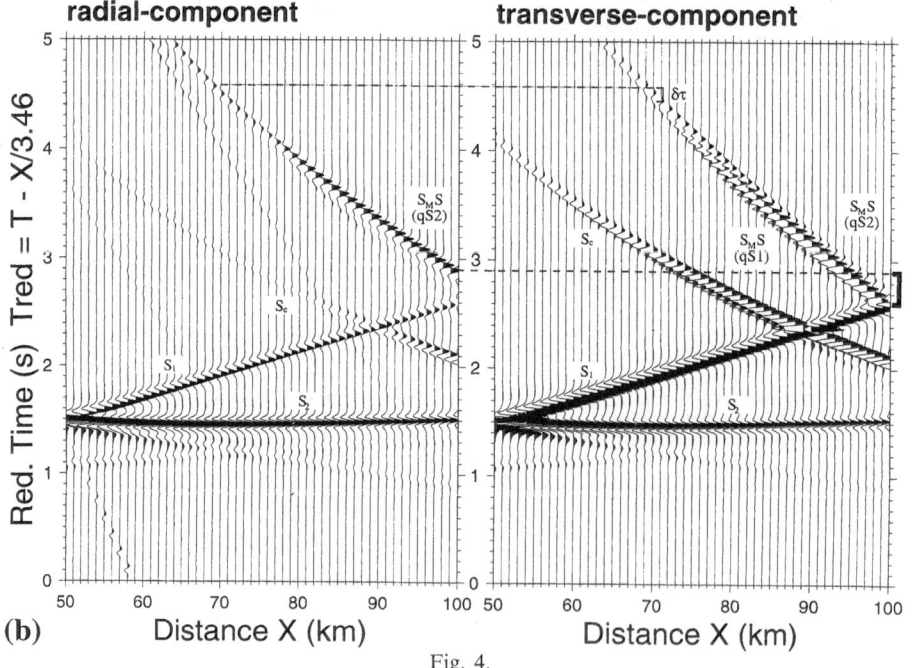

(b)

Fig. 4.

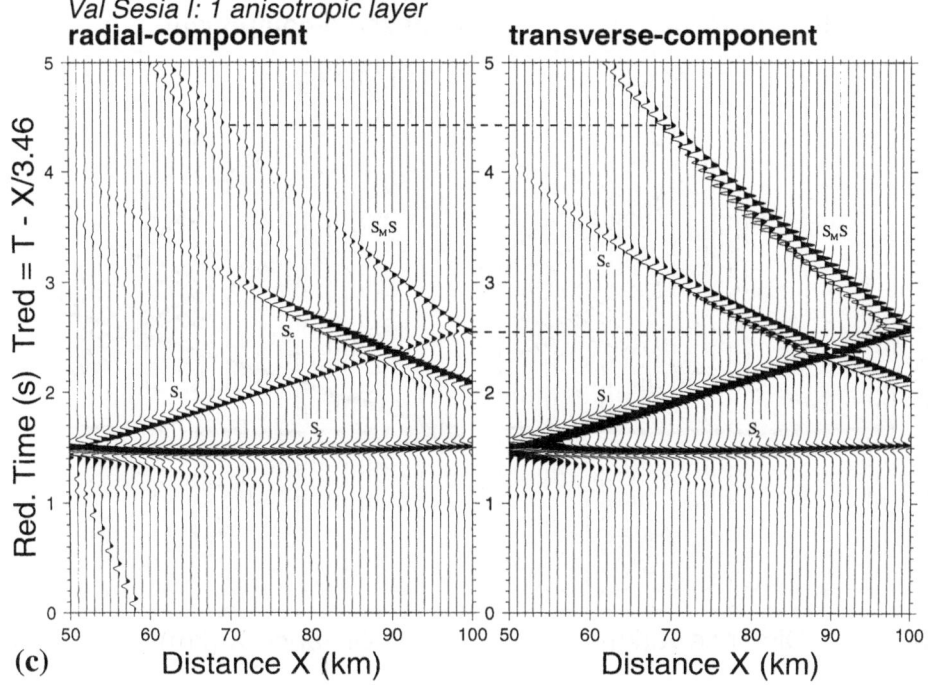

Figure 4
True amplitude synthetics of model type I. Crustal S-to-S reflections are labelled S_1, S_2 and S_c. The Moho reflection is denoted S_MS. (a) For Val Strona the S_MS in the radial component is delayed by at least 500 ms at a distance of 100 km compared to the transverse component. The transverse component displays both split shear waves, $qS1$ and $qS2$. The arrows mark the onset of $qS1$ on the radial component and a $qS2$-to-$qS1$-converted phase on the transverse component. (b) Calabria shows shear-wave splitting up to 300 ms in 100 km distance. Although the S_MS reflections are of mixed polarization, the leading shear wave ($qS1$) is predominantly polarized in the horizontal plane. (c) No birefringence is evident in the Val Sesia synthetics.

horizontally polarized and followed by a mixed polarized $qS2$-wave. At offsets of 70 km the $qS2$-wave is delayed by more than 200 ms. Between both phases runs a $qS2$-to-$qS1$-converted phase (indicated by the arrow in the radial component). On the radial component $qS1$ is very weak (indicated by arrows). With 300 ms at 100 km, the delay time in the Calabria section is about half of Val Strona (Fig. 4b). No evidence of birefringence is found in the Val Sesia synthetics (Fig. 4c).

Compositional layering of model type II produces a complex wavefield with interacting primary, multiple and converted reflections generated within the lower crust (Figs. 5a–c). Especially for Val Strona and Calabria, multiples and mode conversions within the anisotropic layers interfere severely with the lower crustal reflections in the radial component. Despite the more complicated pattern the S_MS remains a distinct phase at the end of the reflective band. Figure 5a shows the

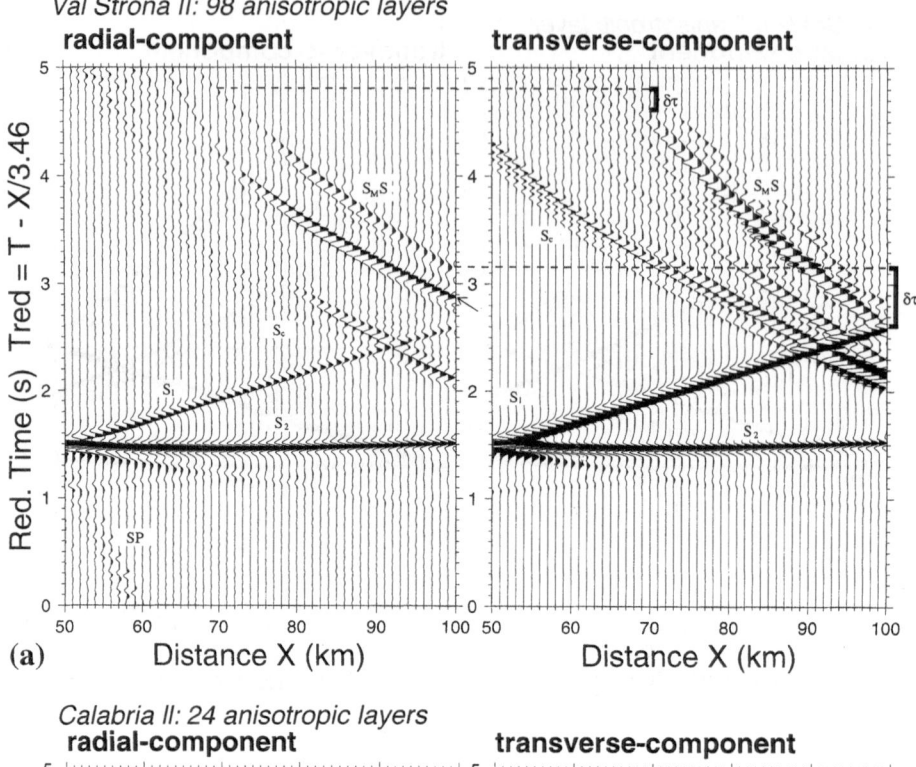

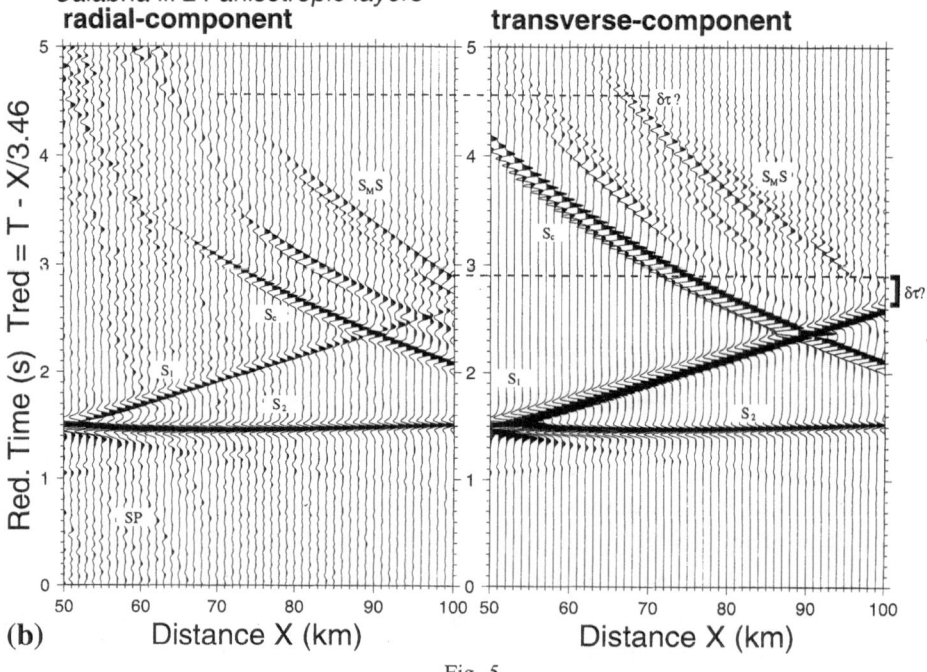

Fig. 5.

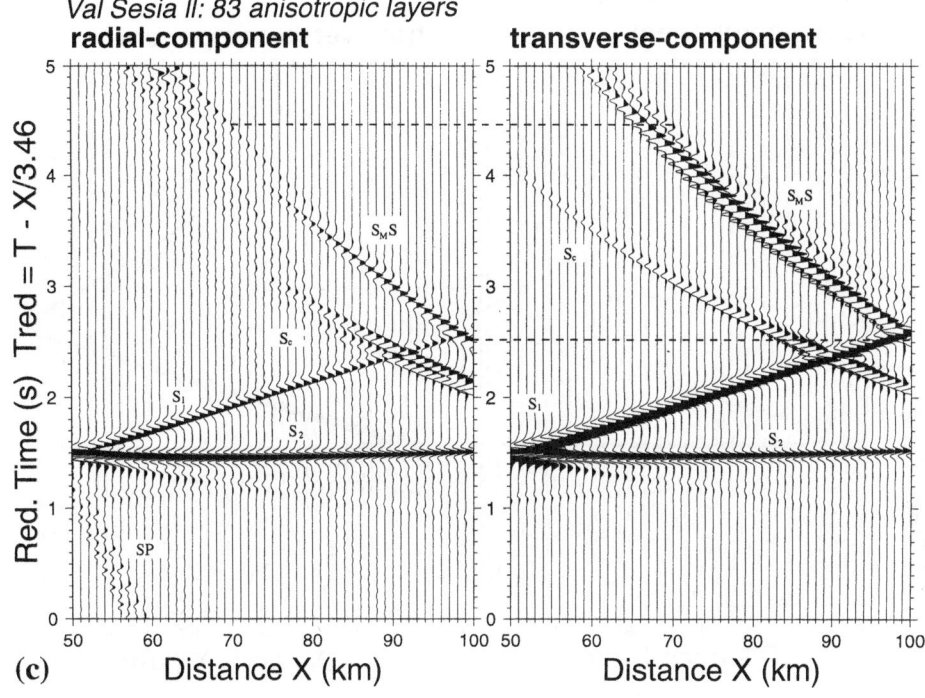

Figure 5
Synthetics calculated for anisotropic compositional layering (model type II). In contrast to Figure 4, the lower crust is now represented by a reflective band. (a) For Val Strona we find the same birefringence in the S_MS as predicted by the averaged tensor. (b) Identification of the S_MS in the transverse component of the Calabria section is difficult due to superimposition of internal multiples at offsets greater than 85 km. (c) Val Sesia shows no indications for shear-wave splitting.

synthetics for the Val Strona model. Here the S_MS appears brightly for offsets beyond 70 km. The travel-time shift between the Moho responses in the horizontal components increases from 200 ms at 70 km to 500 ms at 100 km. The strong lower crustal phase on the radial component (indicated by an arrow) is generated by the change in petrology at 24 km depth associated with an average S-wave velocity increase from 3.8 to 4.25 km/s. For Calabria (Fig. 5b) phase interferences do not allow a clear Moho signal in the wide-angle offset range in the transverse component. Finally, for Val Sesia (Fig. 5c) the lower crust appears transparent rather than reflective. The S_MS appears brightly and is generated by the fine layered stack (average thickness 20 m) in the lowermost part of the lower crust (see Fig. 1a). As in Figure 4c, S_MS travel times coincide for both components.

Although the models of type III contain only isotropic layers, the layering may result in anisotropic wave propagation through fine layering that can gen-

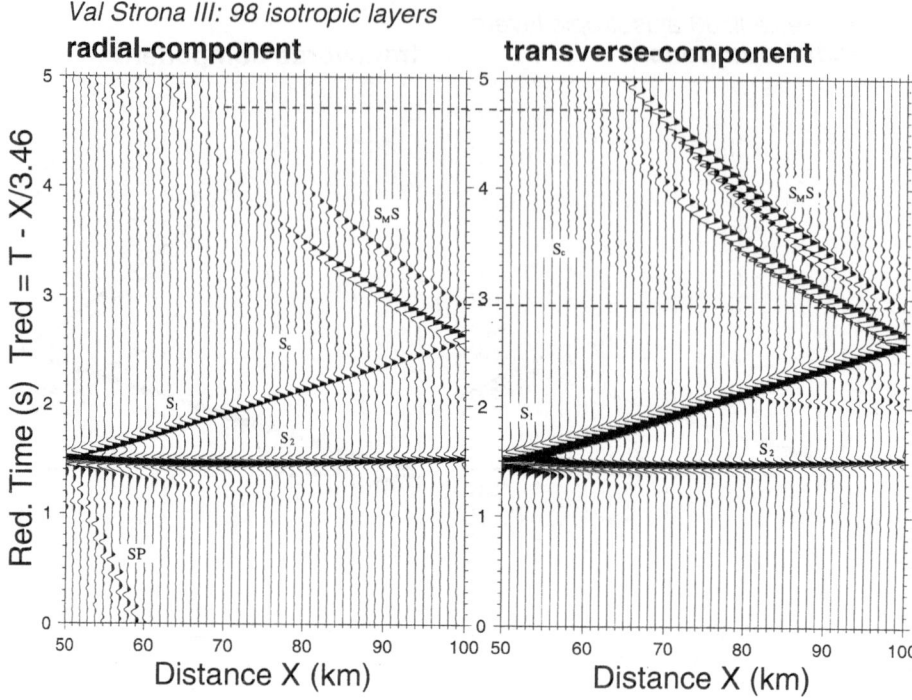

Figure 6

Sections computed for isotropic compositional layering (here shown for Val Strona): (1) There is no evidence of shear-wave splitting at large offsets. (2) The reflectivity pattern is similar to model type II although travel-times and reflection coefficients are altered due to the modified velocities.

erate effective anisotropy. However, modelling results—here shown for Val Strona in Figure 6—indicate no evidence of shear-wave splitting due to thin layering. The $S_M S$ traveltimes coincide for both components. Compared to the anisotropic case (Fig. 4a), amplitudes are smaller for most phases.

Conclusion

We used representative elastic lower crustal models based on exposed lower crustal sections in Calabria and Ivrea as input to anisotropic seismic modelling. The sections represent different types of lower crust varying from felsic to mafic composition. The elastic tensors were obtained from laboratory measurements. Our models are comparable to models suggested by other authors. e.g., SANDMEIER and WENZEL (1990) in terms of alternating high and low velocity layers, increasing velocity with depth and high reflection coefficients. The reflective signature of the synthetic data is comparable to seismic observations of *in situ* lower crust (HOL-BROOK *et al.*, 1988; LÜSCHEN *et al.*, 1990).

Shear-wave splitting is evident in the $S_M S$ reflection for offsets larger than 60 km and provides travel-time differences up to 500 ms between both S-waves. The amount of shear-wave splitting obtained for the Calabria and Val Strona profiles is in good agreement with the observations in the Urach shear-wave experiment in 1990 (RABBEL and LÜSCHEN, 1996). However, the direction of maximum birefringence was estimated to be at 45° incidence.

Isotropic versus anisotropic modelling revealed that shear-wave splitting is related to intrinsic anisotropy of lower crustal rocks rather than to thin layering effects.

Averaging the tensors of elasticity of lower crustal layers gives a good approximation of the lower crustal anisotropy. Differences in the $S_M S$ travel times computed from models I and II are small. Our study shows that seismic anisotropy should seriously be taken into account in investigations of the laminated lower crust.

Acknowledgements

The authors are thankful to M. Bopp and an anonymous reviewer for constructive comments. The study was funded by the Deutsche Forschungsgemeinschaft, grant Ra 496/5-1.

REFERENCES

BABUŠKA, V., and CARA, M., *Seismic Anisotropy in the Earth* (Kluwer Academic Publishers 1990).

BACKUS, G. E. (1962), *Long-wave Elastic Anisotropy Produced by Horizontal Layering*, J. Geophys. Res. *67*, 4427–4440.

BARRUOL, G., and KERN, H. (1996), *Seismic Anisotropy and Shear-wave Splitting in Lower-crustal and Upper-mantle Rocks from the Ivrea Zone—Experimental and Calculated Data*, Phys. Earth Planet. Int. *95*, 175–194.

BARRUOL, G., and MAINPRICE, D. (1993), *3-D Seismic Velocities Calculated from Lattice-preferred Orientation and Reflectivity of a Lower Crustal Section: Examples of the Val Sesia Section (Ivrea Zone, Northern Italy)*, Geophys. J. Int. *115*, 1169–1188.

BARTELSEN, H., LÜSCHEN, E., KREY, TH., MEISSNER, R., SCHMOLL, H., and WALTHER, C., *The combined seismic reflection-refraction investigation of the Urach Geothermal Anomaly*. In *The Urach Geothermal Project* (ed. Händel, R.) (Schweizerbart, Stuttgart 1982) pp. 247–262.

BOOTH, D. C., and CRAMPIN, S. (1983), *The Anisotropic Reflectivity Technique: Theory*, Geophys. J. R. Astr. Soc. *72*, 755–766.

CHRISTENSEN, N. I. (1971), *Shear-wave Propagation in Rocks*, Nature *229*, 549–550.

CHRISTENSEN, N. I. (1989), *Reflectivity and Seismic Properties of the Deep Continental Crust*, J. Geophys. Res. *94*, 17,793–17,804.

CRAMPIN, S. (1987), *Geological and Industrial Implications of Extensive-dilatancy Anisotropy*, Nature *328*, 491–496.

HALE, L. D., and THOMPSON, G. A. (1982), *The Seismic Reflection Character of the Continental Mohorovicic Discontinuity*, J. Geophys. Res. *87*, 4625–4635.

HOLBROOK, W. S., GAJEWSKI, D., KRAMMER, A., and PRODEHL, C. (1988), *An Interpretation of Wide-angle Compressional and Shear Wave Data in Southwest Germany: Poisson's Ratio and Petrological Implications*, J. Geophys. Res. *93*, 12,081–12,106.

HURICH, C. A., and SMITHSON, S. B. (1987), *Compositional Variation and the Origin of Deep Crustal Reflections*, Earth Planet. Sci. Lett. *85*, 416–426.

KERN, H., *P- and S-wave velocities in crustal and mantle rocks under the simultaneous action of high confining pressure and high temperature and the effect of rock microstructure*. In *High-pressure Researches in Geoscience* (ed. Schreyer, W.) (Schweizerbart, Stuttgart 1982) pp. 15–45.

KERN, H., and SCHENK, V. (1985), *Elastic Wave Velocities in Rocks from a Lower Crustal Section in Southern Calabria (Italy)*, Phys. Earth Planet. Int. *40*, 147–160.

LE GALL, B. (1990), *Evidence of an Imbricate Crustal Thrust Belt in the Southern British Variscides: Contributions of Southwestern Approaches Travers (SWAT) Deep Seismic Reflection Profiling Recorded through the English Channel and the Celtic Sea*, Tectonics *9*, 283–302.

LÜSCHEN, E., WENZEL, F., SANDMEIER, K.-J., MENGES, D., RÜHL, TH., STILLER, M., JANOTH, W., KELLER, F., SÖLLNER, W., THOMAS, R., KROHE, A., STENGER, R., FUCHS, K., WILHELM, H., and EISBACHER, G. (1987), *Near-vertical and Wide-angle Seismic Surveys in the Black Forest, SW Germany*, J. Geophys. *62*, 1–30.

LÜSCHEN, E., NOLTE, B., and FUCHS, K. (1990), *Shear-wave Evidence for an Anisotropic Lower Crust Beneath the Black Forest, Southwest Germany*, Tectonophysics *173*, 483–493.

MATTHEWS, D. H., *Seismic reflections from the lower crust around Britain*. In *The Nature of the Lower Continental Crust* (eds. Dawson, J. B., Carswell, D. A., Hall, J. and Wedepohl, K. H.) (Geol. Soc. London, Spec. Publ. 1986) pp. 11–21.

McCARTHY, J., and THOMPSON, G. (1988), *Seismic Imaging of Extended Crust with Emphasis on the Western United States*, Geol. Soc. Am. Bull. *100*, 1361–1374.

MOONEY, W., and MEISSNER, R., *Multi-genetic origin of crustal reflectivity: A review of seismic reflection profiling of the continental crust and Moho*. In *Continental Lower Crust* (eds. Fountain, D. M., Arculus, R., and Kay, R.) (Elsevier Science Publishers, Netherlands 1992) pp. 45–79.

NOLTE, B. (1988), *Erweiterung und Anwendung des Reflektivitätsprogrammes für anisotrope Medien* (in German), Diploma Thesis, Karlsruhe University.

RABBEL, W., and LÜSCHEN, E. (1996), *Shear-wave Anisotropy in Laminated Lower Crust at the Urach Geothermal Anomaly*, Tectonophysics *264*, 219–233.

SANDMEIER, K. J., and WENZEL, F. (1986), *Synthetic Seismograms for a Complex Crustal Model*, Geophys. Res. Lett. *13*, 22–25.

SANDMEIER, K.-J., and WENZEL, F. (1990), *Lower crustal petrology from wide-angle P- and S-wave measurements in the Black Forest*. In *Seismic Probing of Continents and their Margins* (eds. Leven, J., Finlayson, D., Wright, C., Dooley, J., and Kennett, B. N. L.) Tectonophysics *173*, 495–505.

SCHENK, V. (1981), *Synchronous Uplift of the Lower Crust of the Ivrea Zone and of Southern Calabria and its Possible Consequence for the Hercynian Orogeny in Southern Europe*, Earth Planet. Sci. Lett. *56*, 305–320.

SCHENK, V., *The exposed crustal cross section of southern Calabria, Italy: Structure and evolution of a segment of hercynian crust*. In *Exposed Cross Sections of the Continental Crust* (eds. Salisbury, M. and Fountain, D.) (Kluwer Academic Publishers, Netherlands 1990) pp. 21–42.

SIEGESMUND, S., TAKESHITA, T., and KERN, H. (1989), *Anisotropy of v_p and v_s in an amphibolite of the deeper crust and its relationship to the mineralogical, microstructural and textural characteristics of the rock*. In *Evolution of the European Continental Crust: Deep Drilling, Geophysics, Geology and Geochemistry* (eds. Meissner, R. and Gebauer, D.) Tectonophysics *157*, 25–39.

SIEGESMUND, S., KRUHL, J. H., and LÜSCHEN, E. (1996), *Petrophysical and Seismic Features of the Exposed Lower Continental Crust in Calabria (Italy): Field Observations versus Modelling*, Geotekt. Forschungen *85*, 125–163.

SMITHSON, S. B., *A physical model of the lower crust from North America based on seismic reflection data.* In *Nature of the Lower Continental Crust* (eds. Dawson, J. B., Carswell, D. A., Hall, J., and Wedepohl, K. H.) (Geol. Soc. London, Spec. Publ. 1986) pp. 23–34.

WEISS, T. (1997), *Gefügeanisotropie und ihre Auswirkung auf das seismische Erscheinungsbild: Fallbeispiele aus der Lithosphäre Süddeutschlands* (in German), Ph.D. Thesis, University of Göttingen.

(Received February 2, 1998, revised October 2, 1998, accepted October 28, 1998)

 To access this journal online:
http://www.birkhauser.ch

Pure appl. geophys. 156 (1999) 157–171
0033–4553/99/010157–15 $ 1.50 + 0.20/0

❚ Pure and Applied Geophysics

3-D Prestack Kirchhoff Migration of the ISO89-3D Data Set

STEFAN BUSKE[1,2]

Abstract—This paper presents an overview of the results obtained from a 3-D prestack depth migration of the ISO89-3D data set. The algorithm is implemented as a Kirchhoff-type migration, in which the migrated image is generated by weighted summation along diffraction surfaces through the shot record section. The diffraction surfaces are computed by a 3-D finite difference solution of the eikonal equation. A 3-D macro-velocity model derived mainly from wide-angle tomographic inversion served as input for the travel-time calculations. The results of the migration are presented as slices through a volume covering an area of 21 km × 21 km in the horizontal and 15 km in the vertical direction, centered around the KTB drill hole. In these slices the continuation of the Franconian Lineament or SE1 reflector, respectively, can be identified over most of the survey area as a northeast dipping reflector plane. Its signature appears partly curved and discontinuous and with different strength of reflection down to a maximum depth of 9 km. About 5 km to the south-southeast of the KTB drill hole the uppermost top reflection of the Erbendorf body (EB) can be recognized at approximately the same depth. The slices clearly show its complicated internal structure consisting of several apparently separated reflective parts. Moreover, the geometry and the shape of a few other subsurface structures are described.

Key words: Crustal structure, deep seismic reflection, explosion seismology, 3-D Kirchhoff migration.

Introduction

3-D prestack Kirchhoff migration is a leading edge technology in seismic data processing. Its computational requirements regarding memory and CPU time are considerable. Nevertheless, it has proven to yield significant improvements in imaging complex structures, both in the exploration industry and in the case of deep seismic reflection surveys (SIMON *et al.*, 1996).

The main task in the realization of a 3-D prestack Kirchhoff migration algorithm is the computation of the diffraction surfaces along which the weighted summation through the shot record sections is performed. This step requires the calculation of the travel times from the source and every receiver location to every

---

[1] Institute of Meteorology and Geophysics, Feldbergstr. 47, 60323 Frankfurt, Germany. E-mail: buske@geophysik.uni-frankfurt.de

[2] Now at Ensign Geophysics Limited, Brighton Road, Addlestone, Surrey KT15 1PU, United Kingdom. E-mail: stefanb@ensigngeo.com

subsurface point within the volume to be migrated. With the help of modern finite difference eikonal solvers this is feasible. Here the method of PODVIN and LECOMTE (1991) is used, which yields first-arrival travel times on a cubic grid for a given source location and a predefined velocity model. Details regarding the incorporation of the travel-time computation procedure into the migration algorithm as well as the parallel implementation and the required computing times can be found in BUSKE (1999).

This paper focuses on the results of the application of this prestack migration technique to the ISO89-3D data set. As this data set has been the subject of many studies, the following introductory remarks are kept short. Further information can be found in the undermentioned references.

The ISO89-3D Data Set

Overview

This data set was recorded in 1989 in the vicinity of the German Continental Deep Drill hole (KTB) as part of the program *Integrated Seismics Oberpfalz* (*ISO*) (DÜRBAUM et al., 1990, 1992). The field parameters and the shot-receiver setup of this 3-D survey have been described comprehensively in STILLER (1991). For clarity and for the definition of the coordinate axes, Figure 1 shows the shot and receiver locations within the survey area of 21 km × 21 km.

About 3327 Vibroseis shots were carried out and a maximum of 7 seconds recording time was used for the migration. All 3327 shot records were preprocessed separately and for each shot record the whole volume (21 km × 21 km × 15 km) was migrated. Subsequently, a single final migrated image was generated by stacking of the absolute values of these individual migrated volumes.

Examples of preprocessed shot records as well as comparisons with modeled shot records computed for an analogous model can also be found in BUSKE (1999).

The Macro-velocity Model and the Geological Setting

The 3-D interval velocity model used here was mainly obtained from tomographic inversion of wide-angle reflection data (SCHWARZENBÖCK, 1993). Figure 2 shows slices through this velocity model.

The northeastern part, which belongs to the crystalline basement of the Bohemian massif, consists mainly of gneissic-granitic rocks with P velocities of 5–6 km/s up to shallow depths, whereas the southwestern part is composed of permomesozoic foreland sediments of the South German Platform with moderate velocities of 3–4 km/s. Both parts are separated by a major crustal fault zone called the

Franconian Lineament (FL), which can be traced along its outcrop through the entire survey area (see Fig. 1). It is regarded as a suture zone formed by closure of an early Paleozoic ocean basin during the Variscan collision in Devonian/Carbonifereous times (400–330 Ma). From an earlier 2-D survey (KTB8502) oriented approximately perpendicular to the strike of this Lineament (see Fig. 1) it is known that the FL continues through the whole upper crust as a relatively sharp northeast dipping reflector (SE1). Figure 3 displays the relevant part of profile KTB8502 after a 2-D prestack Kirchhoff migration.

From this image the dip angle of the SE1 reflector can be estimated as about 55 degrees. Directly beneath the SE1 reflector in the depth range between 12 and 13 km a highly reflective part known as the Erbendorf body (EB) can be observed. The origin and the nature of this body is unknown. Based on wide-angle reflection seismics it corresponds to a high-velocity zone ($v_p > 7$ km/s) and may be a metabasic relic of a paleosubduction zone, or a sliver of dense rock emplaced tectonically during the Variscan collision (EMMERMANN and LAUTERJUNG, 1997).

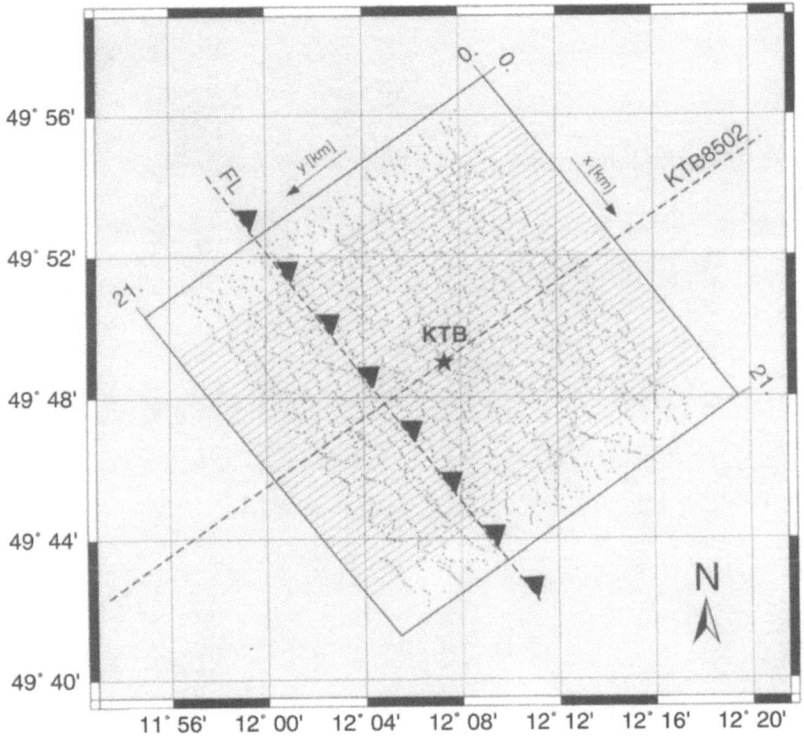

Figure 1

Geographical location of the ISO89-3D experiment including all shot and receiver points marked with a cross or a dot, respectively. The approximate locations of the KTB drill site, the northeast dipping Franconian Lineament (FL) and the 2-D profile KTB8502 are indicated.

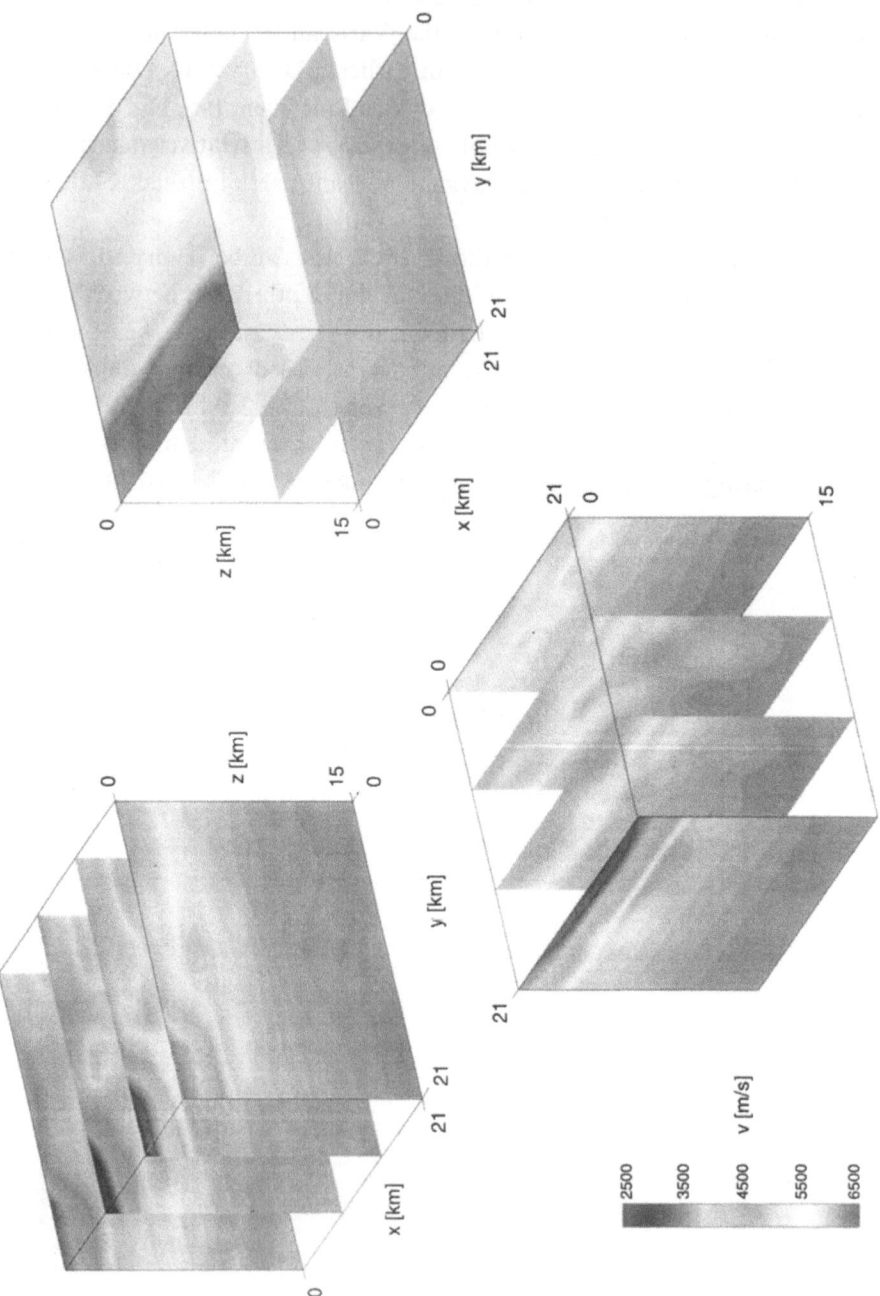

One of the main aims of a 3-D migration of the ISO89-3D data set is to obtain information about the geometry and lateral extent of the SE1 reflector as well as of the EB. While for the SE1 reflector a continuous signature can be expected from the surface geology, the exact location, size and shape of the EB is questionable. The expected location of the SE1 reflector coincides with the transition zone between the two crustal units inherent in the velocity model. It is not clear whether the smoothness of the transition zone in this model is only due to the limited spatial resolution of the tomographic inversion or whether a significant velocity jump is not present at all. Results from borehole measurements suggest that the SE1 reflector at a depth of about 7 km is a cataclastic zone of several hundred meters width rather than a sharp boundary in the velocity function. On the other hand, the outcrop of the SE1 reflector as well as the first arrivals of those shot records covering both crustal units support a sharp discontinuity near the surface. The velocity model is constructed to give both, a high

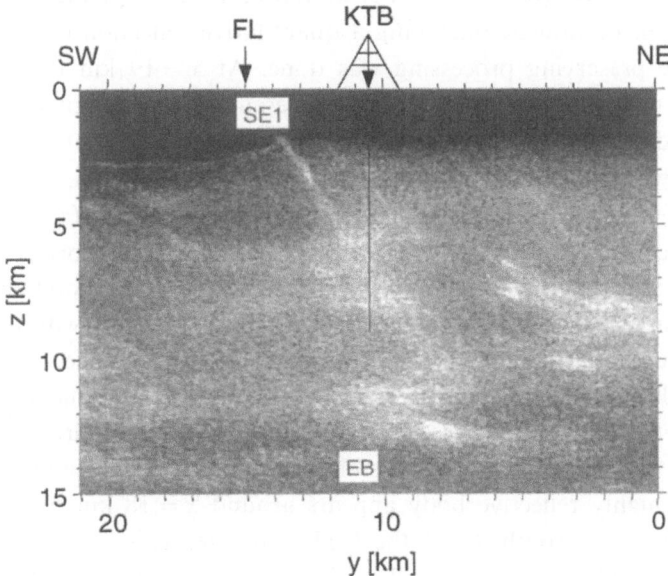

Figure 3

Part of the 2-D prestack migration of the profile KTB8502. As can be seen from Figure 1 this profile is oriented perpendicular to the strike of the Franconian Lineament (FL) and passes through the KTB location. The SE1 reflector appears as a steeply dipping event and the EB appears strongly at a depth of $z = 12$–13 km.

Figure 2

3-D macro-velocity model $v(x, y, z)$ based on tomographic inversion of wide-angle measurements. Low velocities near the surface in the southwestern part are due to the presence of sediments, while the northeastern part contains crystalline rocks with relatively high velocities extending to shallow depths. The transition zone between those two crustal units is smooth.

velocity contrast near the surface as well as a smooth transition zone without significant velocity changes at greater depths. All in all, the velocity model seems to be appropriate for a prestack migration.

Migration Results

In this section slices are shown through the final migrated and stacked volume of all 3327 shot records (Fig. 4).

The uppermost slices parallel to the x-y plane clearly show the continuation of the FL to depth, the SE1 reflector (Figs. 4a,b). Its intersections with the horizontal planes of constant depth exhibit a relatively sharp continuous signature parallel to the x axis between $x = 6$ km and $x = 19$ km. In the x range between 13 and 16 km it appears as a curved plane and reflectivity seems to increase with increasing x coordinate. Here the expression reflectivity is used to characterize the strength of the imaged reflector in a qualitative manner. No conclusions can be drawn concerning a quantitative reflection coefficient, because no amplitude preserving processing was done. At $x = 19$ km the signature vanishes, which is probably due to the low coverage at the marginal parts. According to its dip to the northeast, the SE1 reflector is cut at decreasing y coordinates with increasing z ($y = 12.5$–13.5 km at $z = 3$ km, $y = 12$–13 km at $z = 4$ km, $y = 11.5$–12.5 km at $z = 5$ km). In deeper parts ($z = 6, 7, 8$ km) the signature becomes more diffusive. This observation is in correspondence with borehole results, which characterize the SE1 reflector at its borehole intersection point as a cataclastic zone of several hundred meters width without any significant impedance contrasts. Compared to the upper parts, the curvature of the intersection line seems to have reversed at $z = 8$ km, but due to the diffusive character of the reflector at these depths this is not well confirmed. At $z = 9$ km parts of the SE1 reflector are still visible between $x = 11$ km and $x = 15$ km. At this depth a highly reflective body appears around $x = 18$ km, $y = 10$ km, which is about 5 km south-southeast of the KTB drill site. This strongly reflective part can be interpreted as the top of the EB. Following the slices down to greater depths it appears as a complicated structure covering the entire central part and vanishing at $z = 14$ km. The following vertical depth slices will show its shape more clearly than can be seen here. Several point-diffractor-like signatures can also be identified ($z = 9$ km: $x = 11.5$ km, $y = 2.5$ km; $z = 10$ km: a pattern consisting of three small separated areas between $x = 8$–10 km and $y = 2.5$–6 km; $z = 11$ km: $x = 14$ km, $y = 4$ km and $x = 12.5$ km, $y = 19$ km). These signatures result from the horizontal intersection of also predominantly horizontal oriented structures, whose position and dimension will become more clearly from the following vertical depth slices.

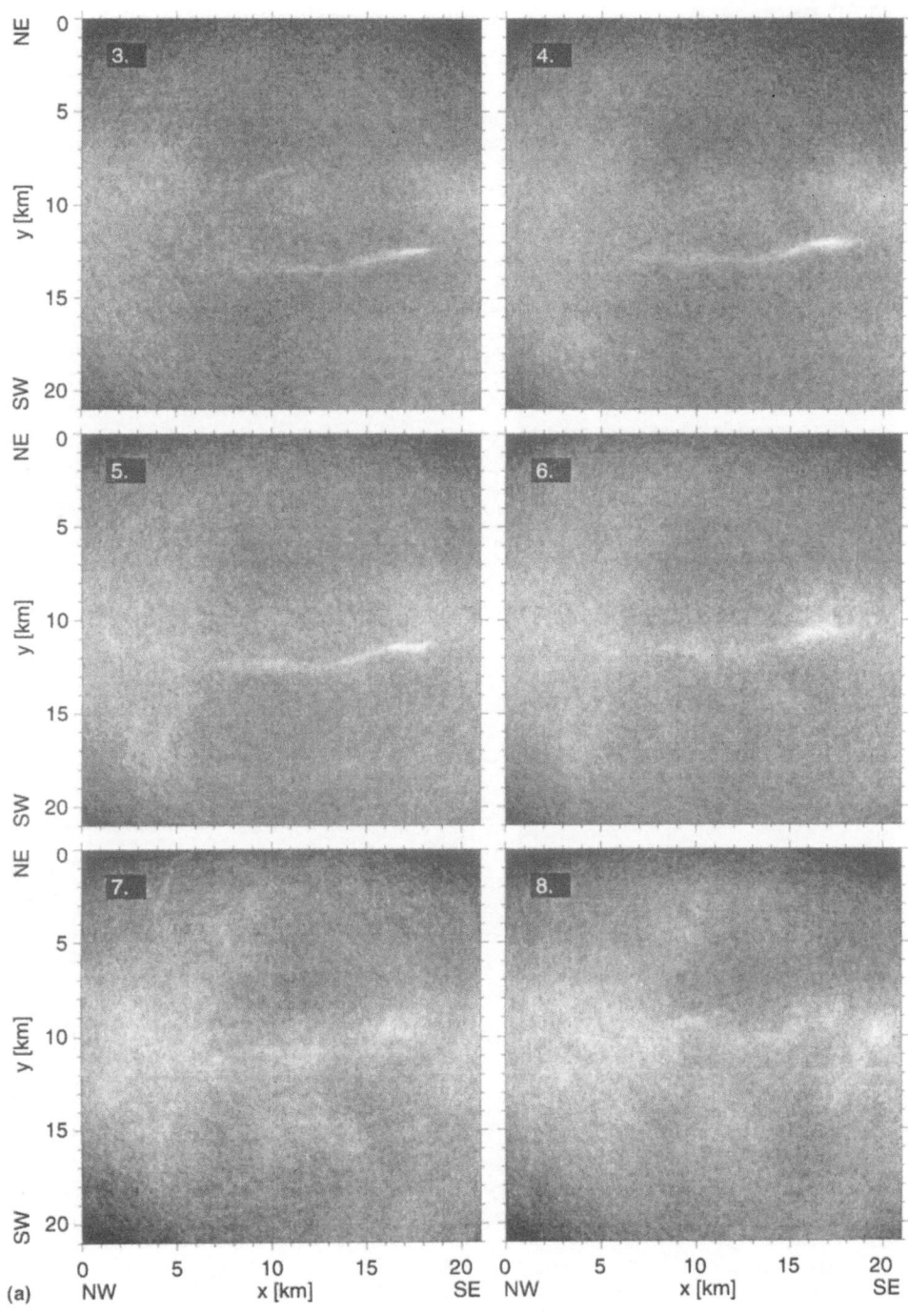

Fig. 4a.

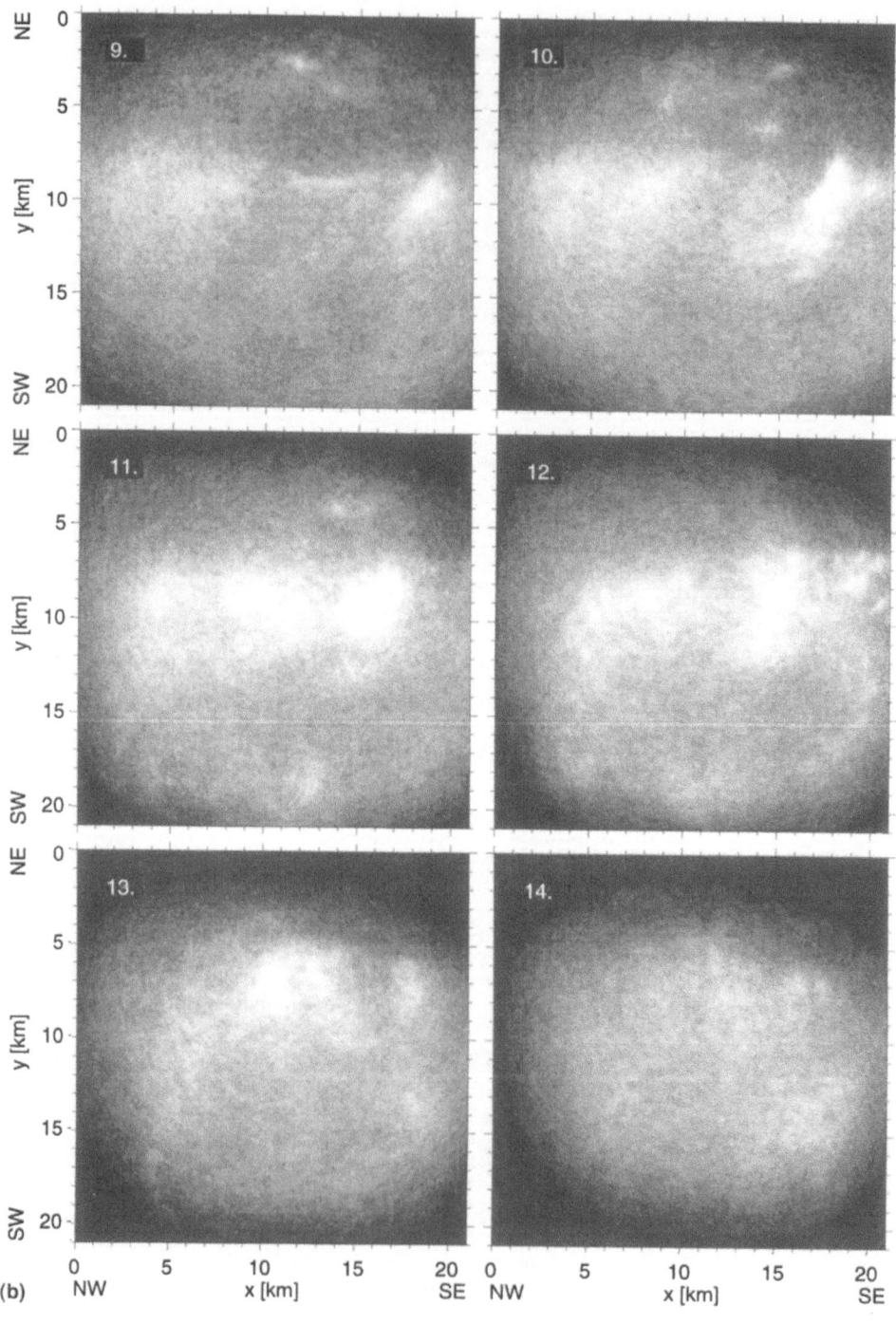

Fig. 4b.

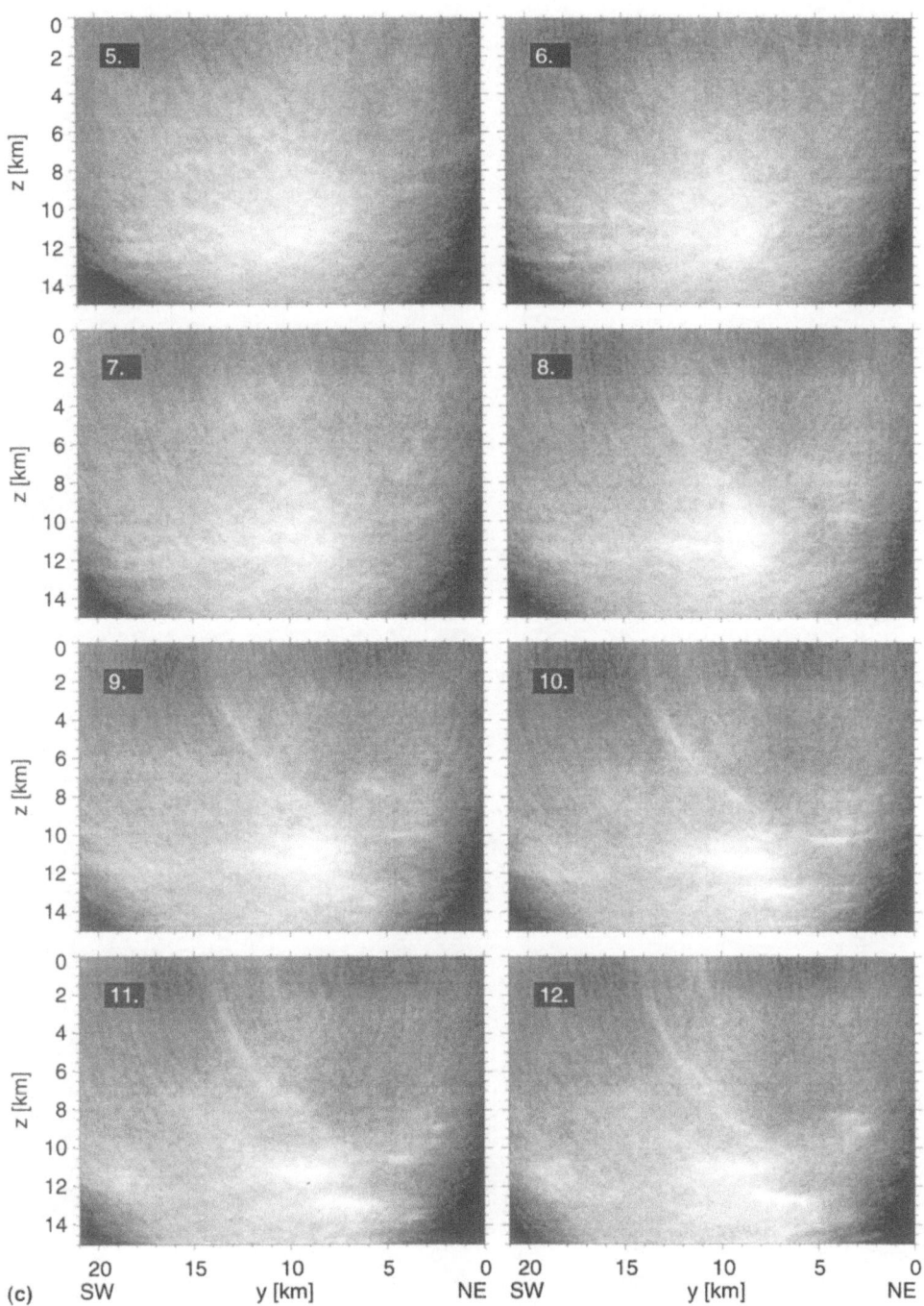

Fig. 4c.

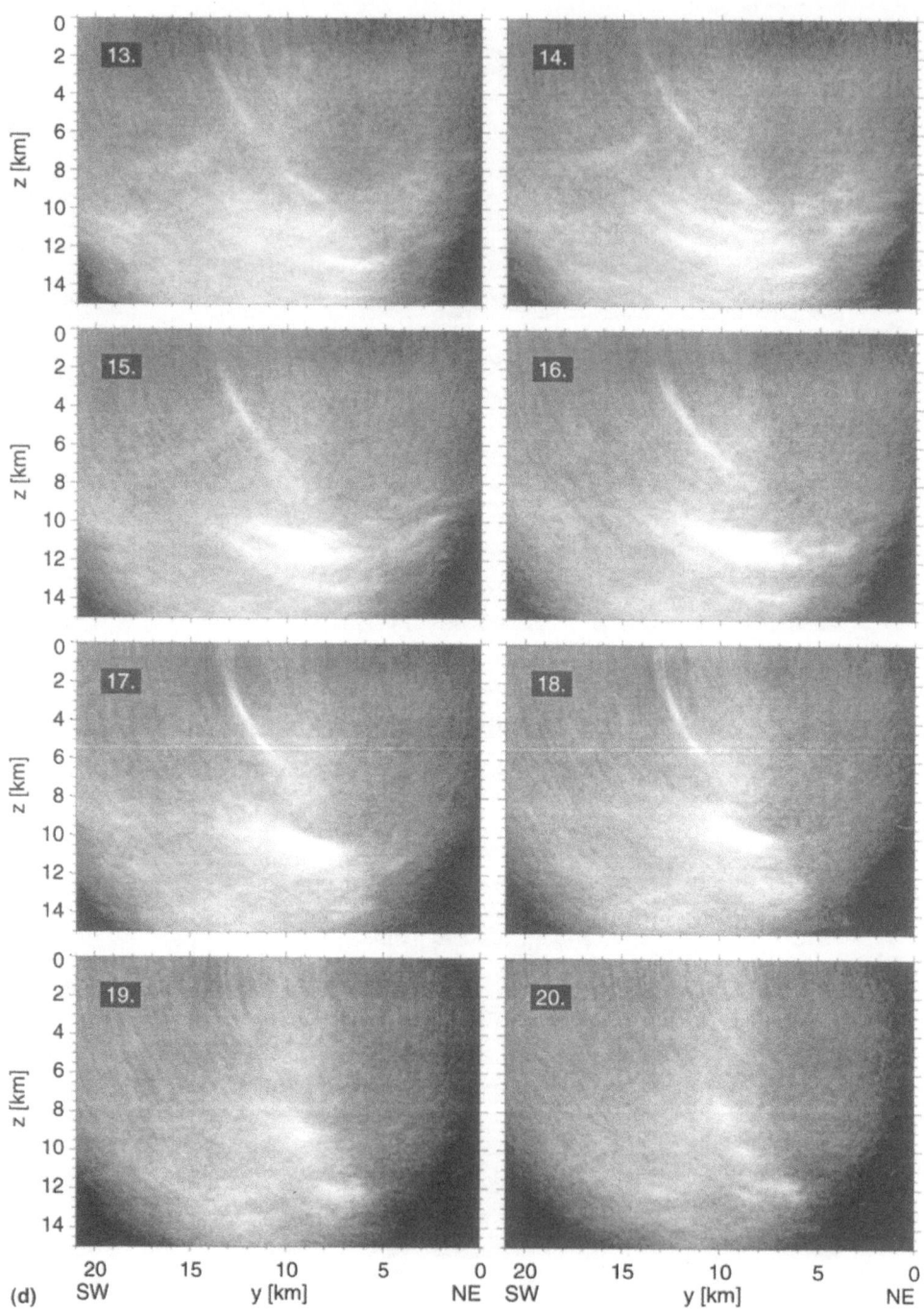

(d)

Fig. 4d.

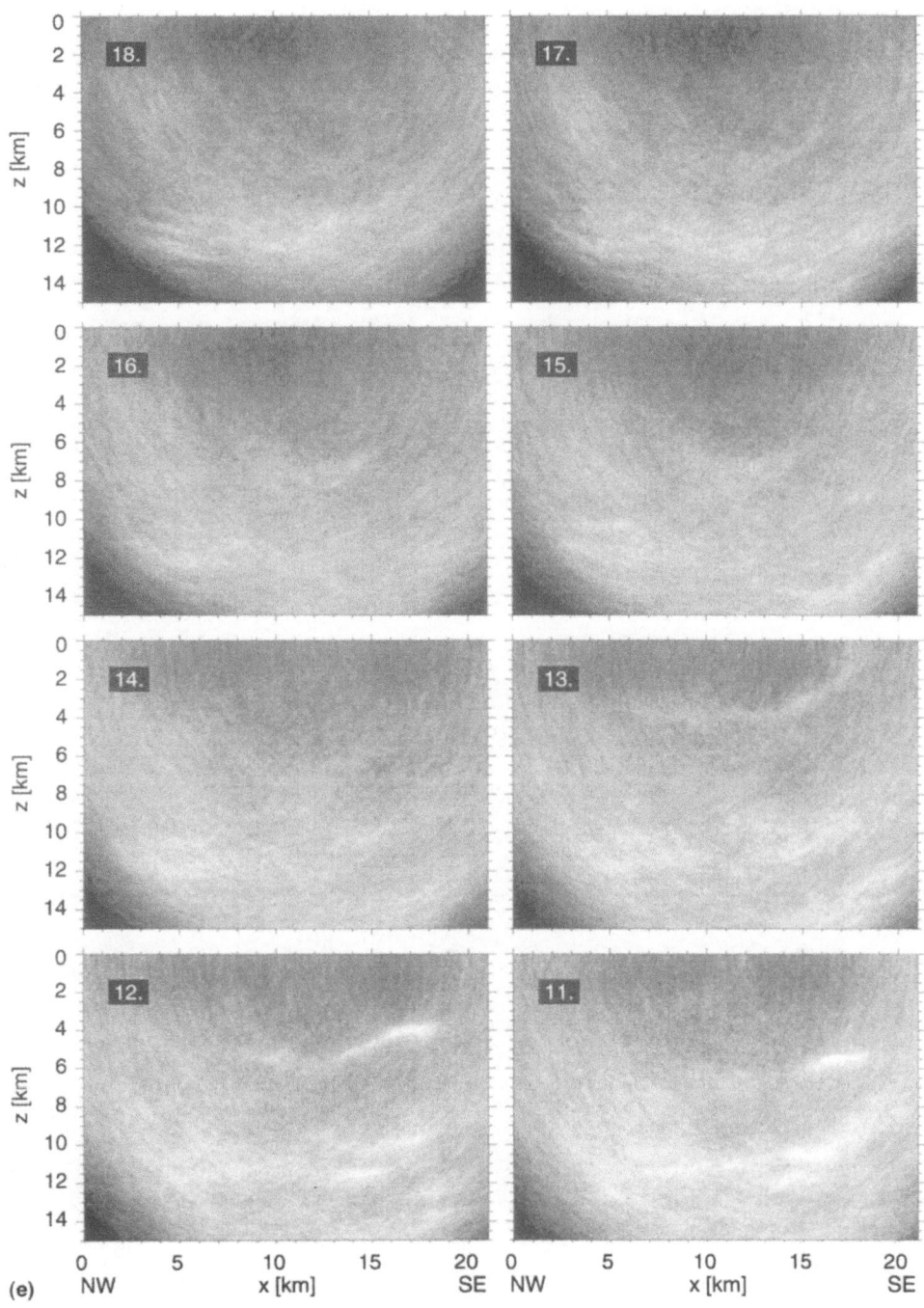

Fig. 4e.

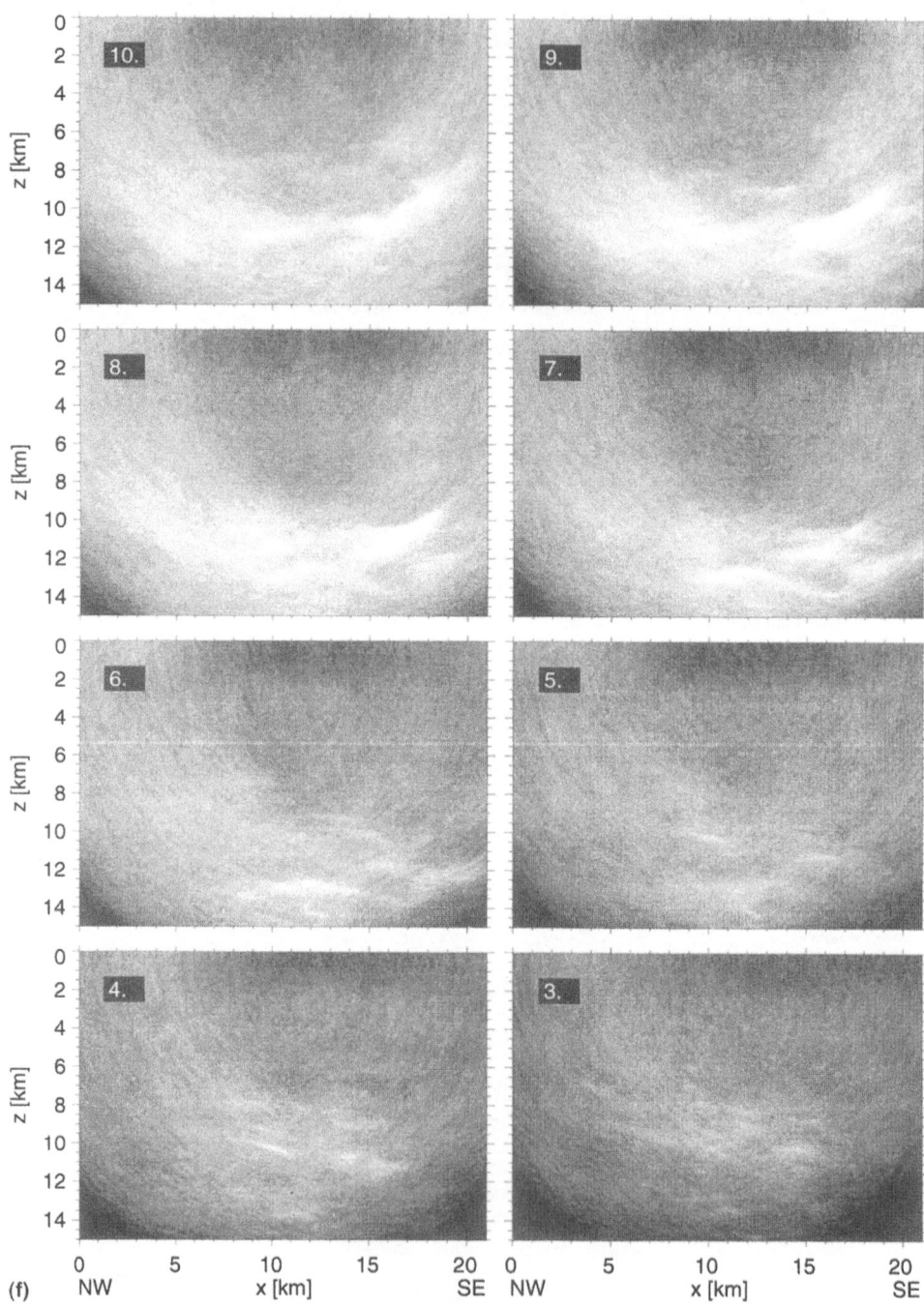

Figure 4f

Slices through the migrated volume of the ISO89-3D data set parallel to the *x*-*y* (a,b), *y*-*z* (c,d) and *x*-*z* (e,f) plane. The value of the constant coordinate is plotted in the upper left corner. For a detailed description see text.

Figures 4c and 4d show vertical depth slices, which are oriented parallel to the y-z plane. In most of these slices ($x = 7$–19 km) the SE1 reflector can be identified as a northeast dipping plane between $z = 2$ km and $z = 9$ km depth. In the central part ($z = 3$–6 km) its dip angle varies slightly around 55 degrees. In some slices this central part appears as a diffusive weakly reflective region. As recognized in the previous x-y slices, the reflection strength reaches its maximum value around $x = 16$ km. Below a depth of about $z = 10$ km the SE1 reflection plane cannot be identified; either it is not visible due to the limited aperture or it is completely absent. The latter is more likely the case because it also cannot be seen in the 2-D profile KTB8502 (see Fig. 2) which has a considerably larger aperture. Some slices, for instance at $x = 10$ km, show a detachment-like horizontal signature at $z = 10$ km. Below this depth the EB is the dominant feature in most of these slices. While it has a two-fold signature at $x = 10$–12 km, it seems to converge into a single strongly reflective body at $x = 15$–18 km. The true character and the origin of this apparent convergence will be resolved in the next slices parallel to the x-z plane. In the southwestern part at depths between $z = 10$ km and $z = 13$ km some subhorizontal reflectors can be identified at $x = 5$ km and $x = 6$ km as well as in those slices located further northwest ($x < 5$ km) although not shown here. The slices at $x = 13$–15 km reveal another interesting feature in the southwestern part. At depths of $z = 5$–8 km a weakly reflective zone seems to run into the SE1 at right angles. The origin and the possible tectonic history of this reflection band is the subject of further investigations.

Finally, the slices parallel to the x-z plane (Figs. 4e,f) show the lateral extension in x direction of the reflecting elements. These slices are oriented approximately parallel to the strike of the FL. At $y = 18$ km the above-mentioned subhorizontal reflectors can be observed between $x = 3$ km and $x = 7$ km at depths of $z = 10$–13 km. Also the weakly reflective zone related to the SE1 appears between $x = 9$–15 km at a depth of 7–8 km ($y = 15$ km and $y = 16$ km). From $y = 13$ km to $y = 9$ km the SE1 reflector is cut at increasing depths following its dip. In correspondence with the x-y slices the curvature of the SE1 reflection plane can be seen as intersections at different depths. The signature itself is continuous and manifests its largest reflection strength again in the southeastern part between $x = 15$ km and $x = 19$ km. At $y = 11$ km the central part becomes more diffusive and at $y = 10$–9 km the curvature has reversed. Now the intersection of the SE1 reflector with the x-z plane appears as two discontinuous subhorizontal reflectors at $z = 7$–9 km depth between $x = 7$ km and $x = 15$ km. Additionally, starting at $y = 12$ km the complex structure of the EB can be traced over a wide x range. Its reflectivity increases strongly with decreasing y coordinate and reaches its maximum approximately at $y = 9$ km. From the slices at $y = 8$ km and $y = 6$ km it becomes clear that the EB consists of several parts. The lowest one at a depth range $z = 12$–14 km between $x = 9$ km and $x = 16$ km shows an upcoming at about $x = 12$ km. This part is separated from the superimposed parts by a nonreflective band of about one kilometer width. From the combination of these y-z slices and the corresponding x-z slices it should be possible to construct a vague image of the complicated structure of the EB. Finally the x-z slices at $y = 5$ km and $y = 4$ km show the above-mentioned possible detachment zone as an accumulation of southeast dipping reflectors at $z = 10$ km between $x = 7$ km and $x = 11$ km.

From the location of the KTB at the center of the survey area ($x = y = 10.5$ km, see Fig. 1) and the corresponding depth slices it can be concluded that the borehole has intersected a rather diffusive and weakly reflective part of the SE1 reflector. This is in good agreement with the borehole results mentioned earlier.

When looking at the depth slices it is important to keep in mind that the volume resulted from the post migration stack of many single migrated volumes. Due to errors in the velocity model used or artificial noise perturbations, destructive stacking of weak or less coherent signals is unavoidable. Although the usage of the absolute values is a very robust way of obtaining a stacked image, it can compensate these errors only to a certain degree. Careful selection and stacking of a few single migrated volumes may produce a clearer image for certain subsurface structures of interest. This has been verified for this data set, too. Nevertheless, it is the main aim of this section to provide an overview of the dominant subsurface structures rather than to present single reflection elements at its best possible resolution.

Conclusions

3-D prestack migration has proven to be applicable to deep seismic reflection data. The ISO89-3D data set used here has only a low coverage (maximum 15-fold), compared with similar 2-D profiles (KTB8502: 80-fold). In spite of the corresponding low S/N ratio the slices through the migrated volume show the geometry and the shape of the dominant structures (SE1, EB) clearly. This is probably due to the used Kirchhoff-type migration, which handles each subsurface point as a possible diffractor. In combination with a postmigration stack of the absolute values this is known (SIMON et al., 1996) and has proven as an appropriate and robust technique yielding good results, especially in a crystalline environment.

This paper has demonstrated the prestack migration of the 3-D survey. In general, the results are in good agreement with those obtained by others, e.g., the poststack migration by KÖRBE et al. (1997).

Attractive and intriguing questions still remain open, for instance the estimation of a reflection coefficient for the SE1 reflector. Further investigations must incorporate the results of other experiments from the same area, e.g., the IN-STRUCT experiment for which a first lithological interpretation of the SE1 reflector has been given by HARJES et al. (1997). The final goal is to yield as much information as possible for a detailed qualitative and quantitative understanding of the geological and tectonic evolution within this area.

Acknowledgements

I am grateful to Manfred Stiller (GeoForschungsZentrum Potsdam) for his valued cooperation in preparing the ISO89-3D data set, to Michael Simon (Universität München) for making available the 3-D velocity model, and to Gerhard Müller for helpful discussions on the manuscript and the subject of this paper.

Further slices through the migrated volume corresponding to Figure 4 can be found at http://www.geophysik.uni-frankfurt.de/~stefan/work/ISO89-3D.

This work was supported by the Bundesminister für Bildung und Forschung (Grant 03GT94026, DEKORP Project).

REFERENCES

BUSKE, S. (1999), *Three-dimensional Prestack Kirchhoff Migration of Deep Seismic Reflection Data*, Geophys. J. Int. *137*, 243–260.

DÜRBAUM, H.J., REICHERT, C., and BRAM, K. (1990), *Integrated Seismics Oberpfalz 1989*, KTB Report 90–6b.

DÜRBAUM, H.J., REICHERT, C., SADOWIAK, P., and BRAM, K. (1992), *Integrated Seismics Oberpfalz 1989*, KTB Report 92–5.

EMMERMANN, R., and LAUTERJUNG, J. (1997), *The German Continental Deep Drilling Program KTB: Overview and Major Results*, J. Geophys. Res. *102* (B8), 18,179–18,201.

HARJES, H.P., BRAM, K., DÜRBAUM, H.-J., GEBRANDE, H., HIRSCHMANN, G., JANIK, M., KLÖCKNER, M., LÜSCHEN, E., RABBEL, W., SIMON, M., THOMAS, R., TORMANN, M., and WENZEL, F. (1997), *Origin and Nature of Crustal Reflections: Results from Integrated Seismic Measurements at the KTB Superdeep Drilling Site*, J. Geophys. Res. *102* (B8), 18,267–18,288.

KÖRBE, M., STILLER, M., HORSTMEYER, H., and RÜHL, T. (1997), *Migration of the 3-D Deep Seismic Reflection Survey at the KTB Location, Oberpfalz, Germany*, Tectonophysics *271*, 135–156.

PODVIN, P., and LECOMTE, I. (1991), *Finite Difference Computation of Travel Times in Very Contrasted Velocity Models: A Massively Parallel Approach and its Associated Tools*, Geophys. J. Int. *105*, 271–284.

SCHWARZENBÖCK, A. (1993), *Tomographische Laufzeitinversion, eine Methode der 3-D Geschwindigkeitsanalyse und deren Anwendung auf ISO89-Daten*, Master's Thesis, Institut für Allgemeine und Angewandte Geophysik, Universität München.

SIMON, M., GEBRANDE, H., and BOPP, M. (1996), *Prestack Migration and True-amplitude Processing of DEKORP Near-normal Incidence and Wide-angle Reflection Measurements*, Tectonophysics *264*, 381–392.

STILLER, M., *3-D vertical incidence seismic reflection survey at the KTB location, Oberpfalz*, Volume 22 of *Continental Litosphere: Deep Seismic Reflections*, Geodynamic Series (AGU, Washington, D.C. 1991) pp. 101–113.

(Received December 23, 1997, revised August 24, 1998, accepted September 22, 1998)

To access this journal online:
http://www.birkhauser.ch

Pure appl. geophys. 156 (1999) 173–186
0033–4553/99/010173–14 $ 1.50 + 0.20/0

❘ Pure and Applied Geophysics

3-D Prestack-migration of Wide-angle Data from a Variscan Transition Zone

F. Bleibinhaus,[1] M. Bopp,[1] M. Simon[1] and H. Gebrande[1]

Abstract—In addition to the near normal-incidence observations within the German DEKORP 2 project in 1984, wide-angle observations have been carried out on a parallel profile across the boundary between the Saxothuringian and Moldanubian crust, approximately 50 km NE of the main transect to control three-dimensional variations. Explosion sources have been used for the entire survey, providing excellent conditions for wide-angle registrations. A velocity model has been derived on the basis of in- and off-line refraction measurements using a kinematic raytracer which was extended to three dimensions by interpolation of 2-D velocity fields between parallel sections. Although prestack-migration of the data led to aliasing effects due to large shot and geophone spacing, stable results were obtained by forming envelopes after single-shot migration. The migrated sections reveal a strongly reflective Moho at about 31 km depth and a steeply (50°) dipping intracrustal reflector, which seems to be related to the border between the two Variscan units.

Key words: Variscan basement, 3-D wide-angle data, DEKORP, deep seismic sounding, crustal structure, prestack-migration.

Introduction

The West European Variscan belt is part of a Paleozoic orogen, reaching from the Appalachians in the southern USA and the Mauretanides in western Africa to the Bohemian Massif in the Czech Republic and Poland. It originated from the convergence of Laurentia-Baltica in the north and Africa in the south, which succeeded the predominant extension regime of the Early Paleozoic about 450 million years ago (Matte, 1991; Behr *et al.*, 1984). During subduction and the following collision, intervening terranes and microcontinents have been accreted, metamorphosed and partially uplifted. Due to post-Variscan crustal thinning the crust was reequilibrated, and it is difficult to find detailed evidence of the Variscan orogeny in southern Germany.

[1] Institut für Allgemeine und Angewandte Geophysik, Theresienstr. 41/IV, D-80333 München, Germany.

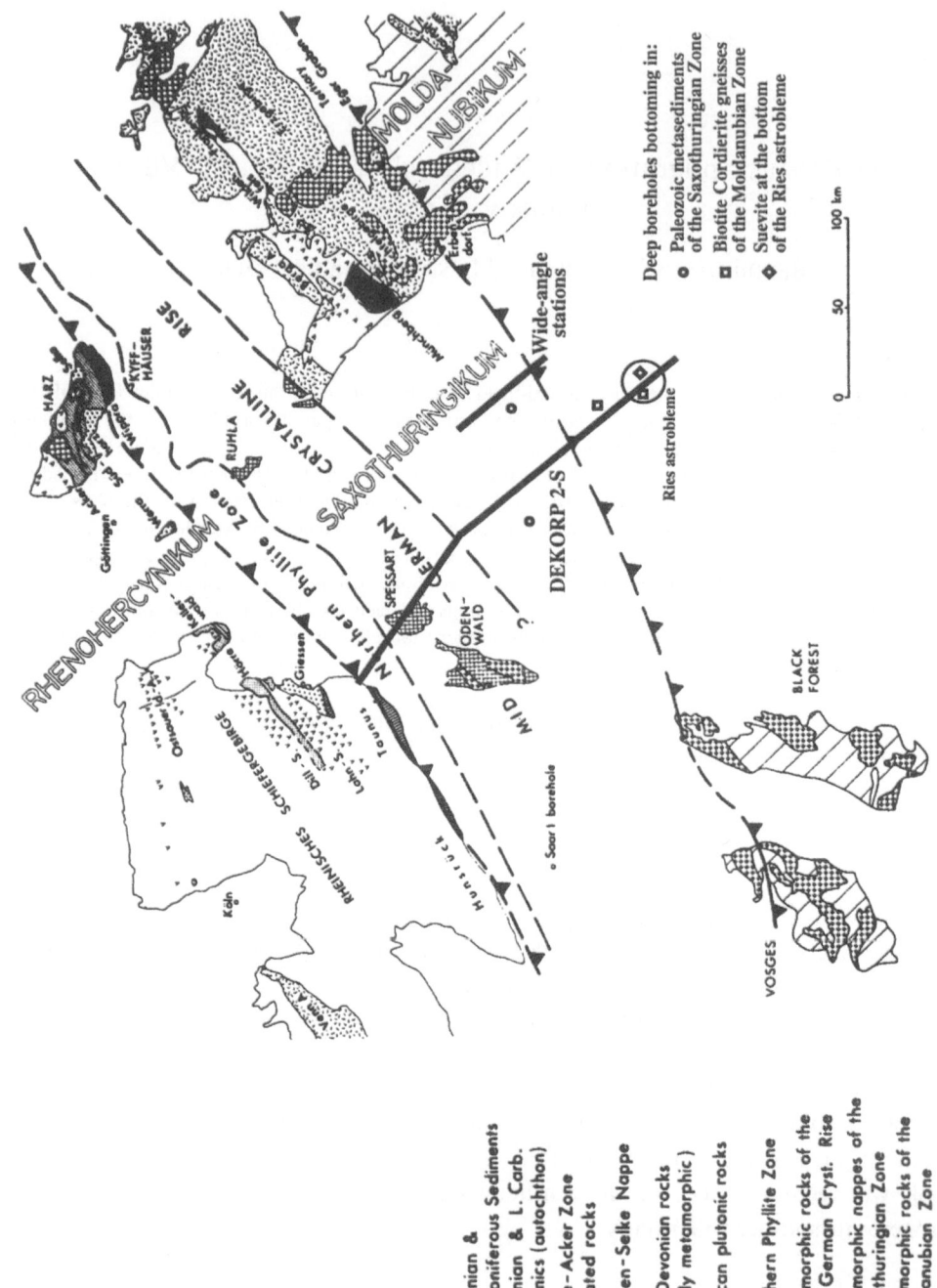

Geological Setting

The Saxothuringian (ST) consists of predominantly neritic metasediments and metavolcanics. The carboniferous HP/LT-metamorphism and the arrangement of

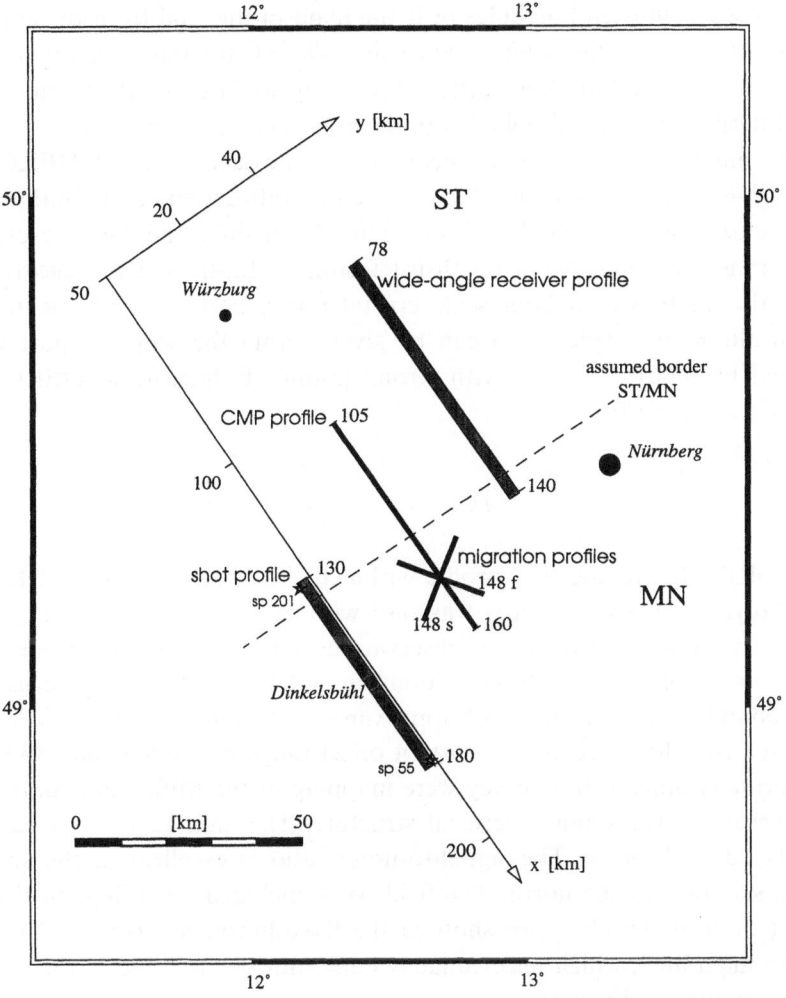

Figure 2
Map of the acquisition geometry.

Figure 1
Outcrops and structural division of the Variscan units in middle Europe (after BEHR *et al.*, 1984). In the region, where DEKORP 2-S crosses the border ST/MN, the basement is covered with Permomesozoic sediments. Deep boreholes indicate the location of this border (after BEHR and HEINRICHS, 1987).

some units indicate an origin by accretion. The Moldanubian (MN) comprises the crystalline core region of the Variscan belt (BEHR *et al.*, 1984; MATTE, 1991).

In the area of investigation, the basement is covered by mesozoic sediments with varying thickness of some hundred meters. The possible location of the ST/MN border can be interpolated from outcrops in the Vosges and the Black Forest in the west and the Bohemian Massif in the east (Fig. 1). More detailed evidence of its location arises from deep boreholes near the main profile and from the gravimetric map (GERKE, 1957) delineating an extensive SW-NE trending gravity rise, which can be related to the Variscan border. According to MATTE (1991), this gradient may be interpreted as a S-dipping slab of mafic rocks and gneisses.

In the migrated section of the near normal-incidence data of DEKORP 2-S (DEKORP RESEARCH GROUP, 1985) numerous diffractions and southward dipping reflection elements can be observed in the middle and lower crust at the ST/MN transition zone. Based on thrust tectonics, BEHR and HEINRICHS (1987) interpret this pattern as a large-scale crustal ramp, part of a continental suture zone. An alternative explanation can be given within the scope of post-Variscan extensional tectonics combined with strong granitoide intrusions (DEKORP RE-SEARCH GROUP, 1985).

Data Acquisition

DEKORP 2-S was the first profile within the German Continental Reflection Seismic Project. The entire survey was shot with explosive charges, thus providing excellent conditions for wide-angle observations These have been carried out with a mobile array of 24 MARS-66 stations in a 60-km long profile crossing the ST/MN boundary. At a distance of approximately 50 km from the main line, they were moved roll-along keeping a constant offset range between 55 and 80 km (Fig. 2). The primary aims of this survey were mapping of the Moho and controlling of three-dimensional variations of crustal structure. The data acquisition parameters are displayed in Table 1. The signal-to-noise ratio is excellent in the south and decreases seriously to the north. The field work included recordings in the Mold-anubian (south of Nürnberg) of shots in the Saxothuringian (between km 80 and 130 of the main line), which unfortunately could not be used due to their very bad signal-to-noise-ratio. Examples of the raw data presented by the DEKORP RE-SEARCH GROUP (1985) show a strong heterogeneity of the crustal structure that might be related to the assumed transition zone.

Preprocessing

The frequency modulated field data of the MARS-66 stations were demodulated and digitized. Subsequently the data were filtered (4/8–35\45 Hz), the spherical

Table 1

Data acquisition and preprocessing

Data acquisition	
Energy source	Dynamite
Hole depth	30 m
Charge	30 kg, every 4th shot 90 kg
Average shot spacing	320 m
Recording instrument	MARS 66
Geophone array	48 vertical
Average geophone spacing	500 m
Eigenfrequency	2.0 Hz
Offset range	55–80 km
Preprocessing	
Trapezoidal bandpass filter	4/8–35\45 Hz
Spherical divergence correction	$\sim t$
Interactive editing	
Static corrections to 400 m a.s.l.	Surface-consistent
Amplitude corrections	Surface-consistent

divergence was corrected by multiplying the amplitude with the traveltime and the data were edited interactively. Shot-statics were calculated using a 2-D velocity model of the uppermost 400 m derived by inversion of the first arrivals of the main line. Intercept times were computed from the true shot positions down to sea level, taking the ray parameter into account (WINKELMANN, 1996). The difference between these local intercept time values and a corresponding intercept time, calculated for a mean 1-D model between the reference datum 400 m above sea level and sea level, was applied as refraction statics (up to 160 ms in the Ries astrobleme). Subsequently, surface-consistent residual statics (up to 40 ms) were calculated for the shotpoints from the remaining delay times of the first arrivals. Due to the lack of information concerning the receiver side, an average velocity for the uppermost meters (1.5 km/s) has been determined correlating the topography with the time shift of the first breaks. This velocity was used to correct the receivers to the reference datum. Due to the large geophone spacing, attempts to calculate residual statics were not successful.

To achieve a quantitative image of crustal reflectivity, surface-consistent amplitude corrections have been applied (SIMON *et al.*, 1996). The average amplitude in a window from 0 to 5 s reduced traveltime was computed and averaged subsequently for common shotpoint gathers (A_i) and for the whole dataset (A_{tot}). To obtain a scaling factor for each shot I, A_{tot} was divided by A_i. After application of this correction factor the data were sorted according to common receivers. Scaling factors were calculated for each receiver j in the same way and applied subsequently. By this means, near-surface effects could be removed, in particular the different coupling of shots and receivers to the ground. As an alternative to the surface-consistent amplitude normalization, traces were scaled with an AGC of 3000 ms to focus on weak intracrustal reflections.

Velocity Model

A 3-D macro velocity model was developed starting from a 2-D model for the main line derived from in-line wide-angle registrations. Gravimetry and the general structure of the Variscan fold belt suggest uniformity of the velocities normal to the main line within several tens of kilometres. Nevertheless, 3-D raytracing in the uniformly extrapolated 2-D model could not duplicate the traveltimes measured on the parallel profile. To satisfy the needs of more realistic 3-D raytracing a new method of velocity modelling has been implemented in the kinematic raytracing software "RAYMUC-3D", developed by GEBRANDE (1975), NIXDORF (1986) and BOPP (1994). It is based on interpolation of velocities between parallel 2-D sections. In the 2-D sections velocities are defined by analytic functions along auxiliary lines and linear interpolation between them in z-direction. 2-D models are extended to the third dimension by a cosine-interpolation between neighbouring 2-D sections (Fig. 3). Outside the 2-D sections the 3-D model is defined by uniform extrapolation of the outermost sections.

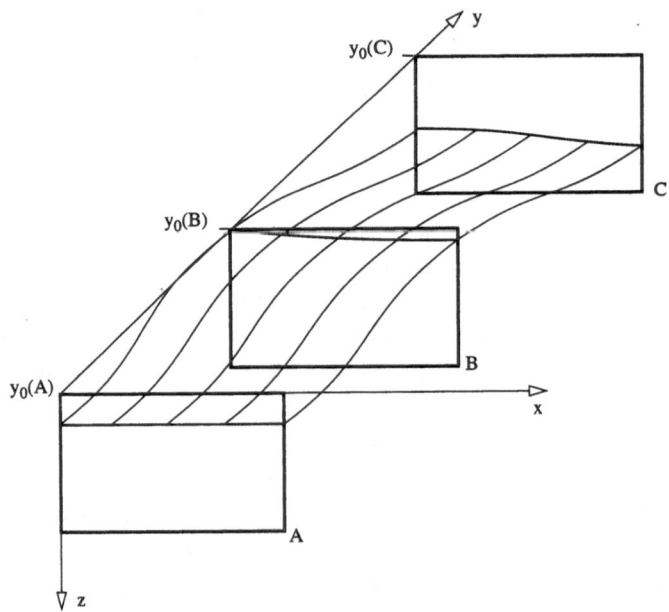

Figure 3

3-D velocity modelling between three parallel 2-D-sections (A, B, C). Each section must have the same number of auxiliary lines (for better visualization, they have only one). Velocities are defined analytically along these lines and interpolated between them.

For the area of investigation, the 3-D model is based on only two 2-D sections, one below the main reflection line DEKORP 2-S and one below the CMP line of the offline wide-angle registrations. The former section was derived from inline (wide-angle) observations and remained unchanged during forward modelling of the offline observations. The latter (Fig. 4), however, was adjusted by 3-D raytracing to match the offline traveltimes. The resulting 3-D model is able to reproduce the first arrivals, the PmP and the most important intracrustal reflections, with less than 100 ms misfit. This is deemed to be satisfactory for a 3-D macro velocity model. Common features of the 3-D model are

- high velocities in the Saxothuringian upper crust (reflected by early first breaks in the northern shot records, Fig. 8a) as compared to rather slow velocities in the Moldanubian zone,
- a relatively strong horizontal velocity gradient in the middle crust at the border between Saxothuringian and Moldanubian.

Lateral variations parallel to the Variscan strike direction are of minor importance: An additional velocity increase in the upper Saxothuringian crust and a deepening of the Moho by 1 km towards the NE are worth mentioning. Quantitatively and locally these features might be influenced by the model building technique, nonetheless their general trend is clearly proven by the offline data.

Therefore we trust that the derived 3-D macro velocity model meets the demands of a migration velocity model and that structural details can be resolved by 3-D migration of the measured wide-angle wavefield.

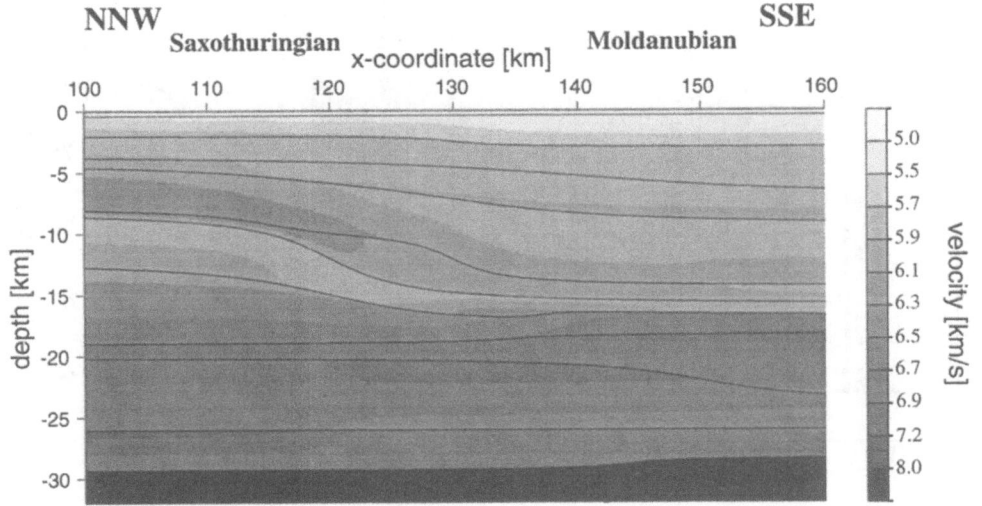

Figure 4
Section through the 3-D velocity model under the CMP line. In this region, high velocities characterize the Saxothuringian upper crust, low velocities the Moldanubian.

Seismic Imaging and Resolution

To obtain a seismic image, 3-D prestack-isochrone migration has been applied according to the principles described by SIMON (1993) and SIMON *et al.* (1996).

The process of seismic imaging requires that the Nyquist condition is satisfied in the time as well as in space. Considering the dominant signal frequency of about 12 Hz, the average incidence angle of 45° and the average angle between the raypaths and the receiver profile of approximately 45° and assuming a *P*-wave velocity of 6 km/s, the effective wavelength can be estimated as 1 km. The average geophone distance of 500 m is therefore not small enough to exclude aliasing effects.

To reduce these effects, forming the envelopes has been included in the migration process as a stabilizing method. It is applied after single-shot migration and before stacking. The danger of destructive interference of reflectors by erroneous shot-statics or velocities is thereby reduced. On the other hand, forming the envelopes causes a reduction of the frequency content and a loss of resolution.

Subsequently the envelope migrations are normalized to simulate a homogeneous subsurface coverage (SIMON *et al.*, 1996; TILLMANNS *et al.*, 1996). Henceforth, all presented migrated sections are normalized envelope-prestack-migrations.

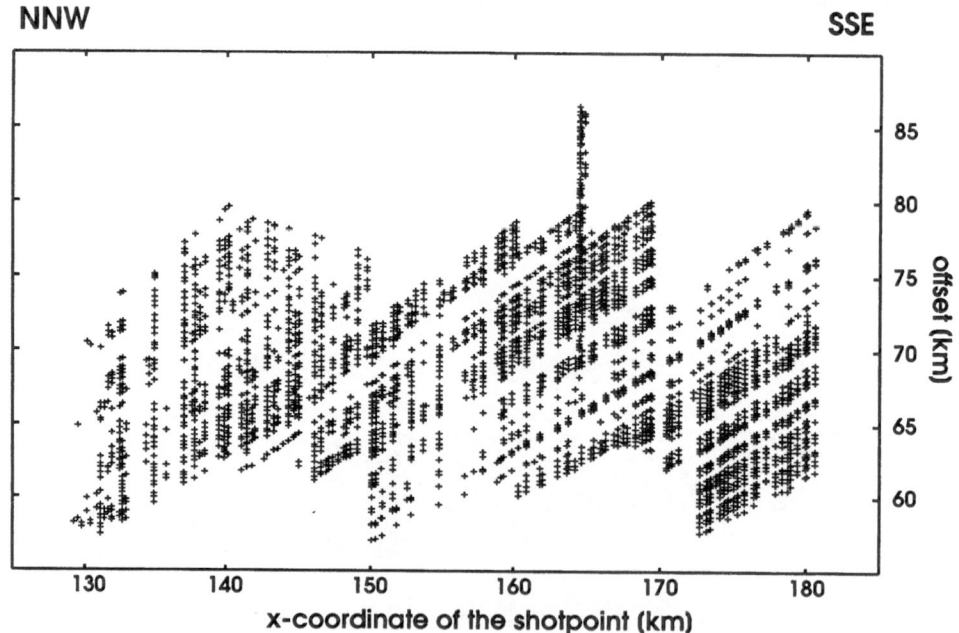

Figure 5

Offset versus *x*-coordinate of the shotpoints. Every migrated trace is marked with a " + ". Large gaps in the covered region may lead to artifacts in the migrated sections.

To test the resolving power of these envelope migrations and possible aliasing effects under the conditions of the real acquisition geometry (Fig. 5), synthetic seismograms have been computed with a Gabor source wavelet of 12 Hz for a constant velocity model with point diffractors below the CMP-line. These data were processed like the real data, but migrated with the known constant velocity model. The focusing of the diffractors in the resulting image (Fig. 6) gives an impression of the spatial resolution. It is poor for all sections above 10 km depth, as it must be expected for wide-angle data with minimum offsets of 55 km. Below this depth the resolution is satisfactory for the CMP and the 148f sections, still improving with

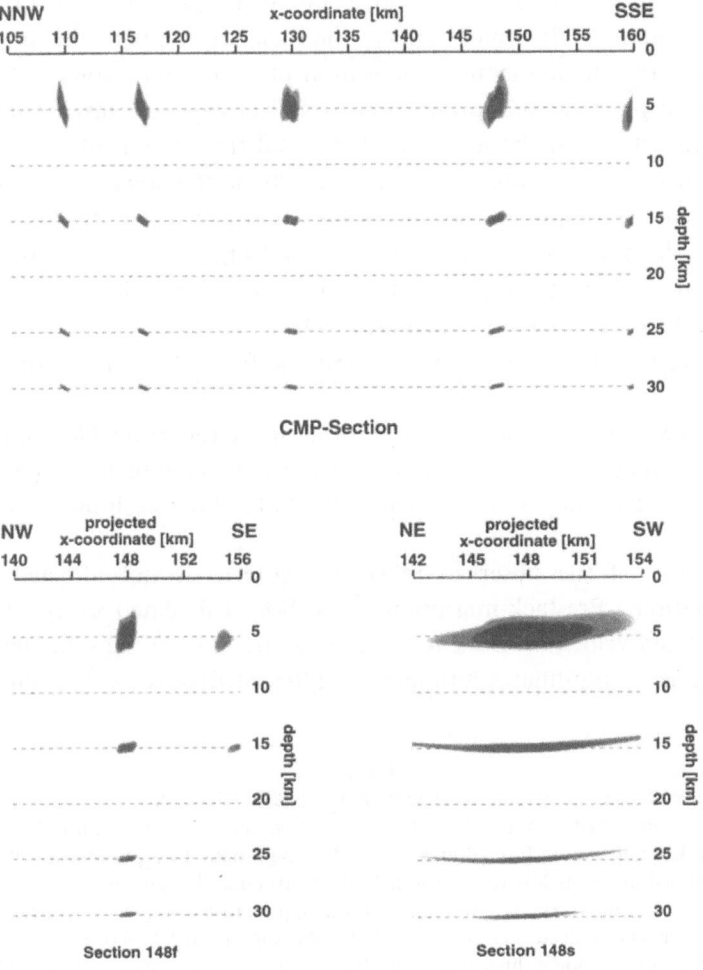

Figure 6

Prestack-migration of synthetic data with envelopes formed after single-shot migration. The diffractors were located under the CMP line at the x coordinates 110, 117, 130, 148 and 160 at depths of 5, 15, 25 and 30 km. Focusing of the diffractors indicates the quality of spatial resolution.

greater depths and to the centre of the sections. At the boundaries of these sections, the migrated diffractors indicate the shape of the isochrones caused by decreasing coverage. Section 148s, which is almost parallel to the average direction of the rays, shows poor horizontal resolution due to the acute angles (below 20°) with which the isochrones intersect. A full 3-D resolution could not be obtained, because the registrations of rays propagating NW-SE could not be included in the analysis due to their poor signal-to-noise ratio.

Figure 7 shows prestack envelope migrations of the real data. In the CMP section (Fig. 7a), amplitudes have been scaled with an AGC of 3000 ms during preprocessing, whereas in section 148f (Fig. 7b) the real amplitudes have been preserved. The transparent lower crust in the CMP section is a typical result of the AGC scaling (the amplitudes above and below strong reflectors, here it is the PmP, are suppressed), which reduces the dynamic range of the amplitudes but thus facilitates the simultaneous visualization of weak and strong reflections. The other section displays the real proportions of reflectivity: A transparent upper crust and higher reflectivity in the lower crust. Crustal thickness is about 30 to 31 km. The high reflectivity beneath x coordinate 120 in the lower crust indicates an inclined reflector (D) dipping into the Moho. Although it may be superimposed by smiles, especially due to the decreasing coverage to the north, it can be regarded as real: One of the northern shotpoints (Fig. 8a) shows a reflection at 2.5–3 s reduced time (distinctly different from the PmP), which is responsible for reflector (D). Below reflector (D), the Moho seems to continue further north at a depth of about 30 km.

Furthermore, one can observe two intracrustal reflectors (A) and (C) and a region of high reflectivity in the upper crust (B) in prolongation of reflector (A). This reflectivity (B) is not well established due to the low resolution in the upper 10 kilometres.

The stability of the observed reflection patterns is not affected by modest velocity variations. Prestack-migrations have been calculated using 1% decreased and 1% increased velocities. Both resulting sections showed the same features A–D below the same x coordinates although in different depths (± 1–2 km).

Figure 7

(a) Prestack-migration of the off-line wide-angle data of DEKORP 2-S in a section under the CMP profile. The data were scaled with an AGC of 3000 ms before migration to enhance the visualization of weak reflectors. Envelopes were formed after single-shot migration. Two prominent reflectors (A) and (C) can be identified in the middle crust. The reflectivity (B) could be regarded as prolongation of the reflector (A), however due to the low resolution in the upper crust it might be a migration artifact as well. The reflectivity (D) indicates an inclined reflector dipping into the Moho, but is superimposed by numerous smiles. Colour coding: high reflectivity (red—blue—green—yellow—white) low reflectivity. (b) Prestack-migration of the southern part of the data in section 148f (see Fig. 2) with real amplitudes. Although the Moho reflectivity had to be clipped for better visibility of the intracrustal reflector (A), this section shows a more reliable distribution of crustal reflectivity than the AGC-scaled section. This section is in the direction of the true dip of the reflector (A) of approximately 50°.

The true dip (50°) of reflector (A) is displayed in Figure 7b. The reflection can be identified in a corresponding shotpoint (Fig. 8b). Attempts were made to verify the reflector by introducing a subsurface plain at its location into the velocity model. The derived traveltime curve proves indeed, that reflector (A) is caused by

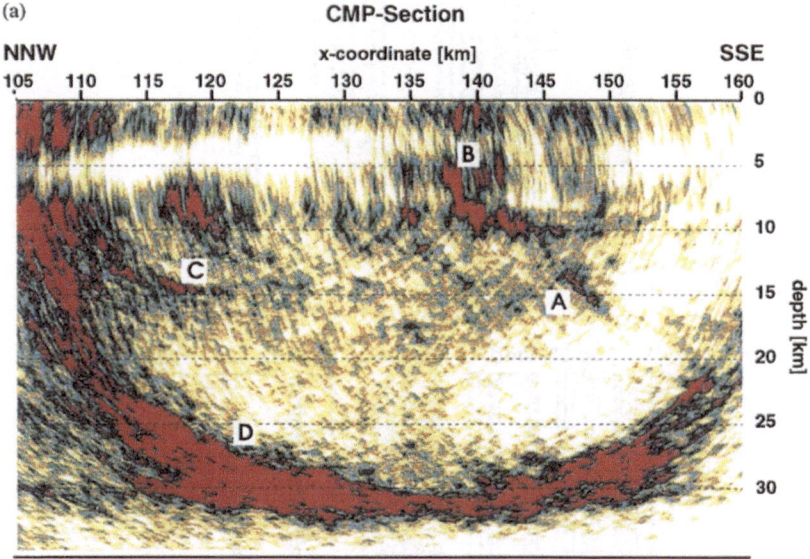

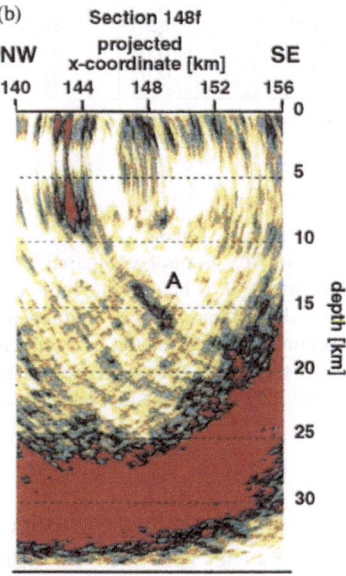

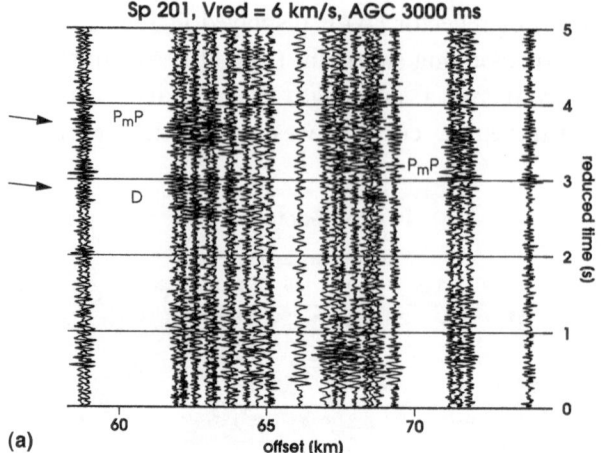

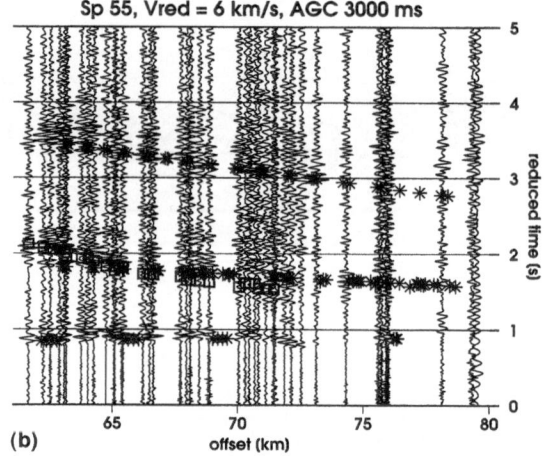

Figure 8

(a) Shotpoint 201 (see Fig. 2). Data quality is poor compared to shotpoints in the south (Fig. 8b). A prominent reflection (D) approximately 1 s before the PmP causes the inclined reflector (D) in the migrated section (Fig. 7a). Calculations with a simplified model proved the traveltimes. (b) Shotpoint 55 (see Fig. 2). Traveltimes were calculated by raytracing in the velocity model as shown in Figure 4 (crosses). Hollow squares symbolize reflections from a subsurface plane at the location of reflector (A) which has been introduced in the velocity model.

that reflection, however the calculated apparent velocity does not favourably fit the observed. This means that the location of optimal superposition of the measured reflection perhaps is not beneath the CMP profile, but some kilometres away.

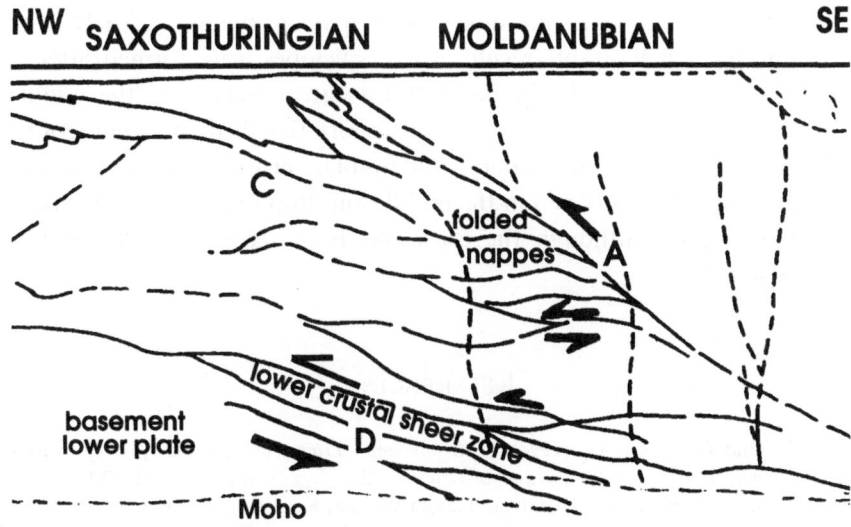

Figure 9

Structural model of the transition zone ST/MN, modified after a preliminary structural model based on a thrust concept derived by the DEKORP RESEARCH GROUP (1985) from an automatic line drawing of the migrated steep-angle data of DEKORP 2-S. The position of the reflection elements A, C and D (see Fig. 7) is projected onto it along the Variscan strike direction.

Geological Interpretation

A possible explanation for the observed reflections can be given in a framework of thrust tectonics. Figure 9 illustrates a slightly altered detail of a preliminary structural model based on that concept (DEKORP RESEARCH GROUP, 1985). It is located approximately 25 km WSW of the CMP profile in Variscan strike direction, and the structures discussed above (A, C, D) have been projected onto it. Within this framework the reflector (A) could be seen as part of a probably tilted thrust plane. The high amplitudes of the reflector (D), which might indicate strong tectonic strain (BEAUMONT and QUINLAN, 1994), can be interpreted as part of a lower crustal shear zone where subduction may have taken place. The observation of similar structures at a distance of 25 km justifies the two-dimensional interpretation of the DEKORP 2-S main line and further supports the idea that the transition zone ST/MN could be considered as a main intracontinental suture zone, remnant of Variscan subduction of Saxothuringian under Moldanubian crust.

Conclusions

The investigated off-line wide-angle arrangement eventuated to provide useful data for velocity analysis as well as for structural imaging. Despite incomplete azimuthal coverage and large shot and receiver spacing, the applied 3-D isochrone prestack-migration led to stable results when envelopes were formed after single-shot migration. Prominent SE-dipping reflectors are observed in the middle and lower crust of the migrated section. They support the thrust-tectonic concept of subduction of Saxothuringian under Moldanubian crust. It is remarkable that steeply dipping reflectors survived in the middle and lower crust of the Variscan belt in Southern Germany, although the crust has been reequilibrated due to post-Variscan extension.

REFERENCES

BEAUMONT, C., and QUINLAN, G. (1994), *A Geodynamic Framework for Interpreting Crustal-scale Seismic-reflectivity Patterns in Compressional Orogens*, Geophys. J. Int. *116*, 754–783.

BEHR, H.-J., ENGEL, W., FRANKE, W., GIESE, P., and WEBER, K. (1984), *The Variscan Belt in Central Europe: Main Structures, Geodynamic Implications, Open Questions*, Tectonophysics *109*, 15–40.

BEHR, H.-J., and HEINRICHS, T. (1987), *Geological Interpretation of DEKORP 2-S: A Deep Seismic Reflection Profile across the Saxothuringian and Possible Implications for the Late Variscan Structural Evolution of Central Europe*, Tectonophysics *142*, 173–202.

BOPP, M. (1994), *RAYMUC 3-D, Benutzeranleitung für ein kinematisches 3-D-Raytracingprogrammpaket*, Internal Report, Ludwig-Maximilians-Universität, München, 58 pp.

DEKORP RESEARCH GROUP (1985), *First Results and Preliminary Interpretation of Deep-reflection Seismic Recordings along Profile DEKORP 2-South*, J. Geophys. *57*, 137–163.

GEBRANDE, H. (1975), *Ein Beitrag zur Theorie thermischer Konvektion im Erdmantel mit besonderer Berücksichtigung der Möglichkeit eines Nachweises mit Methoden der Seismologie*, Ph.D. Thesis, Ludwig-Maximilians-Universität München, 159 pp.

GERKE, K. (1957), *Die Karte der Bouguer-Isanomalen 1:1.000.000 von Westdeutschland*, Frankfurt/Main, Institut für Angewandte Geodäsie.

MATTE, PH. (1991), *Accretionary History and Crustal Evolution of the Variscan Belt in Western Europe*, Tectonophysics *196*, 309–337.

NIXDORF, U. (1986), *Teleseismische Laufzeitresiduen in Bayern von 1979 bis 1984 und ihre Auswertung mittels strahlenseismischer Modellrechnung*, Diploma Thesis, Ludwig-Maximilians-Universität München, 218 pp.

SIMON, M. (1993), *Entwicklung eines 3-D-Migrationsverfahrens mit Anwendungen auf seismische Daten aus dem Umfeld der KTB Oberpfalz*, Ph.D. Thesis, Ludwig-Maximilians-Universität München, 186 pp.

SIMON, M., GEBRANDE, H., and BOPP, M. (1996), *Prestack Migration and True-amplitude Processing of DEKORP Near-normal Incidence and Wide-angle Reflection Measurements*, Tectonophysics *264*, 318–392.

TILLMANNS, M., SIMON, M., ZITZELSBERGER, A., and GEBRANDE, H. (1996), *Neue Seismographien aus dem Umfeld der Kontinentalen Tiefbohrung (KTB), Oberpfalz*, Geologica Bavarica *101*, 291–314.

WINKELMANN, R. (1996), *Entwicklung und Anwendung eines Wellenfeldverfahrens zur Auswertung von CMP-sortierten Refraktionseinsätzen*, Ph.D. Thesis, Ludwig-Maximilians-Universität München, 150 pp.

(Received February 12, 1998, revised December 18, 1998, accepted December 28, 1998)

 To access this journal online:
http://www.birkhauser.ch

Pure appl. geophys. 156 (1999) 187–206
0033–4553/99/010187–20 $ 1.50 + 0.20/0

Pure and Applied Geophysics

Focusing in Prestack Isochrone Migration Using Instantaneous Slowness Information

MARK TILLMANNS[1] and HELMUT GEBRANDE[1]

Abstract—Prestack migration finds increasing application in processing crustal seismic data. However, less effort has been made to incorporate slowness information in the imaging process. The combination of slowness information with migration leads to an improved image in the depth domain, especially by reducing migration artefacts and noise. A slowness-driven isochrone migration scheme is introduced for migration of 2-D seismic data. Instantaneous slowness information $p(x, t)$ is extracted from the data using correlation analysis in moving time and space windows. Slowness values resulting from spatial coherent energy (signal) and incoherent background noise are distinguished by the simultaneous evaluation of an instantaneous coherence criterion $g(x, t)$. In slowness-driven isochrone migration this information is used for locally weighting the amplitude $A(x, t)$ smearing on the isochrone surface. In particular, slowness p and coherence criterion g determine position and sharpness of a Gaussian weighting function. The method is demonstrated using two synthetic data examples and is subsequently applied to two deep crustal data sets, one wide-angle (along DEKORP4) and one steep-angle reflection seismic observation (KTB8506). Both data sets were collected in the surroundings of the KTB drill site, Oberpfalz, as part of the German DEKORP project.

Key words: Prestack migration, instantaneous slowness, migration artefact, crystalline crust, DEKORP.

Introduction

Modern prestack migration algorithms can treat data from areas of relatively complex structures. Besides on correct velocities, their success strongly depends on a sufficient coverage of the illuminated subsurface. The migration algorithms highly depend on the superposition principle such that the reflection amplitudes add up at the correct subsurface points and destructively interfere at all other locations. Migration images of low-fold data or data with limited observation geometry are often corrupted by migration artefacts. Low amplitude reflections are obscured with smeared noise and noise bursts migrate into artificial smiles.

In this paper we demonstrate how to improve the superposition condition by including directional characteristics in a migration process based on the isochrone

---
[1] Institut für Allgemeine und Angewandte Geophysik, Universität München, Theresienstr. 41, D-80333 München, Germany. E-mail: gebrande@seismik.geophysik.uni-muenchen.de

principle. In a first step the observed wavefield is analysed for its instantaneous slowness content by means of a multichannel cross-correlation analysis which yields a slowness and coherence value for each data sample. The main prestack depth migration algorithm uses slowness and coherence information for individually weighting each amplitude smearing along the isochrone surface. Coherent signal energy will be focused back to its reflecting or diffracting subsurface location, while incoherent energy will be defocused and rejected by weighting.

Integrating directional properties into the imaging process is nearly as old as the concept of prestack migration itself. JAIN and WREN (1980) suggested a combination of Kirchhoff migration and likelihood weighting to improve the signal-to-noise ratio of common shot migration schemes. MILKEREIT (1987a) combined multichannel coherence filtering by means of shift-and-sum processes with diffraction stack migration. This method uses normalized local slant stack slownesses to weight the conventional diffraction stack. Other migration algorithms are based on slowness transformed data sets (PHINNEY and JURDY, 1979; HUBRAL, 1981; TEMME, 1984) or on Gaussian beam transformed data sets (HILL, 1990; ALKHALIFAH, 1995). These approaches use tomographic back projection by mapping beams of slowness components back into the subsurface in accordance with their directional properties. Similar approaches in array seismology are described by HEDLIN et al. (1991, 1994) and SCHERBAUM et al. (1997). JACKSON et al. (1991) and TAKAHASHI (1995) presented migration schemes for multicomponent data in which polarization information is combined into the imaging process. These multicomponent methods do not make use of spatial coherence information and thus do not distinguish between the focusing of signal or noise energy.

The imaging procedure we propose uses two independent processing steps: Slowness analysis followed by slowness-driven isochrone migration. Their principle concepts will be discussed in the subsequent sections. The main advantage of the intermediate processing state is its possibility for quality control and/or further processing steps. The underlying isochrone concept allows mapping reflections of any dip and curvature and may handle any velocity field. The integration of instantaneous slowness and instantaneous coherence information into migration yields strong suppression of migration artefacts while simultaneously providing excellent noise rejection. This is demonstrated using synthetic data followed by real data examples which were collected as part of the German DEKORP project.

Concept of Instantaneous Slowness-analysis

Instantaneous slowness (ray parameter or reciprocal apparent velocity) information $p(x, t)$ can be extracted from spatially dense, multichannel seismic data. We may think of $p(x, t)$ as additional information on the actual data $A(x, t)$ likewise a time and space varying seismic attribute. MILKEREIT (1987b) introduced the application of localized slant stacks, weighted by coherence, to produce a decomposition of the seismic data into single-trace instantaneous slowness $p(x, t)$ components. We propose the application of multichannel cross correlation followed by linear regression analysis. This approach, elaborated by HASLINGER (1994), avoids the discretization of the range of slowness values and avoids by this discrete directions of signal focusing in slowness-driven isochrone migration. We will briefly summarize its basic principles (Fig. 1):

Consider a data sample A at time t from trace x, called the reference trace. Define a time window of length δt approximately equal to the dominant signal wavelength which is centred around t. We call it the reference window. Calculate the cross-correlation functions of the reference window with each trace segment of a multichannel (N) dip-limited data subset centred around the reference trace (marked bold in Fig. 1a). The time-shifts, which are associated with the maxima of

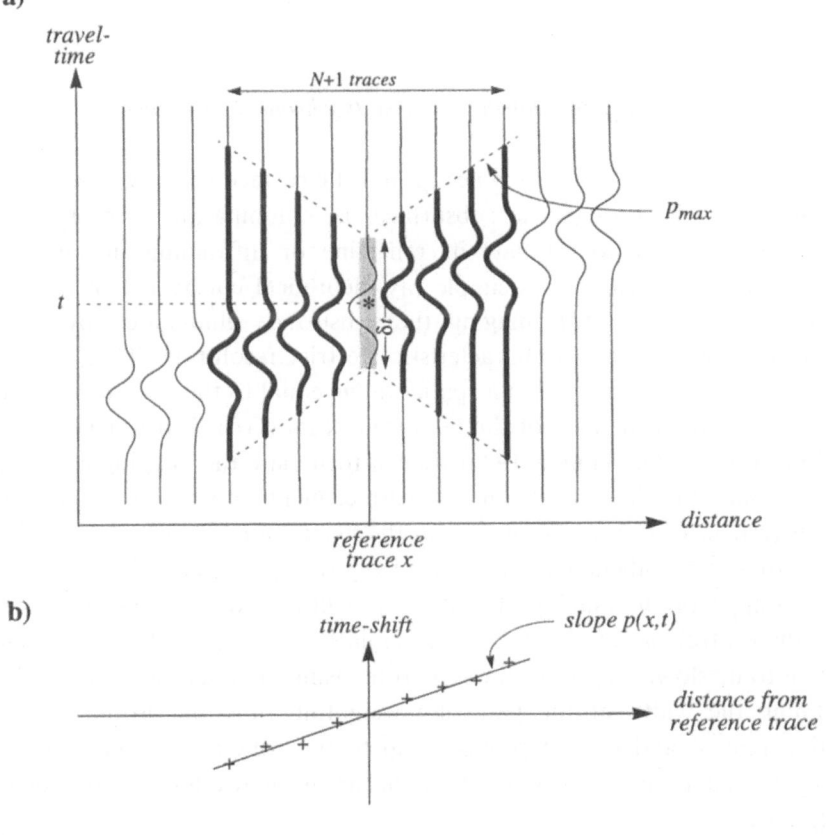

Figure 1
Principle of slowness-analysis by multichannel cross correlation (a) and linear regression (b). The shaded region along the reference trace marks the extent of the reference window, the neighbouring trace segments marked bold indicate the dip-limited sequences for cross correlation, respectively.

the cross-correlation functions and plotted against the trace distances, are input to a linear least-squares regression analysis (Fig. 1b). The slope of the best regression line will give the local slowness component $p(x, t)$ for the present analysed data sample $A(x, t)$. Coherent signal slownesses will be distinguished from incoherent noisy slownesses by means of a coherence (quality) criterion $g(x, t)$ $(0 \leq g \leq 1)$, which is calculated using the correlation and regression results (see Appendix A). The process is repeated for each data sample. The user must specify the number of traces N around the reference trace (depending on the spatial extent of linear signal coherence in the data), the maximal searchable dip in terms of a maximal slowness value p_{max} and the length of the reference window δt. The output consists of three traces (data, slowness and quality) for each seismic channel. At the moment, in case of crossing events we evaluate only a single instantaneous slowness; the slowness of the dominant reflection branch. A multimodal slowness function with respect to each time step might be evaluated, if we would use further (local) maxima of the cross-correlation function as input to further linear least-squares regression analysis.

Concept of Slowness-driven Isochrone Migration

With the knowledge of the slowness p and the surface velocity v we may follow a seismic ray back through the subsurface. In combination with the isochrone condition we are able to estimate its reflecting or diffracting subsurface point, assuming single scattering. This simple ray-theoretical consideration may be suitable for modelling, not for imaging the subsurface illuminated by wavefields. Likewise in optics, wavefield characteristics restrict resolution. Thus we can only locate the subsurface point within a spatial range equal to the resolution limit of the used wavefield. With the help of the Gaussian beam (GB) theory (CERVENÝ et al., 1982; CERVENÝ, 1985; WEBER, 1986) we can formulate the imaging condition. GB theory combines ray theoretical concepts with elements of wave theory by considering the wavefield from a beam of rays in the neighbourhood of a central ray. The ray amplitude perpendicular to the central ray decreases according to a Gaussian bell. In an appropriate imaging procedure we will consider a beam of rays centred around the central ray of slowness p or similarly, focus the reflected amplitude A according to its slowness p onto the isochrone using Gaussian weights W (Fig. 2a). By defining the width of the Gaussian weighting function proportional to the resolution limit σ and inverse proportional to the coherence value g, we achieve focusing of signal energy $(g \rightarrow 1)$ within the limits of resolution while defocusing noise $(g \rightarrow 0)$.

The method is based on the isochrone prestack migration algorithm discussed by SIMON et al. (1996) and in detail by SIMON (1993). The traveltime t of the sample to be added into a specific underground bin $\tilde{x}_0$ is geometrically calculated

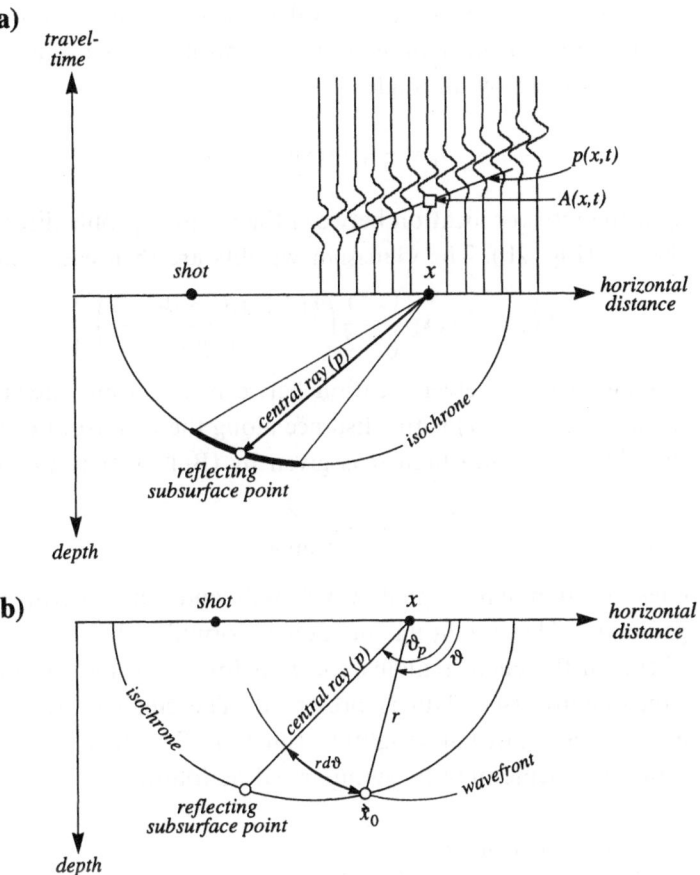

Figure 2

(a) Principle of slowness-driven isochrone migration, depicted for homogeneous subsurface. A sample A will be focused onto the isochrone according to its slowness p using Gaussian weights. The lateral extent of significant weights is marked schematically with bold pen size. (b) Parameters involved in computing the weights along the isochrone.

using an average velocity v_{mig} (migration velocity), hence, avoiding expensive ray tracing:

$$t = \frac{1}{v_{mig}} \cdot (R + r). \tag{1}$$

R and r denote the distances from $\check{x}_0$ to the shot and receiver, respectively. The migration velocities, as defined in the work by SIMON (1993), are expressed prior to migration as a mean function of space. They are calculated with the help of migration velocity analysis and/or ray tracing on a sparse grid, and are subsequently interpolated onto the binned underground. Robustness of migration with straight rays (equation (1))—even in strongly inhomogeneous velocity fields—has been proven several times (e.g., SIMON et al., 1996). In slowness-driven isochrone migration we can further use the migration velocities to replace (without significant departure, see Appendix B) the actually curved central ray associated with p by a straight central ray defined by an angle ϑ_p:

$$\vartheta_p = \frac{\pi}{2} + \arcsin(pv_{\mathrm{mig}}). \tag{2}$$

ϑ_p is measured at the receiver location between the positive profile direction and the direction of the ray (Fig. 2b). The Gaussian weights are then easily computed by:

$$W(\check{x}_0) = \frac{g}{\sigma} \cdot \exp\left\{ -\frac{1}{2}\left(\frac{r(\check{x}_0) \cdot (\vartheta_p - \vartheta(\check{x}_0))}{\sigma/g} \right)^2 \right\} \tag{3}$$

ϑ denotes the angle at the receiver location between the profile direction and the direction to $\check{x}_0$. $rd\vartheta = r \cdot (\vartheta p - \vartheta)$ is the distance along the wavefront of the fictitious beams (Fig. 2b). The resolution limit σ is given by (BERKHOUT, 1984)

$$\sigma = \frac{1}{2} \frac{\lambda}{\sin \alpha} \tag{4}$$

where λ denotes the dominant signal wavelength and $\sin \alpha$ the aperture of the observation geometry. The weighting function is formulated using the constraint, that the area beneath the curve remains constant for any values of p and g. That is, energy of the preprocessed data is preserved. The smaller the width σ/g the higher and sharper the Gaussian weighting function. The main algorithm of the developed migration scheme may be summarized as follows:

For all traces $\check{x}$ of a seismic section {
    for all subsurface bins $\check{x}_0$ of the migration array {
        calculate traveltime $t = t(\check{x}, \check{x}_0)$ using equation (1)
        calculate weight $W = W(\check{x}, \check{x}_0, t)$ using equations (2) and (3)
        add weighted amplitude $W(\check{x}, \check{x}_0, t) \cdot A(\check{x}, t)$ to the value of the bin $\check{x}_0$
    end}
  end}

Application to Synthetic Data

To demonstrate the method a common steep- and wide-angle seismogram section (Fig. 3a) was calculated using dynamic ray tracing (CERVENÝ et al., 1977). The laterally homogeneous velocity-depth model (shown in Fig. 3a, top right) represents typical, crystalline basement, including high- and low-velocity zones. The impulse response of the model has been convolved with a simple 10 Hz Ricker wavelet. PS-converted waves from the discontinuities at 7, 10 and 16 km are simulated. Trace spacing of the 501-channel section is 100 m, giving a maximal offset of 50 km.

The result of the slowness-analysis ($N = 6$, $\delta t = 120$ ms and $p_{max} = 0.2$ s/km) is shown in Figure 3b. The slownesses are displayed using a grey-scale density plot, values ranging from 0 s/km (light grey) to 0.2 s/km (dark grey). No slowness

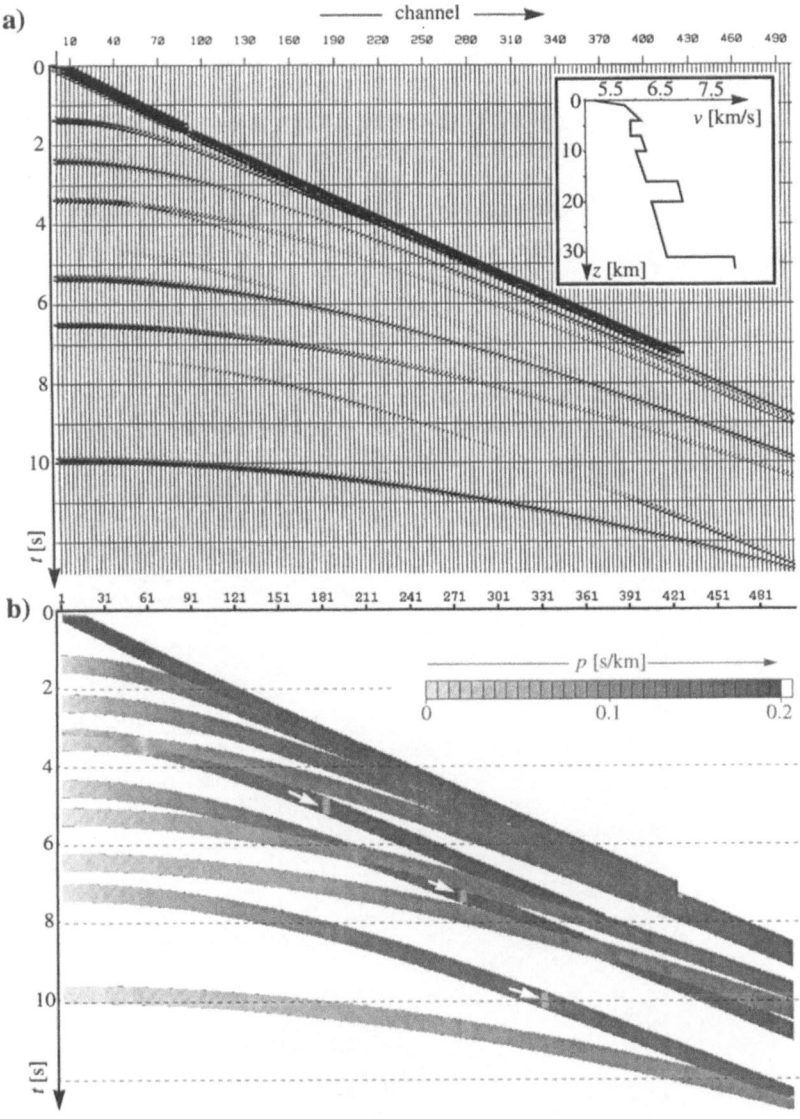

Figure 3

(a) Synthetic seismogram section of the velocity curve shown top right. (b) Result of the slowness-analysis with parameters $N = 6$, $\delta t = 120$ m s and $p_{max} = 0.2$ s/km. Increasing slowness values are displayed from light to dark grey. No slowness information is obtainable from zero data trace segments (white parts). Phase changes of the PS-converted waves (marked with arrows) cause distortion of slowness values.

information is obtainable from zero data trace segments. Those parts are displayed in white colour. Lateral slowness variations are correctly recovered as is indicated by the smooth colour transition along the reflection curves. Cross-overs of travel-time branches do not severely damage the overall image, while phase changes do (arrows in Fig. 3b).

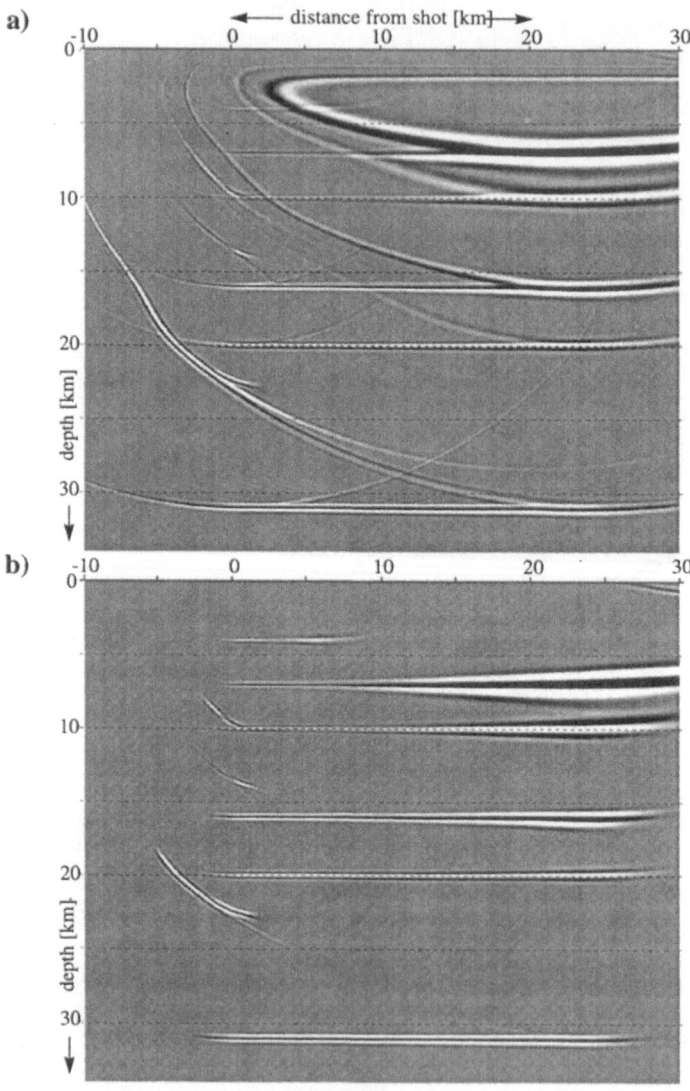

Figure 4
Migration of synthetic data of Figure 3a. The migration is performed using isochrone migration (a) and slowness-driven isochrone migration (b). Results are displayed as variable density plot, negative to positive values from white to black. The images are plotted for better comparison with same average rms-amplitude. Migration artefacts are impressively suppressed in (b).

Migration results are shown in Figure 4. Processing prior to migration included correction for spherical divergence, removal of first breaks and in the case of Figure 4b, analysing for the slowness content as depicted in Figure 3b. Migration velocities were calculated out of the given velocity-depth model. Migration artefacts (smiles), which heavily superimpose the actual subsurface structure on the simple isochrone migration image (Fig. 4a), are impressively suppressed by slowness-driven isochrone migration (Fig. 4b). In both cases converted waves are wrongly imaged and stretching effects degrade the image of wide-angle reflections. The result of Figure 4b implies further that the method may be applied successfully to steep- as well as wide-angle configurations.

The second synthetic example is more sophisticated and closer to the real data example of DEKORP4 in terms of velocity variation and acquisition geometry. Again dynamic ray tracing (Cervený et al., 1977) was used to calculate wide angle data (Fig. 5b) along a laterally heterogeneous velocity-depth model (Fig. 5a). Shot locations move from profile coordinate 5.6 km to profile coordinate 45.6 km at intervals of 2.5 km. Each shot is observed to the right in an offset range from 30 to 50 km. Trace spacing is 100 m and again a 10 Hz Ricker wavelet is used to convolve the impulse response of the model for direct and reflected P waves.

Processing of the synthetic data set included removal of first breaks and correction for spherical divergence. Slowness-analysis ($N = 6$, $\delta t = 100$ ms and $p_{max} = 0.2$ s/km) was performed on the preprocessed data set. Migration velocities (shown in Fig. 6c) were calculated out of the given velocity-depth model (Fig. 5a). All preprocessed shots have then been migrated into depths from 5 to 35 km using isochrone migration and slowness-driven isochrone migration. The subsequent stack of all single shot migrations is shown in Figures 6a and 6b, respectively. The result of Figure 6b shows that slowness-driven isochrone migration is able to handle laterally heterogeneous velocity fields and to image correctly strongly dipping events (C in Fig. 5a). Slowness-weighting still favours the specular part of the isochrone. The image quality is, despite increased subsurface coverage, superior to the result of simple isochrone migration (Fig. 6a). Stretching effects associated with the migration of wide-angle data along with the phase difference between reflection branches A and B (Fig. 5b) decrease the imaging resolution for reflectors A and B (Fig. 5a) equally in both migration results. Slight undulations of the Moho reflector image (E in Fig. 5) are due to migration velocity approximations. The influence of an erroneous velocity model on the migration result appears to be dominated by the isochrone condition (equation (1)). A study of these effects is undertaken by SIMON (1993).

Application to Field Data

In 1985 the German DEKORP project conducted several deep steep- and wide-angle observations in the Oberpfalz area in order to explore a central part of

a)

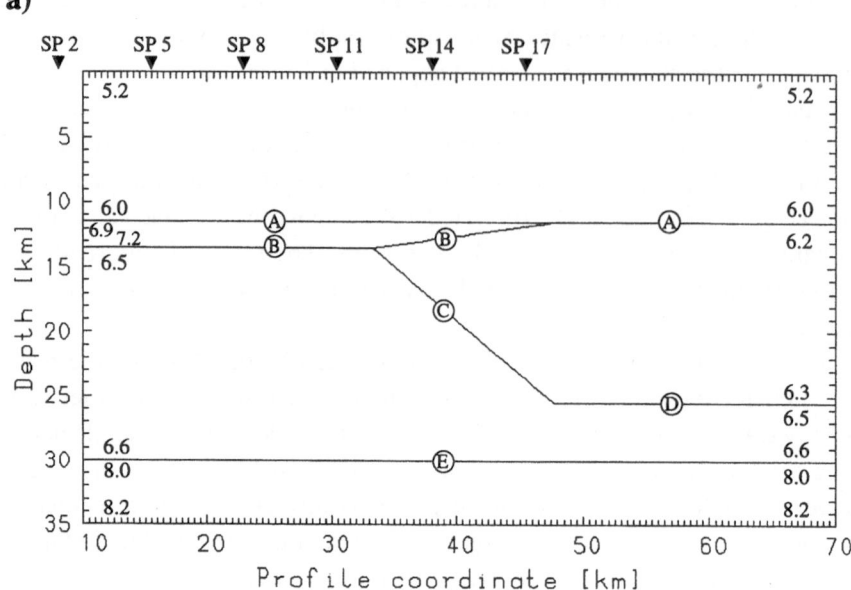

b)

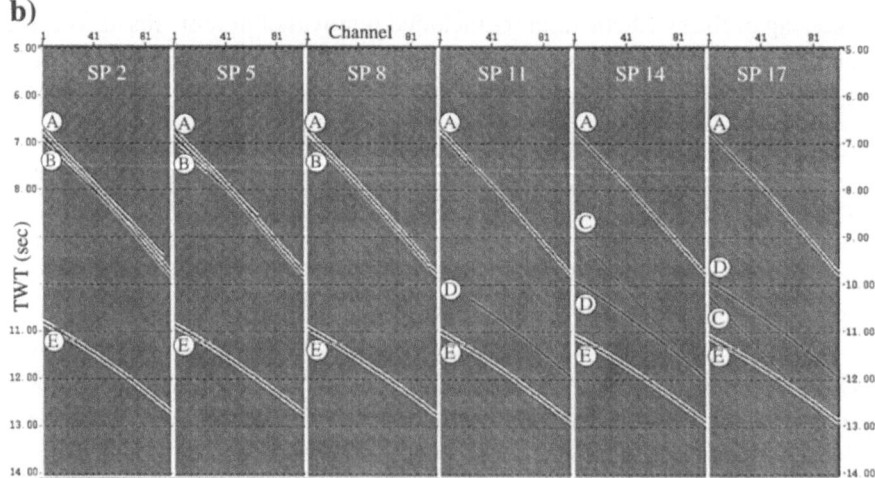

Figure 5
Laterally heterogeneous velocity model (a) and subset of the modelled synthetic data (b). (a) Velocities
within the layers increase linearly between the given top and bottom values (in km/s). The model extends
laterally homogeneous beyond the right and left margins. Discontinuities are labelled from A to E.
Characteristic features of the model are: a high velocity zone between A and B, a dipping reflector C,
the Moho E. (b) Variable density plot of a subset of the synthetic shot records. Shot locations are
marked in (a) (geophon locations are to the right of each shot, from 30 to 50 km offset). Reflection
branches are labelled with the same letters as the reflecting interfaces of (a). First breaks are muted.

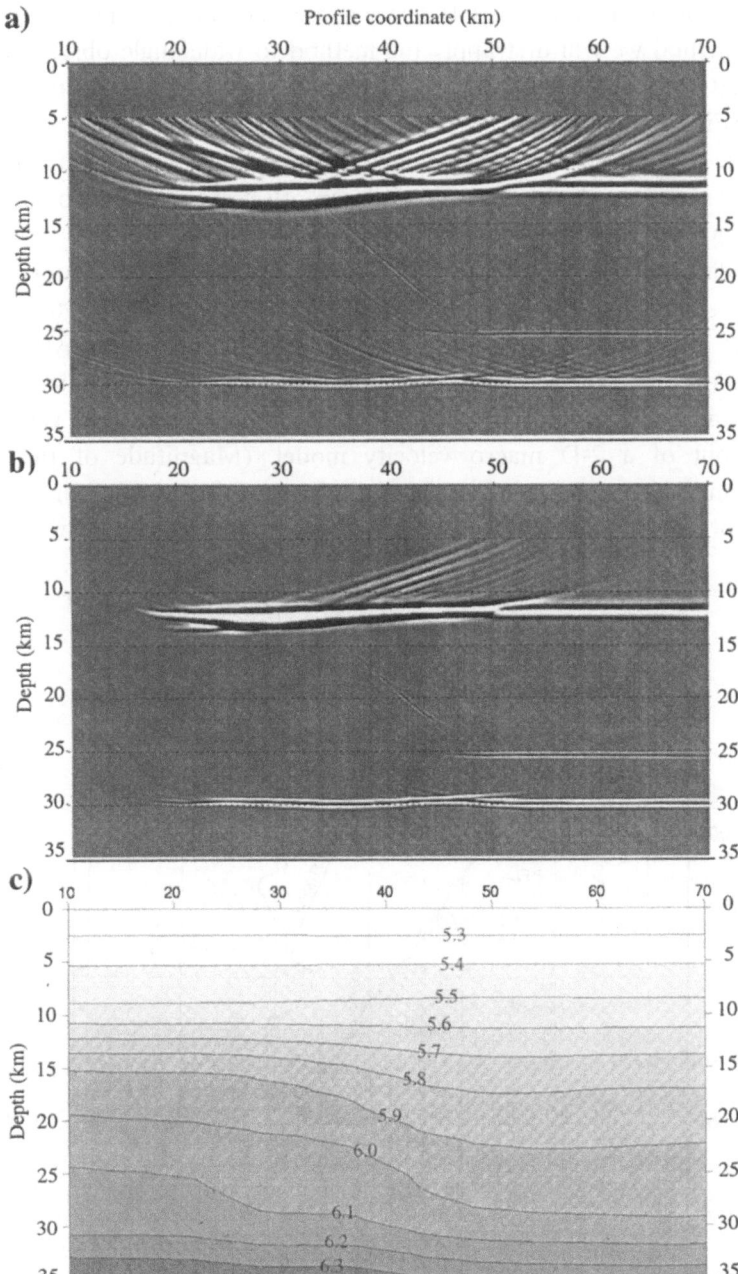

Figure 6

Results of isochrone migration (a) and slowness-driven isochrone migration (b) for the synthetic data of Figure 5, plotted in variable density (negative to positive values from white to black). Migrations were performed in the depth range from 5 to 35 km. (c) Migration velocity field (contours in km/s) used for migration.

the Variscan Belt of Europe. To demonstrate the imaging power of the new migration method we will first apply the method to wide-angle observations along the DEKORP4 profile and afterwards to a portion of the steep-angle profile KTB8506 (Fig. 7).

The wide-angle measurements included 96 explosions, each separated by about 1 km. Every shot was observed southeast in an offset range of approximately 42 to 58 km using a 200-channel equipment, giving a nominal fold of 16. Amplitude preserving data processing successively included careful editing, bandpass filtering (10/40 Hz), spherical divergence correction, amplitude corrections to account for the coupling of sources and receivers to the ground, predictive deconvolution, static corrections and trace interpolation. First breaks have not been muted in order to keep wide-angle reflections at quite similar traveltimes. Migration velocities were calculated out of a 2-D macro velocity model. (Magnitude of the migration velocities, smoothness and trend of the variations are like that of Fig. 6c.) Details of the pre-processing can be found in SIMON (1993). A typical slowness image is

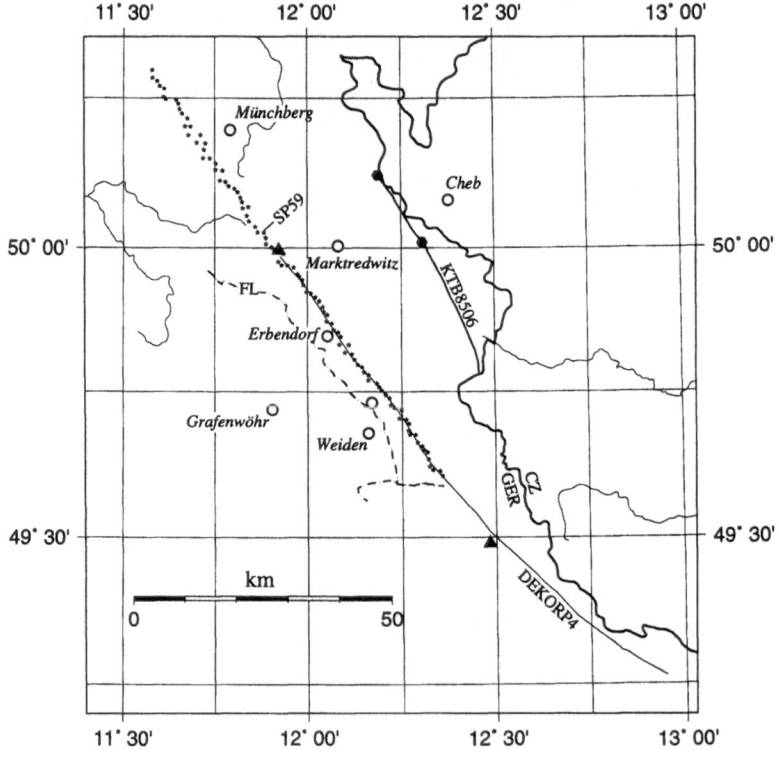

Figure 7

Map showing location of steep-angle reflection profile KTB8506 and wide-angle reflection profile observed on the main DEKORP4 traverse. Stars mark wide-angle shotpoints. The main geological fault zone of the Franconian Line (FL) is sketched for orientation. Filled triangles and hexagons indicate the horizontal extent of the migration arrays used later.

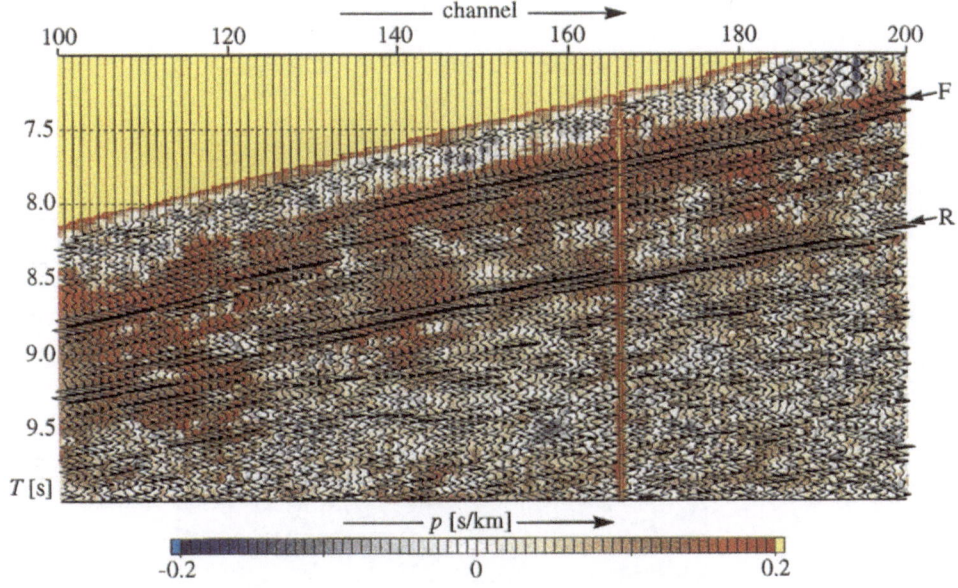

Figure 8

Colour coded slowness plot of the wide-angle shot 59 (marked in Fig. 7) with a wiggle plot of the pre-processed data (AGC-scaled for displaying purposes) superimposed. The display is confined to 100 channels and trace segments from 7 to 10 s including first breaks (F) and prominent wide-angle reflections (R), which originate from the well known but less understood Erbendorf body (e.g., GEBRANDE et al., 1989; HIRSCHMANN, 1996). Slowness value range from negative values in blue to positive values in red shades. No slowness information is obtainable from dead data trace segments (yellow colour).

shown in Figure 8 ($N = 4$, $\delta t = 104$ ms, $p_{max} = 0.2$ s/km). Slownesses are colour coded, negative values in blue, positive values in red shades. No slowness information is obtainable from edited or muted dead data trace segments (yellow colour). A wiggle plot of the pre-processed data is superimposed. Coherent signals are well analysed, each colour transition reflects changes in the slope. Noisy parts are characterised by frequent colour fluctuations.

80 of the 96 shots have been migrated and subsequently stacked using isochrone migration (Fig. 9) and slowness-driven isochrone migration (Fig. 10). The horizontal extent of the migration array is shown in Figure 7 (triangles). The depth of the migration images starts at 5 km, as we cannot expect structural resolution at lower depths due to grazing incidence of the wide-angle observations and associated stretching effects. To compare the performance of the two migration schemes, the images are scaled to identical average rms-amplitudes. Most striking are the significantly different migration noise levels. In Figure 9 considerable structural information is blurred by noise. Interpretation becomes difficult. On the other hand, Figure 10 easily allows identification of less prominent but quite important reflections. Typical examples are the SE dipping events (D) in the depth range of 15

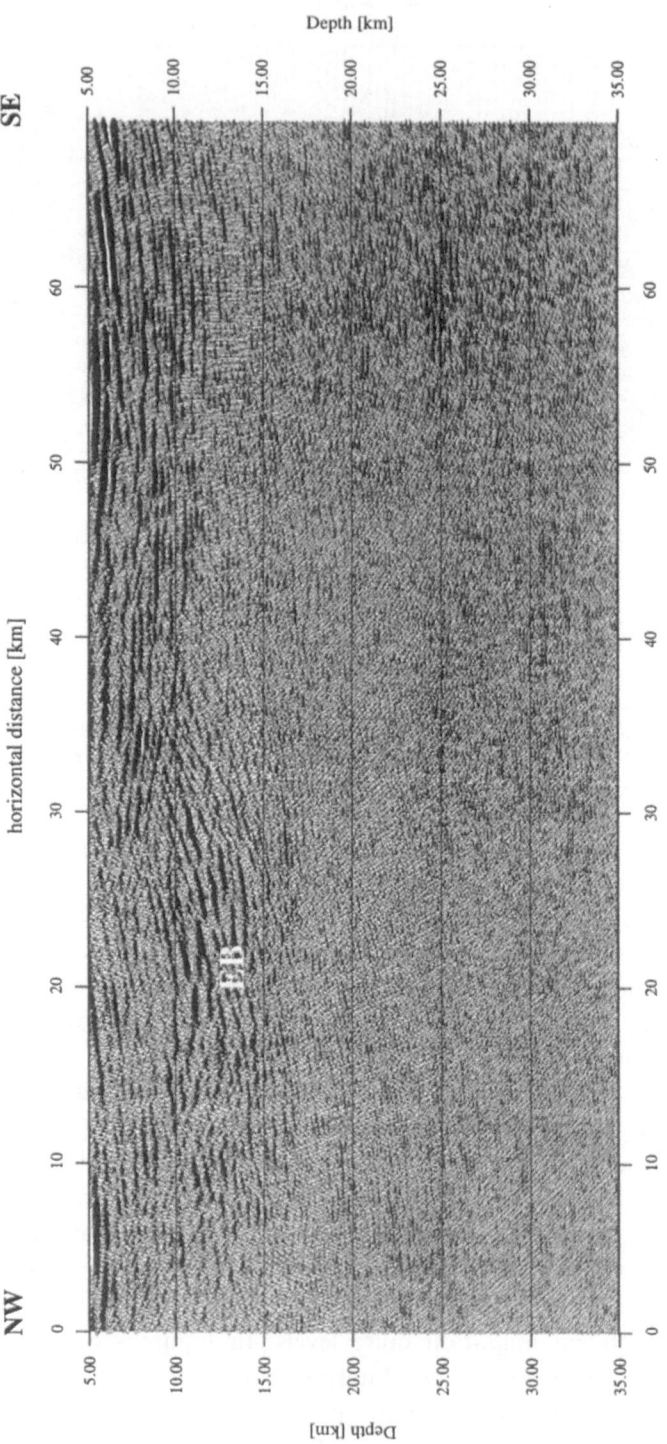

Figure 9

Isochrone migration of the wide-angle observations along DEKORP4. The horizontal extent of the migration array is depicted by triangles in Figure 7. EB: Erbendorf body.

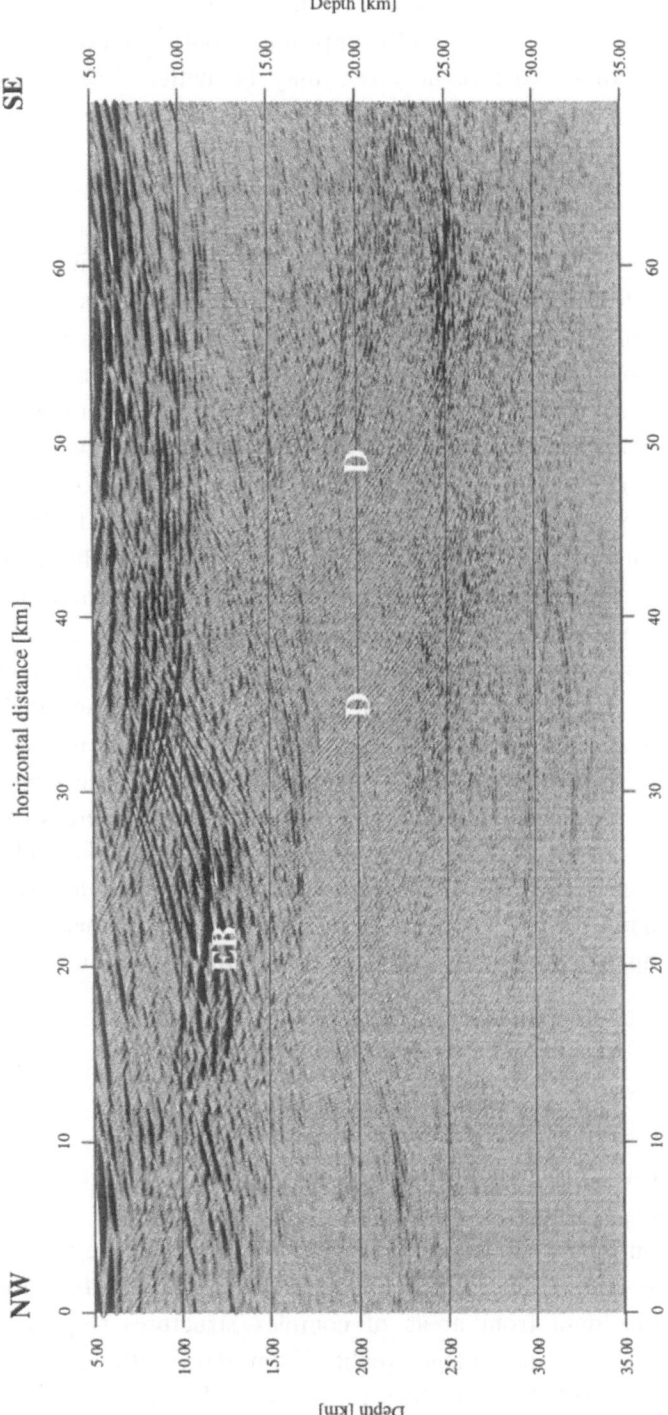

Figure 10

Slowness-driven isochrone migration of the wide-angle observations along DEKORP4. Same migration array as in Figure 9. EB: Erbendorf body; D: dipping events.

to 25 km. The crustal segment at a depth of 8 to 13 km, the Erbendorf body (EB) (e.g., GEBRANDE et al., 1989), appears in both migrations as a dominant feature. Its internal structure however, may be better deduced from slowness-driven isochrone migration (Fig. 10), as reflection elements are more continuously imaged.

The second example demonstrates that the new method is still superior, even if the data are spatially closely sampled with high coverage of the subsurface. The data used were collected with equal source and receiver spacing (80 m) along KTB8506 (Fig. 7). Each vibroseis shot was observed on a 200-channel asymmetric split-spread configuration (offsets from -4 to $+12$ km) giving a nominal fold of 100. The amplitude preserving pre-processing is described in TILLMANNS et al. (1996). It successively included editing, muting of first breaks, amplitude correction for source and receiver coupling to the ground, bandpass filtering, static corrections and spherical divergence correction. Migration velocities were derived primarily by migration velocity analysis (AL-YAHYA, 1989).

Prestack migration was performed on 131 shots at the NW end of the profile, again using isochrone and slowness-driven isochrone migration after slowness-analysis ($N = 6$, $\delta t = 72$ m s, $p_{max} = 0.2$ s/km). The horizontal extent of the migration array is depicted in Figure 7 (hexagons). Trace envelopes (TANER and SHERIFF, 1977) have been computed prior to stacking single-shot migrations, yielding a more robust but less resolved subsurface image after stacking (Fig. 11). The migration images are presented as grey-scale density plots. The slowness-driven isochrone image (Fig. 11b) exhibits spatially stronger confined reflecting elements, as is the case in isochrone migration (Fig. 11a). The reflection element of lystric course in Figure 11a (mid-image) appears rather as a simple SE dipping event in Figure 11b and may be further connected to the equally dipping structure at the NW margin of the migration array (not seen on Fig. 11a).

Conclusions

Slowness-driven isochrone migration avoids migration artefacts (Fig. 4b, Fig. 11b), reducing by this the possibility of misinterpretation. It further strongly rejects noise on migration images (Fig. 10), giving a more detailed and differentiated structure image. The method is an especially powerful tool in mapping low-fold seismic data from areas of complex structures (Fig. 10), and it also significantly improves the image quality from data with high coverage of the subsurface (Fig. 11b). We therefore recommend the method, especially in situations in which structural interpretation is difficult, as is mostly the case in mapping crystalline crust.

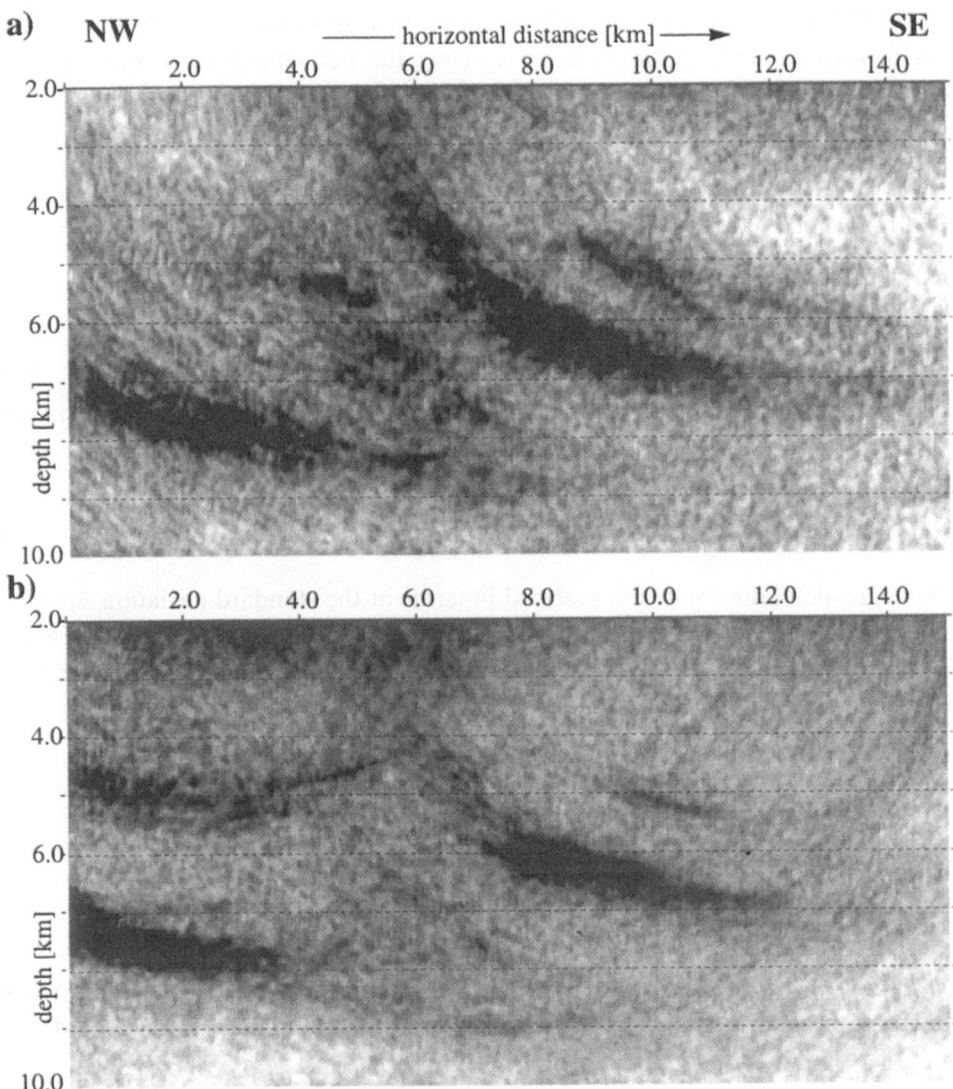

Figure 11
Prestack migration section from steep-angle profile KTB8506 with envelope stack after single shot migration. Migration is performed using isochrone migration (a) and slowness-driven isochrone migration (b). Amplitudes are presented as grey-scaled density plot, low reflectivities in white to high reflectivities in black. The images are scaled to identical average rms-amplitudes. The horizontal extent of the migration array is shown in Figure 7 (hexagons).

Acknowledgements

This study was carried out within the framework of the DEKORP program funded by the German Federal Ministry of Science and Technology under grants 03

GT 94067. We are grateful to Dipl.-Geophys. F. Haslinger, now ETH Zürich, who has performed most of the work in realizing the new approach in instantaneous slowness-analysis.

Appendix A

The quality g is multiplicatively composed of a correlation criterion g_{cor} and a regression criterion g_{reg}

$$g = g_{cor} \cdot g_{reg} \tag{A1}$$

both falling in the range $[0, 1]$. The correlation criterion evaluates the average of the N maxima $r_{max,n}$ of the normalised cross-correlation functions:

$$g_{cor} = \frac{1}{N} \sum_{n=1}^{N} r_{max,n}. \tag{A2}$$

The regression criterion g_{reg} is specified in terms of the standard deviation Δp of the slope of the regression line. Normalisation onto the interval $[0, 1]$ is difficult. Evaluating the relative error $\Delta p/p$ indicates unstable behaviour if $p \to 0$. Instead, g_{reg} is defined as follows

$$g_{reg} = \begin{cases} \dfrac{\Delta p_{min}}{\Delta p} & \text{if} \quad \Delta p > \Delta p_{min} \\ \\ 1 & \text{if} \quad \Delta p \leq \Delta p_{min} \end{cases} \tag{A3}$$

where Δp_{min} is evaluated as the standard error in slope, which statistically results from the discretization error associated with the discrete time-shifts provided by cross correlation.

Appendix B

With the help of migration velocities (defined by SIMON, 1993) we want to replace the actual curved central ray by a straight ray. Thus we must ask whether we can identify the observed slowness p with the slowness $p_{mig} = \sin i_{mig}/v_{mig}$ of the straight ray (Fig. 12). For the 1-D-case we can analytically derive the following inequality

$$p_{mig} = \frac{\Delta \cdot t}{\Delta^2 + z^2} \geq p \tag{A4}$$

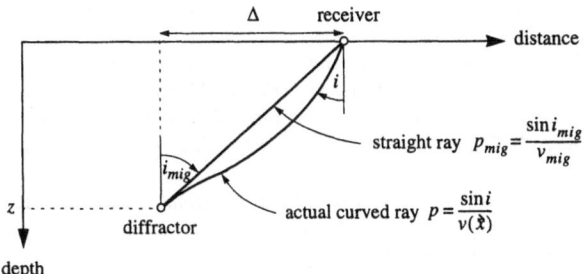

Figure 12
Replacing the actually curved central ray by a straight ray, using migration velocities.

where t is the traveltime of the curved ray and Δ and z are defined as in Figure 12. This inequality solved for z yields:

$$|z| \lesssim \Delta \cdot \sqrt{\frac{t}{p\Delta} - 1}. \tag{A5}$$

From this inequality we know (GIESE, 1976) that it provides very good estimates even in cases of strong velocity variations. Thus, we may expect sharp estimates $p_{mig} \cong p$.

REFERENCES

ALKHALIFAH, T. (1995), *Gaussian Beam Depth Migration for Anisotropic Media*, Geophysics *60*, 1474–1484.

AL-YAHYA, K. (1989), *Velocity Analysis by Iterative Profile Migration*, Geophysics *54*, 718–729.

BERKHOUT, A. J., *Handbook of Geophysical Exploration. Section 1: Seismic Exploration. Vol. 12, Seismic Resolution* (Geophysical Press, London 1984).

CERVENÝ, V., MOLOTKOV, I. A., and PSENCIK, I., *Ray Method in Seismology* (Univerzita Karlova, Praha 1977).

CERVENÝ, V. (1985), *Gaussian Beam Synthetic Seismograms*, J. Geophys. *58*, 44–72.

CERVENÝ, V., POPOV, M. M., and PSENCIK, I. (1982), *Computation of Wave Fields in Inhomogeneous Media—Gaussian Beam Approach*, Geophys. J. Roy. Astr. Soc. *70*, 109–128.

GEBRANDE, H., BOPP, M., NEURIEDER, P., and SCHMIDT, T., *Crustal structure in the surroundings of the KTB drill site as derived from refraction and wide-angle seismic observations*. In *The German Continental Deep Drilling Program (KTB)* (eds. EMMERMANN, R., and WOHLENBERG, R.) (Springer-Verlag, Berlin 1989) pp. 151–179.

GIESE, P., *Depth calculation*. In *Explosion Seismology in Central Europe* (eds. GIESE, P., PRODEHL, C., and STEIN, A.) (Springer-Verlag, Berlin 1976) pp. 146–161.

HASLINGER, F. (1994), *Ein Verfahren zur automatisierten Bestimmung von Scheingeschwindigkeiten aus Mehrspurseismogrammen*, Diploma thesis, Ludwig-Maximilians-Universität München, 80 pp.

HEDLIN, M. A. H., MINSTER, J.-B., and ORCUTT, J. A. (1991), *Beam-stack Imaging Using a Small Aperture Array*, Geophys. Res. Lett. *18*, 1771–1774.

HEDLIN, M. A. H., MINSTER, J.-B., and ORCUTT, J. A. (1994), *Resolution of Prominent Crustal Scatterers near the NORESS Small-aperture Array*, Geophys. J. Int. *119*, 101–115.

HILL, N. R. (1990), *Gaussian Beam Migration*, Geophysics *55*, 1416–1428.

HIRSCHMANN, G. (1996), *KTB—The Structure of a Variscan Terrane Boundary: Seismic Investigation— Drilling—Models*, Tectonophysics *264*, 327–339.

HUBRAL, P. (1981), *Seismic Slant Stack Migration*, Geol. Jb. *E21*, 5–11.

JACKSON, G. M., MASON, I. M., and LEE, D. (1991), *Multicomponent Common-receiver Gather Migration of Single-level Walk-away Profiles*, Geophys. Prosp. *39*, 1015–1029.

JAIN, S., and WREN, A. E. (1980), *Migration before Stack—Procedure and Significance*, Geophysics *45*, 204–212.

MILKEREIT, B. (1987a), *Migration of Noisy Crustal Seismic Data*, J. Geophys. Res. *92*, 7916–7930.

MILKEREIT, B. (1987b), *Decomposition and Inversion of Seismic Data—An Instantaneous Slowness Approach*, Geophys. Prosp. *35*, 875–894.

PHINNEY, R. A., and JURDY, D. M. (1979), *Seismic imaging of deep crust*, Geophysics *44*, 1637–1660.

SCHERBAUM, F., KRÜGER, F., and WEBER, M. (1997), *Double Beam Imaging: Mapping Lower Mantle Heterogeneities using Combinations of Source and Receiver Arrays*, J. Geophys. Res. *102*, 507–522.

SIMON, M. (1993), *Entwicklung eines 3-D-Migrationsverfahrens mit Anwendung auf seismische Daten aus dem Umfeld der Kontinentalen Tiefbohrung Oberpfalz*, Ph.D. Thesis, Ludwig-Maximilians-Universität München, 186 pp.

SIMON, M., GEBRANDE, H., and BOPP, M. (1996), *Pre-stack Migration and True-amplitude Processing of DEKORP Near-normal Incidence and Wide-angle Reflection Measurements*, Tectonophysics *264*, 381–392.

TAKAHASHI, T. (1995), *Prestack Migration Using Arrival Angle Information*, Geophysics *60*, 154–163.

TANER, M. T., and SHERIFF, R. E., *Application of amplitude, frequency and other attributes of stratigraphic and hydrocarbon determination. In Seismic Stratigraphy, AAPG Mem. 26* (ed. PAYTON, C. E.) (American Association of Petroleum Geologists, Tulsa 1977) pp. 301–327.

TEMME, P. (1984), *A Comparison of Common-midpoint, Single-shot, and Plane Wave Depth Migration*, Geophysics *49*, 1896–1907.

TILLMANNS, M., SIMON, M., ZITZELSBERGER, A., and GEBRANDE, H. (1996), *Neue Seismographien aus dem Umfeld der Kontinentalen Tiefbohrung KTB Oberpfalz*, Geologica Bavarica *101*, 291–314.

WEBER, M. (1986), *Die Gauß-Beam Methode zur Berechnung theoretischer Seismogramme in absorbierenden inhomogenen Medien: Test und Anwendung*, Ph.D. thesis, Universität Frankfurt/Main, 141 pp.

(Received April 12, 1999, accepted April 14, 1999)

 To access this journal online:
http://www.birkhauser.ch

Pure appl. geophys. 156 (1999) 207–232
0033–4553/99/010207–26 $ 1.50 + 0.20/0

© Birkhäuser Verlag, Basel, 1999

⌐ Pure and Applied Geophysics

An Attempt to Integrate Reflection Seismics and Balanced Profiles

D. GAJEWSKI,[1] A. LAMBRECHT[1,2] and O. ONCKEN[3]

Abstract—Different techniques in Geophysics and Geology are used to derive the structure of the subsurface. They are based on different data sets, i.e., seismic and geological data, and a combination of these techniques should produce better earth models. The case study presented in this paper is based on data of the German Continental Reflection program (DEKORP) collected in the Münsterland basin and the Rhenish Massif located at the northern border of the Rhenohercynian fold and thrust belt of the Mid-European Variscides. In this study we present an attempt to integrate balanced profiles, i.e., structural geology, and reflection seismics. The integration is performed by synthetically modelling seismic waves according to the acquisition of the field data, where the velocity model is based on the balanced profile. The synthetic data are compared with the field observations. Differences between observed data and field data are either caused by velocity errors in the model or by errors in the balanced profile. Criteria are developed to interpret these differences in order to improve the joint model of geologists and geophysicists. The case study presented in this paper shows that the combination of balanced profiles and reflection seismics may lead to shortcuts in the determination of seismic velocities of the subsurface. These shortcuts can reduce processing times and processing costs of reflection seismic data.

Key words: Balancing, reflection seismic, modelling, prestack depth migration, velocity determination.

Introduction

In this study we investigate the possibility of obtaining a subsurface model by combining different methods from different disciplines. Therefore, an attempt (carried out within the framework of the German Continental Reflection Program DEKORP) was made to integrate reflection seismics and structural geology. The general procedure of the integration is as follows (Fig. 1). A balanced geological profile is used as input for seismic modelling and subsequent processing of synthetic seismic data is carried out according to seismic field experiments. Synthetic stacked or migrated sections are compared with observed data, and differences are used to improve the geological balancing and the processing of the field data. There are

---

[1] Institut for Geophysics, University of Hamburg, Bundesstr. 55, 20146 Hamburg, Germany.
[2] now at: Trappe Erdöl Erdgas Consultant, Burgwedelerstr. 89, 30916 Isernhagen, Germany.
[3] GeoForschungszentrum Potsdam, Telegraphenberg C2, 14473 Potsdam, Germany.

always several possible geological models which fulfill the rules of balancing
(DAHLSTROM, 1969; WOODWARD *et al.*, 1989) and therefore the procedure can be
carried out in an iterative loop (Fig. 1) until a sufficient fit between modelling and
field data is obtained. Finally a model of the subsurface should be obtained which
represents an improved model in comparison to the true subsurface structure.

Geological balancing is used to develop structural models. The geological model
is derived from geological data obtained at the earth's surface and in boreholes. If
also available, seismic data are used to obtain the structural model. For the
balanced profile used in this paper, stacked sections of the reflection data were used
to support the construction of the structural model. However, a balanced profile is
a depth section and, therefore, an unmigrated time section is of limited value in the
balancing process.

A balanced profile is a structural section which accounts for conservation of
mass and bed length during structural deformation. Balanced cross sections of the
earth's crust should be retro-deformable to a reasonable initial depositional position without having voids or overlaps (criterion of strain compatibility,
DAHLSTROM, 1969). Based on this assumption it is possible to obtain geometrical

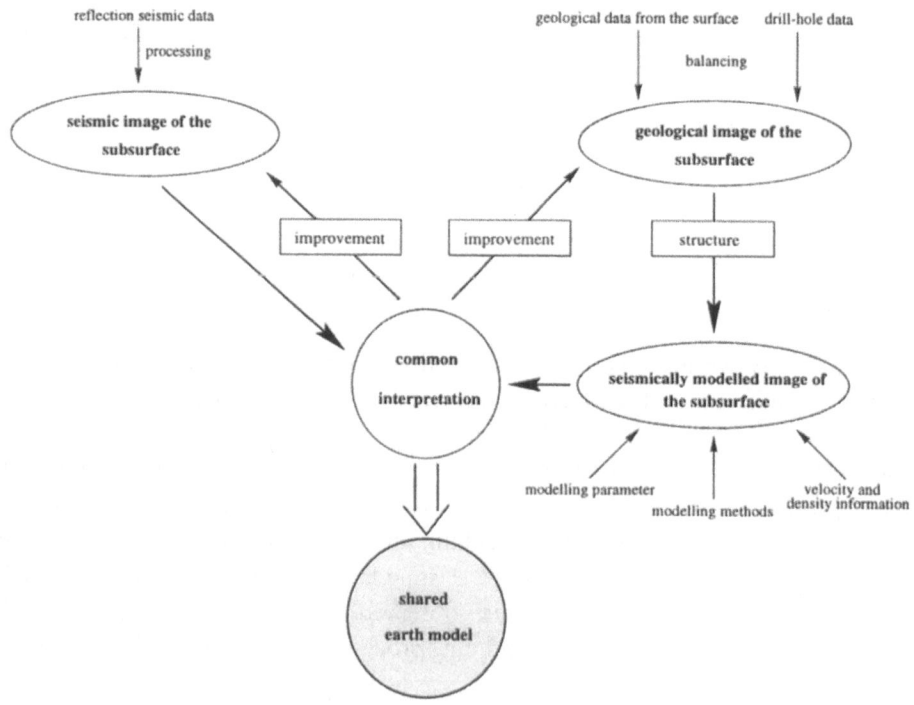

Figure 1
Sketch of the process to integrate reflection seismics and structural geology.

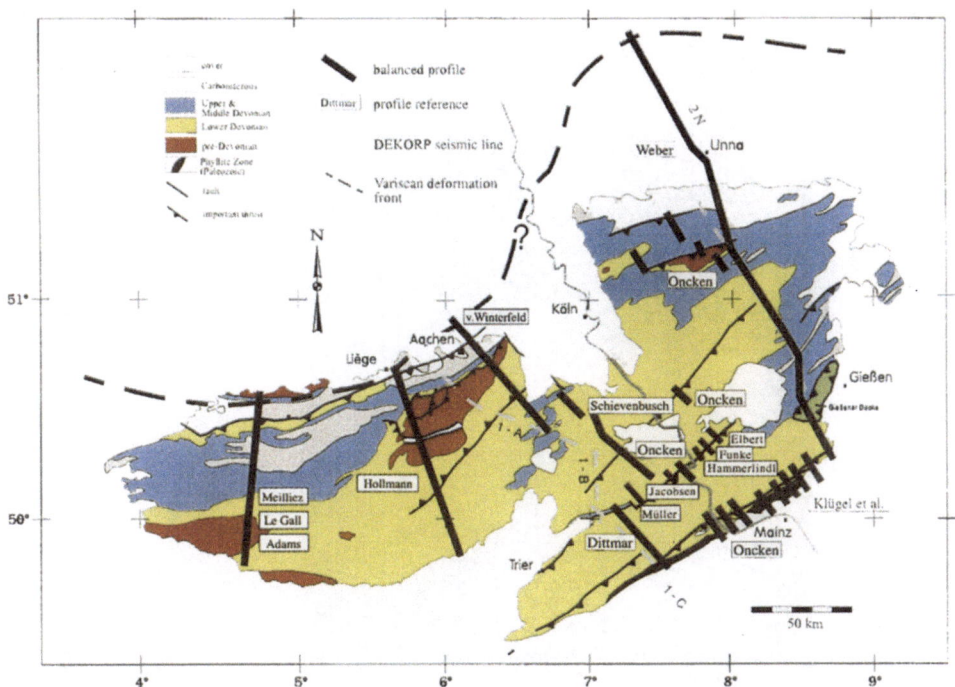

Figure 2
Map of Geology and balanced profiles across the Rhenish Massif (Plesch, A., personal communication).

rules to test the consistency of geological sections. It should be emphasized that a cross section which passes the geometric tests is not necessarily correct, since completely different cross sections can fulfill the law of conservation of mass and bed length. The virtue of crustal balancing is not to produce a unique, essentially correct explanation of a tectonic problem, but to eliminate impossible solutions.

The attempt to integrate a balanced profile and seismics is carried out using a section of the Münsterland basin and the Rhenish Massif located at the northern border of the Rhenohercynian fold and thrust belt of the Mid-European Variscides.

Geology

The balanced profile crosses the Upper Cretaceous syncline of the Münsterland basin and the Rhenish Massif (Figs. 2 and 3). The extension of the profile is 245 km, it strikes in NNW-SSE direction and crosses the Middle European Variscides. The external zones of the Middle European Variscan orogene are the Rhenohercynian and the Subvariscan Zones. Both zones were developed during

Figure 3

Balanced profile of the Upper Cretaceous Syncline of the Münsterland basin and the Rhenish Massif (after Weber, personal communication).

Paleozoic times as intra-continental basins at the northern border of the Variscan orogene (DROZDZEWSKI and WREDE, 1994). The southern boundary of the basin is formed by the Saxothuringian Zone with its crystalline border, the Middle European Crystalline Rise. In the northeast the basin extends to the East-European platform with the Tornquist-Teisseyre Zone, and in the northwest it extends to the Caledonian Iapetus-Suture of the British Islands. In the Rhenohercynian Zone thick Devonian basin fillings with regional magmatism are present. Synorogenic plutonism and regional metamorphism are only of minor importance. The Rheno-hercynian Zone consists of the Ardennen, the Rhenish Massif and the Harz mountains (FRANKE et al., 1990).

Modelling and Processing

Construction of the Velocity Model

Due to limited computer resources only a part of the balanced profile (see the indicated area in Fig. 3) was considered. This part was selected by inspecting the quality of the reflection seismic data of profile DEKORP 2-North (see Fig. 2 for location of profile). To a depth of 10 km, i.e., about 5 s TWT (two-way travel time), the best data quality (high amplitude reflections) are observed over a greater part of the Münsterland basin.

The model has a dimension of 49.25 km in horizontal and 8.0 km in vertical direction. The velocity information to be used for the balanced profile was obtained yielding results from publications of the Münsterland I borehold (LICHTENBERG, 1963; DROZDZEWSKI and WREDE, 1994). The interpretation of the Münsterland I borehole displays good agreement of stratigraphic layers and velocity. The balanced profile consists of stratigraphic layers, lithological layers are not present in this model. Therefore, the velocities for these layers could be easily obtained. In some stratigraphic units the average over different velocities due to lithological changes had to be used. Unfortunately, the velocity measurements in the borehole were only carried out down to 5.67 km depth. For greater depths, i.e., for the Devonian parts of the subsurface, the velocities are not well confirmed because the logging tool was used at higher temperature and pressure than permitted. Newer results for this depth range are not available. It was not possible to use the velocities for this stratigraphic unit, derived in the western part of the Rhenish Massif, because the geological facies are different, i.e., for the same geological time different rocks were deposited. Velocities which were determined from reflection seismic data were not used for the modelling because the modelling of the balanced profile had to be independent from the reflection seismic measurements. This procedure allows checking of the reliability of geological balancing. The velocity model for the numerical seismic P-wave modelling which was constructed from the balanced

profile and the borehole velocities is shown in Figure 4. This model is called
G-(geological) model. The distances of the following figures are specified with
respect to the distance axis of the *G*-model. The position of the Münsterland I
borehole is at about profile kilometer 11.

Velocities of the *G*-model range between 3.2 km/s close to the surface and
5.6 km/s at greater depth (6.6 km). At the base of the Cretaceous layer at 1.3 km
depth a velocity jump of 800 m/s (from 5.7 to 4.5 km/s) is present. The *G*-model
shows several syncline structures within the Westfalian and Upper Devonian. Two
of the layers are low velocity layers, i.e., the velocity of the layer is lower than the
velocity of the layer on top. The balanced profile represents only structural units
which correspond to a certain stratigraphic sequence. The velocities of this sequence
were chosen according to the velocities obtained for the same stratigraphic units in
the borehole. Thus, all discontinuities of the seismic velocity model correspond to
structural boundaries. Lithological boundaries within a stratigraphic unit are not
obtained. Some problems appear because in the log information of the borehole
lithological velocity changes also are present, and the velocities must be averaged
over the entire structural unit of the balanced profile.

Seismic Modelling

Seismic modelling of the balanced profile was performed by applying the
pseudo-spectral method. This technique is a direct solution of the equation of
motion for modelling the complete wave field of complex structures (KOSLOFF and
BAYSAL, 1982). Synthetic seismograms were computed for a range of shot positions. Each of the 160 shots was calculated on a 400×400 grid with a grid spacing
of 25 m horizontally and vertically. On each side and at the bottom of the model
an absorbing zone of 40 grid points was introduced to avoid boundary reflections.
At the top of the model free surface boundary conditions were applied.

The shot-receiver configuration was an end-on-spread configuration with maximal offset of 7.975 km. For each shot the wave field was computed at 320 geophone
positions with a geophone spacing of 25 m. Therefore, the CMP spacing was 12.5 m
which is smaller compared to the field experiment. This was necessary to match the
conditions of stability and numerical dispersion for the modelling. For the Fourier
method, there must be at least two grid points per wavelength.

The shot spacing is 250 m, resulting in a 16-fold coverage which is smaller than
the 100-fold coverage of the vibroseis field data. Owing to the restrictions in
computing time and disk space, the modelling of more shots for a higher coverage
was not considered. The propagation time for each synthetic seismogram was 5 s.
The upper cut-off frequency of the source wavelet (Ricker wavelet) in the seismic
modelling was 60 Hz. The sweep used for the field experiment contained frequencies
between 12–48 Hz.

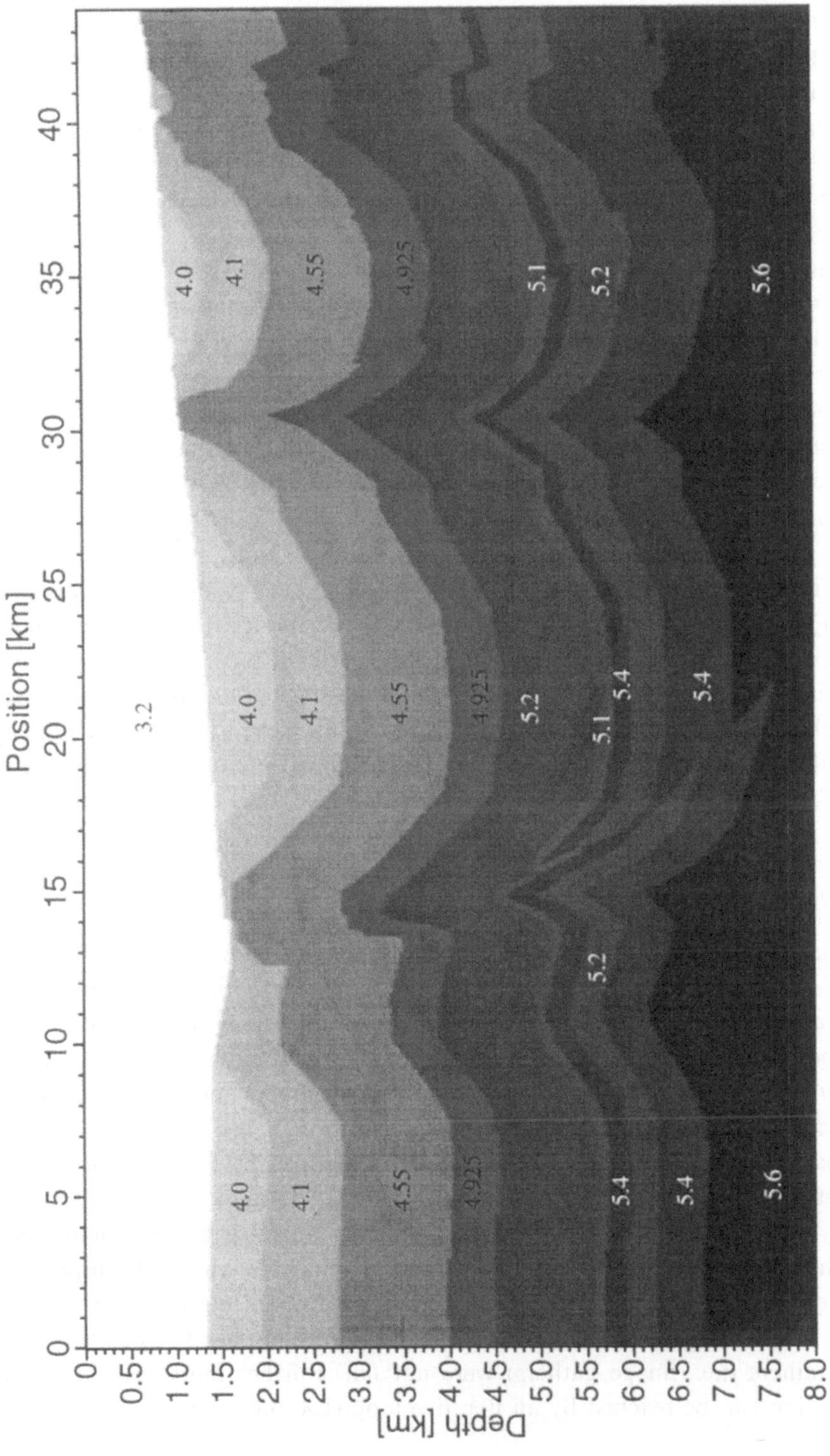

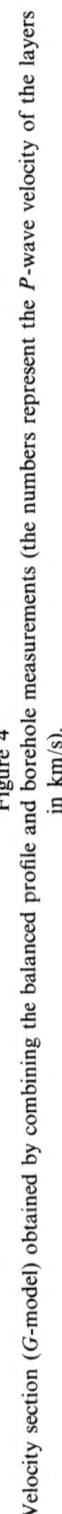

Figure 4

Velocity section (*G*-model) obtained by combining the balanced profile and borehole measurements (the numbers represent the *P*-wave velocity of the layers in km/s).

Processing of Synthetic Data

The synthetic data computed for the *G*-model were acquired and processed according to the DEKORP-2N field experiment. After CMP-sorting the data was loaded into GeoDepth (Trademark of Paradigm Geophysical, Ltd.). Since this is a synthetic data set, the velocity model (i.e., the *G*-model) is known and there is no need for velocity determination. However, we have applied velocity determination techniques on the synthetic data to evaluate the potential of velocity determination tools if ideal data are available.

Semblance analysis and constant velocity analysis with stack were used to determine stacking velocities. The section derived with these stacking velocities was improved by delay-analysis of the NMO corrected gathers. The improved stack section is shown in Figure 5.

The stacked section shows diffractions caused by modelling on a coarse numerical grid (stair-stepping at dipping interfaces). The amplitude of the diffractions are high at the first reflector (boundary Cretaceous-Westfalian) because of the strong velocity contrast at this boundary ($\delta v = 800$ m/s).

Looking at the stacked section two regions (2) and (3) are remarkable due to their low reflectivity. This region corresponds to the steeply dipping layers and faults in Figure 6. From 5 km to the bottom of the model the reflection image shows few reflections from the dipping structures. All other subsurface structures are represented very well with high reflectivity in the stacked section. The horizons in the stacked section are easy to interpret and can be picked for interval velocity determination. Some problems occurred in picking horizons of the thin layers at 2.7 s TWT.

Velocity Determination and Migration

For interval velocity determination 2-D coherency inversion was used. The 2-D coherency inversion is a method for velocity–depth estimation from unstacked data. It is advantageous in that it does not require event picking on unstacked data, and it is not based on curve fitting or hyperbolic approximations of arrival times (LANDA *et al.*, 1988, 1989). The input includes CMP gathers and zero offset travel times for selected reflectors. The inversion is performed in two steps; firstly, for a given velocity function the interface positions are mapped by ray migration using zero offset time information. Then multi-offset travel times are calculated by tracing rays through the obtained depth model, and semblances are computed on a CMP gather. Positions of interfaces and interval velocities within the layers are represented by spline functions (for more details see LANDA *et al.*, 1991).

Using this first interval velocity section for prestack depth migration, the migrated gathers (i.e., image gathers) were not flat. Further improvement of the velocity section can be reached by an iterative loop (LAMBRECHT, 1997). First the

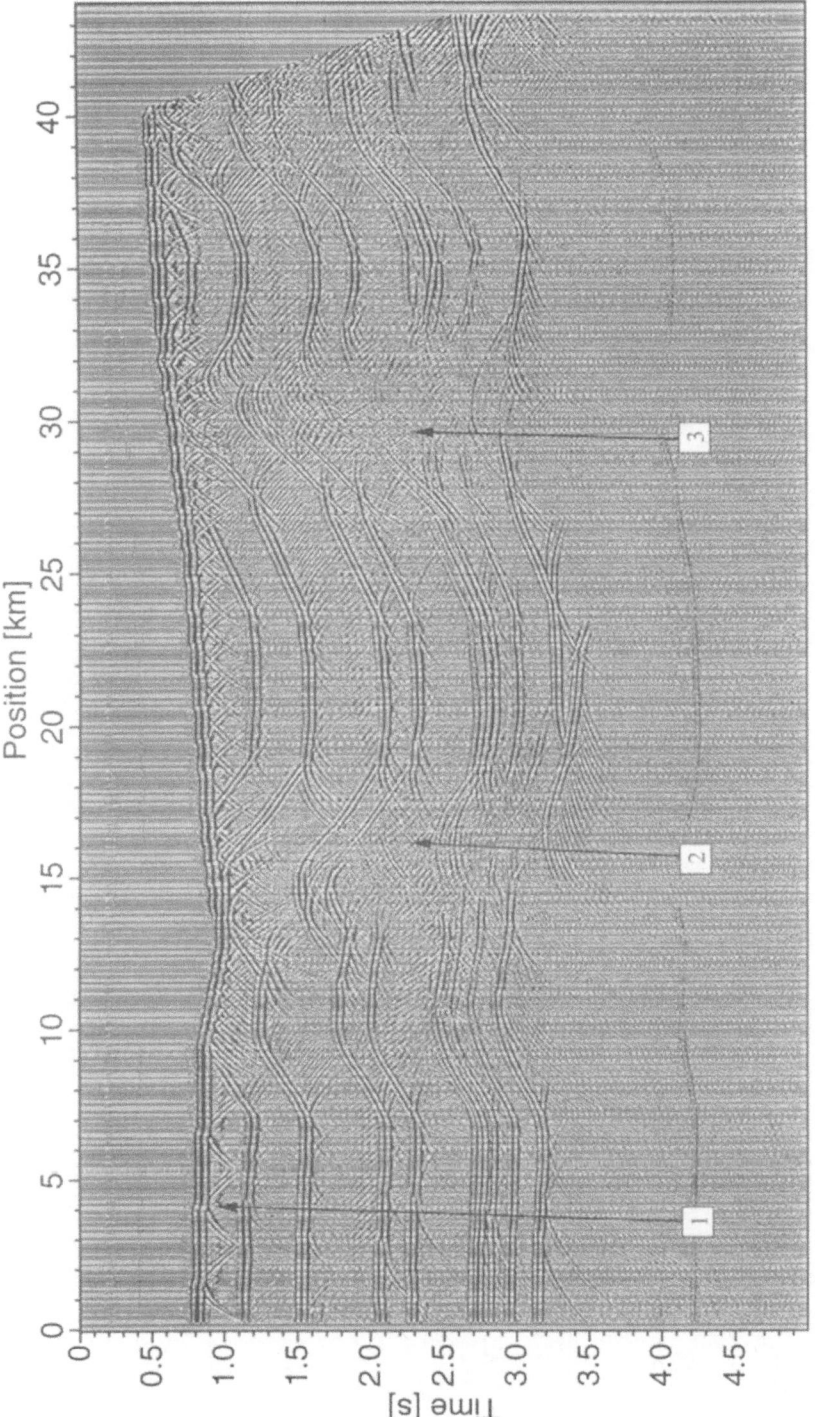

Figure 5

Stacked section of the synthetic data computed for the *G*-model ((1): diffractions, (2) and (3): regions of low reflectivity).

D. Gajewski *et al.*

Pure appl. geophys.,

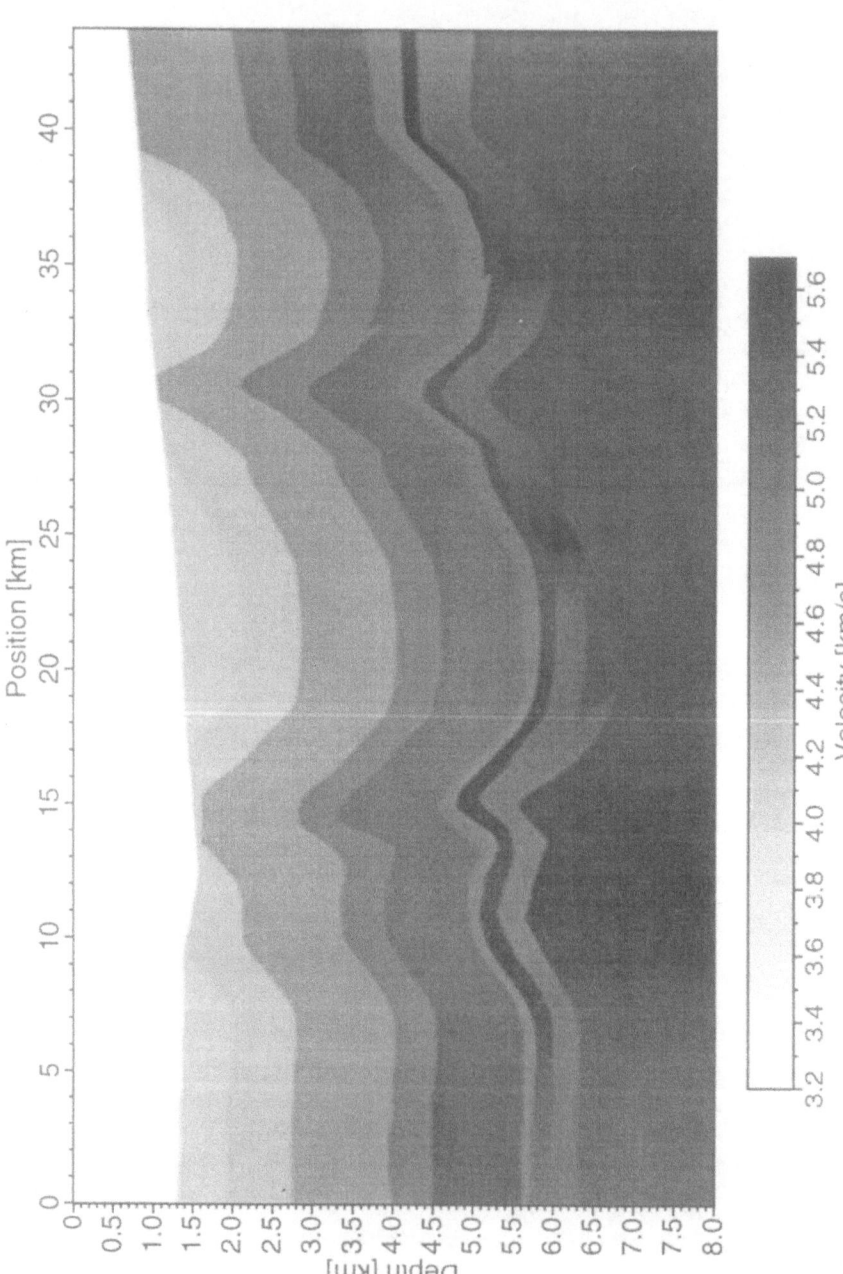

Figure 6

Interval velocity section obtained from the synthetic data of the *G*-model (after delay analysis).

interval velocities were computed by using 2-D coherency inversion. Then prestack depth migration with the velocities from coherency inversion was applied. The migration implemented in GeoDepth is a Kirchhoff type migration based on the numerical solution of the eikonal equation. With a delay analysis on the depth migrated gathers (image gathers), the velocity section was improved. With the new velocity section a new prestack depth migration was carried out and new depth horizons were picked and demigrated into the time domain. Here a better picking was possible, especially in the region of the thin layers where precise picking is very important. This procedure results in a new improved interval velocity section (Fig. 6) which is used for the final prestack depth migration.

The stacked section was also poststack depth migrated. This was a PSPC (phase shift plus correction) migration which works in the (f, k)- and the (f, x)-domain (for details, see STOFFA et al., 1990). The final poststack depth migrated section is shown in Figure 7.

Some of the diffractions caused by the numerical grid are visible in the migrated image, especially for the first horizon. Altogether the subsurface boundaries are imaged very well. For the two regions (2) and (3) showing low reflectivity in the stacked section (at about 2 s TWT), the poststack migration image also displays low reflectivity of the steep dipping structures. The thin layer at a depth of 5.6 km is hardly detectable, especially in the region between x coordinate 18 and 26 km. The reason for this is the inaccuracy of the interval velocity estimation in this region.

The prestack depth migration (Fig. 10) obtained with the improved interval velocity section (Fig. 6) images the thin layers very well along most of the entire profile. In this image some of the diffractions still remain and the zone of low reflectivity (2) is improperly imaged. This could be due to velocity errors in this region or a limited geophone spread not recording the appropriate reflections. The second zone of low reflectivity is imaged very well. In both zones the greatest difference between the imaged and "true" depth is maximal 300 m. For the rest of the profile this deviation is less than 100 m.

As expected, the accuracy of the interval velocities is very important for a good migration result, especially for regions with complex geology. In most cases a simple coherency inversion is not sufficient to derive a reliable interval velocity model. An interactive loop as described above is necessary to improve the results.

Reflection Seismic Data, DEKORP 2-North

General Description

The modelled data correspond to the field reflection seismics of the northern part of the profile DEKORP 2-North. This profile crosses the northern and central part of the Variscan fold and thrust belt. The orientation of the DEKORP 2-North

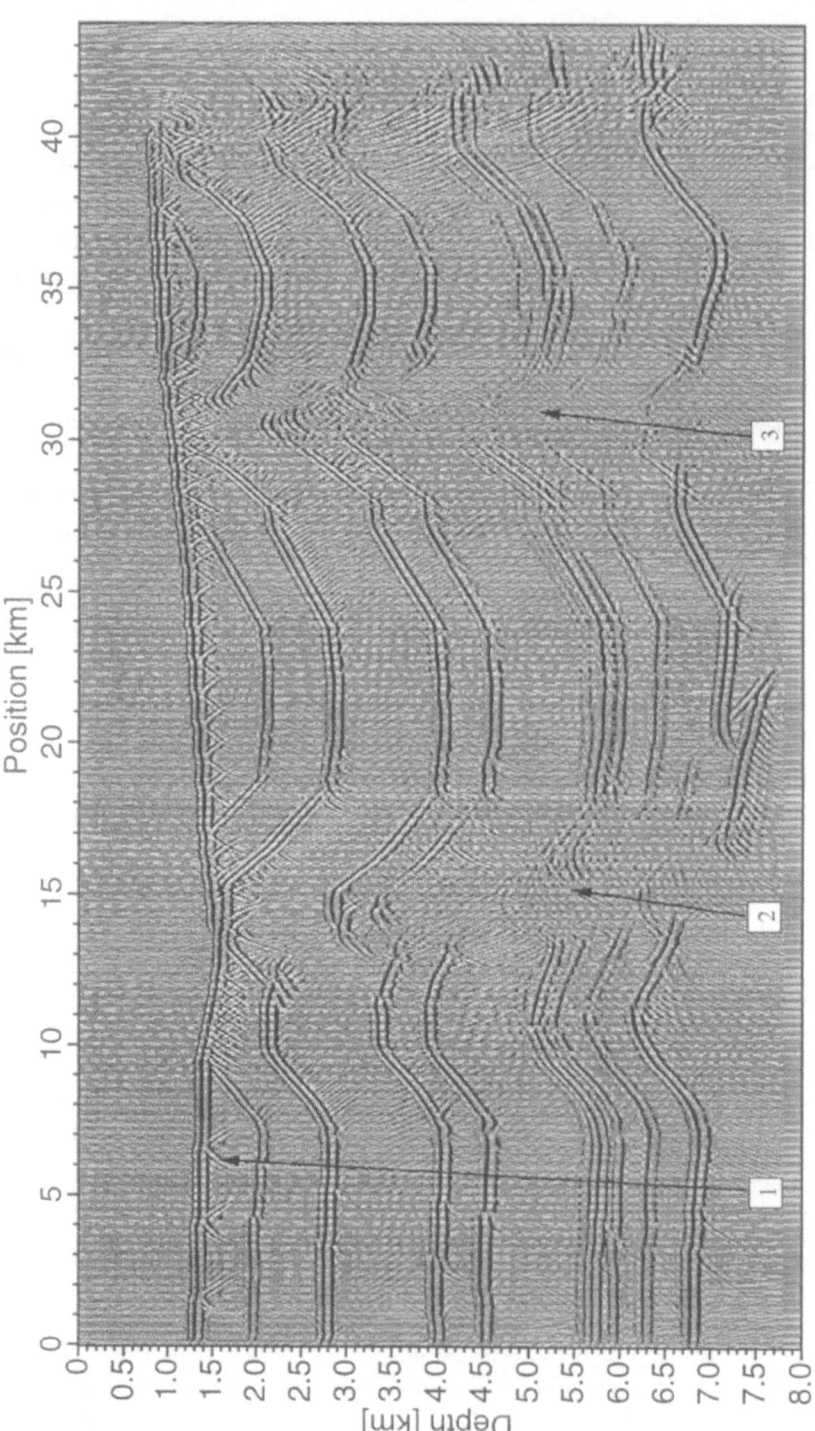

Figure 7

PSPC migration of the stacked section of Figure 5 using the determined interval velocity section (Fig. 6) ((1): diffractions, (2) and (3): zones of low reflectivity).

profile is more or less perpendicular to the important geological structures. Thus, the profile crosses all important fault zones. In the southern part of the profile a 200-fold coverage and a CMP interval of 20 m was chosen because of the complexity of the Rhenish Massif which includes numerous steeply dipping faults. For the northern part through the Münsterland basin a 100-fold coverage was considered to be sufficient since the near-surface structures are less complicated (FRANKE *et al.*, 1990). The reduction of the coverage was combined with a doubling of the geophone spacing from 40 m to 80 m. The shot spacing of 40 m was the same as in the southern part in order to maintain a CMP spacing of 20 m.

An asymmetric split-spread geometry was used to obtain sufficient high velocity resolution and simultaneously to achieve a maximum degree of coverage in the uppermost part of the crust (FRANKE *et al.*, 1990). Due to the general SE-ward dip of the faults, the long part of the split-spread was directed towards NW. On average, 80% of the nominal coverage could be achieved. The split-spread geometry was as follows:

Geophones between 4 km and 0.2 km—vibrator—geophones between 0.2 km and 12 km.

Five vibrators using a sweep length of 20 s were used. Vertical stacking was applied to improve the signal/noise ratio. The recording length was 12 s correlated data with 4 ms sampling. This experiment was particularly designed to investigate the deeper crust and the upper mantle. For the comparison with the modelling, i.e., the integration of structural geology and reflection seismics, only the first 6 s were processed, since this was sufficient for the balanced profile.

Basic Processing

The raw data (geometry and static applied) were processed using ProMAX (Trademark from Advanced Geophysical Corp.) at the GeoforschungsZentrum Potsdam. To improve the first stacking velocity section, residual static corrections were computed. The stacking velocities were improved in an iterative loop, resulting in better focussed semblance clouds. The image quality after stacking was improved with trace mixing, FK-deconvolution with blending and a dip scan stack.

The processing for stacked section comprised 3485 CMPs. This corresponds to a profile length of 69.68 km. After stacking, only the data corresponding to the modelled part were used, processed and interpreted using GeoDepth. Only offsets ranging from -4 km to $+4$ km were considered. Investigations of different offset ranges on a small part of the profile reveal that a reduction of the offset range has no disadvantage with respect to the interval velocity determination or the migration results. In this data set nearly no additional information is detectable in the far offset traces beyond 4 km in the first 6 s TWT. Prestack depth migrations of a small part of the profile using all offsets, and with the reduced offsets shows no differences.

Stacked Section and Interval Velocities

In the stacked section (Fig. 9) the reflector which corresponds to the boundary Cretaceous–Westfalian is very well visible. Above are reflections which probably correspond to the finer stratigraphic units of the Cretaceous or/and to lithological boundaries. The principal syncline structure under the Cretaceous horizon is imaged very well. One can see numerous reflections which correspond to a fine stratification of the subsurface. Such a fine stratification was not present in the interval velocities obtained from the sonic measurements. Another effect displayed in the stacked section is the fact that reflections are not continuous over the entire profile. There are many short non-continuous reflector elements. Hence, the picking of the horizons for the interval velocity determination using coherency inversion is very difficult. The resulting interval velocity section (called D-(data or determined) model in the following) is given in Figure 10.

Strong lateral velocity changes appear in some parts of the profile. It is not possible to establish whether this is caused by real velocity changes or by picking reflector elements which do not correspond to the same horizon. The determination of interval velocities shows that for a medium with fine stratification methods other than coherency inversion may be more suitable for interval velocity analyses. The coherency inversion yields the same coherency values over great velocity ranges. KESSLER *et al.* (1995) showed that for complex geological regions the picking of horizons should be avoided and methods based on migration velocity analysis should be employed. In general, there is a remarkably good fit of determined velocities (D-model) and velocities used for the modelling (i.e., the G-model, compare Figs. 4 and 10).

Poststack Depth Migration

The determined interval velocity section (D-model) was used for PSPC migration of the DEKORP-2N data. The result is illustrated in Figure 11. The depth scale starts at -400 m in this figure since this datum line was used for the DEKORP 2-North data. The datum line of the modelling, however, was at sea level. The profile kilometers used in the following text are not specified according to the original DEKORP 2-North profile, however they are specified according to the G-model, starting with 0 km at the left boundary (for the first 10.3 km of this section no reflection data are available).

The comparison between the PSPC migration results of the synthetic data for the G-model and the field data (Figs. 7 and 11) reveals:
—a difference in the vertical position of the Cretaceous base of approximately 150 m at profile kilometer 15.
—At profile kilometer 35 the difference relative to the vertical position of the Cretaceous base is about 100 m. Over the entire section this horizon is located at greater depth in the field data.

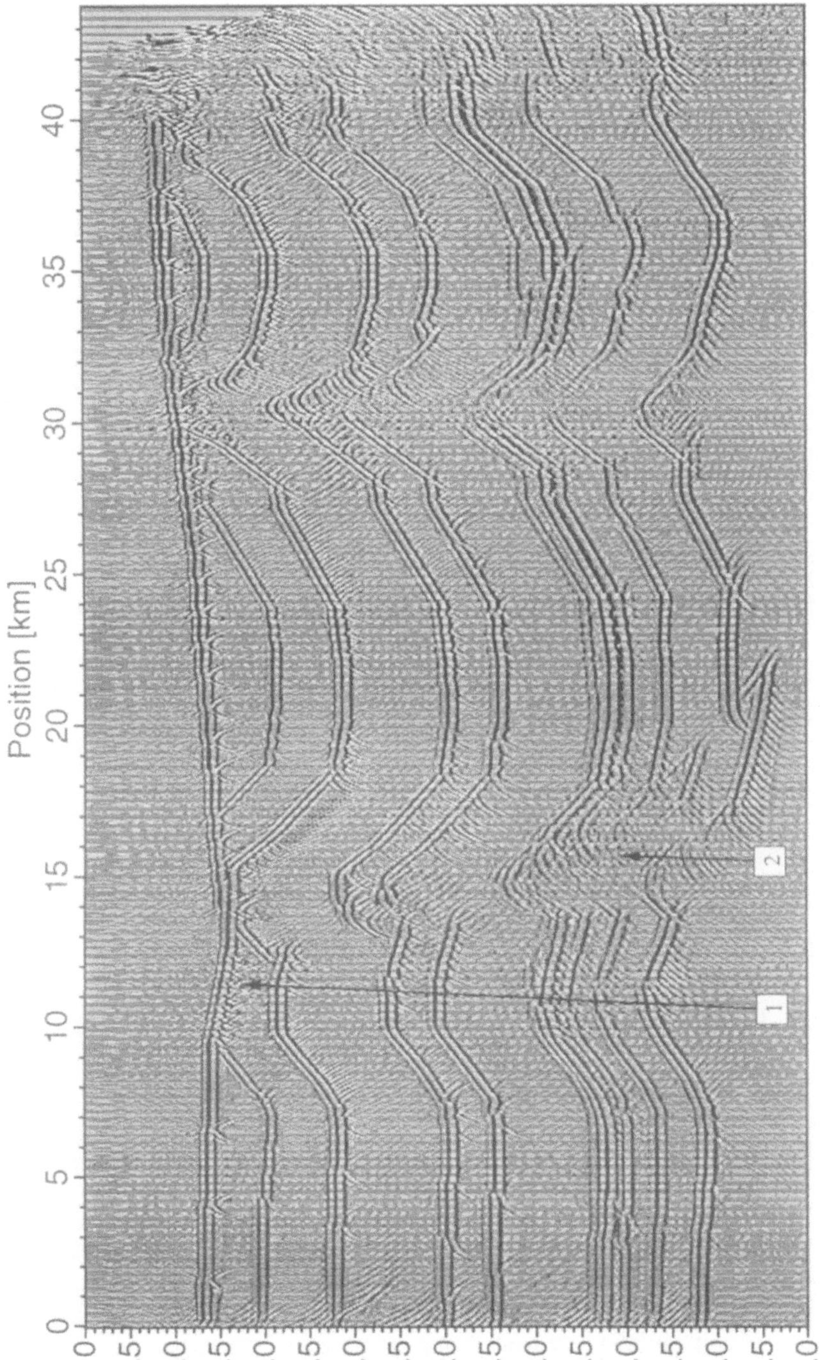

Figure 8

Prestack depth migration of the synthetic data using the determined interval velocity section (Fig. 6) ((1): diffractions, (2): zone of low reflectivity).

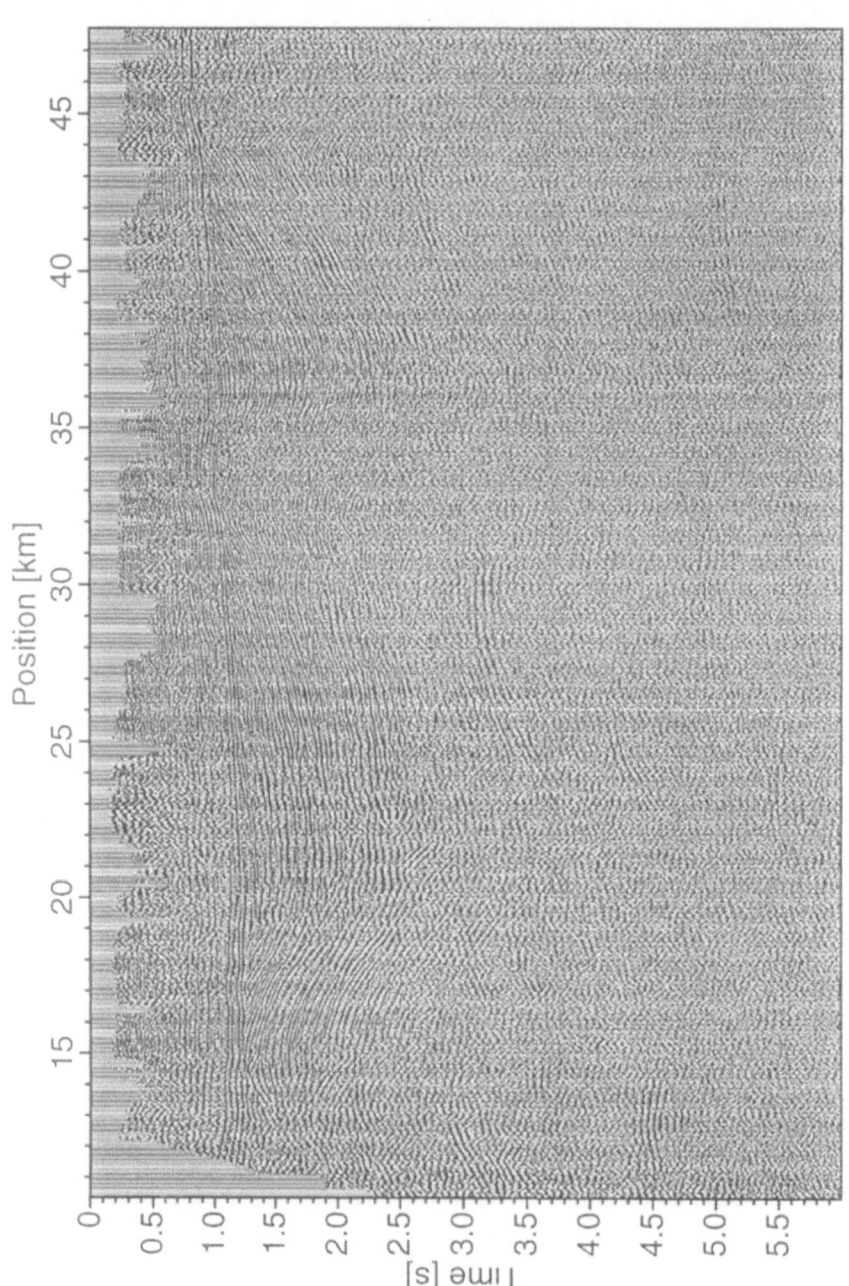

Figure 9

Stacked section of the DEKORP 2-North profile corresponding to the part of the balanced profile considered in this paper (see Fig. 3).

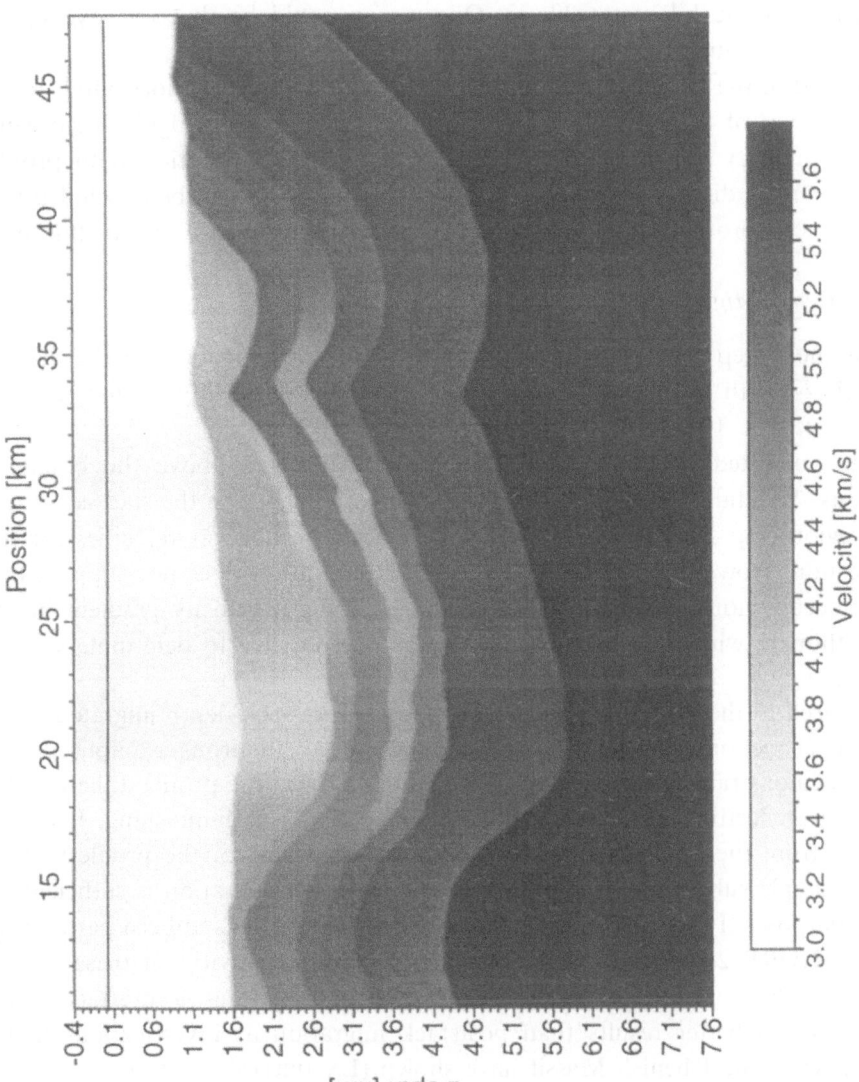

Figure 10

Interval velocity section (*D*-model) of the DEKORP 2-North profile determined with coherency inversion (the horizontal line at 0 km represents the sea level).

—From profile kilometer 10.3 to 14 the DEKORP 2-North data show a zone of low reflectivity. Therefore, it is not possible to see if the syncline as shown in the balanced profile is also present in the real data.

—The second zone of low reflectivity due to the great complexity of the structure is located at profile kilometer 32–35 but in the modelled profile this zone is located at profile kilometer 28–32. One reason could be that the layers were shortened too much during balancing.

—At the end of the modelled profile the horizons became nearly horizontal. In the migrated part of the DEKORP 2-North profile this is not recognizable because of the boundary effects of the migration. In the stacked section both profiles (Figs. 5 and 9) display many similarities in this region. it can be concluded that the balanced profile represents the subsurface structure very well in this part.

Prestack Depth Migration

In the next step a prestack depth migration was carried out with the data of DEKORP 2-North profile using the determined interval velocity section (i.e., the *D*-model, see Fig. 10). The result is given in Figure 12.

In the migrated image few reflectors can be detected above the boundary Cretaceous–Westfalian, although some reflectors are imaged in the stacked section in this region (Fig. 9). The reason might be inaccurate interval velocities for this depth section. However, the result of the PSPC migration does not support this. The Cretaceous horizon might be better imaged using a velocity gradient in this layer. Although, with the methods used it was not possible to determine velocity gradients.

Compared to the PSPC migration (Fig. 11), the prestack depth migrated image of the deeper sections (Fig. 12) shows some reflectors with stronger amplitude and they also demonstrate better lateral continuity along the entire profile. Likewise, the zone of low reflectivity between profile kilometer 32–35 is diminishing. It can be concluded from these results that for the investigated part of the profile with its relatively simple subsurface structure, a poststack depth migration is sufficient for imaging purposes. It might be insufficient for the data in the southern parts of the profile DEKORP 2-North and for the zones of low reflectivity. In these parts a prestack depth migration is justified. In complex structures a prestack depth migration yields better results than poststack migration as investigations in the western part of the Rhenish Massif have shown (LAMBRECHT, 1997).

Cross Migration

In further tests of the correspondence between the balanced profile and reflection seismics, a prestack depth migration of the DEKORP data using the *G*-model was carried out. This procedure is abbreviated as "cross migration" (because the velocity model used for the migration was not derived from the reflection data).

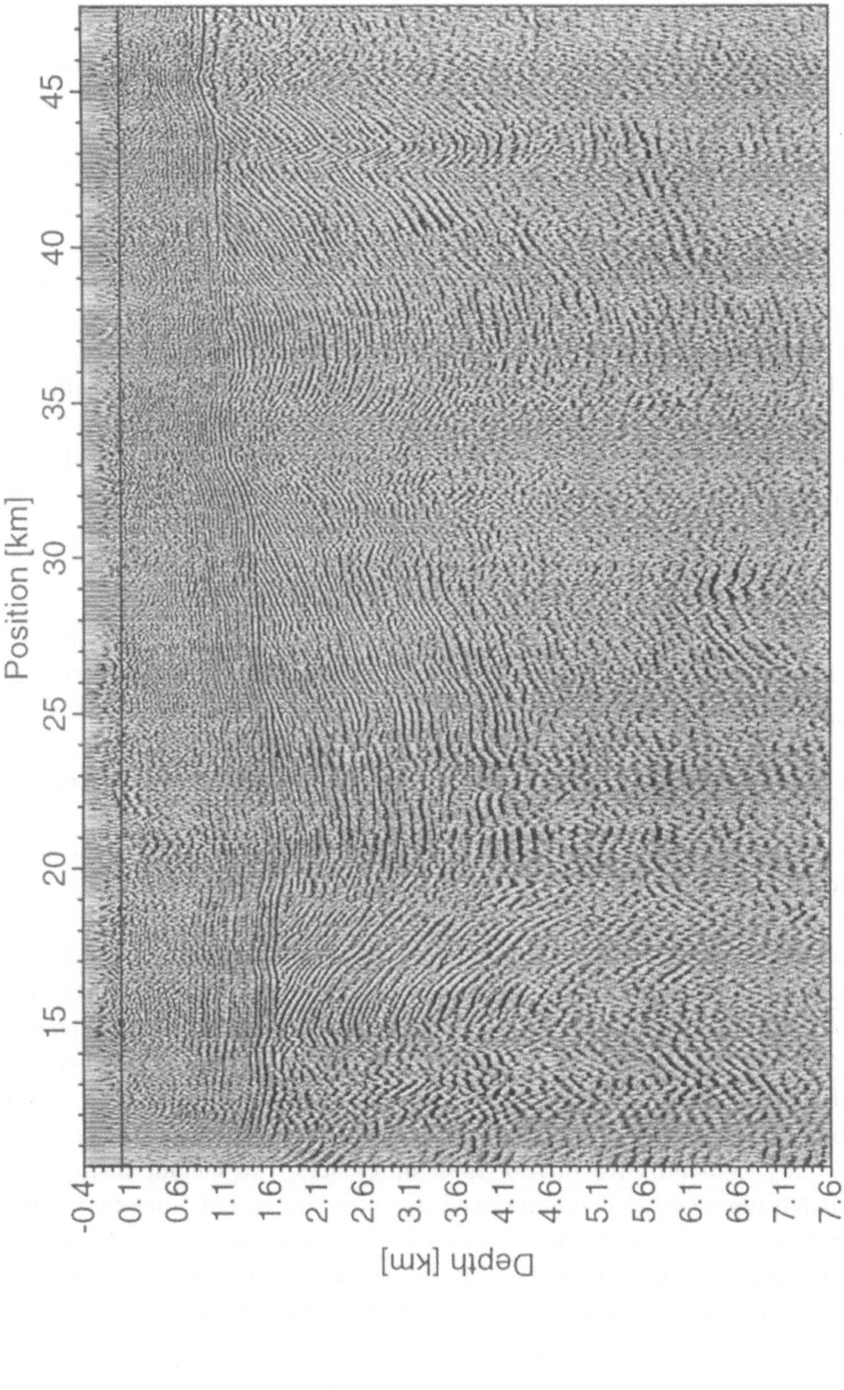

Figure 11

PSPC migration of the DEKORP 2-North profile using the determined interval velocity section (*D*-model, see Fig. 10, the horizontal line at 0 km represents the sea level).

The cross migration (Fig. 13) of the field data using the G-model (Fig. 4) is nearly as good as the migration result (Fig. 12) using the determined velocities, i.e., the D-model (Fig. 10). Some differences appear at the base of the Cretaceous. This reflector is not imaged well because the change in velocity between Cretaceous and Westfalian is 150 m too high in the G-model.

All other differences between the two migration results are negligible. The dip of the reflector element between profile kilometer 39 and 44 at a depth of approximately 6 km is different between the migrated images obtained with the different velocity sections. The differences in depth of some reflector elements reach 100 m.

In summary: A good image of the subsurface could be reached by migrating the field data using the G-model. Please note, that for the construction of this velocity model we have used the balanced profile (i.e., structure) and the velocities obtained from the sonic log of just one drillhole.

The results indicate that structural geology and seismic interpretation complement each other in a very efficient way. The combination of balanced profiles with velocities from borehole measurements produces good starting models for migration or other processing techniques which need a first velocity model (e.g., tomography).

Integration of Reflection Seismics and Structural Geology

In the last section a first comparison between the results of the synthetic and the measured data is carried out in order to integrate reflection seismics and structural geology. It can be concluded from the results presented above that the structure of the balanced profile (Fig. 3) and the migrated image obtained from the real data (Fig. 12) agree rather well. However, some differences have been described in the previous section which might be influenced by four aspects:

1. The interval velocities derived from the reflection data do not represent the true subsurface velocities.
2. The quality of the field data degrade with depth.
3. The geological balancing does not result in the proper subsurface model.
4. Modelling and balancing were performed in 2-D, however, 3-D effects might influence the results, e.g., if dipping reflectors and strong lateral changes of geology occur transverse to the profile (in the case of the Münsterland basin the lateral changes are very small and the dip of reflectors is in most cases small, therefore, the 3-D influence is expected to be negligible or not present).

One possibility to identify reasons for these differences is to include them as input into the balancing process. If the obtained new structure is not consistent with the rules of balancing then the difference may indicate errors in interval velocities and insufficient data quality. If the rules of balancing are fulfilled, a second modelling with the new structure is necessary. The results of the modelling are then again compared to the field data (see Fig. 1).

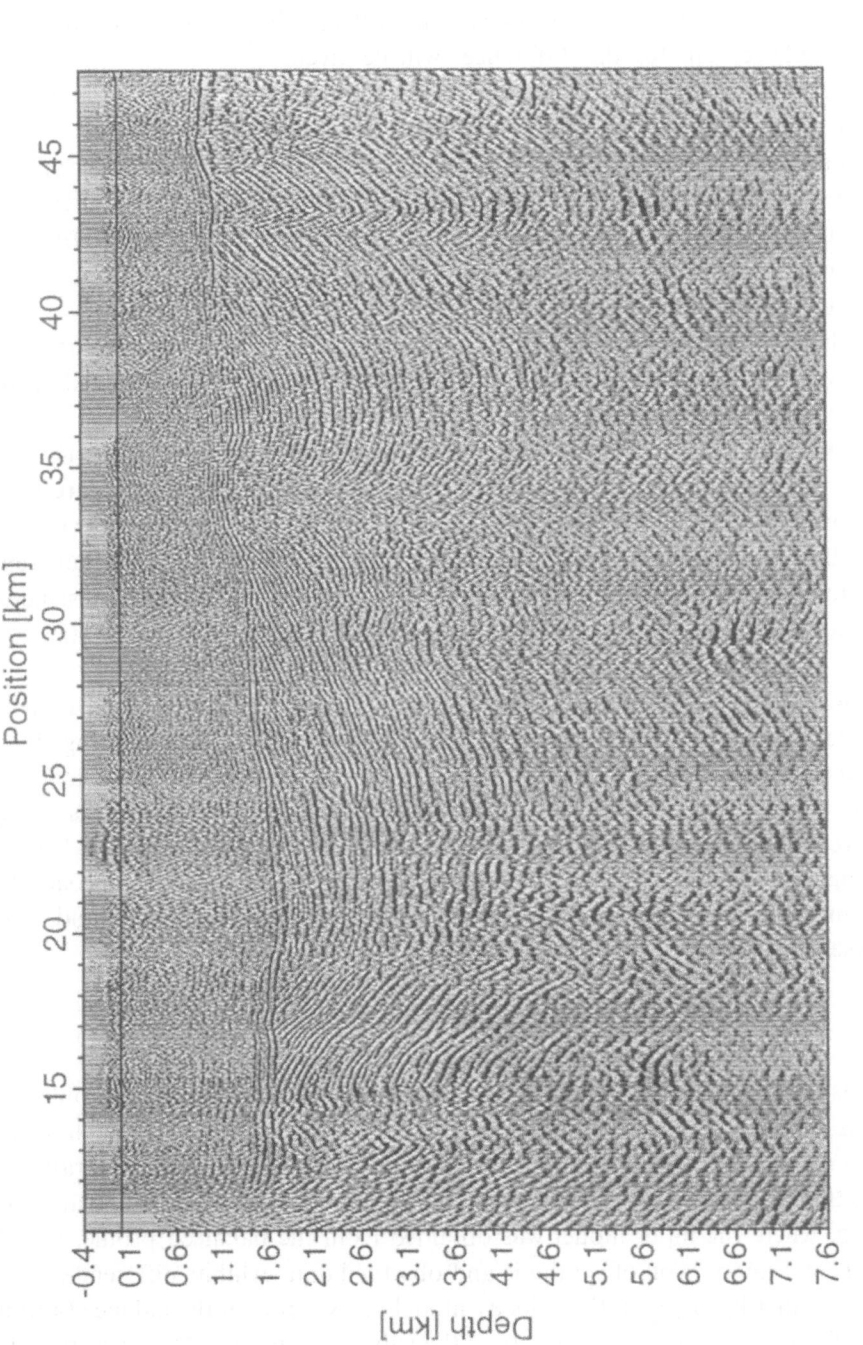

Figure 12

Prestack depth migration of the DEKORP 2-North profile using the determined interval velocity section (*D*-model, see Fig. 10, the horizontal line at 0 km represents the sea level).

The major differences between the migrated synthetic data (Fig. 8) computed for the G-model (Fig. 4) and the migrated reflection data (Fig. 12) using the determined velocities (i.e., the D-model, Fig. 10) are discussed in the following, where possible reasons for the differences will be given:

—The position for the base Cretaceous shows differences of maximal 200 m. The reason for this could be the interval velocity section or the balancing.

—The position of the zone of low reflectivity is laterally mispositioned by approximately 3 km. This zone was imaged in the migrated section of the synthetic data between profile kilometer 28 and 32 and the migrated section of the DEKORP 2-North data between profile kilometer 32 and 35. The reason for this is an error in the balanced profile.

Note that the apparently low reflectivity is caused by steeply dipping structures. These steeply dipping structures are very well represented in the balanced profile, however, it is shifted laterally by 3 km.

—In the migration images of the DEKORP 2-North data the syncline structure is hardly visible for depths greater than 5.5 km. The reason for this feature might be a lower data quality with increasing depth but also wrong interval velocities could be considered, since determination of interval velocities using coherency inversion was difficult in this region. It is also possible that the syncline structure does not exist in this form at this depth. Other possible subsurface structures should be tested and proved by balancing.

—The profile DEKORP 2-North displays considerably more reflections than the modelled data. These reflections could be caused by lithological layering. Introducing lithological layer boundaries into the model would allow for conclusions on the origin of the reflections, i.e., lithological or structural reflections. So far however, without further information, it is not possible to distinguish between lithological, stratigraphical and tectonical or other reflections. To approach this goal more information such as different borehole measurements and ΛVO analyses is needed.

Discussion and Conclusions

In this study we have presented a first attempt to integrate balanced profiles and reflection seismics. Both techniques provide subsurface models, which, however, may not coincide since they are based on different data sets. The integration of both methods was performed by using seismic modelling. Synthetic data were generated, where the input model was obtained from the balanced profile, i.e., the G-model. The comparison of synthetic and observed data exhibits differences which are caused either by errors in the velocity model or by errors in the balanced profile. These differences can be used to improve the subsurface model. The case study has proved that balanced profiles combined with velocity information from boreholes can provide reliable starting models for subsurface imaging.

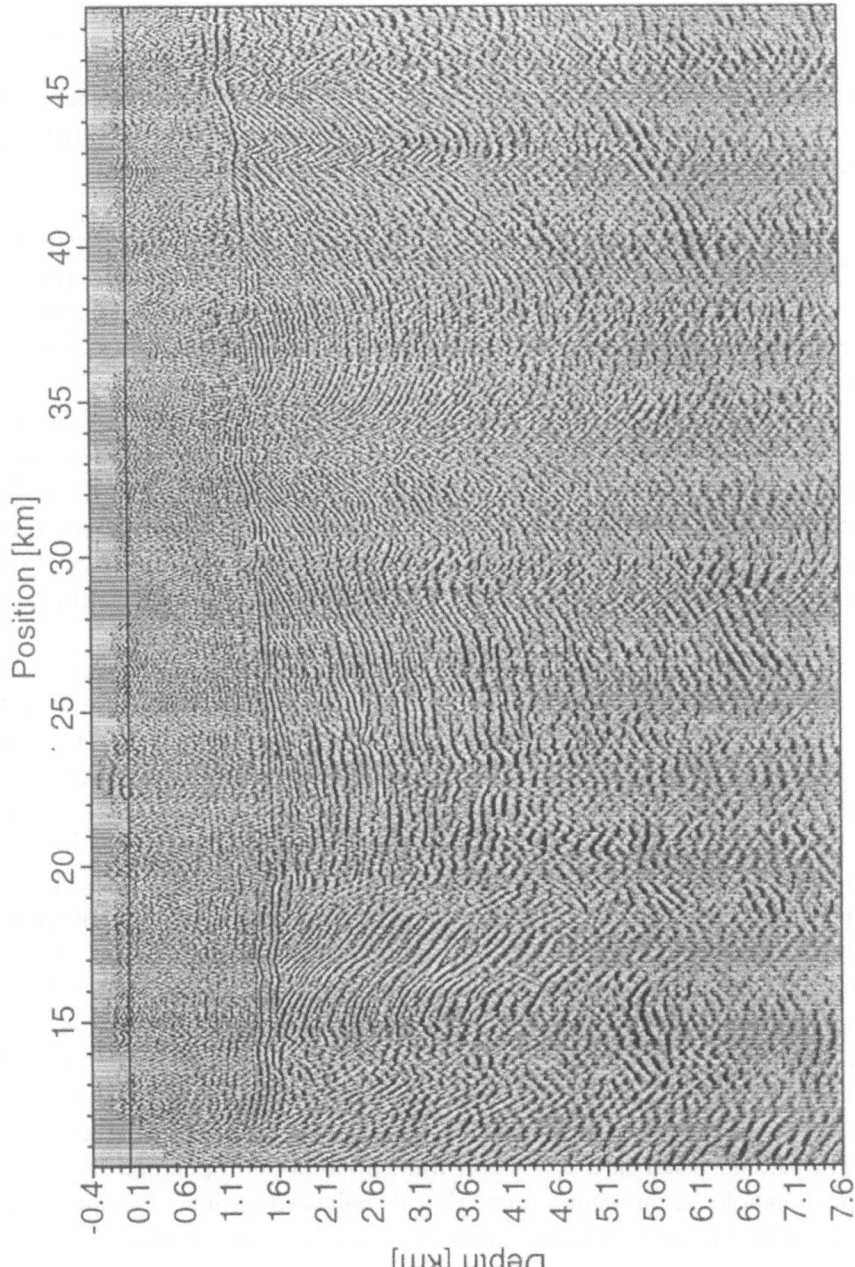

Figure 13

Prestack depth migration of the DEKORP 2-North data using the *G*-model (Fig. 4, the horizontal line at 0 km represents the sea level).

For the seismic modelling, i.e., for the generation of synthetic data, a velocity model is needed. It is obtained by combining the balanced profile (which provides only structural details) with velocity information from borehole measurements on or near the profile. The velocities are laterally extrapolated into the corresponding stratigraphic units of the balanced profile. Obviously, the quality of the velocity model improves with the number of boreholes available. In this study no corrections for pressure and temperature were applied to the velocities measured in the borehole. Such a correction, however, can improve the velocity model, if the stratigraphic units are located at different depths along the profile.

The modelling revealed that the so-called "transparent zones," i.e., zones of low reflectivity in the data, may not necessarily indicate a homogeneous subsurface. On the contrary, transparent zones may result from a subsurface of strong heterogeneity. Therefore, from the geological point of view, transparent zones may also indicate strong tectonics.

Another important conclusion derived from the case study concerns the image quality of migrated seismic sections below strong heterogeneities. In these regions even small errors in velocity degrade image quality considerably. Continuous reflectors may appear as a broken structure, misleading the geological interpretation and resulting in wrongly determined reflector lengths. This, again, will influence the structural model, i.e., the balanced profile, since the correct length of a structure is important for balancing.

A remarkably good migration image of the DEKORP 2-North data was obtained by combining the structure of the balanced profile with the velocities of borehole measurements (resulting in the G-model). This image showed a rather good comparison with the image of the migrated section, using the velocity model developed from the field data (the D-model). Considering the great number of processing steps involved to obtain the D-model, the study indicates that the use of the G-model may lead to savings in processing time and processing costs. This can be of particular importance to the hydrocarbon industry (3-D-seismics). Since balancing is a standard technique for oil exploration and there is usually access to a number of borehole data, the described technique could provide a new tool to develop velocity models faster and less expensively.

It was mentioned above that the quality of the G-model can be improved by incorporating temperature and pressure effects. A further improvement of the G-model is possible by applying reflection tomography (see, e.g., KOSLOFF *et al.*, 1996). This should result in a better constrained lateral extrapolation of velocities from the borehole or boreholes through the balanced profile. In general, if a balanced profile and velocity information from boreholes are available, the resulting G-model appears to be a reasonable starting model. The G-model may provide shortcuts in velocity processing of reflection seismic data.

The above conclusions clearly indicate the benefit of integrating balanced profiles and reflection seismics. To further improve this process, an iterative loop of balancing and synthetic seismic modelling may further improve the results. In this study we only analyzed the differences between synthetic and observed data. In the future these differences should be quantified by modifying the balanced profile according to the observed errors. This will produce an improved G-model.

Obviously the modelling of synthetic seismic data using numerical solutions of the elastodynamic equation is computational intensive. The synthetic modelling is used in this study to test the reliability of the G-model compared to seismic observations of the subsurface. For the development of the G-model alone, no synthetic modelling is necessary. The reliability of the G-model can also be tested if coincident reflection data are available. A prestack depth migration of these data and the analyses of depth focusing panels or the investigation of residual curvatures provide information pertinent to the quality of the velocities of the G-model.

Acknowledgements

The authors are indebted to Johannes Weber for providing the balanced profile prior to its publication. The authors also thank Manfred Stiller for his assistance processing the DEKORP-2N data. We are grateful to H.-J. Brink and G. Müller for their constructive reviews which improved the manuscript. This research was supported by the Bundesministerium für Bildung und Forschung (BMBF).

REFERENCES

DAHLSTROM, C. D. A. (1969), *Balanced Cross Section*, Can. J. Earth Sci. *6*, 743–757.

FRANKE, W., BORTHFELD, R. K., and BRIX, M. *et al.* (1990), *Crustal Structure of the Rhenish Massif: Results of Deep Seismic Reflection Lines DEKORP 2-North and DEKORP 2-North-Q*, Geologische Rundschau *79*(3), 523–566.

DROZDZEWSKI, G., and WREDE, V. (1994), *Faltung und Bruchtektonik—Analyse der Tektonik im Subvariscikum*, Fortschr. Geol. Rheinld. u. West. *38*, 7–187.

KESSLER, D., RESHEF, M., CRASE, E., CHAN, W.-K., TSINGAS, C., and HUBBARD, J. (1995), *Depth Processing: An Example*, The Leading Edge, 949–953.

KOSLOFF, D., and BAYSAL, E. (1982), *Forward Modelling by a Fourier Method*, Geophysics *47*, 1402–1412.

KOSLOFF, D., SHERWOOD, J., KOREN, Z., MACHET, E., and FALKOVITZ, Y. (1966), *Velocity and Interface Depth Determination by Tomography of Depth Migrated Gathers*, Geophysics *61*, 1511–1523.

LAMBRECHT, A. (1997), *Integration of Reflection Seismics and Geological Balanced Profiles with Seismic Modelling (In German)*, Ph.D. Thesis, University of Hamburg.

LANDA, E., KOSLOFF, D., KEYDAR, S., KOREN, Z., and RESHEF, M. (1988), *A Method for Determination of Velocity and Depth from Seismic Reflection Data*, Geophys. Prospect. *36*, 223–243.

LANDA, E., BEYDOUN, W., and TARANTOLA, A. (1989), *Reference Velocity Model Estimation from Prestack Wave Forms: Coherency Optimization by Simulated Annealing*, Geophysics *54*, 984–990.

LANDA, E., THORE, P., SORIN, V., and KOREN, Z. (1991), *Interpretation of Velocity Estimates from Coherency Inversion*, Geophysics *36*, 1377–1383.

LICHTENBERG, K. (1963), *Die Geschwindigkeitsmessung in der Bohrung Münsterland 1*, Fortschr. Geol. Rheinld u. Westf. *11*, 387–388.

STOFFA, P. L., FOKKEMA, J. T., DE LUNA FREIRE, R. M., and KESSINGER, W. P. (1990), *Split-step Fourier Migration*, Geophysics *55*, 410–421.
WOODWARD, N. B., BOYER, S. E., and SUPPE, J. (1989), *Balanced Geological Cross Sections: An Essential Technique in Geological Research and Exploration*, AGU, Short Course in Geology, *6*, Washington.

(Received October 31, 1997, revised October 8, 1998, accepted October 13, 1998)

 To access this journal online:
http://www.birkhauser.ch

Pure appl. geophys. 156 (1999) 233–251
0033–4553/99/010233–19 $ 1.50 + 0.20/0

❙ **Pure and Applied Geophysics**

Incorporation of the Effects of Absorption into Kirchhoff Migration

CHARLES TOUTOU[1]

Abstract—The effects of absorption are incorporated into Kirchhoff migration. The aim is to reconstruct the structures with true-amplitude seismic data and to increase the resolution of migrated data. A complex wave velocity is introduced into the solution of the Helmholtz equation, the starting point of Kirchhoff migration. This leads to an additional filter, the antidissipation operator, which is convolved with the wave field. The general structure of Kirchhoff migration remains unchanged. The effects of the antidissipation operator are illustrated on synthetic data. The new operator is valid for complex media with varying velocity and varying quality factor Q. Moreover there is no limitation to constant-Q, frequency-dependent Q can also be handled. The success of anelastic migration depends on how well the Q macro model is known.

Key words: Kirchhoff migration, absorption.

1. Introduction

The aim of seismic processing is to extract reflection coefficients from seismic data. A successful quantitative interpretation (lithological, petrophysical or stratigraphic) of the data requires the knowledge of reflection coefficients. The extraction of realistic reflection coefficients has been the subject of past and present research, in which effects of second order like absorption and anisotropy are playing an increasing role. It has been observed that those effects sometimes have a greater influence on the (lateral) variation of the amplitude of the wave field than the classical elastic parameters like the velocity or/and the density (WINKLER and NUR, 1979; URSIN and DAHL, 1992).

Migration is originally known as a method to determine the depth and the form of reflectors from seismic data. Presently, however, one is going beyond the original task of migration, and the amplitudes of migrated data are increasingly used for

[1] Institute of Meteorology and Geophysics, University of Frankfurt, Feldbergstraße 47, D-60323 Frankfurt, Germany. Present address: 38 Princeton Terrace, Brampton, Ontario L6S 3S8, Canada.

interpretation. In order to obtain realistic amplitudes, which are of paramount importance for the determination of realistic reflection coefficients, it is necessary to also include the above-mentioned second-order effects. This is already a standard process in forward modeling, but in the inversion process in general and in migration in particular the effects of second order have not gained great attention. It is known that absorption, besides geometrical spreading, is the factor most affecting the amplitudes of the wave field. Additionally to the effect of amplitude attenuation, absorption also affects the wave field by dispersion, i.e., by the frequency dependence of the wave velocity.

There have been attempts to compensate the effects of absorption during the processing of seismic data. Most of the work dealing with this matter merely considers the dispersion effect, which is removed with the aim to improve the resolution of the data. So far, compensation is achieved in a filter-theoretical manner. A seismic trace is considered as the result of the convolution of a source signal with several filters characterizing a linear earth model. Absorption, one of the filters, is removed through deconvolution. In this approach, absorption is generally supposed to vary only in one dimension, depth or time (BICKEL and NATARAJAN, 1985; GELIUS, 1987; HARGREAVES and CALVERT, 1991).

Absorption is, like the velocity or the density, a spatially varying effect, which acts on the propagating wave field like transmission or reflection. Therefore, its effects can also be compensated through backward propagation in a wave-theoretical way. Studies in which absorption is compensated through a backward-propagation procedure like migration are very recent. HARGREAVES and CALVERT (1991) consider a wave-propagation approach to absorption and reverse its effects (inverse Q filtering) through inverse propagation in a Stolt-like migration algorithm (STOLT, 1978). In their approach, the quality factor Q is presumed to be constant or depth-dependent in order to take advantage of the fastness of the Stolt algorithm. The main result of their work was the compensation of the dispersion effect. The amplitudes did not play a key role. MITTET et al. (1995) include absorption in a space-frequency algorithm to improve the quality (resolution) of migrated data. Still, the removal of dispersion played the key role.

In this paper, absorption is incorporated into Kirchhoff migration. This method is known to be a flexible method, allowing the velocity to vary both vertically and laterally. Similarly, no restriction is imposed on the spatial distribution of the quality factor Q. Attention will not only be focussed on the resolution of migrated data, but the amplitudes also play an important role. The amplitudes of migrated data are better suited for AVO interpretation than those of unmigrated sections, since migrated data are free of propagation effects and therefore represent a realistic image of the medium contrasts (MOSHER et al., 1996). In this sense the amplitudes of migrated data, freed from absorption effects, can be a more powerful tool for AVO analysis.

The next section is on the incorporation of absorption into Kirchhoff migration. A new operator is obtained whose behavior is discussed in the third section. In the fourth section, I discuss how migration, including absorption effects, can be implemented. Synthetic examples are shown in the fifth section to illustrate the effects of incorporation of absorption on the migration results.

2. Absorption in Kirchhoff Migration

The principles of the derivation of Kirchhoff migration do not change. The starting point is the solution of the 2-D Helmholtz equation in a homogeneous medium with constant velocity. The details of the solution can be found in many works on Kirchhoff migration (e.g., SCHNEIDER, 1978; STOLT and BENSON, 1986). The equation is solved in the frequency domain for the relevant boundary conditions, and the result is (e.g., TOUTOU, 1997)

$$\bar{U}_P(x, z, \omega) = \int_{-\infty}^{+\infty} i\omega \bar{U}(x', 0, \omega) \frac{1}{2v} H_1^{(2)}\left(-\frac{r}{v}\omega\right) \cos\varphi \, dx'. \tag{1}$$

It represents the spectrum of the wave field continued from the boundary of the medium (here the earth's surface) to depth. $\bar{U}(x', 0, \omega)$ is the spectrum of the recorded wave field, $H_1^{(2)}$ the first-order Hankel function of the second kind, v the velocity of the medium, $r^2 = (x - x')^2 + z^2$, and $\cos\varphi = z/r$ is a directivity factor. The migrated section is obtained by transforming the spectrum into the time domain and then applying the imaging condition. In practice, the exact expression of the time-domain transform of (1) is replaced by its less CPU time consuming high-frequency approximation:

$$U_P(x, z, t) = \int_{-\infty}^{+\infty} \frac{\partial U(x', 0, t)}{\partial t} * \frac{H(-t)}{\sqrt{-t}} * \delta\left(t + \frac{r}{v}\right) V(x') \, dx'. \tag{2}$$

Here, $H(t)$ is the Heaviside function and

$$V(x') = -\frac{\cos\varphi}{\pi\sqrt{2vr}} \tag{3}$$

a weight function, which depends on the velocity of the medium v, on the distance r between the diffracting point and the receiver point, and on the directivity factor. For the incorporation of absorption I start with the spectrum (1).

The general procedure of including absorption into seismic processing is as follows. Absorption of seismic waves is described by the quality factor Q which is generally frequency-dependent (ANDERSON and MINSTER, 1979; BERCKHEMER et al., 1982; MÜLLER, 1983). The quality factor is included into the processing of seismic data through the velocity, which becomes complex. The real part is slightly dependent on Q and describes the dispersion, the imaginary part is responsible for the attenuation. For the derivation of the migrated section including absorption effects I am going to use a frequency-dependent form of Q to show that the formulation is general. For the application, however, I assume that Q is frequency-independent. This is the usual assumption for the frequency band of seismic exploration (10–100 Hz); many studies have shown that the quality factor does not show a noteworthy variation in this frequency band (LIU et al., 1976; KJARTANSSON, 1979). The Q values used further for illustration are taken from the literature (e.g., CARMICHAEL, 1982; MEISSNER, 1986).

The quality factor Q behaves as a power function of frequency, $Q \propto \omega^\gamma$, with $0 < \gamma < 0.5$ (ANDERSON and MINSTER, 1979; BERCKHEMER et al., 1982). The complex wave velocity of such a dissipating medium is (see, e.g., MÜLLER, 1983 for details)

$$v(\omega) = c(\omega_r) \exp\left\{ \frac{\cot\left(\gamma\frac{\pi}{2}\right)}{2Q(\omega_r)}\left[1 - \left(\frac{\omega_r}{\omega}\right)^\gamma\right]\right\} \exp\left\{\frac{i}{2Q(\omega_r)}\left(\frac{\omega_r}{\omega}\right)^\gamma\right\}. \qquad (4)$$

Therein, ω_r is a reference frequency, and $c(\omega_r)$ is the velocity at the reference frequency, which is normally the phase velocity of the elastic (non-dissipative) medium. $Q(\omega_r)$ is the quality factor at the reference frequency. The real velocity in Equation (1) is substituted by the complex velocity $v(\omega)$. This leads to a new expression for the spectrum of the depth-continued wave field ($c_r = c(\omega_r)$, $Q_r = Q(\omega_r)$):

$$\bar{U}_{Pa}(x, z, \omega) = \int_{-\infty}^{+\infty} i\omega \bar{U}(x', 0, \omega) \frac{1}{2c_r h(\omega, Q_r)} H_1^{(2)}\left(-\frac{r\omega}{c_r h(\omega, Q_r)} \right)\cos \varphi \, dx' \qquad (5)$$

with

$$h(\omega, Q_r) = \exp\left\{ \frac{\cot\left(\gamma\frac{\pi}{2}\right)}{2Q_r}\left[1 - \left(\frac{\omega_r}{\omega}\right)^\gamma\right]\right\} \exp\left\{\frac{i}{2Q_r}\left(\frac{\omega_r}{\omega}\right)^\gamma\right\}. \qquad (6)$$

With the assumption of a weakly dissipative medium ($Q \gg 1$) and using the asymptotic approximation of the Hankel function, Equation (5) can be simplified to (see the appendix for details)

$$\bar{U}_{Pa}(x, z, \omega) = \int_{-\infty}^{+\infty} i\omega \bar{U}(x', 0, \omega) \left(\frac{i\pi}{\omega}\right)^{1/2} \exp\left(i\frac{r\omega}{c_r} \right) \bar{D}_a(\omega) V(x') \, dx'. \qquad (7)$$

Except for the weight function $V(x')$, all terms in (7) are frequency-dependent functions. The form of the new antidissipation filter $\bar{D}_a(\omega)$ is derived in the appendix. The inverse Fourier transform of (7) follows from the inverse Fourier transform of each term, including time-domain convolution:

$$U_{Pa}(x, z, t) = \underbrace{\int_{-\infty}^{+\infty} \frac{\partial U(x', 0, t)}{\partial t} * \frac{H(-t)}{\sqrt{-t}}}_{U_f(x', 0, t)} * \delta\left(t + \frac{r}{c_r}\right) * D_a(t) V(x')\, dx'. \qquad (8)$$

U_{Pa} is the downward extrapolated wave field including the effects of absorption, and U_f is a filtered version of the surface wave field $U(x', 0, t)$. This expression differs from that of a nondissipative medium (Equation 2)) through the new filter term $D_a(t)$, which compensates for the effects of absorption during the upward propagation of the wave field. I call this new term the *antidissipation operator*. It will be discussed in detail in the next section. The weight function $V(x')$ remains unchanged, and the imaging condition with the imaging time $t = t_I(x, z)$ can then be applied in the same way as in the nondissipative case, giving the migrated section

$$M_a(x, z) = \bar{U}_{Pa}(x, z, t = t_I(x, z))$$

$$= \int_{-\infty}^{+\infty} [U_f(x', 0, t) * D_a(t)]_{t = t_I(x, z) + (r/c_r)} \cdot V(x')\, dx'. \qquad (9)$$

$D_a(t)$ is, in principle, an additional weight function which varies with time. The incorporation of absorption into migration leaves the migration procedure itself unchanged, i.e., Equation (9) implies that the summation is done along the diffraction hyperbola $t = t_I(x, z) + r/c_r$ (Fig. 1). The migrated section for a nondissipative medium can be recovered from $M_a(x, z)$, if $D_a(t)$ is replaced by the delta function.

3. The Antidissipation Operator

The new filter term $\bar{D}_a(\omega)$, resulting from Equations (5) and (7), is

$$\bar{D}_a(\omega) = \exp\left\{\frac{\omega t_r^*}{2}\left(\frac{\omega_r}{\omega}\right)^\gamma\right\} \exp\left\{-i\frac{\omega t_r^*}{2}\cot\left(\gamma\frac{\pi}{2}\right)\left[1 - \left(\frac{\omega_r}{\omega}\right)^\gamma\right]\right\}. \qquad (10)$$

The antidissipation operator is a function of the dissipation time $t_r^* = r/(c_r Q_r)$ which is the quotient of the travel time on a given path and the corresponding quality factor. $\bar{D}_a(\omega)$ is exactly the inverse of the dissipation operator as it is used in forward modeling (MÜLLER, 1983, eq. (35)). The first factor is an amplitude term which compensates the amplitude loss, and the second term removes the dispersion effect. For the illustration of the behavior of the operator, I will assume, that Q is frequency independent ($\gamma = 0$):

$$\bar{D}_a(\omega) = \exp\left(\frac{\omega t_r^*}{2}\right) \exp\left(-i\frac{\omega t_r^*}{\pi}\ln\frac{\omega}{\omega_r}\right). \qquad (11)$$

For low frequencies this operator is close to unity and does not considerably influence the amplitudes of the migrated section, also the phase is barely changed. For higher frequencies the amplitude of $\bar{D}_a(\omega)$ increases (Fig. 2), which means that

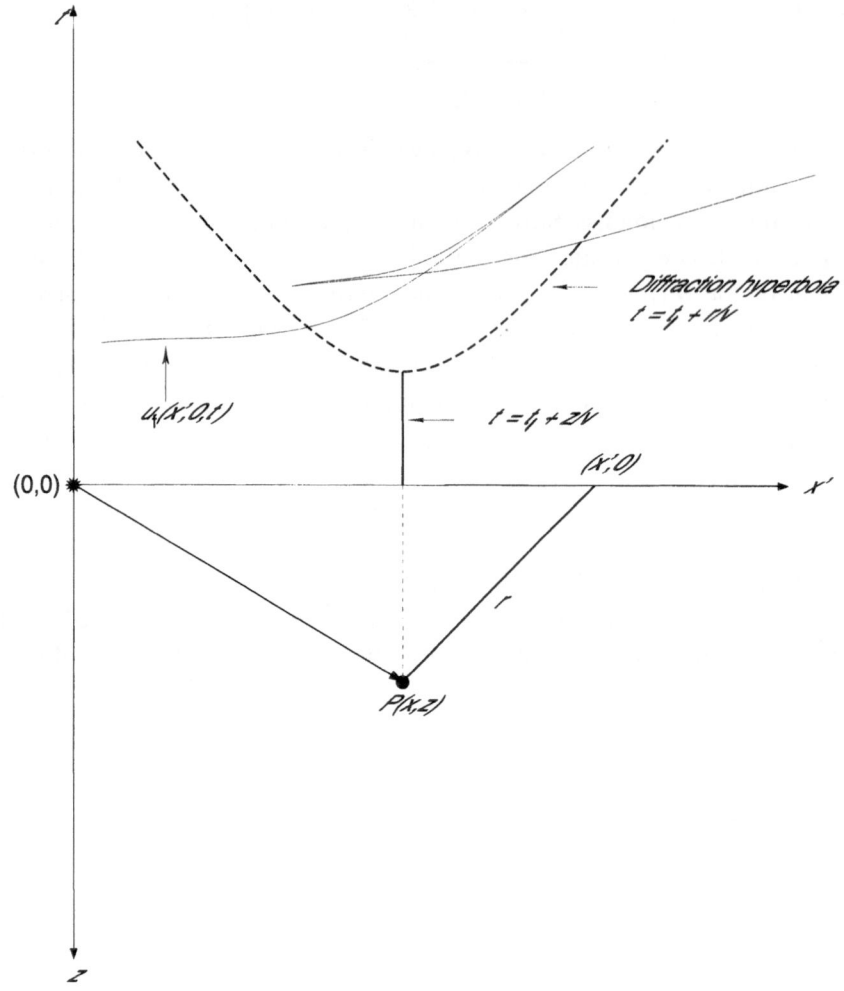

Figure 1

Travel-time curve of the filtered wave field U_f and the diffraction hyperbola for a point $P(x, z)$ in a homogeneous medium. v represents the velocity $c(\omega_r)$.

the compensation for amplitude loss is stronger. The variation of $D_a(t)$ with dissipation time is similar. For low dissipation time t_r^* the operator is almost a δ-function with no compensation for the amplitude loss, whereas it is a strong time-variable pulse for high t_r^* (Fig. 2). In the migration integral (9) the

Figure 2

The antidissipation operator $D_a(t)$ for different dissipation times t_r^* (a: $t_r^* = 3 \cdot 10^{-4}$ s, b: $t_r^* = 10^{-2}$ s), the corresponding amplitude spectrum $\bar{D}_a(\omega)$ and the inverse operator of $D_a(t)$, together with its spectrum. For the sake of comparison the dissipation operator $D(t)$ is also presented (dashed curves). The amplitude spectrum of $D_a(t)$ rises with increasing frequency, and $D_a(t)$ increases with dissipation time.

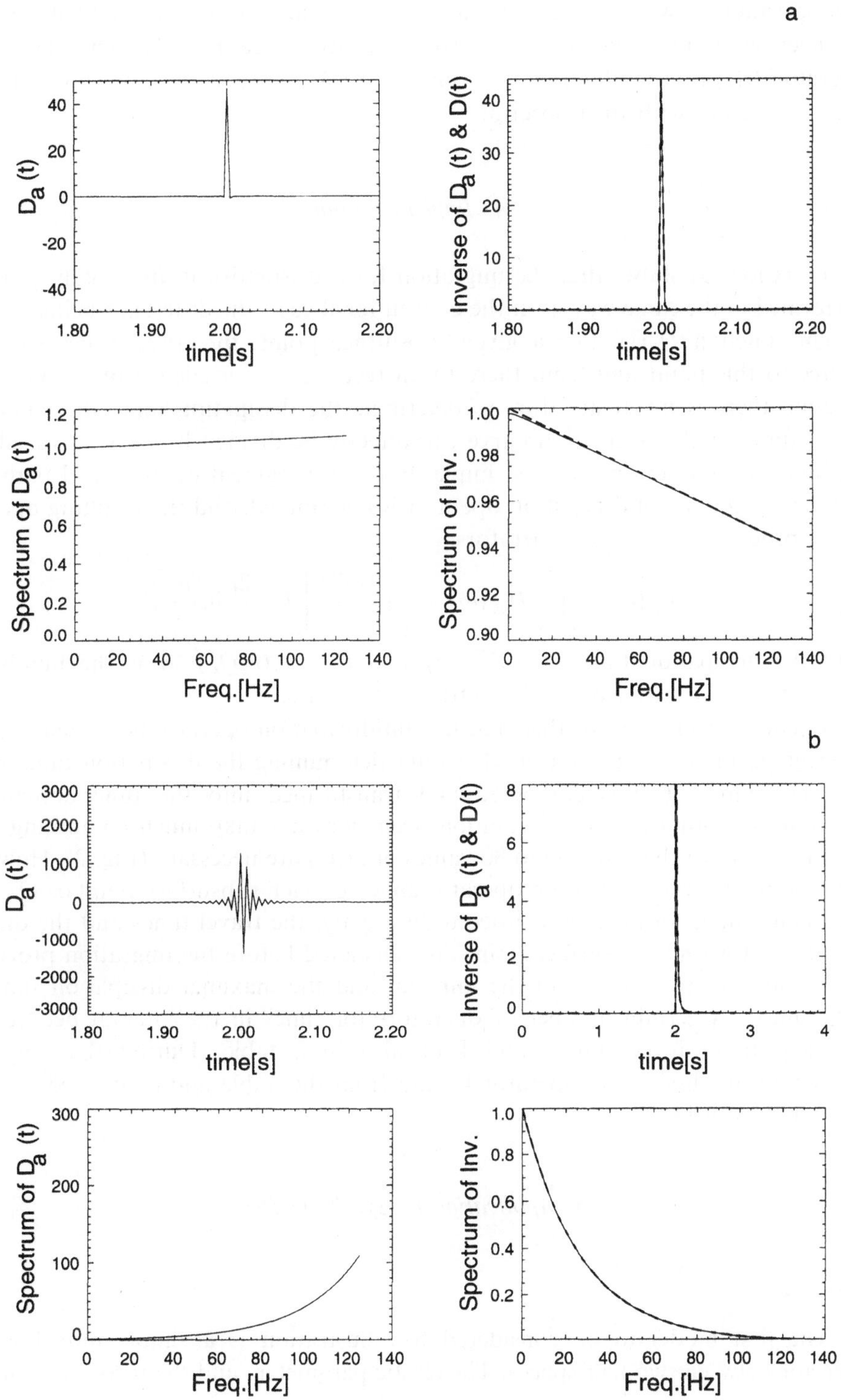
a
D_a (t)
time[s]
Inverse of D_a (t) & D(t)
time[s]
Spectrum of D_a (t)
Freq.[Hz]
Spectrum of Inv.
Freq.[Hz]
b
D_a (t)
time[s]
Inverse of D_a (t) & D(t)
time[s]
Spectrum of D_a (t)
Freq.[Hz]
Spectrum of Inv.
Freq.[Hz]

compensation is weakest near the apex of the summation curve and it becomes stronger with increasing distance from the apex (see Fig. 1). The dissipation operator $D(t)$ (dashed line) and the inverse of $D_a(t)$ are shown for comparison in Fig. 2, together with their spectra.

4. Implementation

Equation (9) shows that the migration for a dissipative medium is in principle performed in the same way as in the conventional case, the difference being the new weight function $D_a(t)$. For a given subsurface point, the travel times from the source to this point and from there to all receivers are needed. For a constant-Q medium there is no great effort to determine the dissipation time; the computed travel time for the diffraction curve can directly be divided by Q. In a variable-Q medium the computation is less simple. For every segment of the travel path with a different Q, the antidissipation operator is determined, and the resulting operator is the product of all single operators:

$$\bar{D}_a(\omega) = \prod_{k=1}^{N} \bar{D}_{ak}(\omega) = \exp\left\{\frac{\omega T^*}{2}\left[1 - \frac{2i}{\pi}\ln\frac{\omega}{\omega_r}\right]\right\} \tag{12}$$

with the dissipation time $T^* = \Sigma_{k=1}^{N} t_k^*$ and $t_k^* = r_k/(v_k Q_k)$. N is the number of segments on the travel path with different Q values.

Equation (9) implies further that the antidissipation operator is needed for each receiver position x'. That means that after determining the dissipation time for x', the spectrum $\bar{D}_a(\omega)$ is computed and transformed into the time domain for convolution with U_f. The convolution itself does not take much computing time; tests have shown that only 50 to 80 values of $D_a(t)$ are necessary (Fig. 2). However, determining the antidissipation operator anew for each subsurface point can be very time consuming. In practice I proceed differently: the travel times and the dissipation times for each subsurface point are calculated before the migration procedure itself and kept in tables, and the minimal and the maximal dissipation time are determined. A certain number of operators for times in the interval between the extremal times is computed and kept also in a table. During the migration procedure the dissipation operator is read from this table and convolved with the filtered wave field U_f.

5. Application to Synthetic Data

Model I

The first model to be considered for illustration is a simple model with a reflector between two half spaces. The elastic parameters and the recording configuration are shown in Figure 3. The distance between two successive receivers is 10 m. The synthetic seismograms are computed with a high-frequency version of generalized ray theory (EMMERICH et al., 1993, Appendix B). The source signal is a Küpper impulse, the dominant frequency is 20 Hz and the sampling rate 1000 Hz. Only the primary reflection is considered.

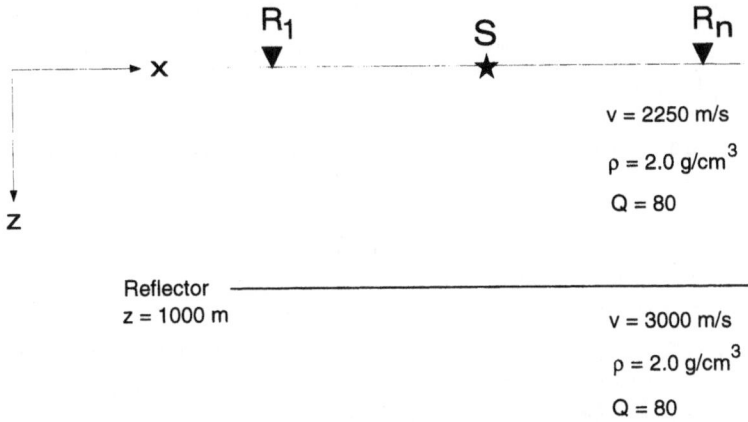

Figure 3
Model *I*: Interface between two half spaces. The quality factor and the density are kept constant, whereas the velocity varies. *S* represents the source position, R_1 and R_n are the positions of the first and the last receiver, at ± 1000 m.

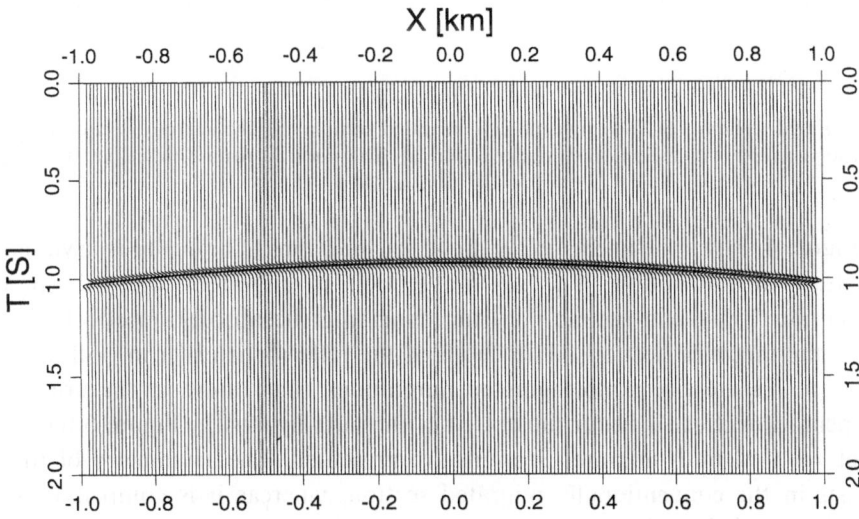

Figure 4
The anelastic synthetic seismograms of model *I*. The signal is asymmetric as a result of the absorption effects.

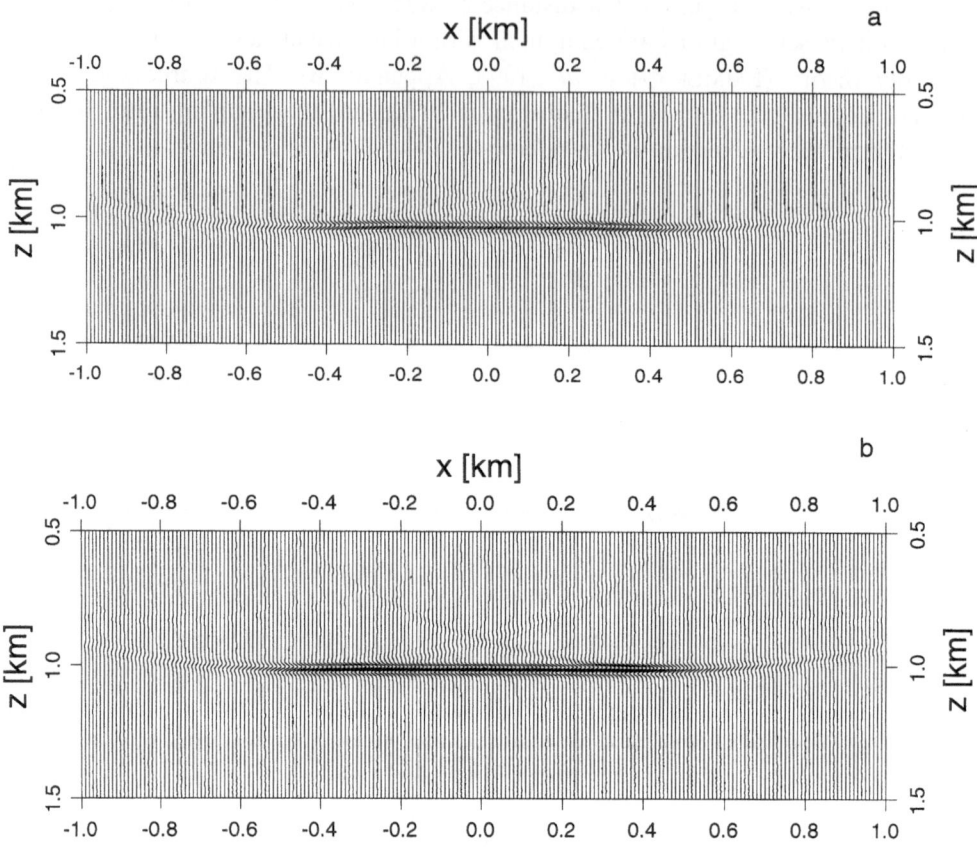

Figure 5
The migrated sections: (a) The conventionally migrated section. (b) The migrated section including the absorption effects (the anelastically migrated section). The traces in (b) are closer, i.e., the amplitudes have increased.

Figure 4 shows the anelastic synthetic seismograms. The signal is asymmetric as an effect of absorption. The seismic section is migrated twice: firstly in the conventional way and secondly by incorporating the absorption effects. Figure 5 shows the migrated sections. The depth of the reflector is exact in both cases. The traces in Figure 5b exhibit greater amplitudes than those in Figure 5a. This compensation effect is better illustrated in the next figures. In Figure 6 the traces at offset $x = 0$ of both migrated sections are presented. The asymmetry of the signal persists in the conventionally migrated section, whereas it is compensated in the anelastically migrated section. Moreover the signal is narrower, this means the dispersion effect is removed. The amplitude increase is also shown in Figure 7, where the AVO curves of the migrated sections are displayed. The curve with the thick dashes is the theoretically estimated reflection amplitude on the reflector.

Disregarding migration artifacts at $x = \pm 0.3$ km due to the limited illumination of the reflector, the agreement of this curve with that of the anelastically migrated section is good. The increase in amplitude due to the incorporation of absorption is almost 30%.

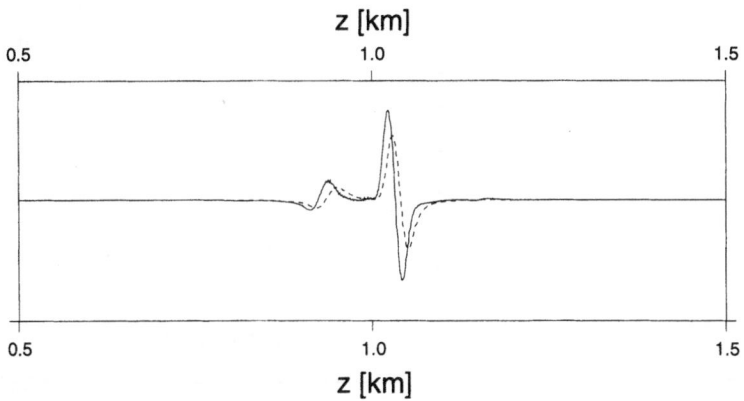

Figure 6

Traces at offset $x = 0$ from both migrated sections. The continuous line is the trace from the anelastically migrated section, and the dashed line follows from the conventionally migrated section. The amplitude increase and the compensation of the dispersion effect are evident. The weak first arrival is a migration artifact (see Fig. 5).

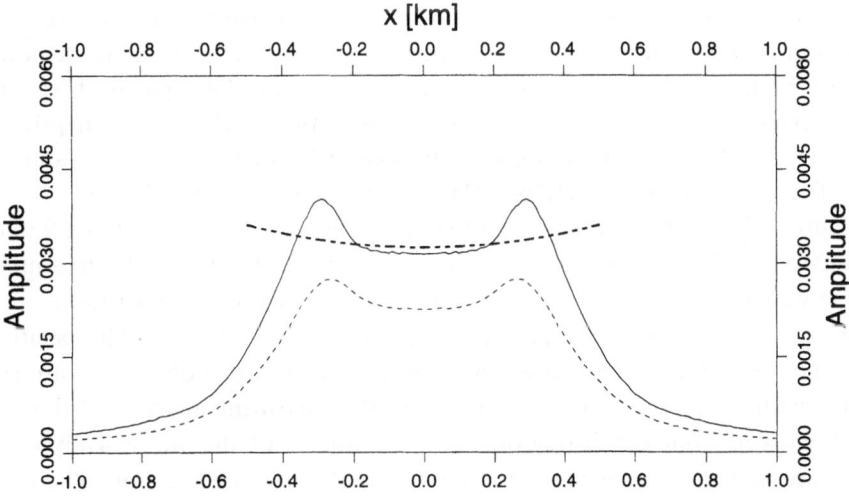

Figure 7

The *AVO* curves of both migrated sections. The maximal amplitude of each trace is picked and displayed against the offset. The line styles have the same meaning as in Figure 6, and the third curve with thick dashes is the theoretically estimated amplitude.

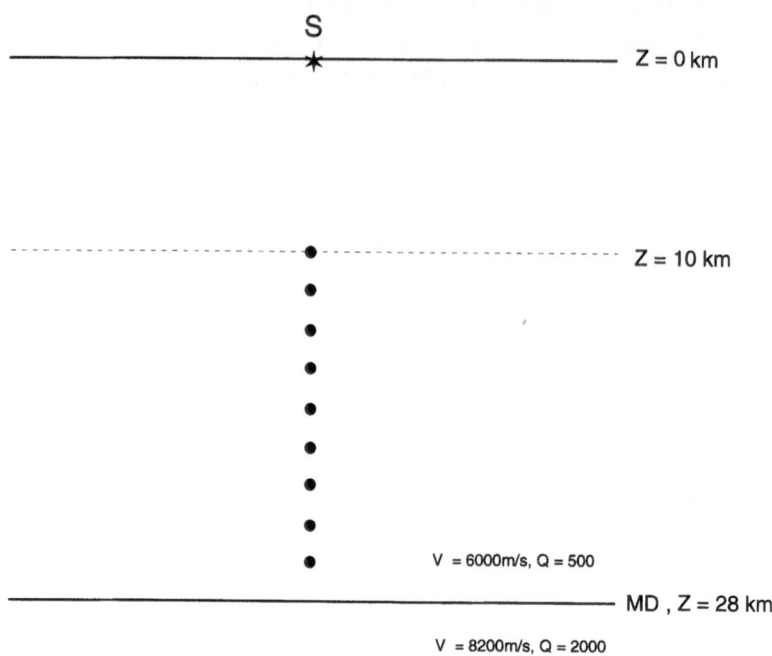

Figure 8
Model *II*: A homogeneous crust with 9 diffractors embedded. The Moho-discontinuity (MD) is at 28 km
depth. *S* is the source position, and the receiver configuration is split-spread with a total length of 16 km.

Model II

The second model is that of a homogeneous earth's crust, where 9 point
diffractors of the same strength are embedded (Fig. 8). The aim is to demonstrate
how the compensation of amplitude varies with depth. The spacing between two
consecutive diffractors is 2 km. The Moho discontinuity (MD) is at depth 28 km.
There are 201 receivers, they are equally spaced between $x = -8$ km and $x = +$
8 km. The dominant frequency is 20 Hz and the sampling rate 250 Hz.

Figure 9 shows the synthetic seismograms. The strong event between 9 and 10 s
is the Moho reflection, the earlier events are the diffractions. Their amplitudes
become weaker with increasing time. This is partially due to absorption.

The migrated sections are presented in Figures 10a and 10b. The positions of
the Moho and the diffractors are exactly reconstructed. The diffractors are slightly
smeared with increasing depth. This is due to the narrowing aperture of the receiver
spread, as seen from the diffractors. The amplitudes of the anelastically migrated
section are higher than those of the conventionally migrated section. The traces at
$x = 0$ for both sections (Fig. 11) establish clearly the amplitude recovery. The
dispersion effect is very slight, and its compensation is barely visible. The artifacts
just ahead of the MD event are also amplified (Fig. 11). This leads to a slightly
noisy image of the Moho.

Figure 12 illustrates the amplitude compensation versus the dissipation time, which represents here the depth of the different diffractors. The squares represent the values picked from the migrated sections, whereas the triangles are theoretically estimated values. The curve of picked values fluctuates, this is probably due to the coarse discretisation with the spacing between two successive diffractors. Both curves rise in the same way with increasing dissipation time or increasing distance from the source, which means that the amplitude compensation increases with depth as expected. The picked values lie systematically above the theoretical curve, which means that the amplitudes of the anelastically migrated section are overamplified or/and those of the conventionally migrated section are underestimated. Because of the finite aperture, and, in this example, its narrowness compared to the depth of diffractors, the impulse cannot be well reconstructed, it is smeared as is seen in Figure 10a. This may lead to amplitude errors, which make the amplitudes stay below their real values. This could be one reason for the systematic effect observed, however a slight overamplification due to the compensation of attenuation is also possible, if for example the frequency band used for the antidissipation

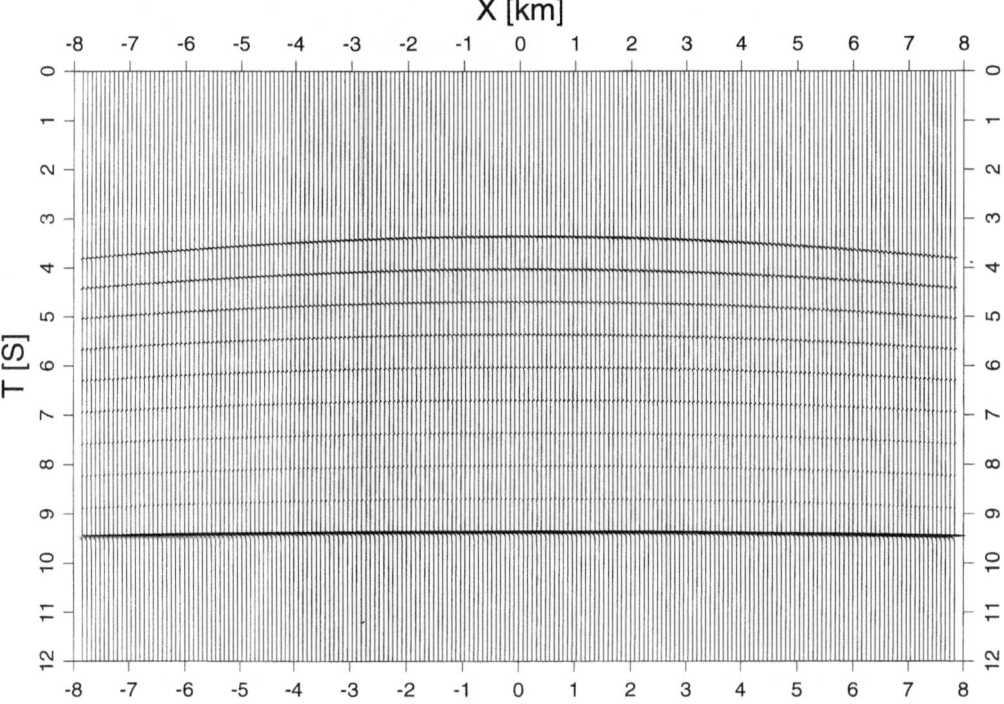

Figure 9

The synthetic seismograms for model *II*. The strongest event is the Moho-reflection, the other events are from the diffractors. They become weaker with increasing travel time.

operator is broader than the frequency band of the signal. This second effect seems to dominate the first.

6. Conclusions

I have shown that the effects of absorption can be incorporated into Kirchhoff migration. The principles of the migration itself do not change, and the absorption effects are compensated during the backpropagation of the observed wave field. A

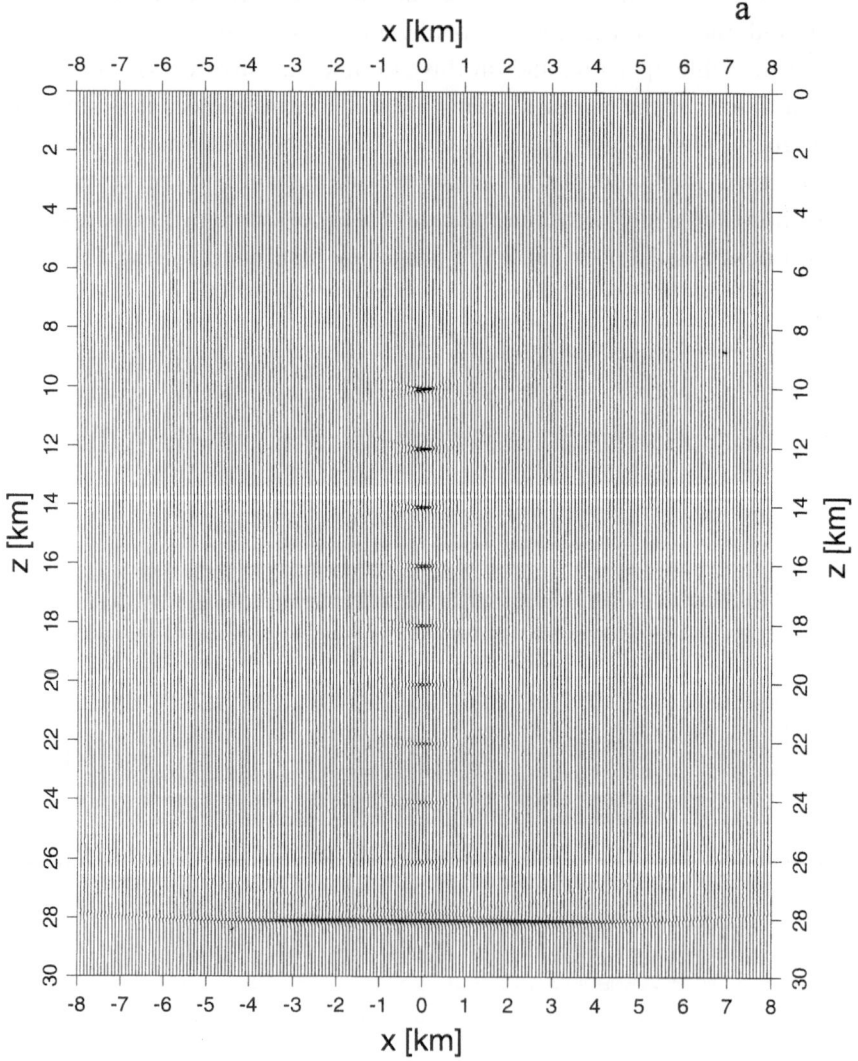

Figure 10 (a)

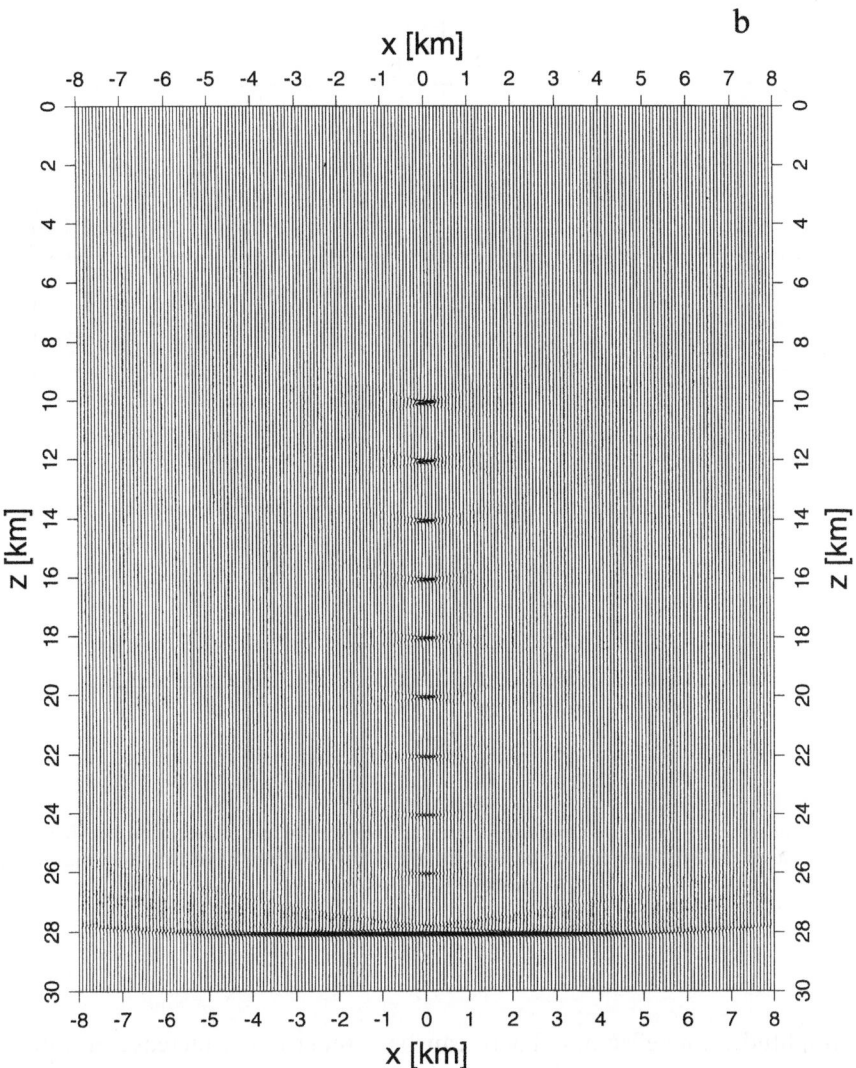

Figure 10
The migrated sections for Figure 9: (a) The conventionally migrated section. (b) The anelastically migrated section. Both sections have the same amplitude scale.

new antidissipation operator has been given. This operator, which is practically the inverse of the known dissipation operator, provides the compensation of the absorption effects through filtering of the observed wave field prior to summation along the diffraction curve of a subsurface point. The models considered have shown that the amplitude attenuation and the dispersion are compensated. The amplitudes in the anelastically migrated section match well the theoretically com-

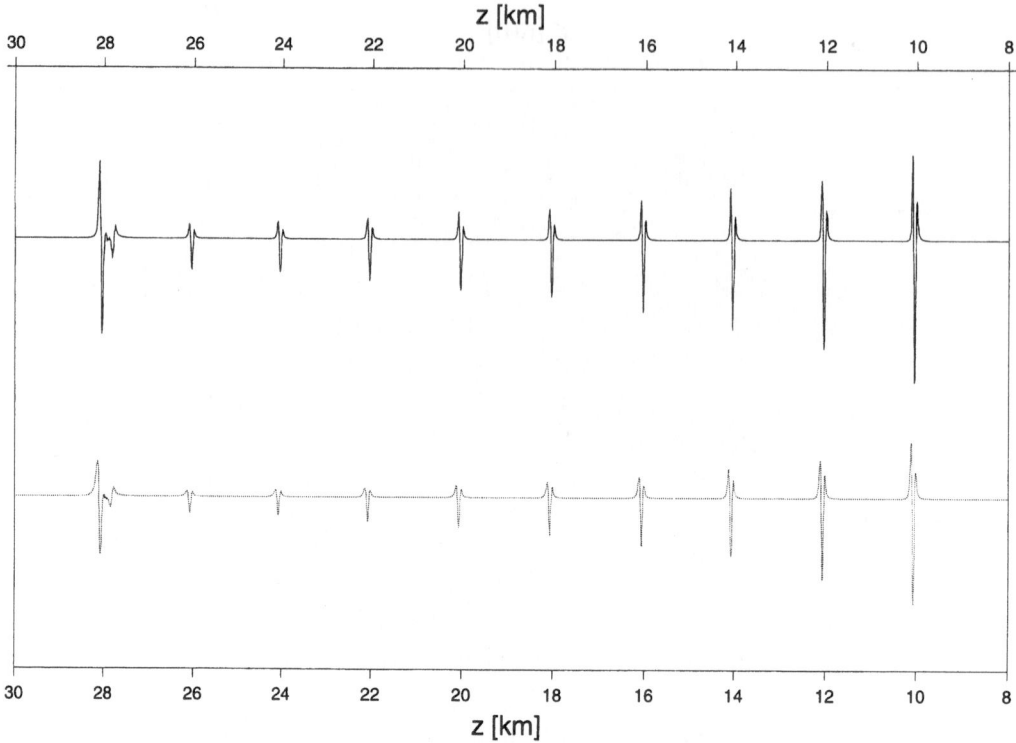

Figure 11

The traces at offset $x = 0$ from both migrated sections. The amplitude compensation is evident, but the dispersion removal is slight and cannot be seen clearly. The line styles have the same meaning as for model *I* (see Fig. 6).

puted amplitudes on reflectors. There can be a remarkable increase in amplitudes compared to the amplitudes of the conventionally migrated section. The increase observed cannot be generalized for all cases, since it depends on the Q structure of the medium and on the frequency content of the signal, but it shows that the lack of attenuation compensation in conventional migration cannot simply be ignored. What this increase quantitatively means for *AVO* analysis has still to be investigated, for *P* waves and for *S* waves as well.

The operator developed here for the compensation of absorption effects has no limitation concerning the macro model of migration. It is also valid for heterogeneous media, only the computation of the dissipation time can be more time-consuming. Just as the macro velocity model determines the success of conventional migration, the success of anelastic migration depends on how well the Q macro model is known.

Acknowledgements

The research of this paper has been supported by the Friedrich–Ebert–Stiftung and by the Bundesministerium für Bildung und Forschung (grant 03GT94026, DEKORP project). I wish to thank Gerhard Müller for his support and many suggestions, which increased the quality of this work. My gratitude also to Dr. Jörg Schleicher and to an anonymous reviewer, whose suggestions helped to improve the quality of this paper, and to Ingrid Hörnchen for typing the manuscript.

Appendix: Derivation of the Antidissipation Operator

The starting point is the spectrum of the depth-continued wave field with the real velocity substituted by the complex velocity (Equation (5)):

$$\bar{U}_{Pa}(x, z, \omega) = \int_{-\infty}^{+\infty} i\omega \bar{U}(x', 0, \omega) \frac{1}{2c_r h} H_1^{(2)}\left(-\frac{r\omega}{c_r h}\right)\cos \varphi \, dx' \qquad \text{(A-1)}$$

with (Equation (6))

$$h = \exp\left\{\frac{\cot\left(\gamma\frac{\pi}{2}\right)}{2Q_r}\left[1 - \left(\frac{\omega_r}{\omega}\right)^{\gamma}\right]\right\}\exp\left\{\frac{i}{2Q_r}\left(\frac{\omega_r}{\omega}\right)^{\gamma}\right\}. \qquad \text{(A-2)}$$

The Hankel function is replaced by its asymptotic approximation (MORSE and FESHBACH, 1953):

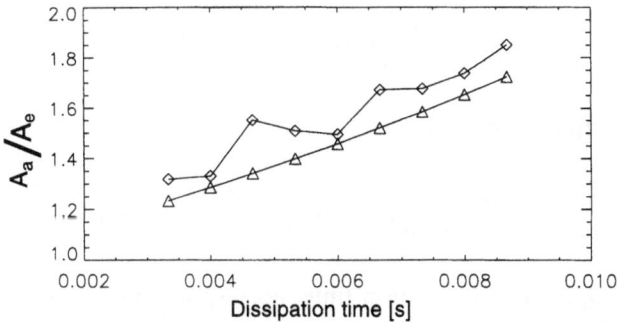

Figure 12

The amplitude compensation as function of depth, which is represented by the corresponding dissipation time. A_a/A_e is the quotient of the maximal amplitude from the anelastically migrated section and the corresponding amplitude from the conventionally migrated section (squares). The triangles are theoretically estimated values. Both curves rise with increasing dissipation time demonstrating that the amplitude compensation increases with depth.

$$H_1^{(2)}(s) \simeq \left(\frac{2}{\pi s}\right)^{1/2} \exp\left[-i\left(s-\frac{3\pi}{4}\right)\right], \quad |s|\gg 1. \tag{A-3}$$

This gives the high-frequency expression

$$H_1^{(2)}\left(-\frac{r\omega}{c_r h}\right) = -\frac{1}{\pi}\left(\frac{2c_r h}{r}\right)^{1/2}\left(\frac{i\pi}{\omega}\right)^{1/2} \exp\left(i\frac{r\omega}{c_r h}\right) \tag{A-4}$$

which is inserted into (A-1):

$$\bar{U}_{Pa}(x, z, \omega) = \int_{-\infty}^{+\infty} i\omega \bar{U}(x', 0, \omega)\left(\frac{i\pi}{\omega}\right)^{1/2} \exp\left(i\frac{r\omega}{c_r h}\right)\frac{-\cos\varphi}{\pi(2c_r hr)^{1/2}}\,dx'. \tag{A-5}$$

For weakly dissipative media ($Q \gg 1, Q_r \gg 1$) the quantity h in (A-2) can be expanded as follows,

$$h = 1 + \frac{\cot\left(\gamma\frac{\pi}{2}\right)}{2Q_r}\left[1-\left(\frac{\omega_r}{\omega}\right)^\gamma\right] + i\frac{1}{2Q_r}\left(\frac{\omega_r}{\omega}\right)^\gamma, \tag{A-6}$$

and differs only slightly from 1. Consequently, the last factor in the integrand of (A-5) can be replaced by the real expression $-\cos\varphi(2c_r r)^{-1/2}/\pi$ which is the weight function $V(x')$ from (3), taken for $v = c_r$, the real reference velocity. The exponential term in (A-5) reads with (A-6):

$$\exp\left(i\frac{r\omega}{c_r h}\right) \approx \exp\left\{i\frac{r\omega}{c_r}\left[1-\frac{\cot\left(\gamma\frac{\pi}{2}\right)}{2Q_r}\left[1-\left(\frac{\omega_r}{\omega}\right)^\gamma\right]-\frac{i}{2Q_r}\left(\frac{\omega_r}{\omega}\right)^\gamma\right]\right\}$$

$$= \exp\left(i\frac{r\omega}{c_r}\right)\bar{D}_a(\omega) \tag{A-7}$$

with the antidissipation operator

$$\bar{D}_a(\omega) = \exp\left\{\frac{\omega t_r^*}{2}\left(\frac{\omega_r}{\omega}\right)^\gamma\right\}\exp\left\{-i\frac{\omega t_r^*}{2}\cot\left(\gamma\frac{\pi}{2}\right)\left[1-\left(\frac{\omega_r}{\omega}\right)^\gamma\right]\right\}. \tag{A-8}$$

Here, $t_r^* = r/(c_r Q_r)$ is the dissipation time, corresponding to the distance r. Finally, inserting (A − 7) into (A-5) yields the depth-continued wavefield spectrum with absorption incorporated:

$$\bar{U}_{Pa}(x, z, \omega) = \int_{-\infty}^{+\infty} i\omega \bar{U}(x', 0, \omega)\left(\frac{i\pi}{\omega}\right)^{1/2} \exp\left(i\frac{r\omega}{c_r}\right)\bar{D}_a(\omega)V(x')\,dx'. \tag{A-9}$$

REFERENCES

ANDERSON, D. L., and MINSTER, J. B. (1979), *The Frequency Dependence of Q in the Earth and the Implication for Mantle Rheology and the Chandler Wobble*, Geophys. J. R. Astron. Soc. *58*, 431–440.

BERCKHEMER, H., KAMPFMANN, W., AULBACH, E., and SCHMELING, H. (1982), *Shear Modulus and Q of Forsterite and Dunite near Partial Melting from Forced-oscillation Experiments*, Phys. Earth Planet. Inter. *29*, 30–41.

BICKEL, S. H., and NATARAJAN, R. R. (1985), *Plane-wave Q Deconvolution*, Geophysics *50*, 1426–1439.

CARMICHAEL, R. S., *Handbook of Physical Properties*, volume III (CRC Press, Boca Raton, Florida 1982).

EMMERICH, H., ZWIELICH, J., and MÜLLER, G. (1983), *Migration of Synthetic Seismograms for Crustal Structures with Random Heterogeneities*, Geophys. J. Int. *113*, 225–238.

GELIUS, L. J. (1987), *Inverse Q-filtering—A Spectral Balancing Technique*, Geophys. Prosp. *35*, 656–667.

HARGREAVES, N. D., and CALVERT, A. J. (1991), *Inverse Q-filtering by Fourier Transform*, Geophysics *56*, 519–527.

KJARTANSSON, E. (1979), *Constant Q Wave Propagation and Attenuation*, J. Geophys. Res. *84*, 4737–4748.

LIU, H. P., ANDERSON, D. L., and KANAMORI, H. (1976), *Velocity Dispersion due to Anelasticity: Implications for Seismology and Mantle Composition*, Geophys. J. R. Astron. Soc. *47*, 41–48.

MEISSNER, R., *The Continental Crust—A Geophysical Approach* (Academic Press, Orlando 1986).

MITTET, R., SOLLIE, R., and HOKSTAD, K. (1995), *Prestack Depth Migration with Compensation for Absorption and Dispersion*, Geophysics *60*, 1485–1494.

MORSE, P. M., and FESHBACH, H., *Methods of Theoretical Physics* (McGraw-Hill Book Company, New York 1953).

MOSHER, C. C., KEHO, T. H., WEGLEIN, A. B., and FORSTER, D. J. (1996), *The Impact of Migration on AVO*, Geophysics *61*, 1603–1615.

MÜLLER, G. (1983), *Rheological Properties and Velocity Dispersion of a Medium with a Power-law Dependence of Q on Frequency*, J. Geophys. *54*, 20–29.

SCHNEIDER, W. A. (1978), *Integral Formulation for Migration in Two and Three Dimensions*, Geophysics *43*, 49–76.

STOLT, R. H. (1978), *Migration by Fourier Transform*, Geophysics *43*, 23–48.

STOLT, R. H., and BENSON, A. K., *Seismic Migration—Theory and Practice* (Geophysical Press, London 1986).

TOUTOU, C. (1997), *Incorporation of the effects of absorption into the migration of seismic data*, Ph.D. Thesis, University of Frankfurt, Germany (in German).

URSIN, B., and DAHL, T. (1992), *Seismic Reflection Amplitudes*, Geophys. Prosp. *40*, 483–512.

WINKLER, K., and NUR, A. (1979), *Pore Fluids and Seismic Attenuation in Rocks*, Geophys. Res. Lett. *6*, 1–4.

(Received January 15, 1998, revised August 27, 1998, accepted September 25, 1998)

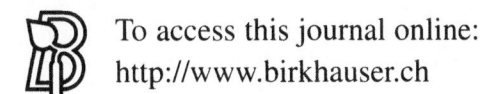

To access this journal online:
http://www.birkhauser.ch

Pure appl. geophys. 156 (1999) 253–278
0033–4553/99/010253–26 $ 1.50 + 0.20/0

❘ Pure and Applied Geophysics

Enhancement of Deep Seismic Reflections in Pre-stack Data by Adaptive Filtering

BURKHARD BUTTKUS[1] and CHRISTIAN BÖNNEMANN[1]

Abstract—Adaptive filters offer advantages over Wiener filters for time-varying processes. They are used for deconvolution of seismic data which exhibit non-stationary behavior, and seldom for noise reduction. Different algorithms for adaptive filtering exist. The least-mean-squares (LMS) algorithm, because of its simplicity, has been widely applied to data from different fields that fall outside geophysics. The application of the LMS algorithm to improve the signal-to-noise ratio in deep reflection seismic pre-stack data is studied in this paper. Synthetic data models and field data from the DEKORP project are used to this end.

Three adaptive filter techniques, one-trace technique, two-trace technique and time-slice technique, are examined closely to establish the merits and demerits of each technique. The one-trace technique does not improve the signal-to-noise ratio in deep reflection seismic data where signal and noise cover the same frequency range. With the two-trace technique, the strongest noise reduction is achieved for small noise on the data. The filter efficiency decreases rapidly with increasing noise. Furthermore, the filter performance is poor upon application to common-midpoint (CMP) gathers with no normal-moveout (NMO) corrections. Application of the two-trace method to seismic traces before dynamic correction results in gaps in the signal along the reflection hyperbolas. The time-slice technique, introduced in this paper, offers the best answer. In this case, the one-trace technique is applied to the NMO-corrected gathers across all traces in each gather at each time to separate the low-wavenumber component of the signal in offset direction from the high-wavenumber noise component. The stacking velocities used for the dynamic correction do not need to be known very accurately because in deep reflection seismics, residual moveouts are small and have only a minor influence on the results of the adaptive time-slice technique. Noise reduction is more significant with the time-slice technique than with the two-trace technique. The superiority of the adaptive time-slice technique is demonstrated with the DEKORP data.

Key words: Prestack signal enhancement, adaptive filtering, deep reflection seismics.

1. Introduction

The primary objective of the geological/geophysical interpretation of DEKORP data is a structural assessment based on stacked sections. Problems beyond this

---

[1] Bundesanstalt für Geowissenschaften und Rohstoffe, Stilleweg 2, D-30655 Hannover, Germany.
E-mail: buttkus@bgr.de; boennemann@bgr.de

scope, for example, distinguishing between different lithological units, determination of the fine structure and the physical properties of vertical or lateral transition zones and boundaries, or the assessment of the structural continuity, require, however, a more extensive analysis of the DEKORP data, especially of the pre-stack data.

Such an interpretation has to be based on dynamic signal parameters of seismic reflections and how they change as a function of reflection time and offset, especially the frequency-dependent amplitude and phase behavior. Determination and analysis of these parameters from pre-stack data require good signal quality. In deep reflection seismics, in most cases reflections are weak and highly disturbed by noise; the quality of the data, therefore, is normally inadequate for such studies and preprocessing is necessary to improve the signal-to-noise ratio.

Digital filter techniques applied in exploration seismics, e.g., conventional frequency filters, frequency-wavenumber filters, and optimum filters based on the Wiener criterion, normally show only limited success in deep reflection seismics, because the signal and noise are nonstationary. Adaptive filters offer an alternative with a large potential for enhancing deep seismic reflections in pre-stack data so that they can be analyzed.

Fundamentals of adaptive filtering are given in several text books, for example, WIDROW and STEARNS (1985), HAYKIN (1991), HONIG and MESSERSCHMITT (1984). The problem of adaptive noise suppression is treated by WIDROW et al. (1975), and the use of adaptive filtering for noise removal in magnetotelluric data by HATTINGH (1989). In reflection seismics, adaptive filters have been used mainly for deconvolution (GRIFFITHS et al., 1977; PRASAD and MAHALANABIS, 1980). Results of first applications in seismics for noise suppression have been published by HATTINGH (1990) and DRAGOSET (1995).

2. Problems of Filtering Deep Reflection Seismic Data

The quality of the pre-stacking DEKORP data has to be improved (*primary goal*) before the dynamic signal parameters can be analyzed. When the data have been improved, better stacking velocity estimates and stacking results (*secondary goal*) may be achieved.

The signal analysis requires not only efficient noise reduction by filtering, but also that the signal shape is not changed by the filtering or at least that the filter effect is the same on signals which are compared during the analysis of dynamic signal parameters (e.g., on neighboring traces). This requirement can be fulfilled strictly only by applying the same filter operator to the complete data set. In this case, optimum noise reduction is not obtained for a signal or noise that varies with time or offset; optimum noise suppression requires that the filter design takes these variations into account. This, however, results in filter operators that vary with time and offset; such operators change the signals differently depending on reflection time and offset. Because in deep reflection seismics, improvement of the signal-to-noise ratio is a prerequisite for pre-stack signal analysis, we have to design time-varying filters and to accept changes in the signal as a result of the filtering. The influence of different filter operators has to be considered when signals are analyzed and compared.

Wiener filters are usually used for signal enhancement in seismics. If the signal is chosen as desired filter output, the desired improvement of the signal-to-noise ratio is obtained with the Wiener filter, while leaving the signal unchanged to a large extent, depending on the signal and noise power spectra. Normally, the filter operators are determined trace by trace for successive time windows to take into account the nonstationary signal and noise properties.

But there are restrictions on the application of the Wiener filter in deep reflection seismics:

Two basic assumptions of the Wiener filter (the filter input is random and stationary) are not fulfilled in deep reflection seismics: There are only a few reflections in the seismogram. Moreover, the signal shape (owing to absorption) and the statistical values of the noise may change with time.

The problem that the input is not stationary can be overcome either by applying a Wiener filter in a blocked form for successive time windows or by using the Booton extension for time-varying processes of the Wiener filter (e.g., BOOTON, 1952 and BUTTKUS, 1991). These techniques for filter design require estimates of the time-varying signal and noise correlation functions. In deep reflection seismics, however, such estimates can seldom be carried out with the necessary accuracy.

The nonrandomness of the reflections is at least as restrictive for the application of the Wiener filter as their not being stationary, because this important condition for the estimation of the correlation functions needed to calculate the filter operator, is not fulfilled. The Wiener filter results, therefore, are not optimal.

Adaptive filtering is an alternative to the Wiener filter. This type of filter offers several advantages over a Wiener filter:

(1) Adaptive filters are able to handle time-varying processes. These filters have the ability to adjust their coefficients automatically to changing data. The filter operator is continuously updated with time.

(2) Algorithms are available which do not require *a priori* information like a Wiener filter (correlation functions) and which can easily be implemented.

Owing to these advantages, the application of adaptive filters to data of deep reflection seismic measurements is attractive. But surprisingly, adaptive filters have seldom been applied for signal-to-noise improvement in seismics and have not even been studied systematically for this purpose. Therefore, the primary objective of this paper is to illustrate some strengths and weaknesses of adaptive filtering of seismic data, using examples of their application to synthetic seismic data as well as to field data.

3. Introduction to Adaptive Filtering

The efficiency of an adaptive filter depends on the relationships between signal and noise (especially significant is the manner in which they correlate), its structure, and algorithms.

There are two filter structures that are most universally utilized for adaptive filters: the transversal and lattice structures. The most commonly employed structure for an adaptive filter is the transversal structure. This filter can be represented by the equation of discrete convolution:

$$y_k = \sum_{i=0}^{M-1} w_i x_{k-i}. \tag{1}$$

Lattice structure is equivalent to the transversal structure, in the sense that any transfer function that can be represented by a transversal structure can also be represented by a lattice structure. The coefficients of a lattice structure are commonly called partial correlation coefficients or reflection coefficients. A review of lattice structures and their use for adaptive prediction of time series is given by FRIEDLANDER (1982).

Adaptation algorithms can be classified as block or recursive algorithms (see, e.g., HONIG and MESSERSCHMITT, 1982). In block algorithms, the input signal is divided into blocks and each block is processed independently. In recursive algorithms, the adaptive algorithm is implemented as a continually operative set of recursive equations that are continually updated. A new set of filter parameters is determined for each input value.

A very simple approach to recursive adaptation algorithms is the least-mean-squares (LMS) technique. An explicit solution of a system of linear equations is avoided by a gradient approximation approach. The LMS algorithm is the most widely applied adaptation algorithm and is also the one used in this study.

Our adaptive filter approach is a joint process estimator. In this case, there are two inputs, x_k and d_k. The x_k input is filtered and the result is subtracted from d_k to yield the error e_k. The objective is to minimize the size of e_k. This is equivalent to the objective that the adaptive filter generates an estimate of d_k based on a filtered version of x_k.

4. The Concept of Adaptive Noise Cancellation

4.1 Principles of Adaptive Noise Cancellation

The adaptive noise cancelling (ANC) method goes back to WIDROW et al. (1975, 1976). It was developed for real-time elimination of interference. A block diagram of the principle of noise cancellation is shown in Figure 1. The primary trace d contains the signal superimposed by noise n_0, which is assumed to be uncorrelated

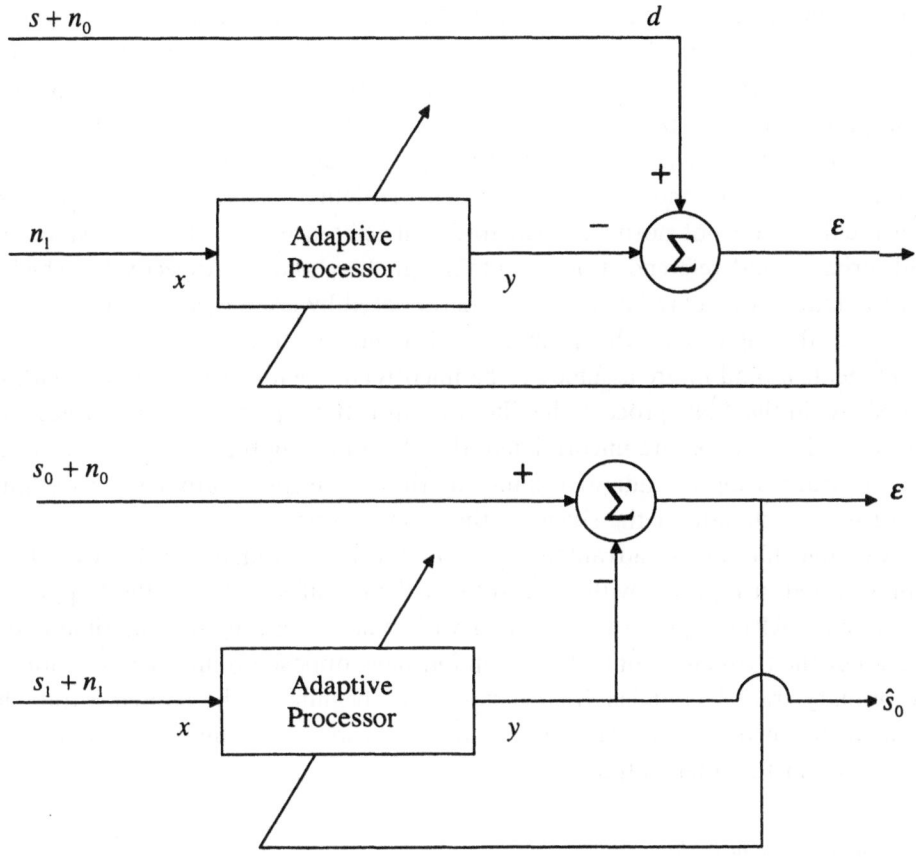

Figure 1
Block diagram of the adaptive noise-cancelling method.

with the signal. The reference trace x contains the noise n_1, uncorrelated with the signal, however correlated with n_0, where the exact nature of this correlation does not need to be known.

The reference input is adaptively filtered to match n_0 as close as possible. It is then subtracted from the primary input $s + n_0$ to produce the system output e. This system output contains the signal s plus residual noise. The system output is fed back to the adaptive filter and the filter coefficients are adjusted by a gradient algorithm to minimize the residual noise. The better the noise components n_0 and n_1 correlate, the stronger the noise reduction.

Figure 2 illustrates an example of the application of the ANC method to synthetic data. Signals, n_0 and n_1 are different second-order autoregressive processes (AR(2)), which can be described by

$$x_k = \alpha_1 x_{k-1} + \alpha_2 x_{k-2} + r_k, \tag{2}$$

where α_1 and α_2 are constants and r_k is white noise. The example shows the general principle of adaptive filtering: Besides the trace which we want to filter (the so-called primary trace (c)), a reference trace (d) is needed. The selection of an appropriate reference trace is usually the main problem in adaptive filter design.

For applications in geophysics, the principle of adaptive noise suppression shown in Figure 1 must be modified as illustrated in Figure 3, because signal-free noise measurements are not normally available as a reference trace. Let us assume that both primary and reference traces contain signals s_0 and s_1, respectively, which in some unknown way correlate and are superimposed by noise n_0 and n_1, respectively.

While the signals in the primary and reference traces correlate, the noise components n_0 and n_1 are assumed to be uncorrelated with each other and with the signals. As in the ANC process, the filter output is those parts of the two traces that correlates. If n_0 and n_1 are uncorrelated, then by adjusting the reference trace to the primary trace using the adaptive filter algorithm, the filter output is an optimum least-squares estimate of the signal in the primary trace.

Adaptive filtering is advantageous only for nonstationary processes. This is demonstrated in Figure 4 with the results of the modified ANC method applied to a stationary AR(2) signal superimposed with time-varying noise. The time dependencies of the frequency content of n_0 and n_1 have opposite trends, i.e., the noise of the primary trace has higher frequencies at the beginning, the reference trace has them at the end of the trace. Even in this worse case, most of the noise is suppressed in the filter output.

4.2 The LMS Algorithm

The LMS (least-mean-squares) algorithm goes back to WIDROW and HOFF (1960). The algorithm is an approximation of the exact solution of minimizing the mean-square error. This approximation is the price to be paid for not requiring a stationary input and knowledge of the signal and noise statistics.

The filter operator is updated in the following form:

$$\vec{W}_{k+1} = \vec{W}_k - \mu \nabla[E[e_k^2]]. \tag{3}$$

$\vec{W}_{k+1}$ is the new filter operator at time $k + 1$; it is determined by updating $\vec{W}_k$, the previous operator. e_k is the difference (or error) between the filtered reference trace and the primary trace at time k. The correction term is an estimate of the derivative of the mean-squared error with respect to the filter coefficients. μ is the adaptation parameter which determines the convergence and stability of the filter.

Figure 2

Numerical example of noise cancellation by adaptive filtering. The signal shown in (a) is estimated from the primary trace (c) using the reference trace (d). The reference trace and the noise of the primary trace (e) are two different high-frequency AR(2) processes; the signal component (b) is an AR(2) process with lower frequencies.

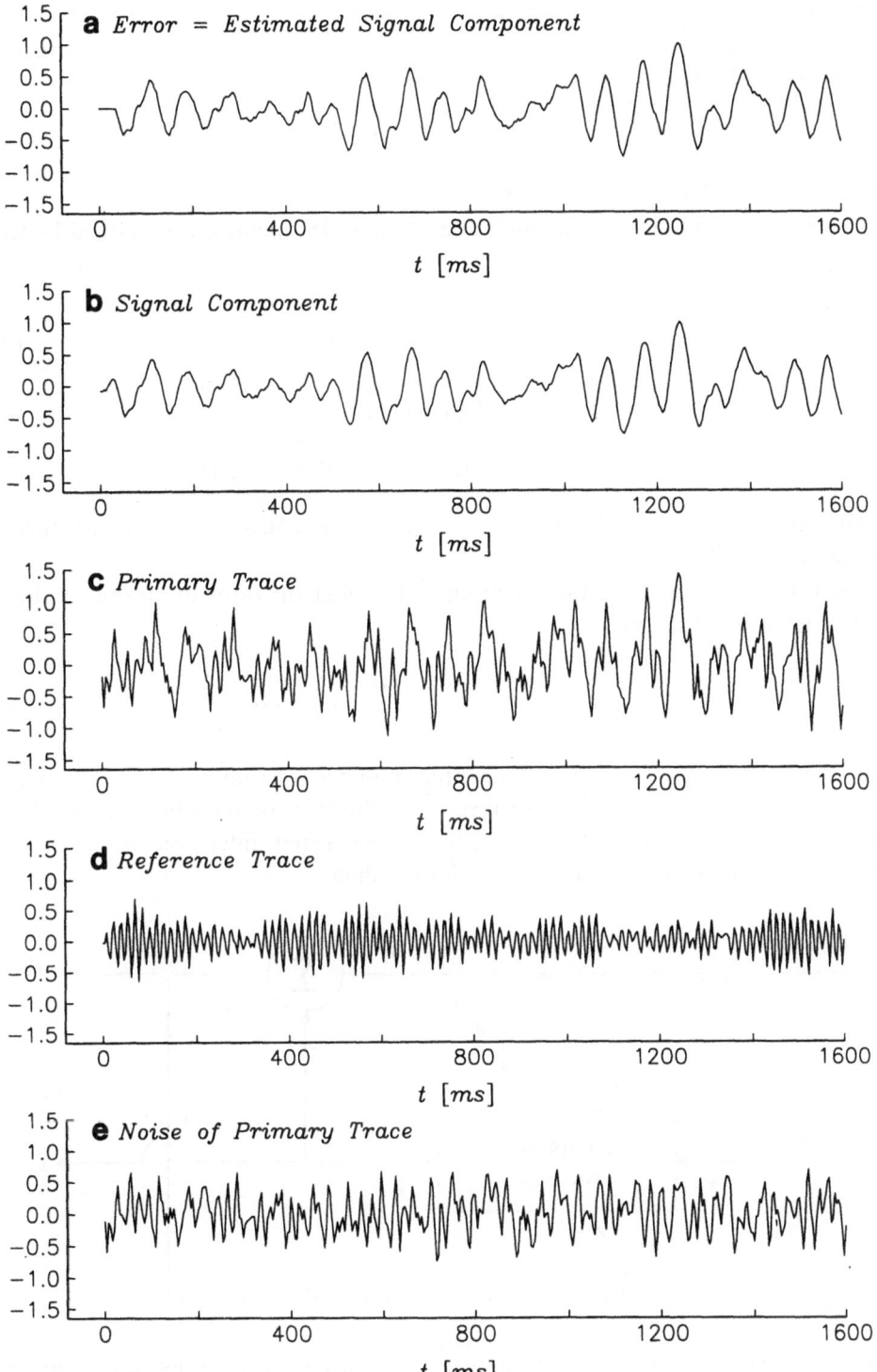

In the LMS algorithm, instantaneous estimates are used to approximate the expectation of the error square e_k^2. Equation (3) then yields the following recursive relation for updating the operator:

$$\vec{W}_{k+1} = \vec{W}_k - \mu \nabla[e_k^2] = \vec{W}_k + 2\mu e_k \vec{x}_k, \tag{4}$$

where $\vec{x}_k$ is the reference trace at time k.

Using expanded notation, the updating of the LMS operator is described by the following equations:

$$y(k) = \sum_{i=0}^{M-1} w_k(i) x(k-i), \tag{5}$$

$$e_k = d(k) - y(k), \tag{6}$$

$$W_{k+1}(i) = W_k(i) - 2\mu e_k x(k-i), \quad i = 0, 1, \ldots, M-1. \tag{7}$$

The procedure starts with an initial estimate $\vec{W}_0$. A convenient choice for this is the zero vector, $\vec{W}_0 = 0$.

The error resulting from the updating of the filter operator in all our applications is defined as follows:

$$e_k = d(k) - y(k) = x\left(k - \frac{M}{2} - 1\right) - y(k). \tag{8}$$

Thus, the error is calculated by taking equal numbers of samples of the reference trace forwards and backwards with respect to the time of the filter output. This definition is advantageous in that data prior to the current filter time are used, and this improves the adaptation for nonstationary data.

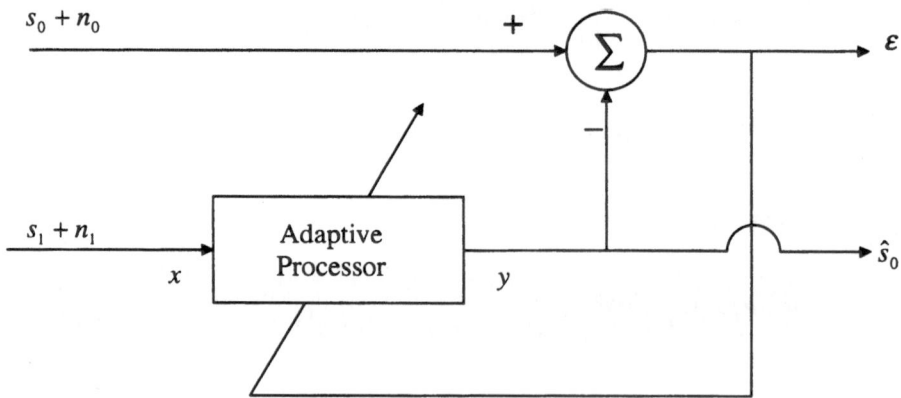

Figure 3
Principle of the modified ANC process. At the filter output the correlating part between the primary trace and the reference trace is estimated.

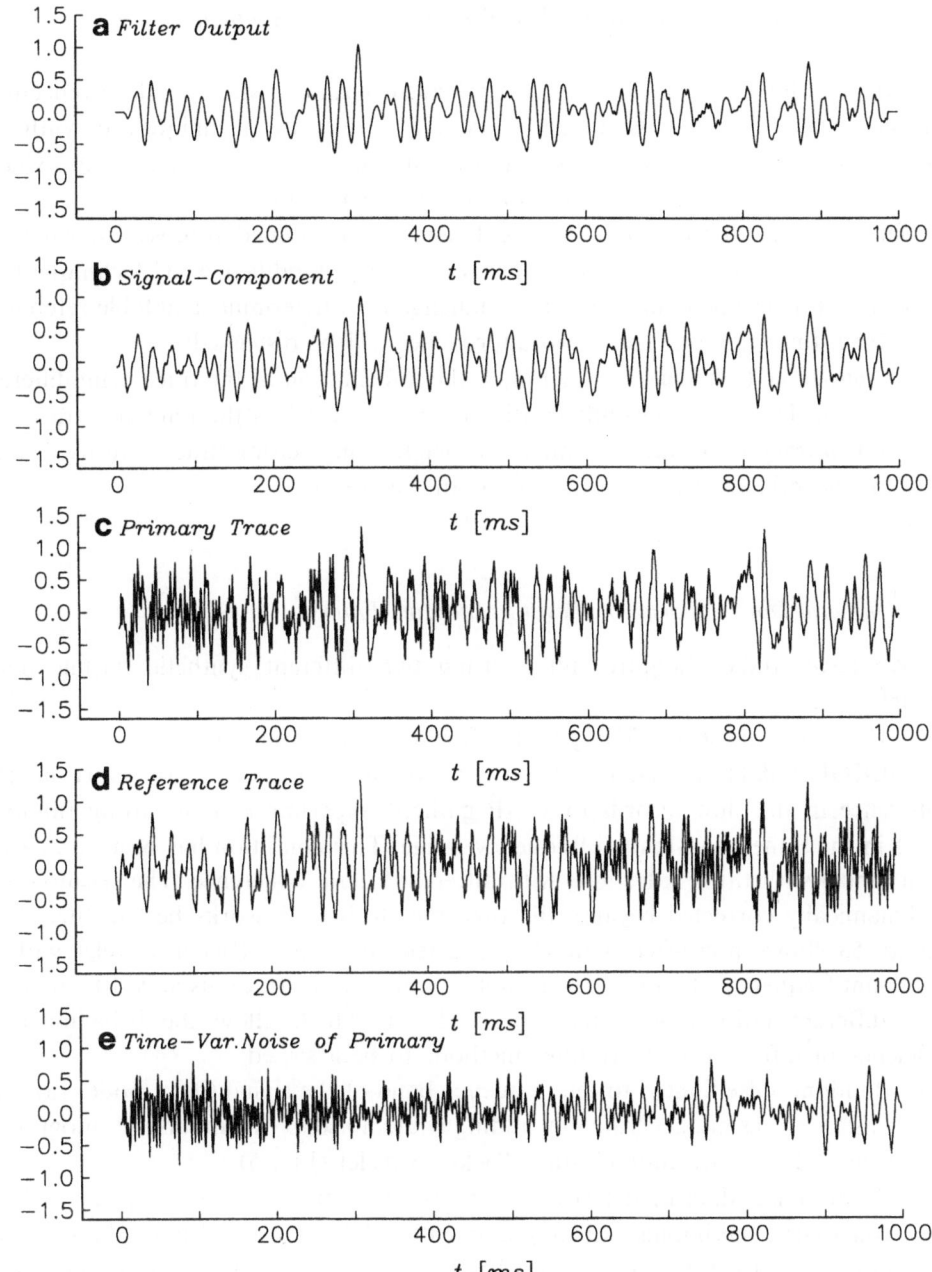

Figure 4

Noise suppression by adaptive filtering for time-varying noise, s, n_0, n_1 are different AR(2) processes. n_1 is the time-reverse of the time-varying noise n_0. The noise of the primary trace (e) is generated by modifying the coefficients of the AR(2) process.

5. Tests of Adaptive Filter Techniques Using Synthetic Data

Adaptive filters seldom have been applied in seismics, especially deep reflection seismics. Therefore, we have applied various adaptive filter techniques to synthetic data in order to develop a technique most suitable for signal-to-noise ratio (SNR) improvement of deep seismic reflections in pre-stack data.

The type of adaptive filter discussed here is the "joined process" in which a reference trace is used. As discussed in section 4.1, a modified ANC technique has to be used for seismic data. The main challenge is to determine a suitable reference trace from the available data. Two approaches will be discussed:

(1)   A **one-trace technique**, in which only the information of the trace being filtered is used. Here the time-shifted primary trace is used as the reference trace.

(2)   A **two-trace technique**, in which two neighboring seismic traces are used, one for the primary trace, the second as the reference.

5.1 Synthetic Data Models for Testing Adaptive Filters

We have studied adaptive filters using two different synthetic seismic data models.

(1)   *Model 1*: Synthetic CMP gathers after dynamic correction

DEKORP data are characterized by isolated reflections with short lateral coherence in the shot records or CMP gathers, superimposed by strong random noise in the same band-limited frequency range. This condition has been simulated by a model with three reflections with different lateral coherence. The seismogram is dynamically corrected because the two-trace technique works best in this case. Figure 5a shows noise-free data. The selected signal is a Ricker wavelet with a dominant frequency of 20 Hz. The sampling rate is 4 ms, trace distance 100 m. Two very different noise models have been chosen which allow the behavior and efficiency of different adaptive filter methods to be assessed:

 (i)   Random noise limited to the frequency band of the signal (data model 1a); the seismic traces are generated by adding white noise to the impulse seismograms, followed by convolution with a Ricker wavelet (Fig. 5).

(ii)   White noise (data model 1b); in this case, the seismic traces are generated by convolving the impulse seismograms with a Ricker wavelet and adding white noise (Fig. 6). Whereas the signal is band-limited, the noise covers the entire frequency range.

Even if the latter model is unrealistic in deep reflection seismics, the test with white, random noise is important to understand the behavior and efficiency of adaptive filter techniques.

Note: In this paper, the noise level is given by the ratio of the maximum amplitudes of noise and signal.

(2) *Model 2*: Synthetic CMP gathers before dynamic correction

To assess the power of adaptive filters when applied to CMP gathers *before* stacking, a synthetic model (model 2) has been chosen in which the signal component comprises two reflection hyperbolas at 0.65 and 1.0 s without dynamic correction (Fig. 7). The curvatures of the reflections correspond to 4000 and 6000 m/s stacking velocities. Wavelet, sampling rate, and trace separation are the same as in model 1.

Even if the models do not satisfy the non-stationarity criteria, they are ideal for testing the different filter techniques.

5.2 One-trace Adaptive Filter Technique

Only a single seismic trace is used in this filter technique. This trace forms the primary trace, while a time shift of this trace is taken as the reference trace. Signal and noise are assumed to cover different frequency bands. To separate signal and

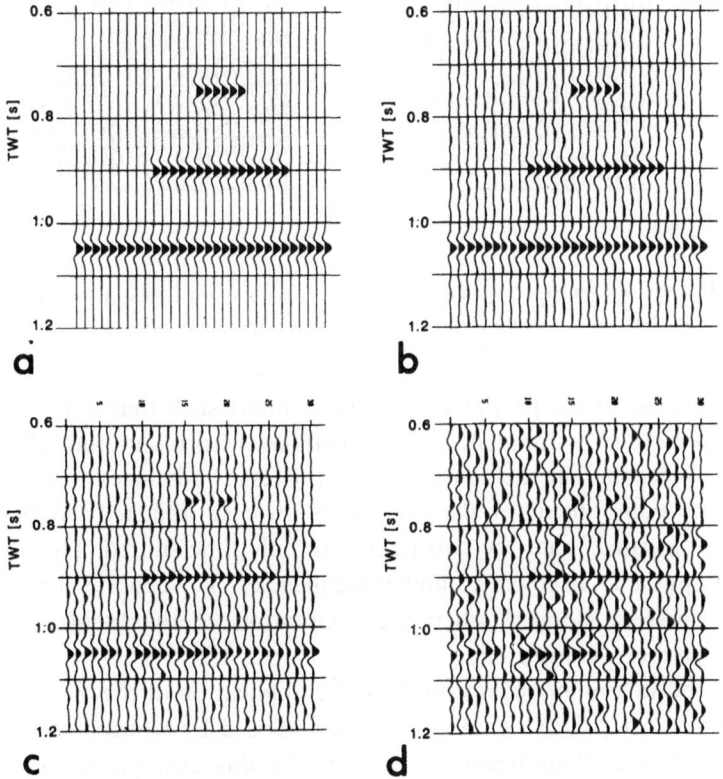

Figure 5
Data model 1a: synthetic CMP gathers after dynamic correction. (b), (c) and (d) show the synthetic seismograms in which the signal component, shown in (a), is superimposed by 10, 20, 50% band-limited noise.

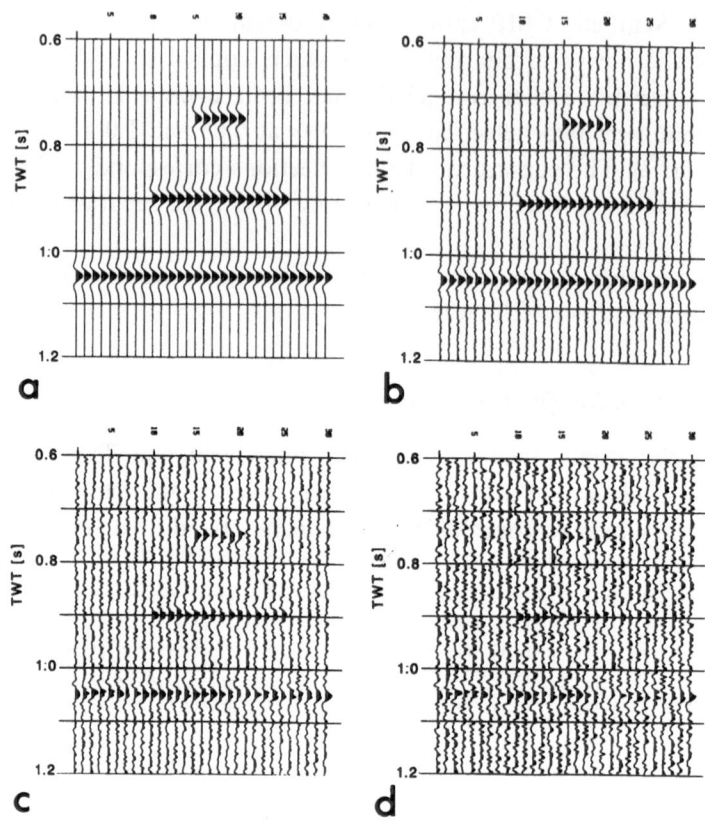

Figure 6
Data model 1b: synthetic CMP gather after dynamic correction. (b), (c) and (d) show the synthetic
seismograms in which the signal component, shown in (a), is superimposed by 20, 50, 100% white noise.

noise, we make use of the property of adaptive filters such that at the beginning of
the filtering only the low frequency components of the primary and reference traces
are fitted to each other. If the adaptive filter process is advanced to later times, the
high frequency components of the two traces are also adjusted. At this stage the
operator has "learned" that the two traces are only time shifted. The adaptation is
conducted stepwise from lower to higher frequencies. To separate signal and noise,
the filter process is stopped when the adaptation of the low-frequency signal
component has been completed.

Figure 8 shows a numerical example of the application of the one-trace adaptive
filter technique. An AR(2) signal (b) is superimposed by white noise (e). The two
components differ in their frequency content. In this example the low frequency
part is defined as signal, the high frequency component as noise. The adaptation
requires the right choice of parameter values, especially of the length of the filter
operator and the value of the adaptation parameter. On the basis of numerical tests
with AR(2) signal/random noise models we derive the following conclusions:

(1) Owing to the error definition in equation (8), the length of the adaptive filter operator must be at least twice the time shift between the primary trace and the reference trace so that the time shifted reference trace can be matched to the primary trace. The reason for this is that a given sample in the primary trace and its equivalent in the reference trace must be used simultaneously for the error calculation. Only then can the filter process match the two traces.

(2) Whether only the low-frequency component can be adjusted at the beginning of the process depends on the signal-to-noise ratio. If the amplitude of the high-frequency noise component is too large relative to the signal, the filter process attempts at the beginning of the procedure to adjust the two components simultaneously, which means that the two components cannot be separated.

(3) The value of the adaptation parameter, which must be defined in advance, usually on the basis of several test runs, is critical for a frequency-dependent adaptation. A value that is too large results in poor filter performance, because

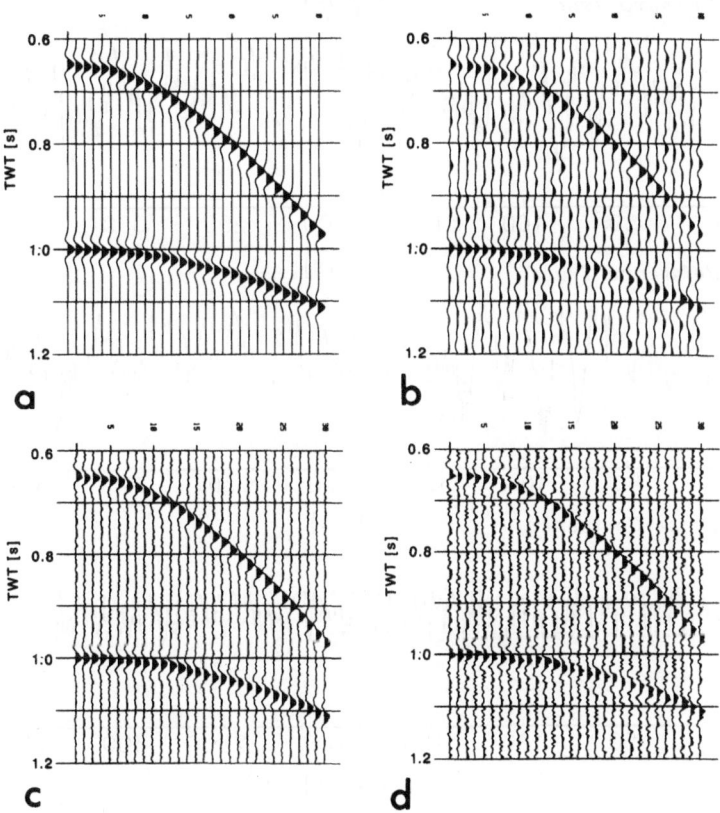

Figure 7

Data model 2: synthetic CMP gathers before dynamic correction. (a) shows the noise-free signal, (b) the signal superimposed by 20% band-limited noise, (c) and (d) the signal superimposed by 20 and 50% white noise, respectively.

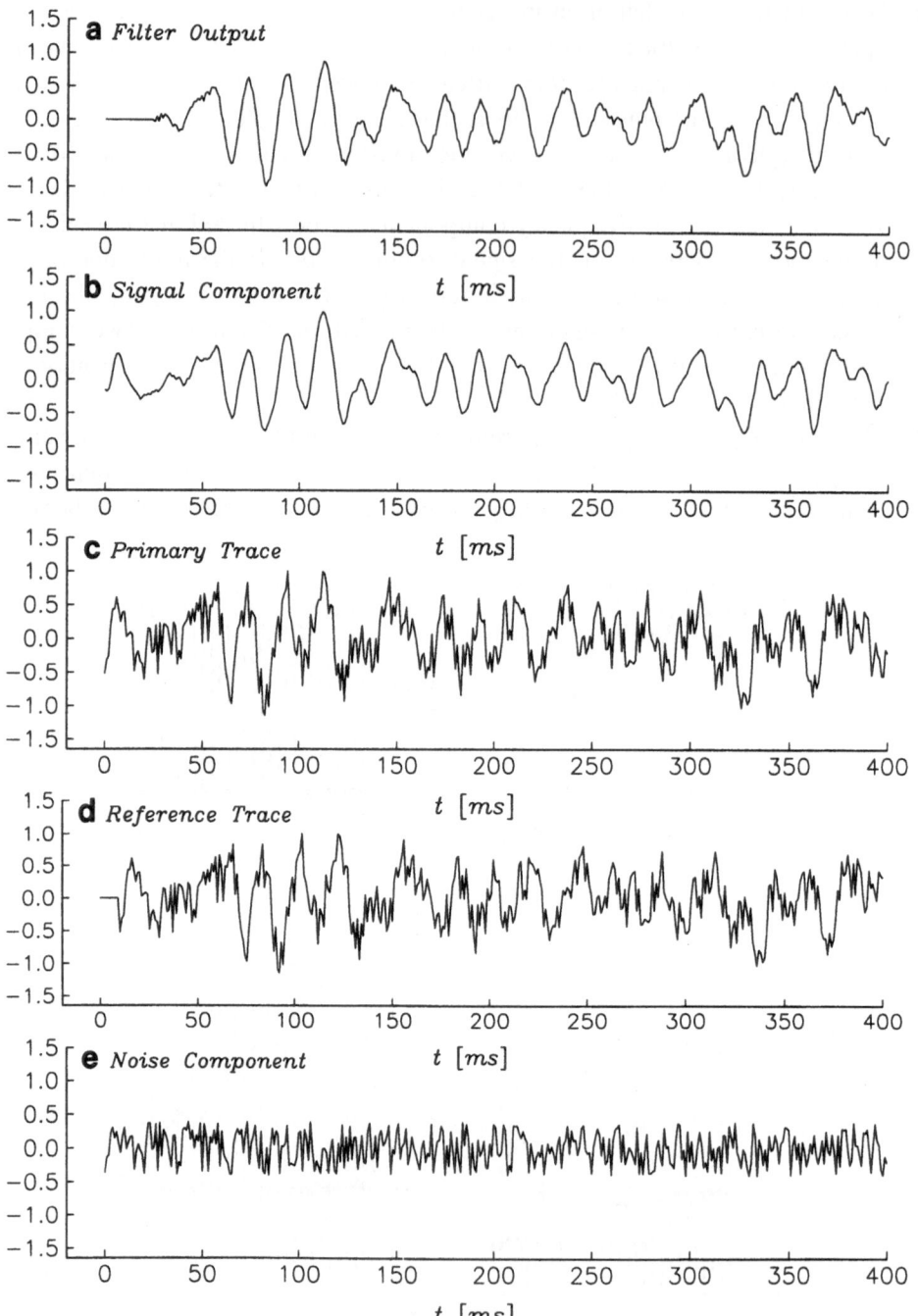

Figure 8

Noise-suppression by one-trace adaptive filtering. The AR(2) signal (b) is overlain by white noise (e). The adaptive filter operator is twice as long as the time shift between the reference trace (d) and the primary trace (c). SNR = 5, time shift = 10 ms, operator length = 21 samples, dominant signal period ≈ 20 ms, sampling rate = 1 ms.

the updating of the filter coefficients in proximity to the minimum of the error function is not accurate enough to adjust to the minimum. If the adaptation factor is too small, convergence towards the minimum of the mean-square error function is slow and the filter may possibly adjust both frequency components simultaneously.

For short records the procedure must be applied in an iterative manner. The adaptive filter process starts with a filter operator whose coefficients are set equal to zero and which are updated continuously with processing time. The filter operator at the end of the trace is the starting operator for the next iteration. The iterative process is stopped before the high frequency noise of the reference trace is adjusted to the primary trace. For long records the iterative procedure must be applied to short time windows.

The one-trace technique has been studied using models that are more similar to seismic data. The objective has been to investigate the dependency of the filter results on the signal-to-noise ratio as well as on the bandwidth of signal and noise. Model 2 (CMP gather before dynamic correction) has been used for these studies, because for the application of the one-trace technique, the dynamic correction is not important.

Figure 9 displays results of the one-trace adaptive filtering process for 20% band-limited noise and 20% white noise. The noise reduction is weak for band-limited noise, and considerably stronger for white noise, although with serious signal distortion in both cases. The degree of signal distortion increases with the magnitude of the noise. For band-limited noise the one-trace technique can be successfully applied only if the noise amplitude is substantially less than 50% of the signal amplitude.

Application to synthetic data confirms that the one-trace technique is efficient only if signal and noise have different frequency contents, while no improvement can be attained if signal and noise are restricted to the same frequency bands. Therefore, application of this method to deep reflection seismic data cannot be expected to improve data quality, because signal and noise have similar frequency content.

5.3 Two-trace Adaptive Filter Technique

This is a modified ANC technique. Two neighboring traces of a seismic CMP gather are used as primary and reference traces, resulting in one output trace. Application of the procedure to a CMP gather with N input traces results in $N - 1$ output traces.

To best satisfy the requirements for this modified ANC technique, the CMP gathers must be dynamically corrected. If the CMP gathers are corrected with the right stacking velocities, the signals are in phase from trace to trace, while the noise, which is assumed to be random, does not correlate between traces.

The two-trace technique has been tested using the dynamically corrected data models 1a (band-limited noise) and 1b (white noise) shown in Figures 5 and 6, respectively. Results of filtering the 20 and 50% band-limited noise models are presented in Figure 10, those of the 20 and 50% white noise models in Figure 11.

In addition, the two-trace method has been applied to CMP gathers before dynamic corrections. Figure 12 illustrates one example with 20% band-limited noise. The signals are now time shifted from trace to trace. Longer operators are needed to take this time shift into consideration, with the consequence that the adaptation capability and the efficiency of the filter process are reduced. While the noise reduction is comparable to that of filtering dynamically corrected traces, the signal component acquires gaps along the reflection hyperbolas.

Our tests yielded the following results:

● For low-noise situations (<20% of the signal), there is little noise reduction between signals because the operator does not adjust fast enough after a signal to the new, signal-free situation.

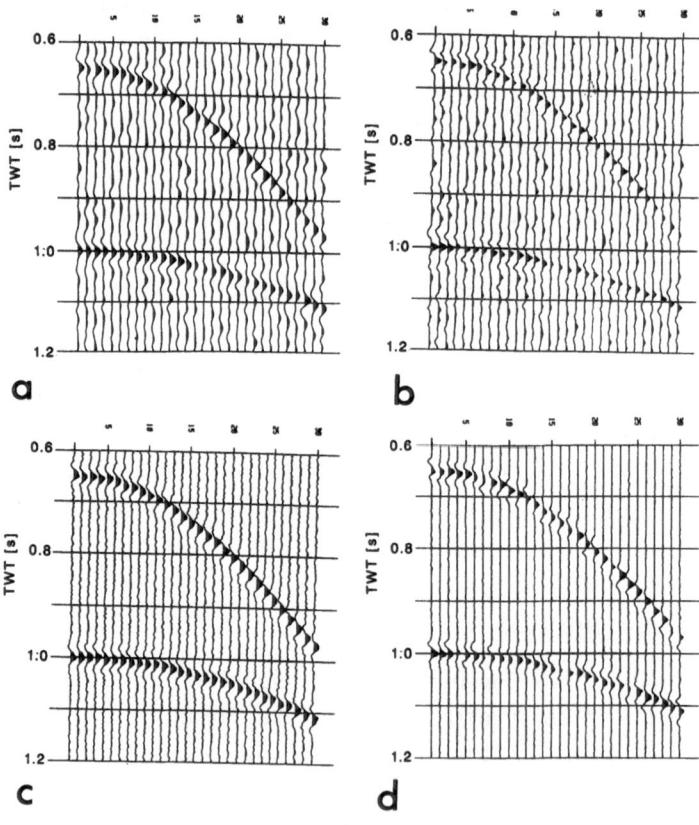

Figure 9

Application of the one-trace adaptive filter technique to synthetic seismic data: a) synthetic CMP gather before dynamic correction (20% band-limited noise), b) results after one-trace adaptive filtering of (a), c) same as (a), however, 20% white noise, d) results after adaptive filtering of (c).

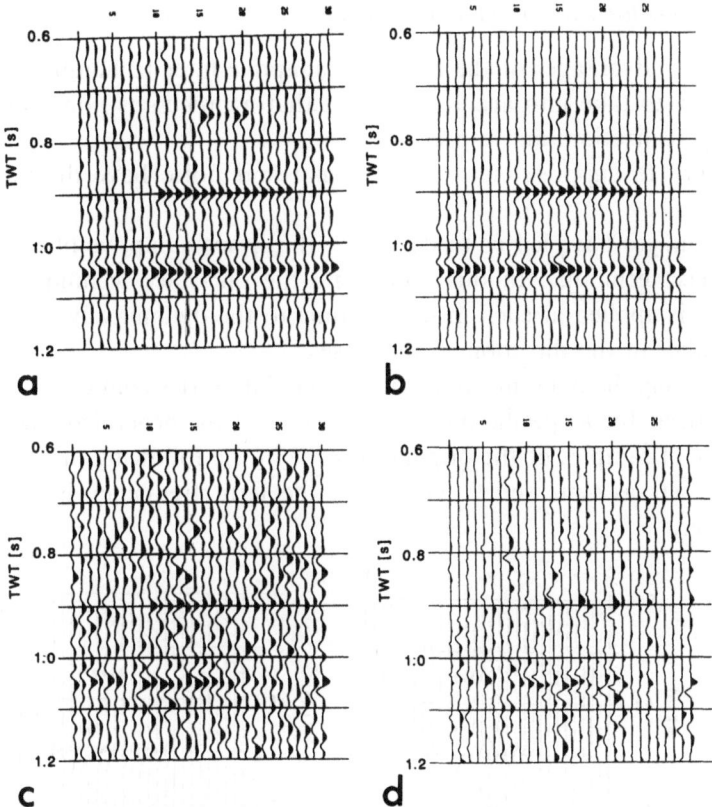

Figure 10
Application of the two-trace adaptive filter technique to data model 1a (band-limited noise). a) synthetic CMP gather after dynamic correction (20% band-limited noise), b) result of two-trace adaptive filtering of the CMP gather shown in (a), c) same as (a), but 50% band-limited noise, d) synthetic seismogram (c) after two-trace adaptive filtering.

- When signal and noise cover the same frequency range, noise reduction is strong only if the noise amplitude is less than 30% of the signal amplitude. With increasing noise the efficiency of the adaptive filter decreases rapidly. For ≥ 50% noise the filter results are not satisfactory. For band-limited signals and broadband noise, noise reduction by the two-trace technique is satisfactory up to 50% noise.
- If dynamic corrections are not applied, signals are distorted and fade out.

No satisfactory results can be expected or achieved by applying the two-trace technique to deep reflection seismic data, because the signal and noise lie in the same frequency range and the noise amplitudes are of the same magnitude as those of the signal or even larger.

5.4 *The Time-slice Adaptive Filter Technique*

The main advantages of the one-trace adaptive filter technique are:
— its ability to take into account the fact that a process is nonstationary,
— its easy realization, and especially
— its ability to suppress noise when signal and noise have different frequency
  contents.

These advantages have led us to develop a third, new adaptive filtering procedure:
the time-slice technique. In this procedure, the one-trace technique, discussed in
section 5.2, is applied at each instant of time to the dynamically corrected data of
a CMP gather in the direction of the offset.

After sorting the data into time slices, each data series contains the values at the
different offsets for a specific time. The number of the generated traces is equal to
the number of samples of the original record. In the following discussion, we treat
these data sets as time series. Therefore, the term frequency is used instead of
wavenumber.

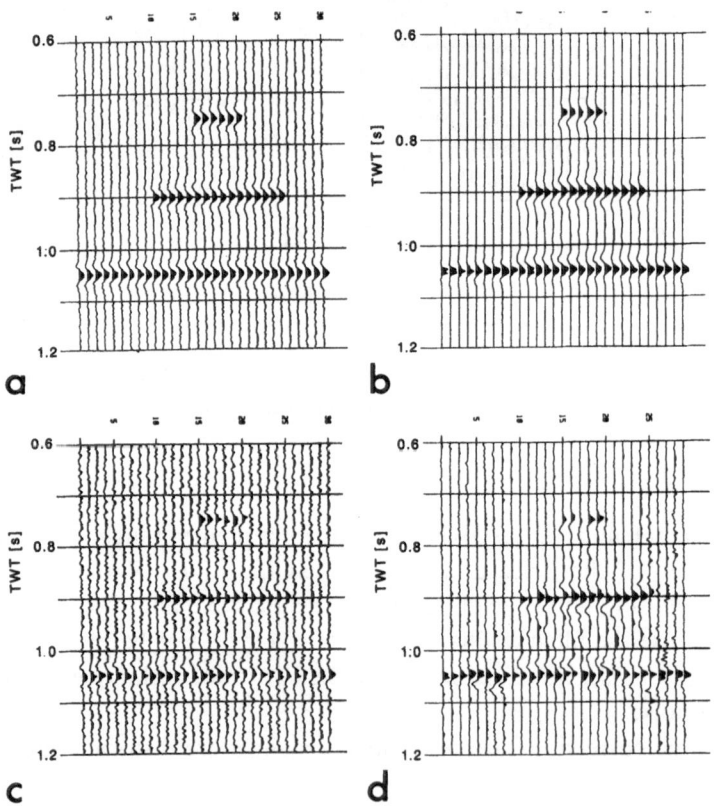

Figure 11

Application of the two-trace adaptive filter technique to data model 1b (white noise). a) synthetic CMP
gather after dynamic correction (20% white noise), b) result of two-trace adaptive filtering of (a), c) same
as (a), but 50% white noise, d) two-trace adaptive filtering result of (c).

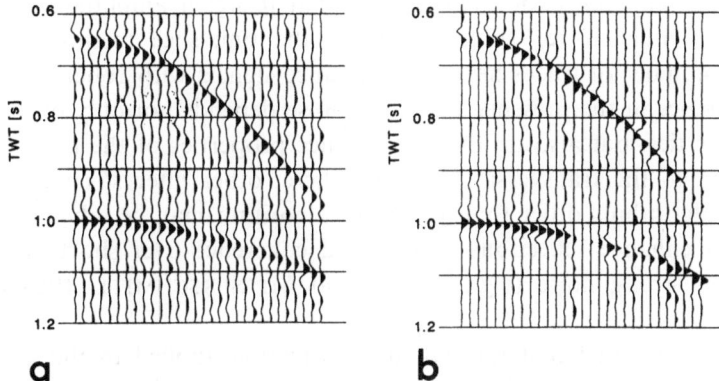

Figure 12
Application of the two-trace technique to synthetic seismic traces (20% band-limited noise) before
dynamic correction.

After this sorting, the signal and noise differ in frequency:
- In the ideal case, the signal has approximately the same amplitude at each offset in the dynamically corrected CMP gather, and therefore, the signal will have very long periods in the rearranged trace (the low frequency part of the time slice).
- Noise—either band-limited or white—that does not correlate in space is random with respect to offset and, therefore, forms the high-frequency part of the time slice.

After sorting, the prerequisites for successful application of the one-trace adaptive filter technique (band-limited, low-frequency signal superimposed by broadband noise) are fulfilled. Separation of the signal from the noise using the one-trace adaptive filter technique is especially easy due to the very low frequency character of the signal component. By resorting the filtered time slices back into time traces, we obtain the filtered seismic traces. In contrast to the two-trace technique, the signal coherence across all traces of the CMP gather is used.

The efficiency of the adaptive time-slice technique is demonstrated by filtering the data of model 1a (band-limited noise) and model 1b (white noise).

For both cases, the results (Figs. 13 and 14) are considerably better than those of the one-trace and two-trace techniques (see Figs. 10 and 11). Even for strong noise, the time-slice technique significantly reduces the noise. The zeroing at the first few and last few traces is produced by (i) the initial operator, which is set equal to zero; (ii) the shifting of the primary trace in order to generate the reference trace, by which the reference trace is zero at either the beginning or the end, depending on the direction of the shift; and (iii) the direction in which the filter is applied.

As can be seen, the short, uppermost reflection is not recognizable in the strong noise example (Figs. 13d–f). A reflection must extend across several traces in order for the filter operator to adjust from noise to signal, which in our example is too short for the top reflection. This means that if signal amplitude variations with offset are significant for some reflectors, the time-slice method may fail. The minimum number of traces needed to enhance signals with coherence over a short lateral distance in the presence of strong noise depends not only on the signal and noise properties, but also on the convergence properties of the filter algorithm. In comparison to the transversal filter used here, other filter structures, e.g., lattice filters, or algorithms other than the LMS algorithm may be advantageous in that they better enhance reflections with lateral coherence over short distances in the presence of strong noise (e.g., HONIG and MESSERSCHMITT, 1984 or HAYKIN, 1991).

We have assumed that the dynamic correction applied to the CMP gather is done with the correct stacking velocities so that the signals are in phase from trace to trace. However, stacking velocities can only be estimated with limited accuracy and the reflections after dynamic correction are often either slightly overcorrected or undercorrected. For the practical application of the time-slice technique, it is important to determine how much the filter results are influenced by stacking-velocity errors, i.e., by normal-moveout residuals.

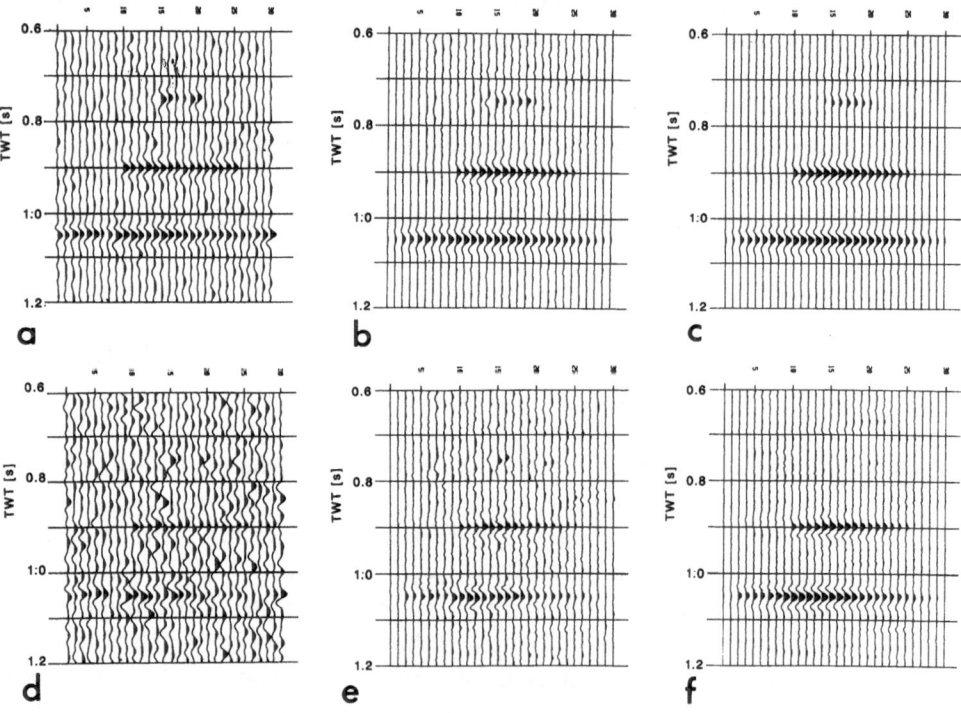

Figure 13

Application of the adaptive time-slice filter process to data model 1a (band-limited noise). a) Synthetic CMP gather after dynmic correction (20% band-limited noise), b) adaptive time-slice filter result of (a), c) median filtering of (b), d) same as (a), but 50% band-limited noise, e) adaptive time-slice filter result of (d), f) median filtering of (e) (3-point filter in time-slice direction).

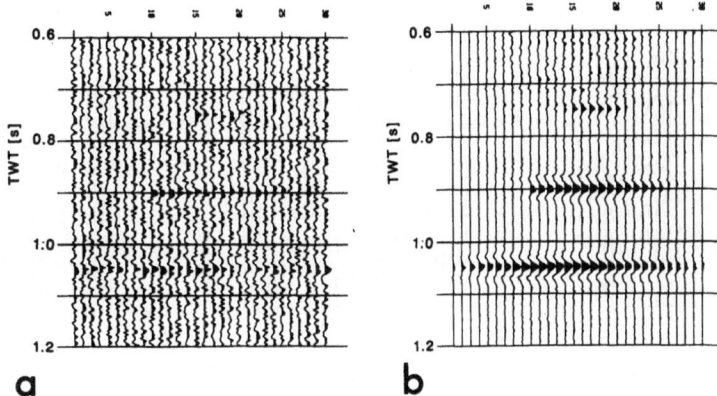

Figure 14
Application of the adaptive time-slice filter process to data model 1b (white noise). a) synthetic CMP gather after dynamic correction (100% white noise), b) adaptive time-slice filter result of (a).

Figure 15 presents one example of the application of the time-slice adaptive filter technique to a synthetic CMP gather after dynamic correction with normal-moveout residuals. It can be seen that stacking velocity errors are not very critical. In deep reflection seismics, the targets are deep seated, the velocities are high with only small variations, and the normal moveouts are also small. Even if the correct stacking velocities can be estimated only roughly, the residual moveouts, ΔT, for deep reflections are only small, i.e., less than or on the order of half of the dominant period T of the reflection. For example, for a reflection with $t_0 = 6$ s,

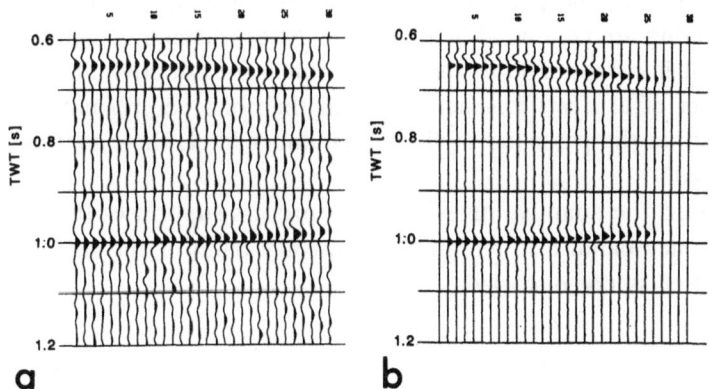

Figure 15
Application of the time-slice adaptive filter technique to a CMP gather after erroneous dynamic correction: a) synthetic CMP gather after dynamic correction (20% band-limited noise). The stacking velocities used for dynamic correction have been too high for the reflection at 0.65 s and too low for the reflection at 1.0 s. The normal-moveout residuals are in the order of 0.5–1.0 the dominant period of the reflection, b) time-slice adaptive filtering result of (a).

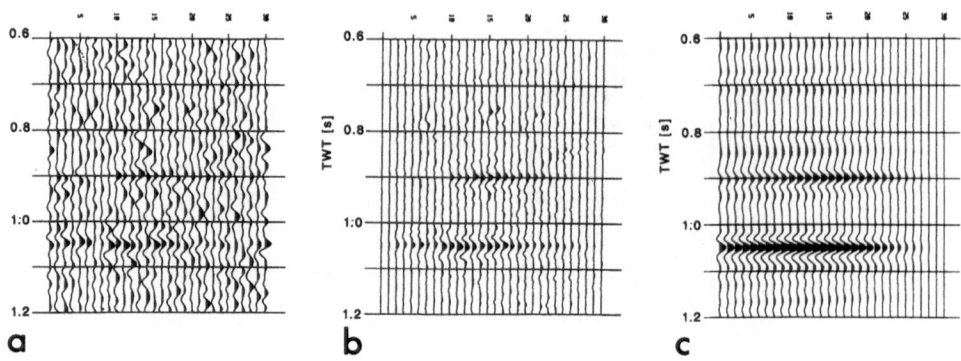

Figure 16
Comparison of adaptive time-slice and wavenumber filtering: a) synthetic CMP gather after dynamic correction (50% band-limited noise), b) time-slice adaptive filtering result of (a), c) wavenumber filtering result of (a).

$V_{RMS} = 5000$ m/s, and maximum offset $= 5000$ m, dynamically corrected with a stacking velocity of 6000 m/s, the residual moveout is on the order of 25 ms. Therefore, we can assume that in deep reflection seismics the condition for successful application of the time-slice technique, i.e., $\Delta T \leq T/2$, is fulfilled. Even for reflections near the earth's surface, the time-slice method can be applied successfully if stacking velocity errors are less than 10%.

In deep reflection seismics, the adaptive time-slice method offers significant advantages versus wavenumber filtering, which also acts in the offset direction. While the wavenumber filter operates with a fixed transfer function, the adaptive filter can adjust to the particular data. Results of adaptive time-slice and wave-number filtering are compared in Figure 16. After wavenumber filtering, the signal enhancement is comparable to that of the time-slice adaptive filtering; however, a large number of artifacts is generated from the noise component by wavenumber filtering. Within the passband of the wavenumber filter, the signal-to-noise ratio is not improved.

In contrast, the adaptive time-slice filter process suppresses the noise across the entire frequency band, also within the frequency band of the signal. Nevertheless, the signal itself is not distorted.

6. Application of Adaptive Filter Techniques to DEKORP Data

The results in section 5 show that of these adaptive filter techniques, the newly developed time-slice technique can be expected to give the best signal-to-noise improvement when applied to DEKORP data, and the one-trace technique will produce hardly any noise reduction, because signal and noise cover approximately the same frequency range.

In a first example, therefore, only the adaptive two-trace and time-slice techniques are applied to a CMP gather (after dynamic correction) of the DEKORP 1C

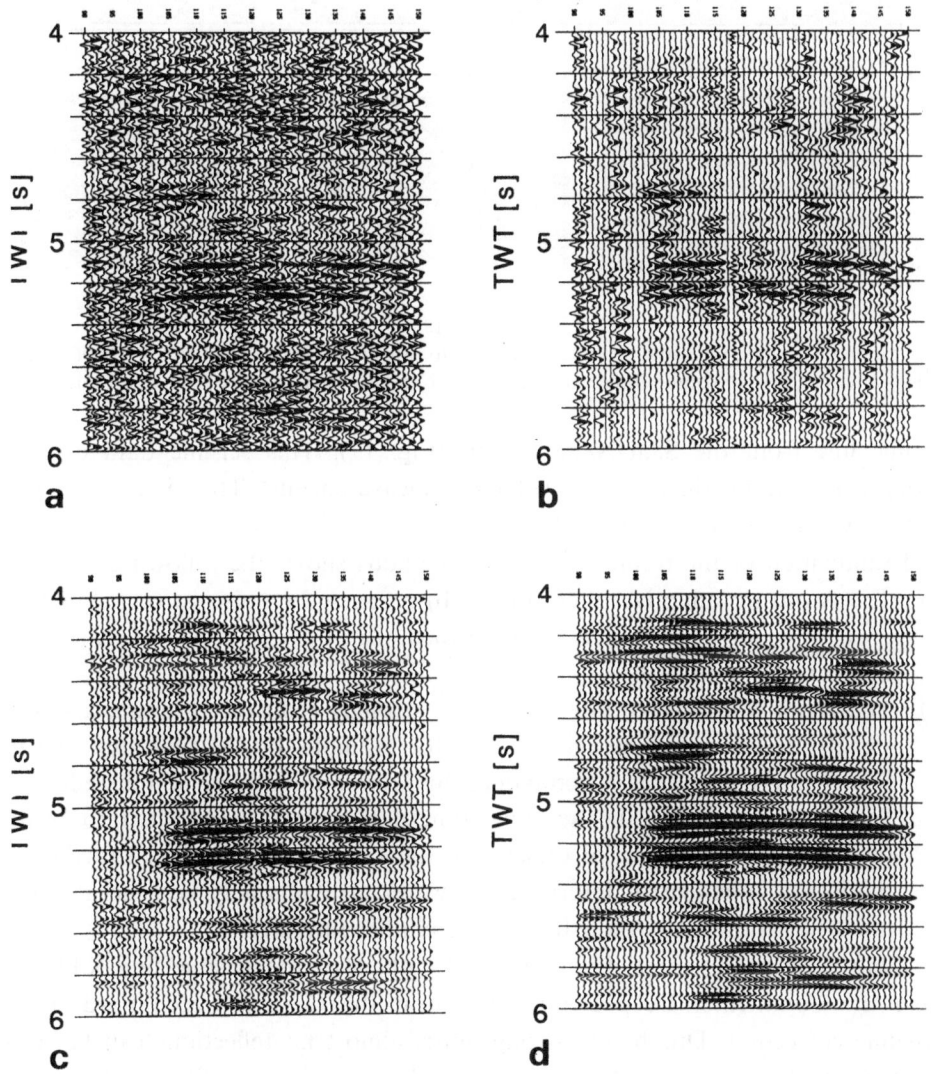

Figure 17
Comparison of adaptive filter techniques to a CMP gather of line DEKORP 1C (Saar-Nahe trough). a) CMP gather after dynamic correction, b) results after two-trace adaptive filtering of (a), c) results after time-slice adaptive filtering of (a), d) results after median filtering of (c) (3-point filter in time-slice direction).

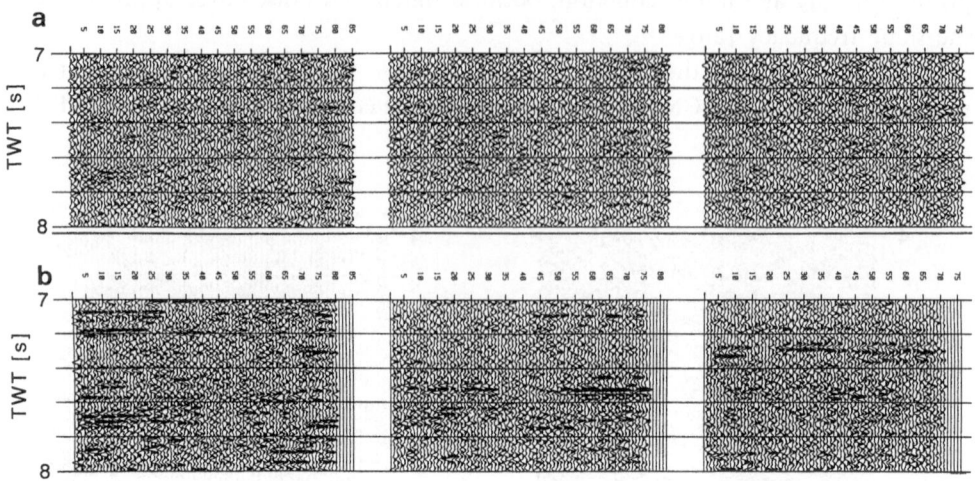

Figure 18

Application of the time-slice adaptive filter technique to three CMP gathers of line KTB 8506. a) CMP gathers after dynamic correction (time window: 7–8 s), b) results after adaptive time-slice filtering of (a).

seismic line from the Saar-Nahe trough (Fig. 17). This seismic data expresses stronger reflections than other DEKORP measurements. The signal and noise frequency contents are comparable.

Examination of the results of the two methods shows the following:

(1) The two-trace adaptive filter (Fig. 17b) reduces the noise; strong signals are preserved, but weak signals are weakened, e.g., the one at 5.9 s between trace numbers 130 and 140.

(2) The adaptive time-slice filter (Fig. 17c) strongly reduces the noise, and weak signals are enhanced, as can be seen, for example, by the reflection at 5.9 s. Signal improvement is accompanied by lateral smoothing of the reflections. When the results of the adaptive time-slice filter are subjected to median filtering in the direction of offset (Fig. 17d), the noise is reduced further and the lateral coherence is improved. However, artifacts may be generated from noise like those that result from wavenumber filtering.

Figure 18a shows three different CMP gathers of the seismic line KTB 8506 (near the Continental Deep Drilling site near the German-Czech border) after dynamic correction. Due to the strong noise, almost no reflections can be recognized. Nevertheless, the application of the adaptive time-slice filter technique has significantly improved the signal-to-noise ratio and the reflections are clearly distinct from the noise (Fig. 18b).

Figure 19 shows a stacked portion of profile KTB 8506 (a) without and (b) with adaptive time-slice filtering of dynamically corrected CMP gathers. Comparison of the two results shows that most reflections are enhanced by adaptive filtering and the surrounding noise is suppressed. Some weak reflections are also suppressed, however. This can be explained by the fact that the adaptive time-slice filter only enhances pre-stack reflections with a minimum lateral coherence (cf. Fig. 16). But these are the reflections that are important for further analysis of dynamic signal parameters. On the other hand, the stacking results show that the enhanced reflections also exist in the unfiltered data set. Thus the filter process does not generate artifacts in pre-stack data.

7. Conclusions

Owing to the signal and noise properties of deep seismic reflection data (same frequency content and small signal-to-noise ratio), conventional adaptive one-trace and two-trace filter techniques do not provide separation of signal and noise. The general disadvantage of these two adaptive filter techniques is that its efficiency decreases rapidly with increasing noise.

In contrast to the one-trace and two-trace filter techniques, the newly developed adaptive time-slice filter method improves very efficiently the signal-to-noise ratio of deep seismic reflections in pre-stack data. Even in the case of strong noise, the signal-to-noise ratio can be improved significantly. The technique takes advantage of the wavenumber difference between signal and noise in the offset direction at each time stage. The method has to be applied to CMP gathers after dynamic

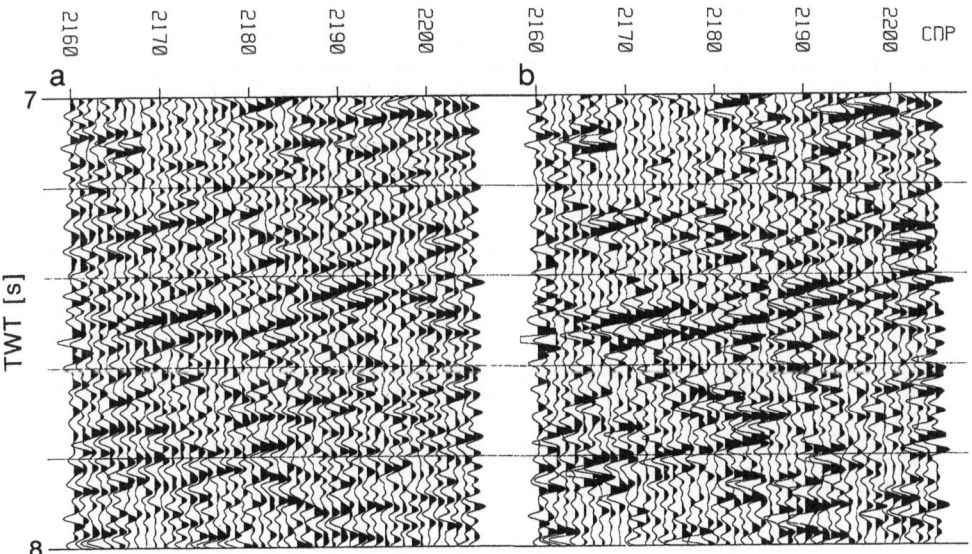

Figure 19
Stacked portion of the DEKORP profile KTB 8506 (a) without and (b) with adaptive time-slice filtering of dynamically corrected CMP-gathers.

correction. Normally, the residual normal moveouts in deep reflection seismics are small and the adaptive time-slice process can be applied successfully.

Acknowledgement

This project was sponsored by the Federal Ministry for Education, Science, Research and Technology (BMBF) within the scope of DEKORP. The authors thank Kris Vasudevan, Calgary, for his critical comments and recommendations for improvements of the paper.

REFERENCES

BOOTON, R. C. (1952), *An Optimization Theory of Time-varying Linear Systems with Nonstationary Statistical Inputs*, Proc. I.R.E. *40*, 977–981.

BUTTKUS, B., *Spektralanalyse und Filtertheorie in der angewandten Geophysik* (Springer-Verlag, Berlin, Heidelberg, New York 1991).

DRAGOSET, B. (1995), *Geophysical Applications of Adaptive-noise Cancellation*, 65th Annual Internat. Mtg., Soc. Expl. Geophys., Expanded Abstracts, 1389–1392.

FRIEDLANDER, B. (1982), *Lattice Filters for Adaptive Processing*, Proceedings of the IEEE *70*, 829–867.

GRIFFITHS, L. J., SMOLKA, F. R., and TREMBLY, L. D. (1977), *Adaptive Deconvolution: A New Technique for Processing Time-varying Seismic Data*, Geophysics *42*, 742–759.

HATTINGH, M. (1990), *Robust, Data-adaptive Filtering Algorithms for Geophysical Noise Problems*, 60th Annual Internat. Mtg., Soc. Expl. Geophys., Expanded Abstracts, 1715–1718.

HATTINGH, M. (1989), *The Use of Data-adaptive Filtering for Noise Removal on Magnetotelluric Data*, Phys. Earth and Planet. Int. *53*, 239–254.

HAYKIN, S., *Adaptive Filter Theory* (Prentice Hall, Englewood Cliffs 1991).

HONIG, M. L., and MESSERSCHMITT, D. G., *Adaptive Filters: Structures, Algorithms, and Applications*, In *Lattice Filters for Adaptive Processing* (Kluwer Academic Publishers, Hingham 1984).

PRASAD, S., and MAHALANABIS, A. K. (1980), *Adaptive Filter Structures for Deconvolution of Seismic Signals*, IEEE Transactions on Geoscience and Remote Sensing GE-18, 267–273.

WIDROW, B., and STEARNS, S. D., *Adaptive Signal Processing* (Prentice-Hall, Inc., Englewood Cliffs 1985).

WIDROW, B., McCOOL, J. M., LARIMORE, M. G., and JOHNSON, Jr., C. R. (1976), *Stationarity and Nonstationarity Learning Characteristics of the LMS Adaptive Filter*, Proceedings of the IEEE *64*, 1151–1162.

WIDROW, B., GLOVER, Jr., J. R., McCOOL, J. M., KAUNITZ, J., WILLIAMS, C. S., HEARN, R. H., ZEIDLER, J. R., DONG, Jr., E., and GOODLIN, R. C. (1975), *Adaptive Noise Cancelling: Principles and Applications*, Proceedings of the IEEE *63*, 1692–1716.

WIDROW, B., and HOFF, M. (1960), *Adaptive Switching Circuits*, in IRE WESCON Conv. Rec., 96–104.

(Received November 13, 1997, revised May 20, 1997, accepted May 25, 1998)

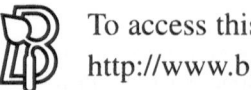

To access this journal online:
http://www.birkhauser.ch

Pure appl. geophys. 156 (1999) 279--301
0033-4553/99/010279-23 $ 1.50 + 0.20/0

Pure and Applied Geophysics

Estimation of Seismic Velocities from Deep Reflections in the τ-p Domain

CHRISTIAN BÖNNEMANN[1] and BURKHARD BUTTKUS[1]

Abstract—In deep reflection seismics the estimation of seismic velocities is hampered in most cases due to the low signal level with respect to noise. In the τ-p domain, it is possible to perform the velocity analysis even under such unfavorable signal conditions. This is achieved by making use of special properties of the transform, which enhance the signal-to-noise ratio. Further noise suppression is realized by incorporating filter procedures into the transform algorithm. The velocity analysis itself is also done in the τ-p domain by calculating and evaluating constant velocity gathers. The results can be directly used in the time domain. A mute algorithm, implemented into the τ-p velocity analysis procedure, further reduces noise. This velocity estimation method is discussed with synthetic data and applied to DEKORP data.

Key words: Deep seismic reflections, velocity estimation, τ-p processing, noise reduction.

1. Introduction

As in all fields of reflection seismics, in deep reflection work the velocity is an important parameter for processing and interpretation. But here the estimation of the velocities faces additional problems: the signal-to-noise ratio is low, the targets are often deep, and the lateral coherence of reflections is usually limited. Because of these conditions, routine velocity analysis methods for single CMP gathers may give unsatisfactory results, particularly considering further velocity studies. Stacking neighboring CMP gathers with a range of constant velocities (constant-velocity stacks) is still one of the main analysis tools in deep reflection seismics. However, this method has some shortcomings:

- Due to maximum offsets, which are small in comparison to the target depth, the mapping of deep reflections is very insensitive to changes in stacking velocities. The stacking process, both in constant-velocity stacks as in the final stack, does not make full use of subtle differences in the shape of the moveout curve.

[1] Bundesanstalt für Geowissenschaften und Rohstoffe, P.O. Box 510153, 30631 Hannover, Germany.

- The reference to single reflection events in CMP gathers is lost because it is not in any case clear which elements of reflections in CMP gathers contribute to the stacked signal.
- The velocity information is laterally smeared because a group of CMP gathers is involved in the analysis.

The aim of this paper is to present a method which allows velocity analyses in single CMP gathers even under the conditions met in deep reflection seismics. At first, the possibilities of the τ-p transform for signal enhancement will be discussed. In the second part, the algorithm for estimating velocities in the τ-p domain is presented. The last part deals with the application of the method to DEKORP data.

2. Improvement of Signal-to-noise Ratio by the τ-p Transform

The parameters τ (intercept time) and p (ray parameter) have been introduced by GERVER and MARKUSHEVICH (1966) into geophysics via the function

$$\tau = t - px, \tag{1}$$

which relates p and τ to travel time t and offset x in seismic data. In wave-propagation theory, these two parameters are sufficient to define the position of a plane wave in the time-offset space.

The τ-p transform of a wave field $u(x, t)$, measured at the surface, is defined by the integral

$$\bar{u}(p, \tau) = \int_{-\infty}^{\infty} dx \, u(x, \tau + px). \tag{2}$$

The use of the τ-p transform in digital seismic signal processing goes back to SCHULTZ and CLAERBOUT (1978). They introduced the term "slant stack" for the transform algorithm, which is numerically realized by stacking seismic data along slanted lines. The slope of this stacking trajectory is determined by the ray parameter p; the intercept time τ denotes the intersection of the trajectory with the time axis.

The slant stack only approximately decomposes a wave field into its plane-wave components. To calculate the exact decomposition, modifications for the slant stack and alternatives have been developed (e.g., CHAPMAN, 1981; TREITEL et al., 1982; BRYSK and McCOWAN, 1986). These modifactions concern essentially the phase and amplitude of the signal. For the determination of seismic velocities, the signal form is not important. Therefore, the slant stack is used for computing the τ-p transform. Furthermore, the slant stack allows an easy implementation of filter procedures for signal enhancement.

In principle, two categories which improve the signal-to-noise ratio by applying the τ-p transform can be distinguished: signal improvement due to properties of the transform itself and signal improvement by implementation of filter algorithms into the transform algorithm. Both categories will be discussed below.

Signal Improvement due to the Properties of the τ-p Transform

Signal and noise are often characterized by different ray parameters. Limiting the transform to the ray-parameter range of the signal reduces noise components, e.g., ground roll, which are characterized by ray parameters lying outside this range. This is the simplest and most straightforward way to use the τ-p transform for signal improvement.

The second property which yields to signal improvement is directly linked to the hyperbolic form of the reflections to be analyzed. Looking at reflection hyperbolas in CMP sorted data, we notice two facts:

1. Due to the spread length the offset is limited.
2. The deeper the reflection, the smaller the curvature of the hyperbolae.

This means that deep reflections are mapped on a small ray-parameter range (Fig. 1). When we compute the τ-p transform with the slant stack, greater parts of the reflection hyperbola are included by the stacking trajectory. Thus the

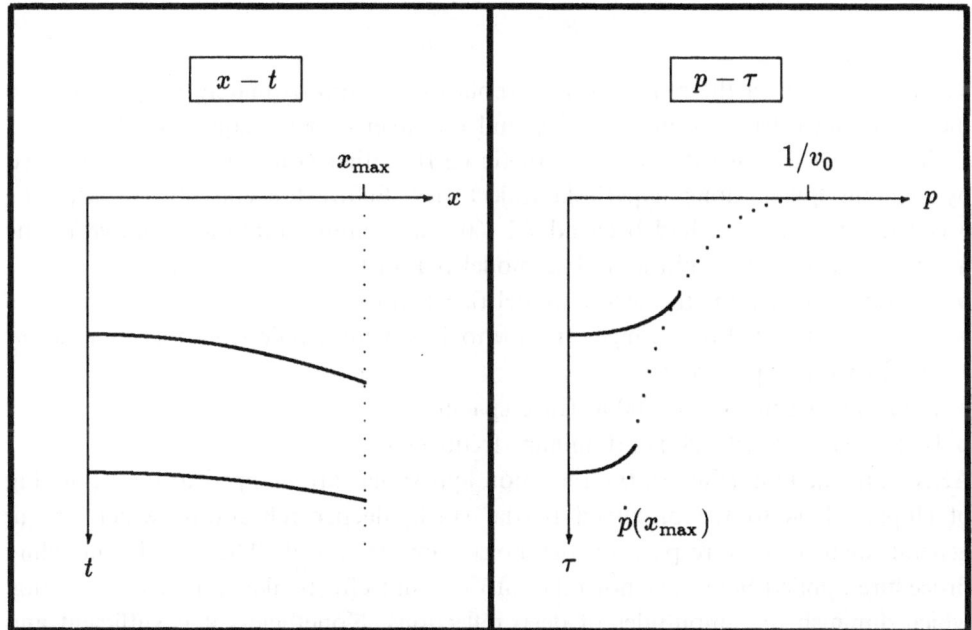

Figure 1

Compression of deep seismic reflections by the τ-p transform, v_0 is the velocity below the surface.

application of the slant-stack results in a partial stacking, which enhances the signal.

For the theoretical understanding of this effect, we start with the equation of the travel-time curve. In the framework of the stacking concept, the two-way travel time $t(x)$ of a seismic reflection in CMP gathers is supposed to follow a hyperbola, with the parameters offset x, stacking velocity v and two-way travel time t_0 at zero offset:

$$t(x) = \sqrt{t_0^2 + \frac{x^2}{v^2}}. \tag{3}$$

The ray parameter p is calculated by differentiating (3) with respect to offset:

$$p(x, t, v) = \frac{dt(x)}{dx} = \frac{x}{v^2 t}. \tag{4}$$

Equation (4) relates p with the time domain parameters x and t and the velocity v. By evaluating (4) at $x = x_{max}$ and inserting (3), the ray parameter at maximum offset is calculated:

$$p(x_{max}) = \frac{x_{max}}{v\sqrt{t_0^2 v^2 + x_{max}^2}}. \tag{5}$$

For $t_0^2 v^2 \gg x_{max}^2$ the steep angle approximation is valid:

$$p(x_{max}) \approx \frac{x_{max}}{v^2 t_0}. \tag{6}$$

The compression of the reflection hyperbola is thus approximately proportional to the maximum offset, the inverse of t_0 and the inverse of the squared velocity.

This effect is illustrated by a synthetic CMP gather (Fig. 2) which is modelled by aligning spikes along hyperbola trajectories. Before the convolution with the wavelet, random noise had been added (60% maximum amplitude relative to the maximum reflection amplitude). The model parameters are:
● constant velocity for the whole model ($v = 6$ km/s);
● the two-way travel time ranges from 0 to 12 s, with a reflection hyperbola every second with respect to t_0;
● 12 km maximum offset, 100 m trace spacing;
● Ricker wavelet with 20 Hz dominant frequency.
After slant stacking (Fig. 3), the reflection hyperbolas are mapped to the beginning of ellipses. Due to the discussed partial stack, deeper reflections, which are in general weaker with respect to the noise, are enhanced. The simple modeling procedure applied here does not take into account effects like spherical spreading, which diminish the amplitudes of deep reflections. Nonetheless it is sufficient and appropriate to demonstrate and evaluate the signal enhancement properties of the τ-p transform. The shallower reflection hyperbolas at 1 and 2 s two-way travel time

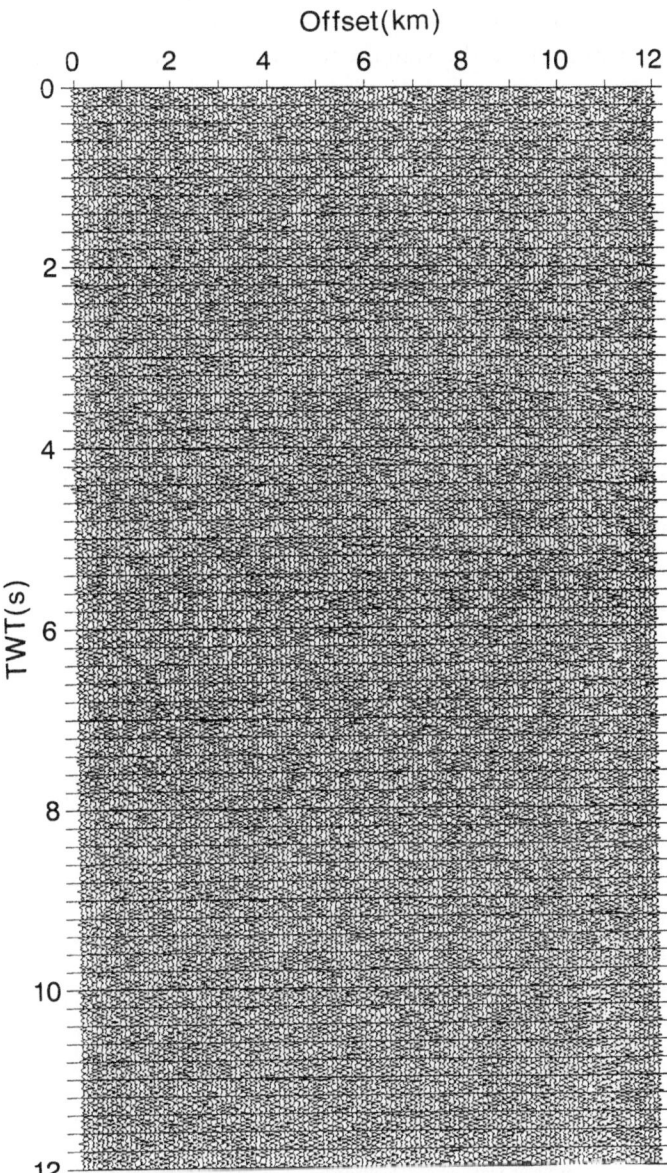

Figure 2
Synthetic CMP gather modelled for 6 km/s constant velocity and 12 km maximum offset with superimposed noise (bandlimited, 60% maximum amplitude relative to the maximum of the signal). Reflection hyperbolas every second with respect to t_0.

are diminished. The transformed noise has no more random character, but a preference direction, which is an effect of the limited offset. At the reflection ellipses it is not clearly visible where the signal ends and where noise portions begin.

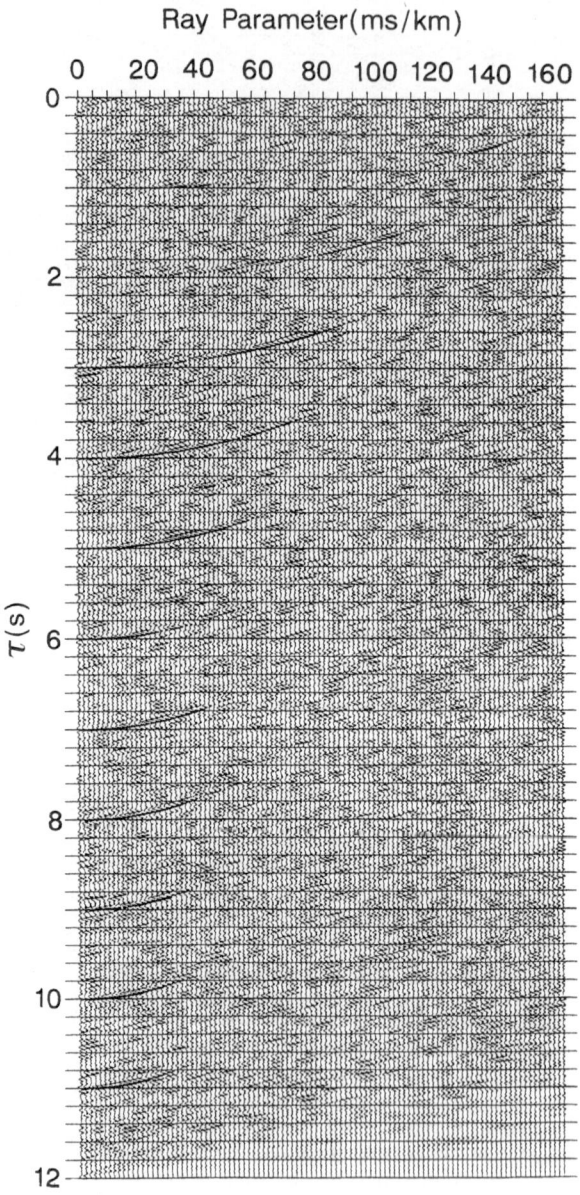

Figure 3
Slant stack of the synthetic CMP gather from Figure 2.

Modification of the τ-p Transform by Filter Algorithms

As in the previous section we make use of the hyperbola shape of the reflection, which suggests the use of hyperbolic velocity filtering, introduced by TATHAM *et al.* (1983). The filter equation is based on the τ-p domain expression for the reflection hyperbola (4), which is used to restrict the transform to a velocity band with the lower limit $v_{\min}$ and the upper limit $v_{\max}$. The filter equation is

$$\frac{x}{v_{\max}^2(t_0)t} \le p \le \frac{x}{v_{\min}^2(t_0)t}. \tag{7}$$

Only $u(x, t)$ in (2) which satisfy Equation (7) contribute to the transform result $\bar{u}(p, \tau)$. Thus, the filter enhances hyperbolic reflections.

In contrast to processes like frequency-wavenumber filtering the passband of the hyperbolic velocity filter can be a function of time. This equation is implemented into the slant stack and the filter process is active during the transform. The main difference relative to the noise suppression mechanism described in the previous section (limitation of the ray-parameter range) is the potential to suppress noise portions, which map into the ray parameter range of the signal as well.

Figure 4 shows the result of transforming the data from Figure 2 with hyperbolic velocity filtering included. The filter passband is wide with limits set to $\pm 30\%$ relative to the model velocity 6 km/s. This demonstrates that this filter can be applied before the velocity analysis: for setting the filter parameters, no exact knowledge about the velocity field is needed. In case of structures with extreme dips, the upper filter limit should be increased to cover the high velocities needed for stacking. In comparison to Figure 3 the reflections are enhanced, especially in the upper part. On the right hand side a mute effect is visible, resulting from the lower velocity limit.

Another class of filters applied for signal enhancement is the implementation of coherence weighting into the slant slack. STOFFA *et al.* (1981) introduced it to reduce aliasing effects caused by wide trace spacing. This technique, which is a nonlinear filter, works by calculating the semblance along each slant stacking trajectory. The stacking result for this trajectory is weighted by the semblance. Applying this method to the data from Figure 2 we see that the deep reflections arc further enhanced and the noise is reduced (Fig. 5). However, shallow reflections are suppressed. This can be explained by the effect that the shallow reflection hyperbolas have stronger curvature, so that each slant stacking trajectory comprises only a small part of the signal. Consequently, the semblance values are low, and multiplication with the slant-stack value further reduces the result. Closely related to the coherence weighted slant stack is the replacement of the pure slant stack by nonlinear stacking (McFADDEN *et al.*, 1986; KRAVIS, 1990). This technique also enhances signals, which are coherent along the stacking trajectory.

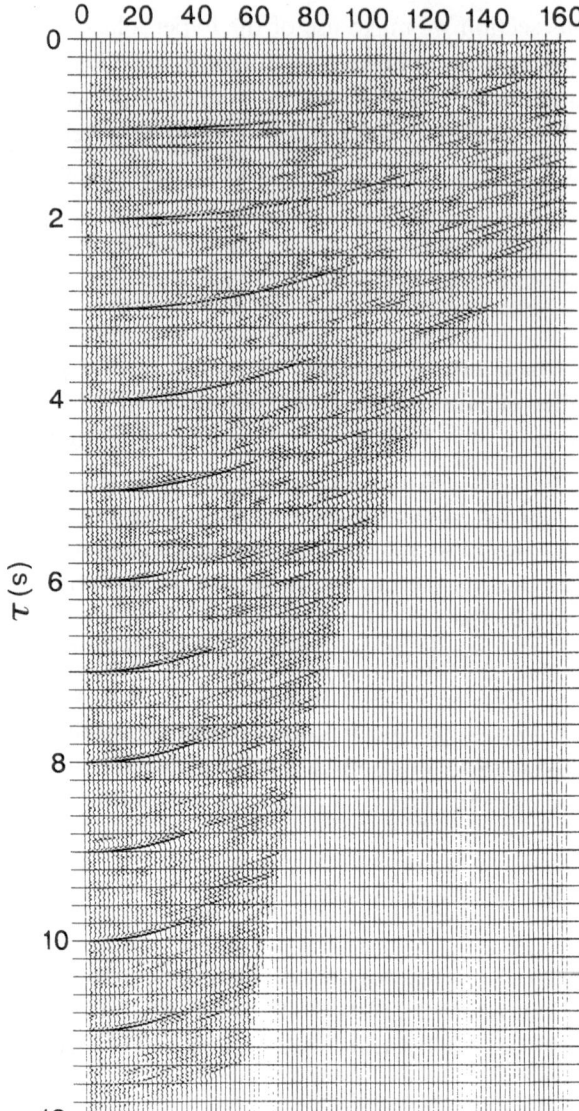

Figure 4
Slant stack of the synthetic CMP gather from Figure 2 with hyperbolic velocity filtering (passband 6 km/s ±30%).

The key idea applied here to enhance hyperbola shaped reflections is the combination of the discussed techniques. To understand the effect we write down the equation of the hyperbolic velocity filter (7) as a function of offset x:

$$pv_{\min}^2(t_0)t \leq x \leq pv_{\max}^2(t_0)t. \tag{8}$$

The slant-stacking trajectory is shortened with respect to the offset. Figure 6 illustrates the effect: the summation is concentrated on seismogram parts which contain hyperbola shaped reflections. On these reduced trajectories enough signal

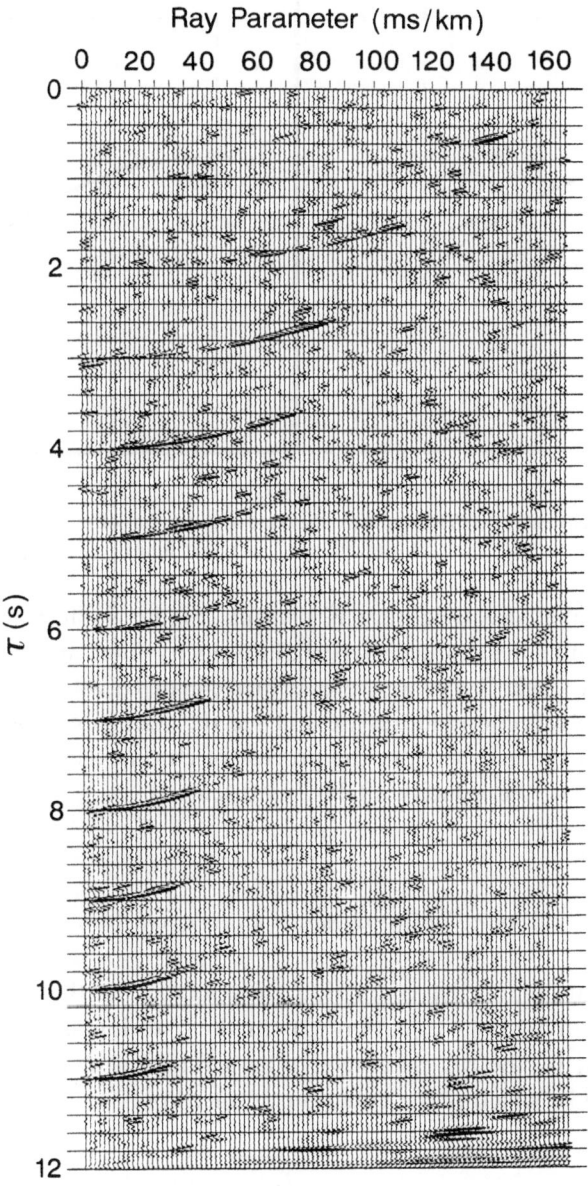

Figure 5
Coherence weighted slant stack of the synthetic CMP gather from Figure 2.

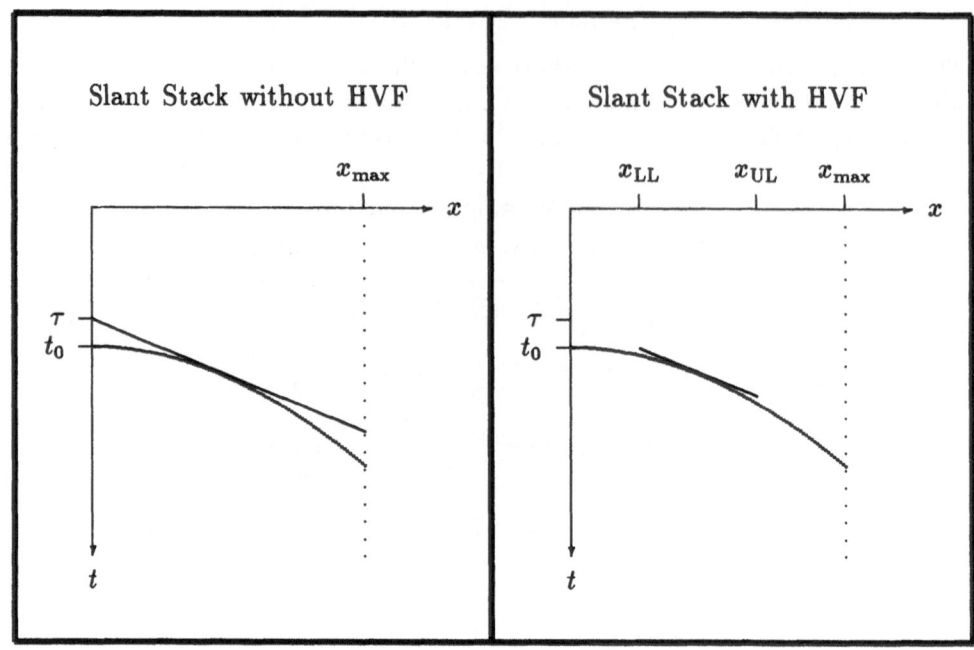

Slant Stack without HVF

Figure 6
The hyperbolic velocity filter (HVF) reduces the stacking to the offset range from x_{LL} to x_{UL} (x_{LL}—lower limit, x_{UL}—upper limit).

energy for the coherence weighting is present. Figure 7 illustrates the result of integrating both filter techniques (hyperbolic velocity filter and coherence weighting) into the slant stack. Reflections are further enhanced with respect to the noise. At the end of the reflection ellipses we still see the transition from signal parts to noise artefacts. In the next section, the process of cutting these artefacts by adding a mute procedure to the velocity analysis is discussed.

3. Velocity Estimation by Dynamic Correction in the τ-p Domain

The key equation for the estimation of velocities in the τ-p domain is the transform of the hyperbola Equation (3). In the τ-p domain the trajectory of a reflection is given by the ray parameter p as a function of intercept time τ (e.g., CLAERBOUT, 1985), which is the equation of an ellipse:

$$\tau(p) = \tau_0 \sqrt{1 - p^2 v^2}, \tag{9}$$

τ_0 is identical with t_0 in (3).

Comparing (3) and (9) we see that the form of the hyperbolas and ellipses, respectively, is determined by the velocity v. Both equations share the same t_0. This opens the way for estimating the seismic velocity by transferring concepts from the

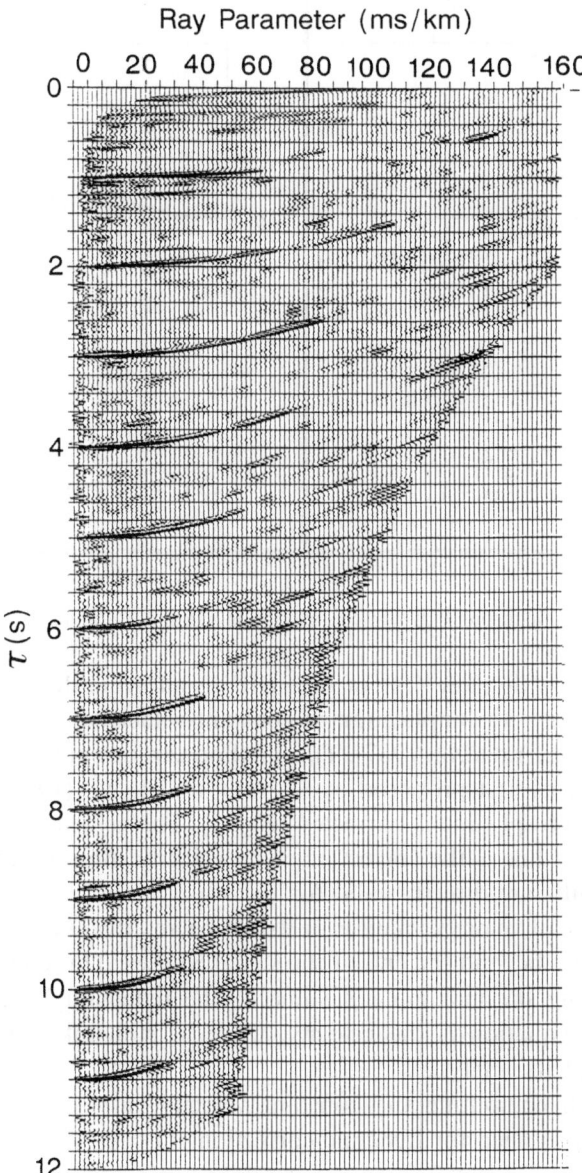

Figure 7
Coherence weighted slant stack of the synthetic CMP gather from Figure 2 with hyperbolic velocity filtering (passband 6 km/s $\pm 30\%$).

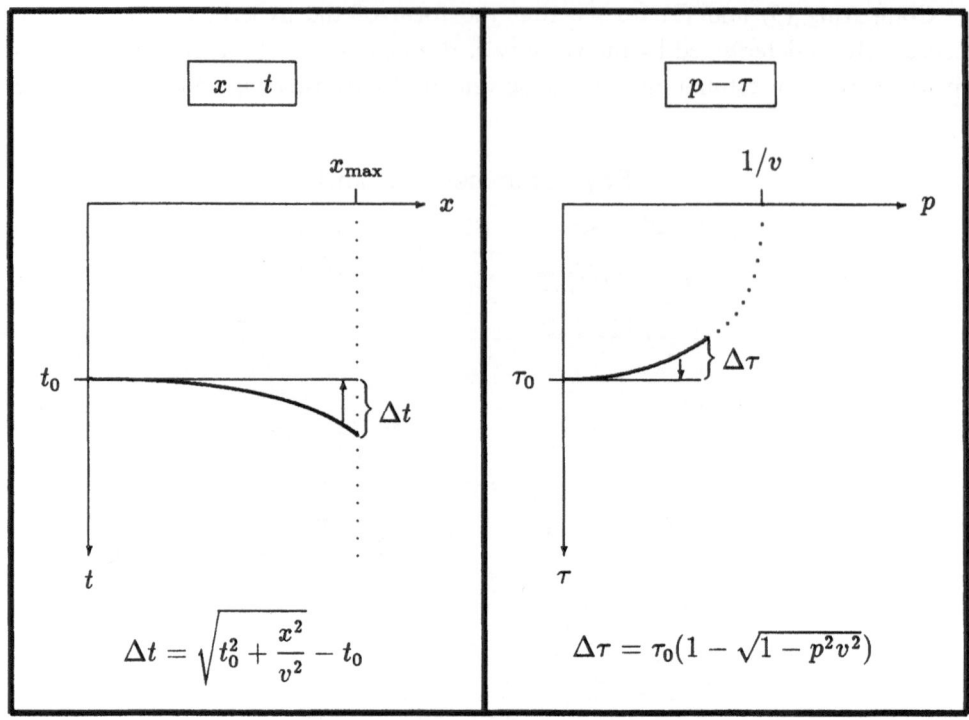

Figure 8
Comparison of dynamic correction in x-t and τ-p domain.

x-t domain to the τ-p domain. One of the methods allowing the velocity analysis in a single CMP gather is the application of constant-velocity moveout corrections to the CMP gather (e.g., YILMAZ, 1987). This method can be applied in a similar form in the τ-p domain, as first proposed by CUTLER and LOVE (1979). Instead of the travel time, the intercept time is corrected (Fig. 8). The velocity which corrects the ellipses to a straight line is the correct stacking velocity. Because of the identity of τ_0 and t_0, the results are directly valid for the x-t domain.

For the purpose of velocity analysis, it would be an advantage to separate signal and noise by muting beyond the end of the signal ellipses. This could be done using Equation (5), giving the maximum p value of the reflection hyperbola. The key parameter in this equation is the velocity, which is still unknown before having performed the velocity analysis.

However, when computing the constant velocity gathers, we know for each gather the velocity applied for dynamic correction. In case of applying the correct velocity, the ellipses are transformed to a straight horizontal line and Equation (5) leads to the correct mute. In neighboring gathers, the mute works approximately (Fig. 9). For the purpose of velocity analysis, this mute procedure is sufficient because the emphasis is on the gather with correct velocity, where the mute is appropriate.

To test the velocity analysis under severe noise conditions, the noise level in the synthetic CMP gather is raised to 100%. Figure 10 presents the result of calculating constant velocity gathers for the deep reflections below 6 s two-way travel time; the velocity range is 5500 m/s–6500 m/s. As given by the model, all reflections are corrected at 6000 m/s to horizontal lines. Short artifacts at small ray-parameter values remain, but they can be distinguished from reflections by the fact that no physically reasonable velocity is able to correct them properly. The example also demonstrates the effect of the muting. For ray-parameter values greater than p_{max} (cf. Equation (5)) the data are set to zero.

The synthetic example shows that even under severe noise conditions the τ-p velocity analysis technique is suitable for the extraction of stacking velocities from deep reflections in single CMP gathers.

4. Application to DEKORP Data

For the application to DEKORP data a CMP gather from the profile DEKORP 1 C has been chosen. The data has been acquired with 400 channels and 40 m source and receiver spacing (DEKORP RESEARCH GROUP, 1991), which results in 20 m CMP spacing and a maximum coverage of 200. The asymmetric split-spread configuration gives 12000 m maximum offset. The profile is located in the region of the Saar-Nahe-Trough (SW Germany). Figure 11 shows an example of a CMP gather to be analyzed. The data contain some reflection bands, mostly at offsets less

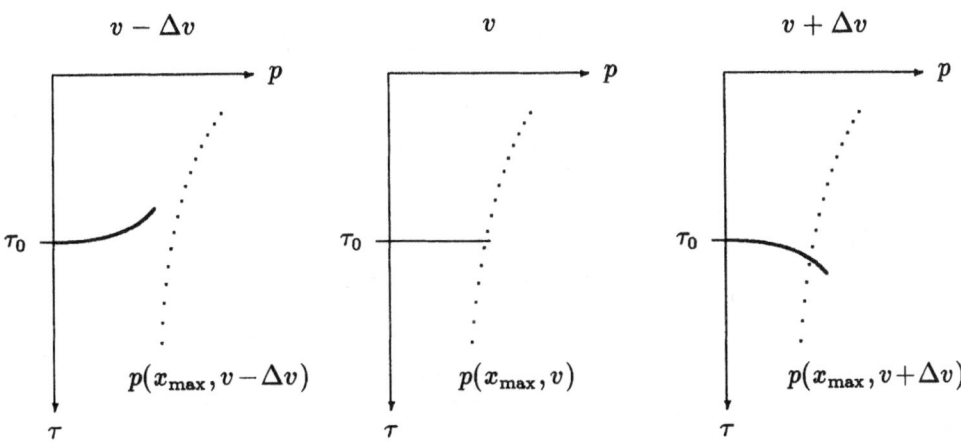

Figure 9

Implementation of velocity-dependent muting into constant-velocity gathers in the τ-p domain. In the case of undercorrection (left) the muting curve is away from the reflection event. The appropriate correction velocity yields the exact mute (center), while in the case of overcorrection (right) the muting curve cuts a portion of the reflection event.

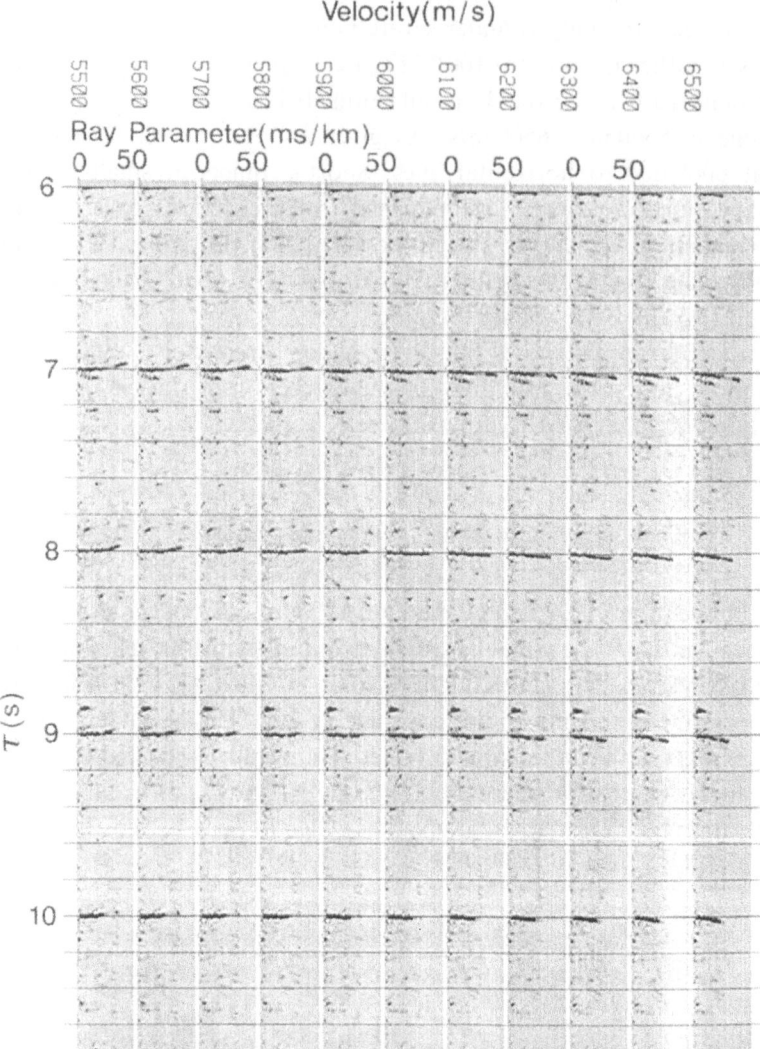

Figure 10
Calculation of constant-velocity gathers from Figure 2 (noise raised to 100% maximum amplitude relative to the maximum of the signal, followed by a slant stack with the same parameters as in Fig. 7); two-way travel time 6 s–12 s, velocity range 5500 m/s–6500 m/s, velocity increment 100 m/s.

than 5 km. This is due to the split-spread configuration which doubles the trace density at offsets less than 4 km. The reflections at 2 s two-way travel time can be assigned to the Holz conglomerate, located in the Upper Carboniferous at the boundary between Stephanian and Westfalian. The reflection band at 3 s two-way

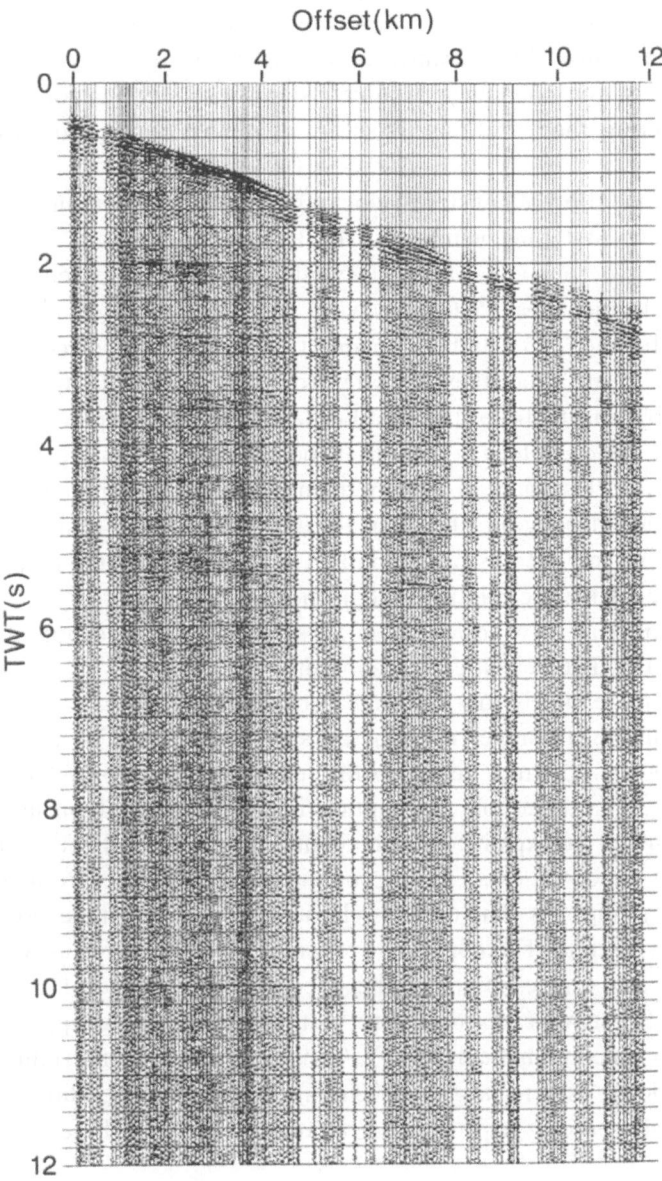

Figure 11
Offset dependent plot of CMP gather 2950, profile DEKORP 1 C; maximum offset 12 km.

travel time originates from the Middle Devonian. At 5 s and 8 s we see reflections from the middle crust and the lower crust, respectively. The reflections from the Mohorovičić discontinuity are located at about 10 s, as indicated by the stacked section.

A first attempt at velocity analysis shows the computation of a velocity spectrum (Fig. 12), as introduced by TANER and KOEHLER (1969) and NEIDELL and TANER (1971). The spectrum shows trends for the velocity, but too many coherence maxima. After transforming the data from Figure 11 into the τ-p domain, using the coherence weighted slant stack with incorporated hyperbolic velocity filtering, the reflections are sufficiently enhanced to allow a velocity analysis (Fig. 13).

The computation of constant-velocity gathers will be demonstrated in three time windows. The first time window to be analyzed covers the reflections from the Upper Carboniferous and Middle Devonian (Fig. 14). The reflections at 2 s show stacking velocities near 5200 m/s with an uncertainty of about 100 m/s. The reflections between 2.6 s and 2.8 s cannot be completely corrected to a horizontal line, which means that they deviate from the hyperbola form. This could be explained by effects from the overburden such as complex structures or large contrasts in seismic velocity. Note that the reflections do not extend entirely to the muting curve. This again is a consequence of the split-spread configuration: the doubled trace density in the first third of the offset range results in higher amplitudes in the τ-p transformed data. The restriction of the reflection to the near offset also explains the relatively high uncertainty.

The next analysis window (4.6 s–5.4 s) covers the reflections from the middle crust (Fig. 15). The velocities are nonuniform, covering a range from 4800 m/s to 6000 m/s. The relatively low stacking velocities could be caused either by multiple reflections inside the upper crust or by side reflections. The uncertainty is in the range of 300 m/s, because here also the reflections cover only the first third of the offset range. The hyperbola assumption is even less true than in the upper crust, which again can be explained by the complex structure of the overburden.

The last window (9.4 s–10.4 s) contains the reflections from the Mohorovičić discontinuity (Fig. 16). The event is characterized by two reflection arrivals at 10 s two-way travel time. The corrected reflection extends only half way to the muting curve. This gives an estimate of the effective offset range of the reflection event from the Mohorovičić discontinuity in this gather, being approximately 6000 m (50% of the maximum field offset). Consequently, the uncertainty range for the stacking velocity is very large (± 500 m/s). Under these conditions stacking velocities can only serve as a processing parameter, not as a base for further velocity studies.

The last data example is taken from the profile KTB 8506 (SE Germany, vicinity of the KTB well) and it demonstrates that the stacking velocity analysis by the τ-p procedure can result in a better stack (Fig. 17). Note the improved lower

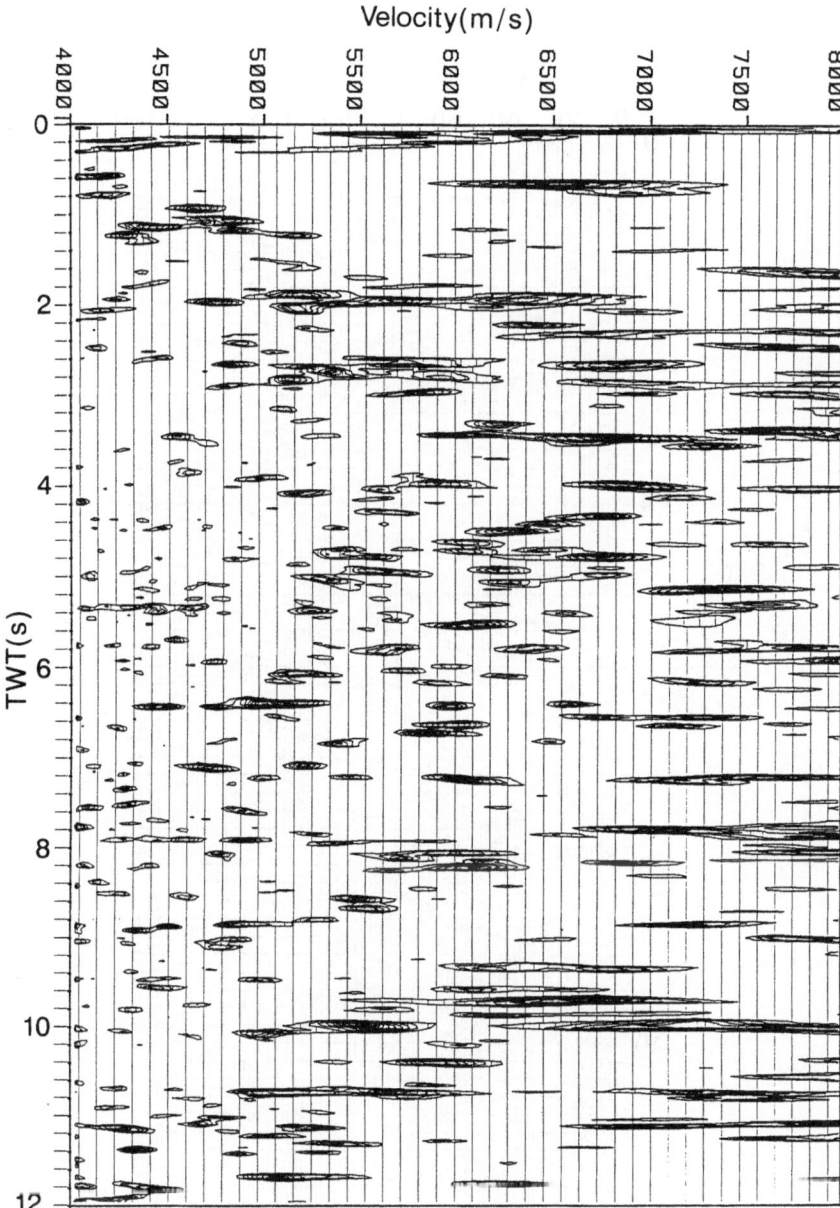

Figure 12
Velocity spectrum computed from Figure 11 for the velocity range 4000 m/s–8000 m/s; two-way travel
time 0 s to 12 s.

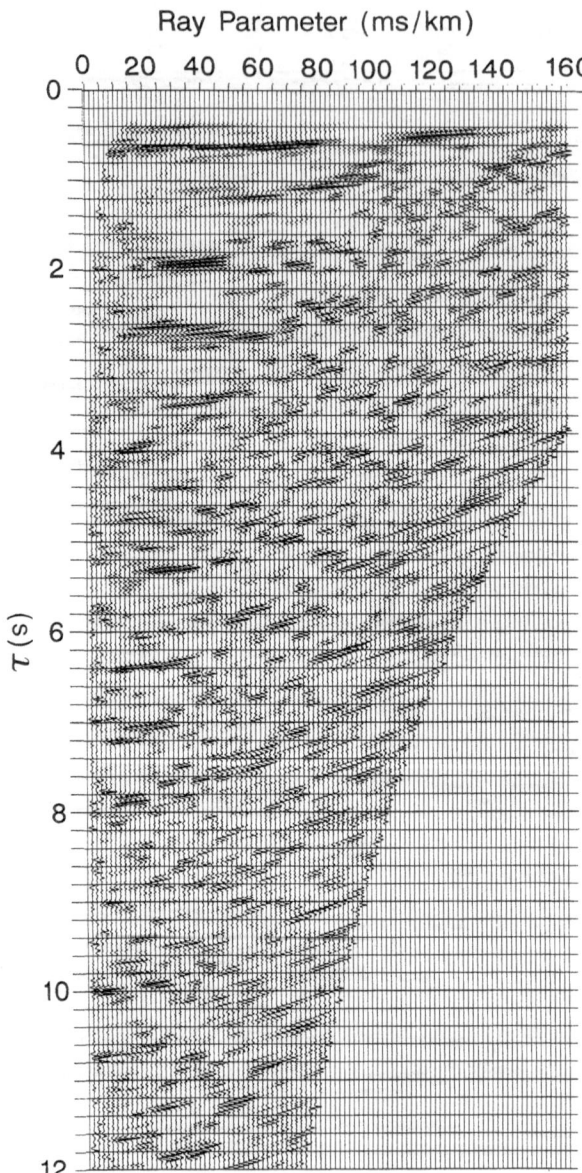

Figure 13
Coherence weighted slant stack of the CMP gather from Figure 11 with hyperbolic velocity filtering
(passband: averaged stacking velocities $\pm 30\%$); maximum ray parameter 167 µs/m.

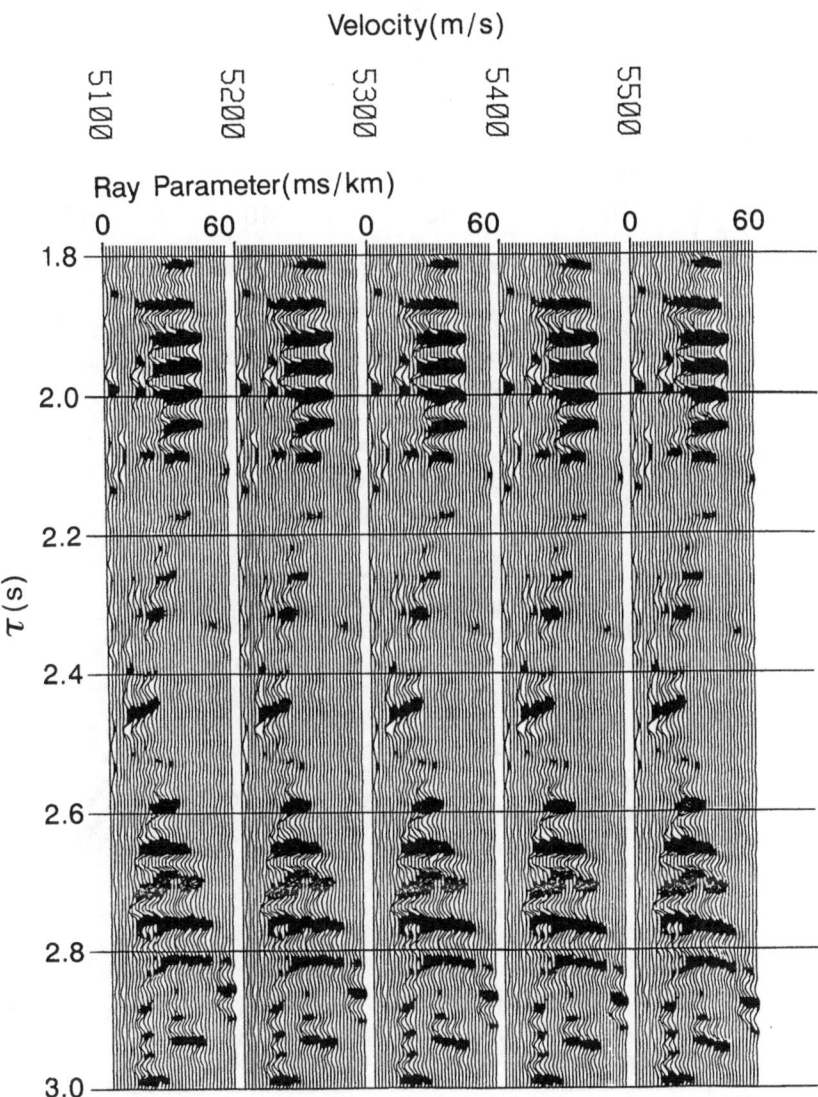

Figure 14
Calculation of constant-velocity gathers from Figure 13; two-way travel time 1.8 s–3 s, velocity range
5100 m/s–5500 m/s, velocity increment 100 m/s.

crust reflector sequence between 6.4 s and 6.6 s after stacking with the velocities
estimated in the τ-p domain.

5. Conclusions

The proposed method for velocity analysis in the τ-p domain is especially
advantageous in deep reflection seismics. The combined effects of the contraction of

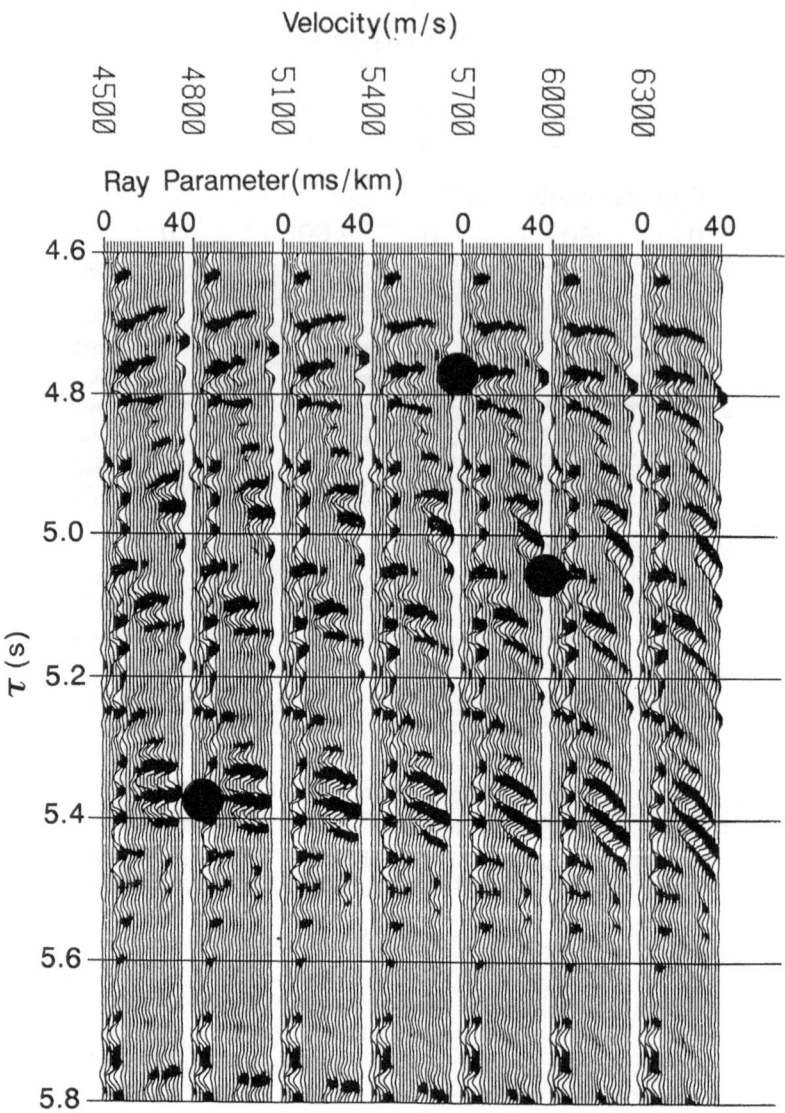

Figure 15

Calculation of constant-velocity gathers from Figure 13; two-way travel time 4.6 s–5.8 s, velocity range 4500 m/s–6300 m/s, velocity increment 300 m/s. The dots mark picked velocities.

deep reflections in the transformed gathers and supplementary filter processes also enhance reflections in the presence of strong noise. This allows the estimation of stacking velocities from single CMP gathers even in the event of low signal-to-noise-ratio and reflections which do not cover the entire offset range. The choice of the passband for the hyperbolic velocity filter is essential for signal improvement. In practice this is no problem, because only a very general knowledge of the stacking

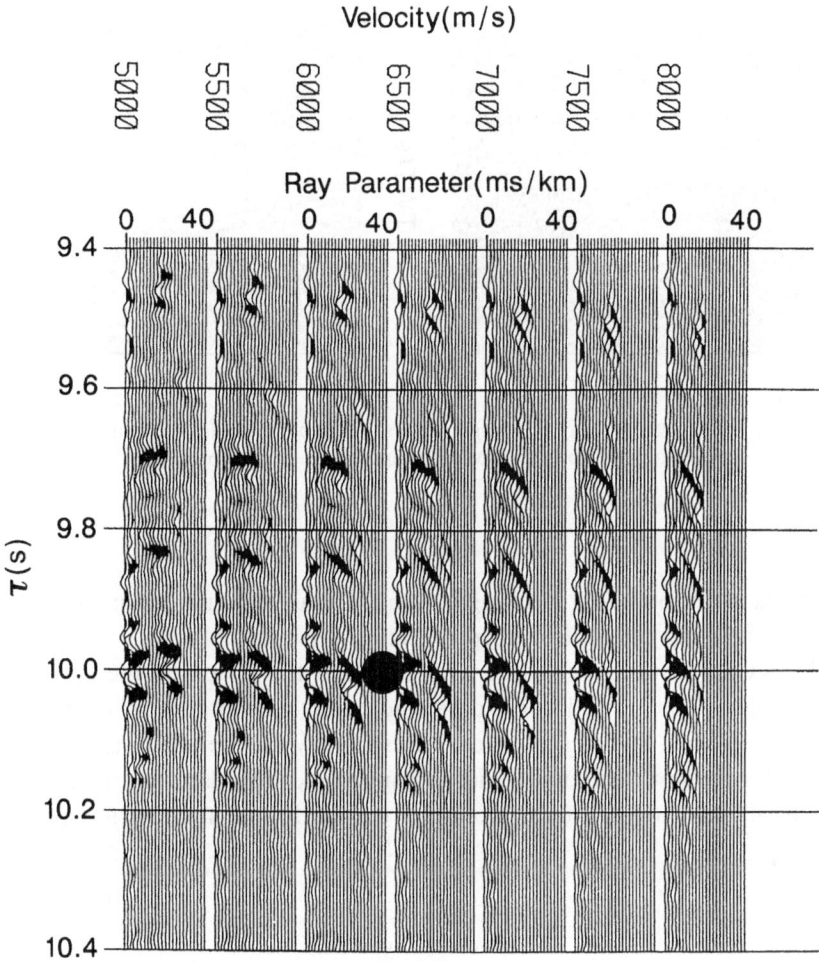

Figure 16

Calculation of constant-velocity gathers from Figure 13; two-way travel time 9.4 s – 10.4 s, velocity range 5000 m/s–8000 m/s, velocity increment 500 m/s. The dot marks a possible velocity pick at the Mohorovičić discontinuity (6500 m/s).

velocities is needed. After the completion of the velocity analysis the stacking also can be performed in the τ-p domain, taking advantage of the signal improvement due to the τ-p transform.

The interpretation of the constant-velocity gathers in the τ-p domain requires no drastic change for the interpreter, because it is carried out largely in analogy to the procedure in the x-t domain. In contrast to velocity analysis by constant velocity stacking the interpreter can be certain that the analysis is based on events which have been identified and chosen in the prestack data. By evaluating the behavior of the seismic event in the τ-p constant-velocity gathers, it is possible to estimate the uncertainty of the velocity estimate. In addition, it can be assessed to which degree

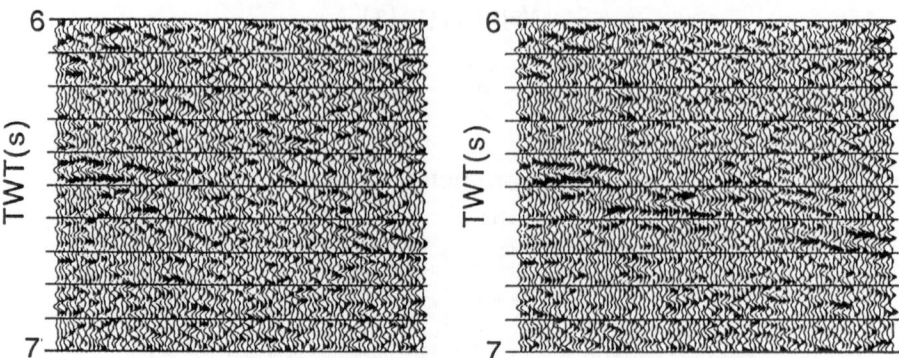

Figure 17
Portion of profile KTB 8506 (two-way travel time 6 s–7 s), stack based on: (a) the velocities from routine processing and (b) the velocities estimated by the τ-p procedure.

the hyperbola assumption for stacking is fulfilled, thus showing the possibilities and limits of the stacking concept in deep reflection seismics.

The DEKORP data examples demonstrate that the proposed method works in practice for deep seismic reflection data and offers advantages in comparison to conventional velocity analysis procedures, especially for detailed studies. The τ-p velocity analysis is not restricted to deep seismic reflection data, but can be applied in all fields of reflection seismics. Examples can be found in BÖNNEMANN (1995).

6. Acknowledgements

The research was funded by the German Federal Ministry of Education, Science, Research and Technology (RG 8709 1) in the frame of the DEKORP project. Professor Dr. H.-P. Harjes, Institute for Geophysics, Ruhr-University Bochum, Germany, is gratefully acknowledged for the academic supervision of the author's Ph.D. thesis. We thank the DEKORP Processing Center, located at the Department of Geophysics at the Technical University Clausthal, for the preparation of the data.

REFERENCES

BÖNNEMANN, C., *Determination of Seismic Velocities from Steep Angle Reflections in the p-τ Domain (in German)* (Ph.D. Thesis, Reports of the Institute for Geophysics, Ruhr-University Bochum, Series A, No. 40, 1995).

BRYSK, H., and MCCOWAN, D. W. (1986), *A Slant-stack Procedure for Point-source Data*, Geophysics *51*, 1370–1386.

CHAPMAN, C. H. (1981), *Generalized Radon Transforms and Slant Stacks*, Geophys. J. R. Astr. Soc. *66*, 445–453.

CLEARBOUT, J., *Imaging the Earth's Interior* (Blackwell Scientific Publications, Oxford, 1985).

CUTLER, R. T., and LOVE, P. L. (1979), *Elliptical Velocity Analysis*, 49th SEG Meeting, New Orleans, Technical Program Abstracts.

DEKORP RESEARCH GROUP (1991), *Results of the DEKORP 1 (BEL-CORP-DEKORP) Deep Seismic Reflection Studies in the Western Parts of the Rhenish Massif*, Geophys. J. Int. *106*, 203–227.

GERVER, M., and MARKUSHEVICH, V. (1966), *Determination of a Seismic Wave Velocity from the Travel-time Curve*, Geophys. J. R. Astr. Soc. *11*, 165–173.

KRAVIS, S. (1990), *The Nth Root Slant Stack—A new Method of Coherency Enhancement*, First Break *8*, 339–344.

MCFADDEN, P. L., DRUMMOND, B. J., and KRAVIS, S. (1986), *The Nth Root Stack: Theory, Applications, and Examples*, Geophysics *51*, 1879–1892.

NEIDELL, N. S., and TANER, M. T. (1971), *Semblance and other Coherency Measures for Multichannel Data*, Geophysics *36*, 482–497.

SCHULTZ, P. S., and CLEARBOUT, J. F. (1978), *Velocity Estimation and Downward Continuation by Wavefront Synthesis*, Geophysics *43*, 691–714.

STOFFA, P. L., BUHL, P., DIEBOLD, J. B., and WENZEL, F. (1981), *Direct Mapping of Seismic Data to the Domain of Intercept Time and Ray Parameter—A Plane Wave Decomposition*, Geophysics *46*, 255–267.

TANER, M. T., and KOEHLER, F. (1969), *Velocity Spectra—Digital Derivation and Application of Velocity Functions*, Geophysics *31*, 859–881.

TATHAM, R. H., KEENEY, J. W., and NOPONEN, I. (1983), *Application of the τ-p Transform (Slant-stack) in Processing Seismic Reflection Data*, Bull. Aust. Soc. Explor. Geophys. *14*, 163–172.

TREITEL, S., GUTOWSKI, P. R., and WAGNER, D. E. (1982), *Plane-wave Decomposition of Seismograms*, Geophysics *47*, 1375–1401.

YILMAZ, Ö., *Seismic Data Processing* (Society of Exploration Geophysicists, Tulsa 1987).

(Received March 3, 1998, revised September 22, 1998, accepted September 25, 1998)

To access this journal online:
http://www.birkhauser.ch

Pure appl. geophys. 156 (1999) 303–318
0033–4553/99/010303–16 $ 1.50 + 0.20/0

| Pure and Applied Geophysics

The Use of Wave-field Directivity for Velocity Estimation: Moving Source Profiling (MSP) Experiments at KTB

J. MÜLLER,[1] M. JANIK[1] and H.-P. HARJES[1]

Abstract—Within the "Integrated Seismics Oberpfalz 1989 (ISO89)" a three-component Moving Source Profiling (MSP) experiment, also named walk-away VSP, was carried out at the drilling site of the "Kontinentales Tiefbohrprogramm der Bundesrepublik Deutschland (KTB)" in Germany. Analysis of transmitted waves traveling from the source locations at the surface down to the receiver array in the borehole reveals velocity information about the illuminated part of the subsurface. Complementary to the widely used evaluation of travel-time perturbations to locate velocity inhomogeneities we suggest the use of the directivity of transmitted wave types down in the borehole. To determine the wave-field directivity we focus on transmitted arrivals by employing principles of "Controlled Directional Reception (CDR)." We calculate local slant-stacks for three different depth positions as a function of the source offset, thus obtaining the variation of the vertical slowness (vertical ray parameter) of incident waves along the horizontal source profile and the vertical receiver array. The slowness data combined with travel times are interpreted by forward modeling taking into account geological information of the survey area. Our findings confirm results from gravity measurements which suggest the existence of large amphibolite/metabasite complexes in the vicinity of the borehole. The described method is also used to identify *P*-to-*S* converted energy originating from fracture zones above the receiver array and to locate the region in which conversion occurs.

Key words: Borehole seismics, velocity estimation, KTB seismic experiments, vertical receiver array, transmitted wave field, *P*-to-*S* conversion.

Introduction

In the framework of the "Integrated Seismics Oberpfalz 1989 (ISO89)" at the site of the "Kontinentales Tiefbohrprogramm der Bundesrepublik Deutschland (KTB)" a Moving Source Profiling (MSP) survey was conducted in the pilot borehole (HARJES *et al.*, 1990).

This experiment aimed at the prediction of structural elements that were to be hit by the later KTB main borehole in the depth range between 4 km and about 10 km. Furthermore it provides a natural link between logging data on the one hand and surface seismics on the other.

The geometry of MSP measurements consists of a vertical receiver spread down in the borehole and a source profile along the surface crossing the drilling site.

[1] Institut für Geophysik, Ruhr-Universität, D-44780 Bochum, Germany.

Apart from travel-time measurements of first arrivals to determine the average velocity of P waves traveling from source to receiver, survey configurations of this kind offer an excellent opportunity to determine the directivity of seismic wave-field components. As such they provide additional information about the spatial distribution of physical properties that cannot be derived from surface seismic measurements.

The idea to use the directional sensitivity of linear receiver arrays to analyze the directivity of interfering wave-field components was first exploited by RIEBER (1936). His efforts to decompose the wave field into its plane wave components by a delay and sum algorithm led to the first variant of beamforming in seismic applications. In the following decades this method was mostly refined in the former Soviet Union where it became known as "Controlled Directional Reception (CDR)" (RIABINKIN, 1957 referred to in GARDNER and LU, 1991; GAL'PERIN, 1974). The more familiar term "Slant-Stack" was introduced by CLAERBOUT (1975) and denotes in a descriptive manner the core of this method, which is widely used currently in seismic data processing.

Whereas the principles of the CDR method can also be used to design offset dependent velocity filters for wave-field separation of MSP data (HARJES *et al.*, 1998; MÜLLER, 1997), here a method for velocity estimation is presented. By calculating local slant-stacks of transmitted phases as a function of source position, we determine the vertical ray parameter at the receiver locations. These data combined with travel times are interpreted by forward modeling, using ray-tracing to test and refine existing crustal models at the KTB.

Data Acquisition

Running from 3 km southwest of the pilot borehole up to 7 km to the northeast, the source profile was chosen perpendicular to the strike of the Franconian Line (FL). The Franconian Line is delineated by a system of steeply northeastward dipping fault zones which separate the eastern crystalline rocks of the Bohemian Massif from Permo-Mesozoic sediments in the west (Fig. 1). The KTB site is situated at the northwestern margin of the Bohemian Massif which represents the largest outcrop of Variscan basement in central Europe (EMMERMANN and WOHLENBERG, 1989). In the vicinity of the drilling site medium-pressure metamorphic basement units prevail, which are mainly formed by paragneisses with intercalations of amphibolite rocks. The entire area was intruded by large volumes of late-to-post Variscan granitoids. One of these, the Falkenberg granite, is crossed at the eastern end of the source profile.

Along the source profile vibroseis sources generated sweeps from 20–80 Hz at distance intervals of 50 m. 5–10 sweeps were vertically stacked and cross-correlated with the sweep, yielding an effective recording time of 12 seconds at a sampling rate of 2 msec. The vertical receiver array in the borehole consisted of a chain of 5

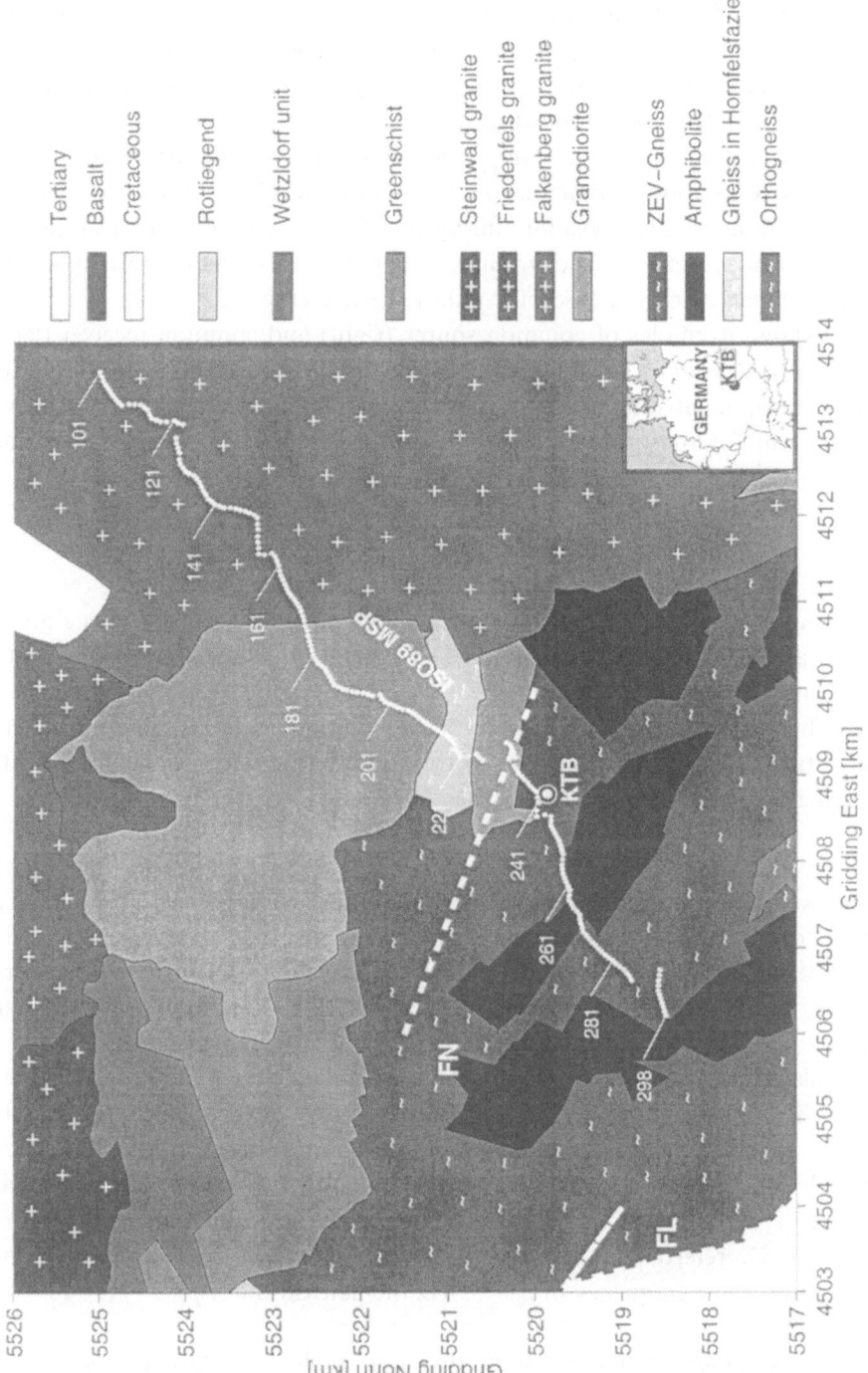

Figure 1

Geological setting at the drilling site of the "Kontinentales Tiefbohrprogramm der Bundesrepublik Deutschland (KTB)" and configuration of the "Integrated Seismics Oberpfalz 1998 (ISO89)" Moving Source Profiling (MSP) experiment. The shot positions are represented by dots and extend from shot number 101 in the northeast to shot number 298 in the southwest. The thick dashed lines denote the Franconian Lineament (FL) and the Fichtelnaab Fault (NF).

three-component geophones separated by intervals of 25 m. The deepest geophone position was at 3685 m. From there the receiver chain was lifted three times by 125 m during the survey so that altogether the depth range between 3210 m and 3685 m was covered. Due to insufficient data quality the lowest 5 receiver locations had to be excluded from the following data processing. The recorded data are available in two different sortings, comprising either all available traces belonging to the same source location, "common source gather (CSG)," or all traces that were recorded at the same receiver position during the survey, "common receiver gather (CRG)". For a homogeneous background model the ray paths of both types of sortings are compared for the case of a reflected wave (Fig. 2).

Figure 3 shows examples of common source (right) and common receiver (left) sorted data for a section of the horizontal profile comprising shot locations 284 to 298 (vertical component only). The data sortings differ in wave-form coherency. This is important for the processing scheme and is discussed in the next section.

Determination of the Vertical Ray-parameter

The estimation of the vertical ray parameter with high accuracy requires a great amount of wave-form coherency in the data. To ensure this, the following aspects have to be taken into account:
1. Signal coherency should not be affected by variation in the weathering layer.
2. The investigated signal must be isolated from interfering parts of the wave field.
3. The estimated vertical ray parameter has to be corrected for borehole dip.

1. It is well known that variation of the weathering layer and topography distort the coherency of seismic wave forms. Borehole registrations avoid this distortion because for data in the CSG sorting the travelpath through the weathering layer is similar (BLIZNETSOV and JUHLIN, 1993). This results in greater coherency of first arrivals and multiples as is obvious in Figure 3.

2. To isolate the event of interest, an offset-dependent velocity filter in the (f, k) domain can be applied. Therefore a theoretical slowness value for a homogeneous model with $v_p = 6100$ m/s was calculated for different source offsets. The filter width was chosen according to the slowness resolution of the vertical array which is controlled by the aperture and the signal frequency according to $\Delta p_f = 1/f_0 \cdot D$ (RIABINKIN, 1957, referred to in GARDNER AND LU, 1991). For an aperture of $D = 350$ m and a mean frequency of $f_0 = 35$ Hz the halfwidth of $\Delta p_f/2 = 40 \cdot 10^{-6}$ s/m is a good choice.

3. The measurement of the apparent slowness implies a determination of the angle of incidence along the receiver profile. Difficulties that arise when the receiver array is fixed down in the borehole are deviations of the borehole from the vertical,

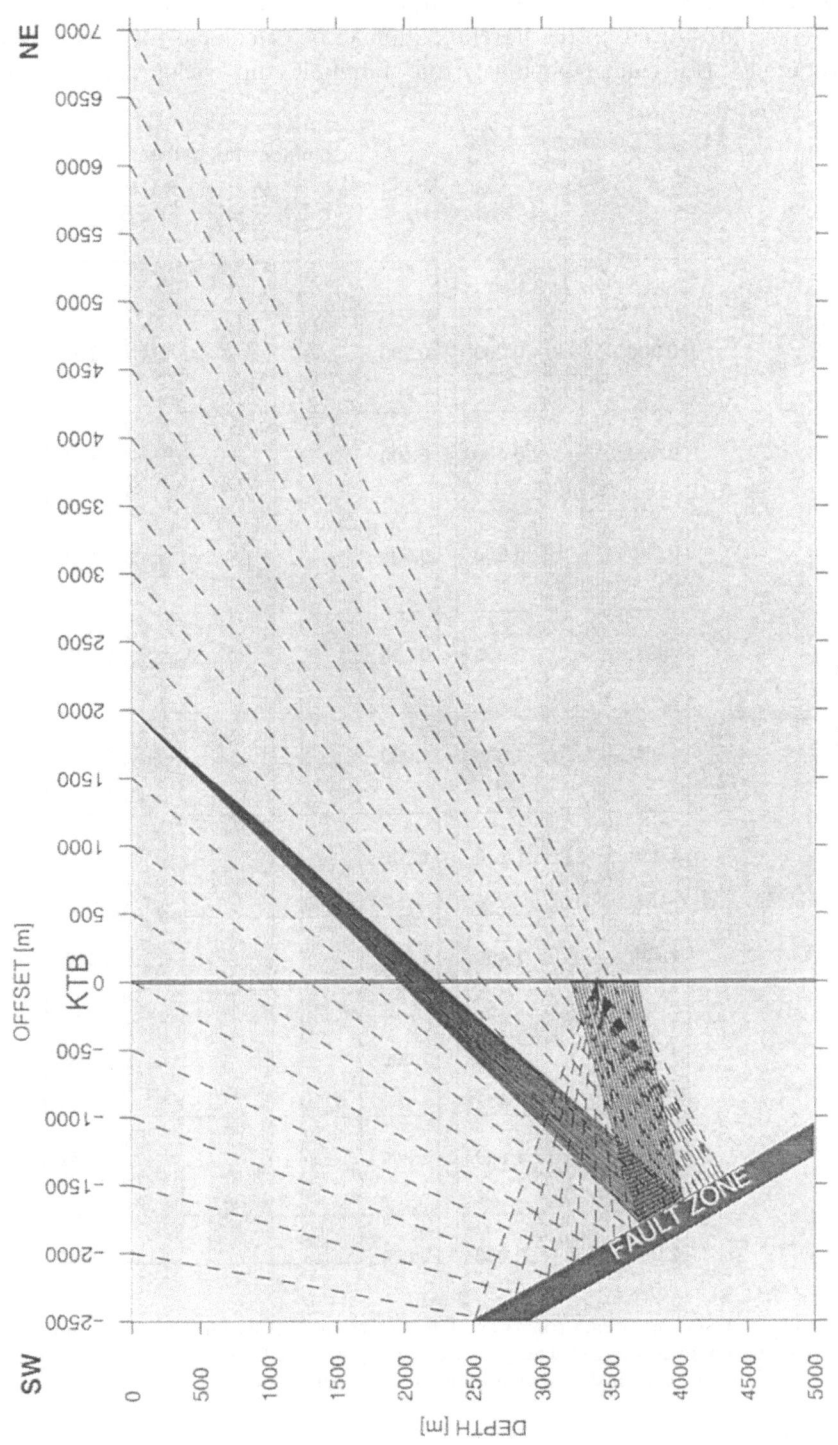

Figure 2

Ray paths for "Common Shot Gather" (CSG) (solid line) and "Common Receiver Gather" (CRG) (dashed lines) sorted data for the case of a reflected wave.

resulting in a wrong offset and dip (Fig. 4a). In the KTB pilot hole, dips up to 3.7 degrees occur dependent on source azimuth and source offset (Fig. 4b).

The error introduced by the borehole dip can be calculated for a homogeneous velocity model (Fig. 4c). Obviously the borehole dip yields higher apparent

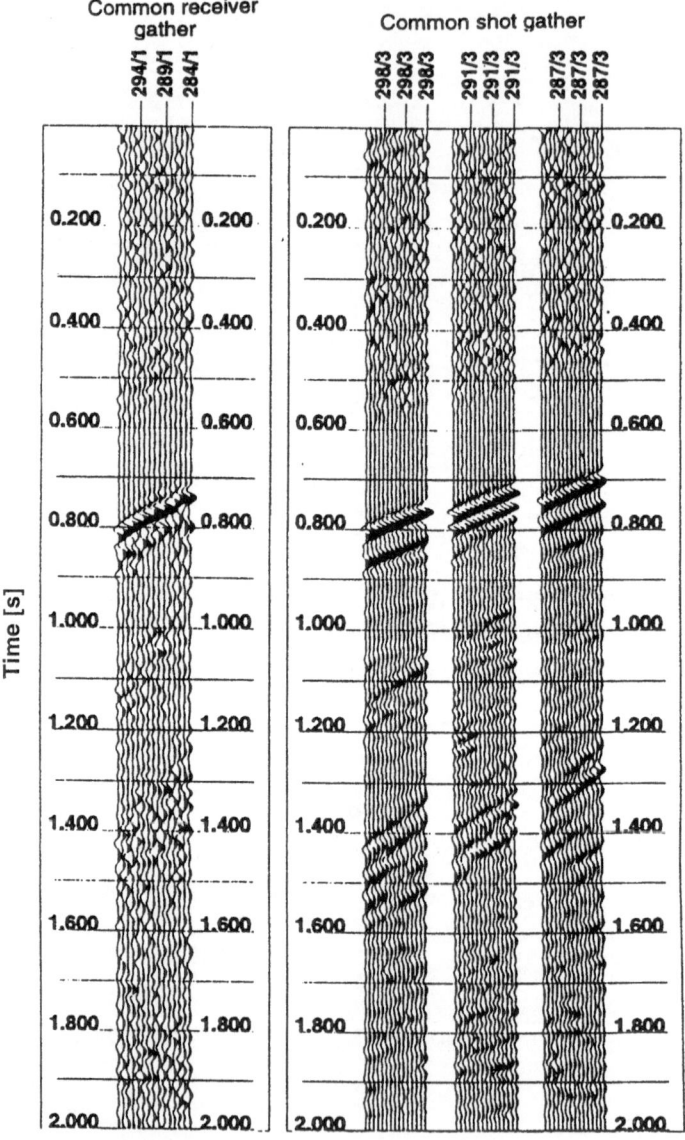

Figure 3
Data example for CSG sorted data (right) and CRG sorted data (left) for a horizontal profile section comprising source points 284 to 298. The greater coherency of wave forms observed in CSG sorted data can be exploited to design optimum processing sequences (see text).

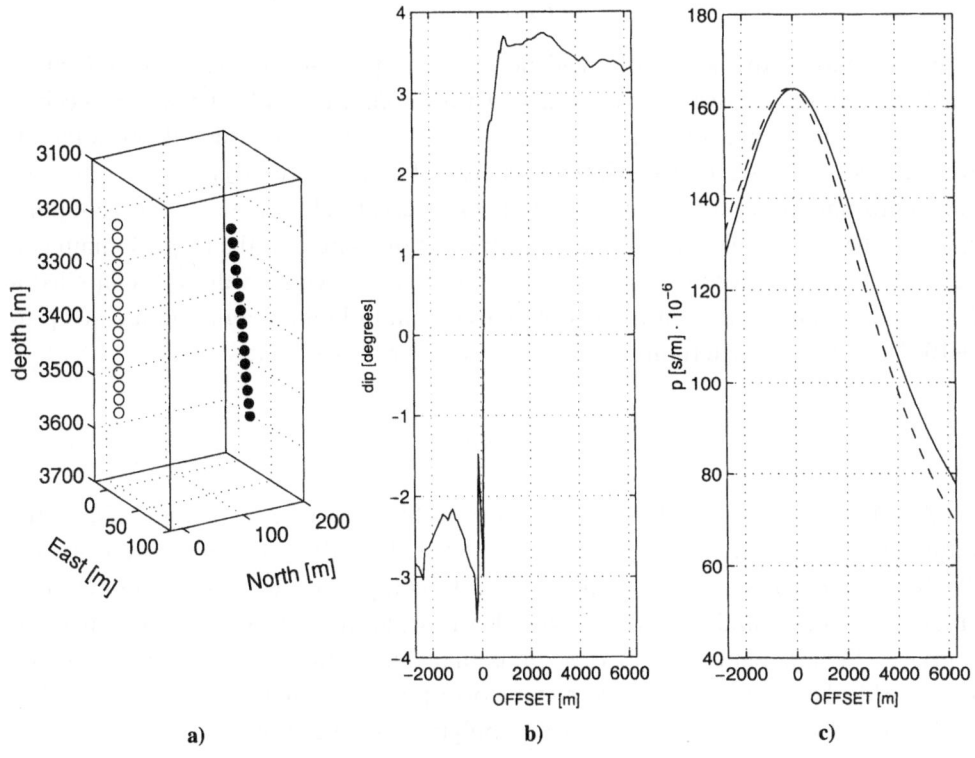

Figure 4

Effect of borehole deviation on measured vertical ray parameters. (a) Real receiver positions (black dots) in comparison to an ideal vertical array (open symbols). (b) Dip of the receiver array in relation to shot point distance. (c) Effect of borehole deviation on the vertical ray parameter for a homogeneous model with $v_p = 6100$ m/s. The dashed line marks the vertical ray-parameter curve that has been corrected for the borehole dip.

velocities in the direction of the dip and lower values in the opposite direction, thus resulting in a tilt of the data curve.

Three-component receivers allow a rotation of the data to P, SV, and SH components, thus achieving maximum reception of the corresponding wave-type energy. After bandpass filtering in the frequency range of 15–45 Hz and trace equalization to balance varying ground coupling of the vibrator sources, we perform an offset dependent extraction of the direct P and S wave in the (f, k) domain. We focus on specific arrivals in the time domain by using time windows centered on arrival times picked from the real data. This facilitates the later determination of the slowness value and saves computation time for the slant-stack operation. To extract the downgoing wave field, the vertical ray parameter p_d for a homogeneous background model with velocity v is determined within the borehole following the equation:

$$p_d = \frac{\cos(i)}{v}$$

where i denotes the angle of incidence with respect to the vertical. To study variations of the vertical ray parameter with depth we divide the whole receiver spread into three subarrays and calculate local slant-stacks for each depth interval. By this procedure we lose resolution because of reduced receiver aperture but we gain information regarding the variation of interval velocity along the geophone spread. The final determination of the amplitude maximum in the (τ, p) domain can be carried out by interpolation between the sampled slowness values. We applied a cubic interpolation between the slowness traces and obtained an accuracy of up to $\pm 10^{-6}$ s/m for the determination of the vertical ray parameter.

Results

A full CRG section along the source profile is shown in Figures 5a–c for different polarization components. Figures 5d–f show the extracted wave fields for P, SV and SH components sorted to CRG. In comparison with the unfiltered data (Figs. 5a–c) the P and S energy is now clearly separated. A strong event is marked on the SH component (CE). It is obvious that this event is part of the downgoing wave field. The vertical ray parameter corresponds to that of the S wave. The coherency of the event is enhanced by applying P-wave statics. It seems that this wave travels with P velocity through the weathering layer. In the following the event is identified as a PS-converted wave.

The vertical ray parameters have been determined for the following wave types and polarization components:
- P wave on the P component,
- S wave on the SV component,
- S wave on the SH component,
- PS-converted wave on the SH component.

Figure 6a illustrates the variation of the P-wave ray parameter along the horizontal profile for each of the three receiver chain positions. Seven source positions have been averaged. For source positions up to ± 2000 m away from the borehole, a decrease of the vertical ray parameter with depth is obvious. This corresponds to an increase of the vertical apparent velocity. For greater distances to the northeast of the borehole, this trend cannot be observed. A possible explanation is the better velocity resolution of the linear receiver array for wave fronts which are incident nearly in the direction of the profile, in contrast to the better directional resolution if the wave is incident perpendicular to it. Thus, for greater angles of incidence the array is more sensitive to variation of the incidence angle due to lateral velocity inhomogeneities. For the nearest shot point (SP 244, offset 60 m) the vertical ray parameter corresponds to the inverse of the interval velocity. The velocities within the depth intervals of interest are listed in Table 1.

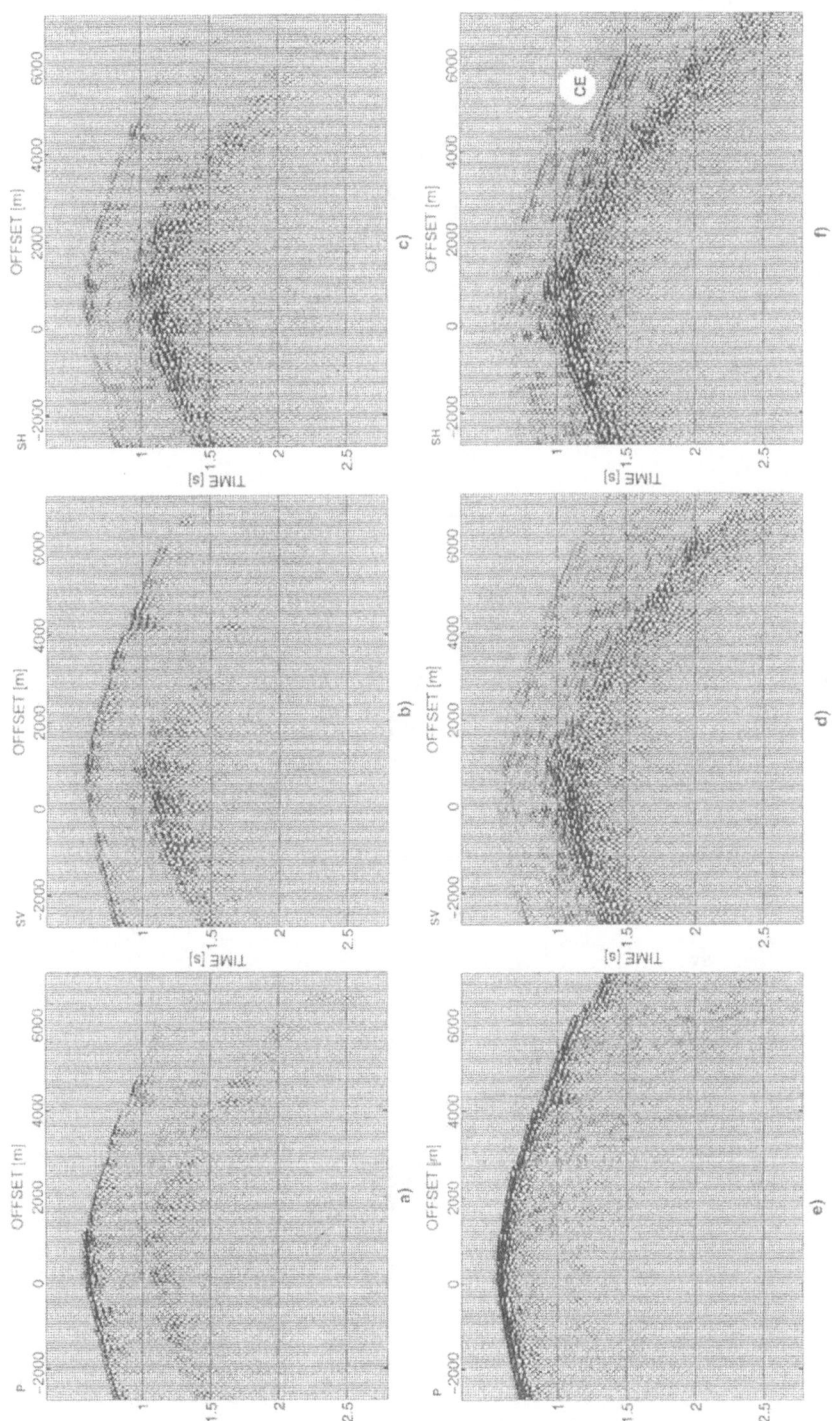

Figure 5

CRG data. (a–c) *P*, *SV*, *SH* component of raw data. (d–f) *P*, *SV*, *SH* components of the extracted transmitted wave fields. All data with 15–45 Hz bandpass filter, trace equalization and static corrections, all data for the receiver at 3385 m depth.

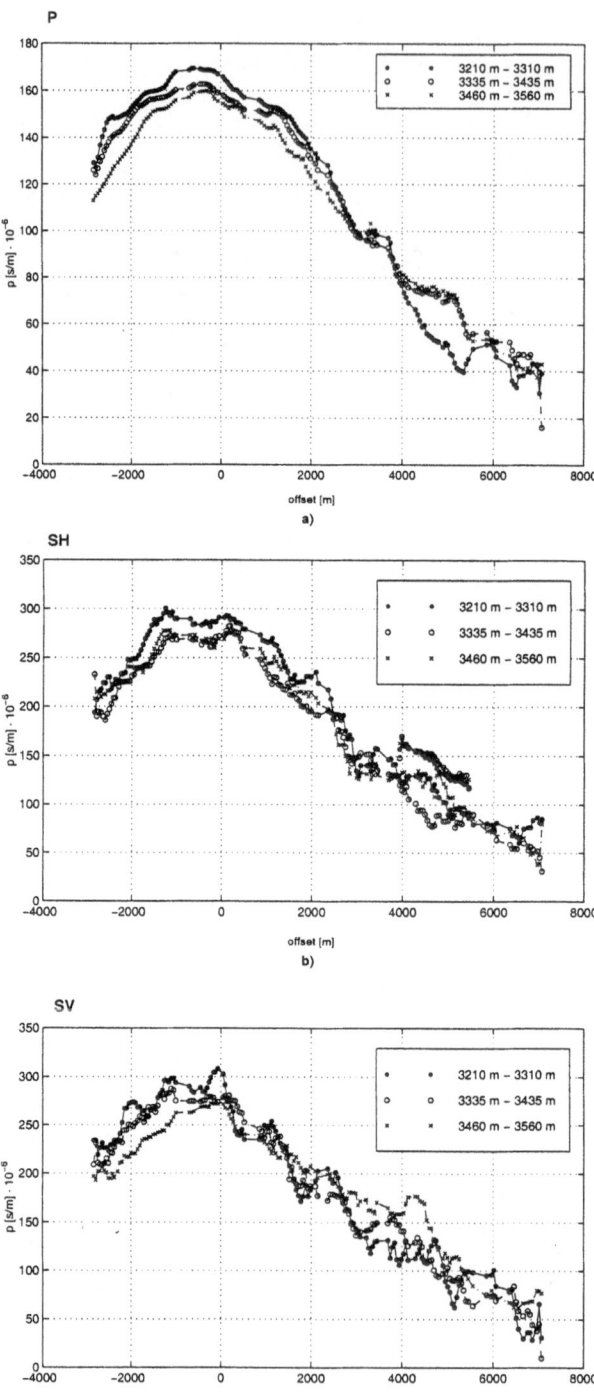

a)

b)

c)

Table 1

Interval velocities from vertical ray-parameter measurements

Depth interval	v_P	v_{SH}	v_{SV}
3210 m–3310 m	5920 m/s	3440 m/s	3310 m/s
3335 m–3435 m	6188 m/s	3660 m/s	3580 m/s
3460 m–3560 m	6281 m/s	3720 m/s	3660 m/s

The increase of the interval velocity has also been observed in the conventional VSP data (RÜHL and HANITZSCH, 1992). The vertical ray parameters of the SV and SH waves show a similar dependency on depth (Figs. 6b,c), although not as extreme as the P-wave ray parameter. Although the data of the horizontal components have been averaged over seven shot points, they contain more local variation in the horizontal direction than the P-wave ray-parameter curve. One possible reason is the anisotropy in the gneisses of the ZEV, caused by the orientation of the foliation, which has more impact on shear waves. Further, it is obvious that the PS-converted wave (CE) in Figure 6a, which runs as a P wave through the weathering layer, does not manifest the strong variation. Therefore it can be

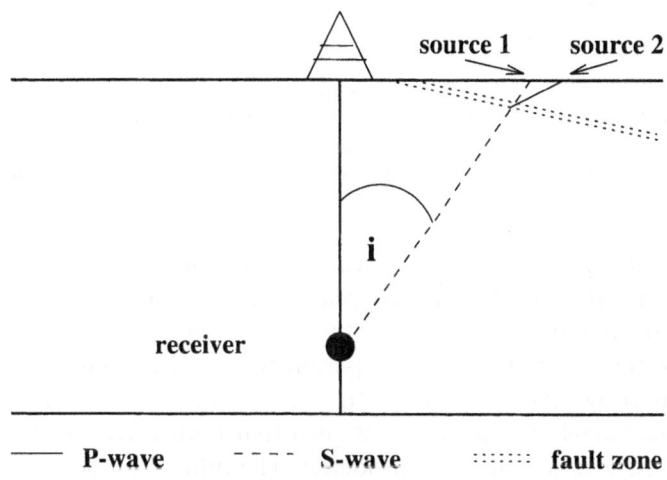

Figure 7

Sketch of the effect of a PS converting interface. Converted and direct wave have the same incidence angle at the receiver array. Source points 1 for the direct wave and 2 for the converted wave differ at the surface.

Figure 6

Vertical ray-parameter values for P (a), SH (b) and SV (c) as a function of offset for the three receiver depth intervals given in the figure. The data are averaged over seven source positions.

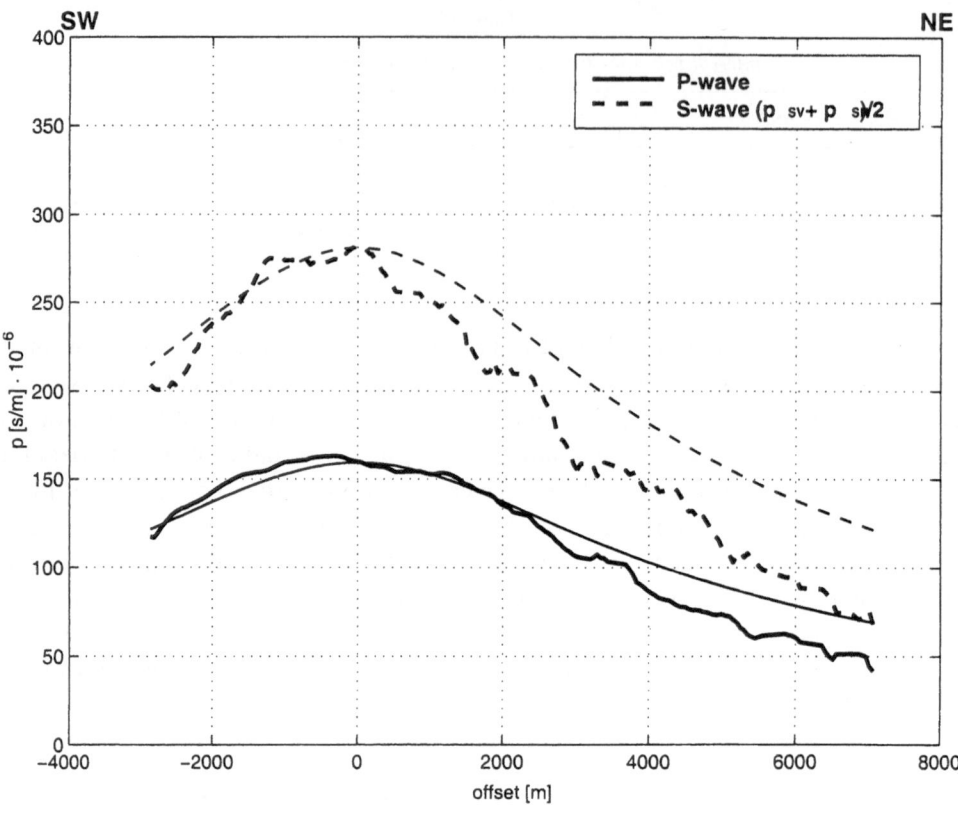

Figure 8

Averaged measured vertical ray-parameter curves for P and S waves (thick lines), together with the vertical ray-parameter curves for a homogeneous model as functions of shot point distance (thin lines). The velocities for the homogeneous model are the mean of the observed interval velocities: $v_P = 6130$ m/ s, $v_S = 3560$ m/s.

concluded that the greatest part of the variation is caused by the weathering layer. The interval velocities of SV- and SH-polarized waves are also listed in Table 1.

The vertical ray-parameter curve of the PS-converted wave (CE) in Figure 6b corresponds to the curve of the SH ray parameter, shifted to greater offsets. Figure 7 schematically shows the ray paths of the direct S wave and a converted wave for a homogeneous model. The geometry is such that both waves hit the array in the borehole with an identical angle of incidence. The different offsets of the sources 1 and 2 at the surface are caused by a change of the ray path of the converted wave and depend on the position and orientation of the converting interface as well as on the v_p/v_s ratio.

The averaged vertical ray parameters of all three depth intervals are combined in Figure 8 and can be compared with theoretically calculated ray parameters for a homogeneous model. The model velocities correspond to the mean of the interval velocities. For convenience, the polarization directions of the S waves have also been averaged.

In contrast to the homogeneous model, an increase of the measured slowness values is observed in the northeast part of the profile. In the southwestern part, an opposite effect is visible for the vertical ray parameter of the P wave. Local effects such as jumps in the vertical ray-parameter curve, caused by rapid small-scale variations of physical parameters or orientation of interfaces, are smoothed by the spatial averaging over 350 m distance in both vertical and horizontal directions. Differences with respect to the homogeneous model are caused by refraction. Reasons are first-order as well as second-order discontinuities, i.e., interfaces and gradient zones, respectively. In a tectonically overprinted region such as the KTB environment, anisotropy is also an important effect. To interpret the measured vertical ray-parameter data we performed a ray-theoretical modeling procedure. The modeling makes use of the measured ray-parameter data and the travel times of the transmitted direct P wave and the P-to-S-converted wave (CE). The aim is to derive a velocity model for the region above the receiver positions and to locate the converting boundary which generates the event CE. To choose an appropriate initial model all available information has been used including

- surface geology,
- a vertical geological profile in the pilot hole (RÖHR et al., 1990),
- results of the 3-D seismic campaign ISO89 (HLUCHY et al., 1992),
- core sample studies of the Falkenberg granite (WÖHRL, 1981),
- velocity model 1001 near the KTB site (WIEDERHOLD, 1992),
- results of gravity modeling (CASTEN et al., 1997).

The most important information from surface geology is the intrusion of the Falkenberg granite and the amphibolite close to the drilling site (compare Fig. 1). The vertical geological profile shows also the gneiss/amphibolite intercalation. The velocity in the granite has been taken from the results of laboratory measurements of WÖHRL (1981). The velocity in the ZEV southwest of the bore-hole is taken from the velocity model 1001 (WIEDERHOLD, 1992). The near-surface information obtained from refraction statics supplied the velocities of the weathering zone. The gravity field shows that the drill hole occurs at the edge of a large amphibolite complex dipping to the southwest, truncated by the Franconian fault zone. Figure 9 illustrates the result of the modeling. The model mostly conforms with the results of 3-D seismics and gravity modeling. The increase of the vertical ray parameter southwest of the hole is caused by the amphibolite body. It shows strong differences with respect to the homogeneous model. The northeast dipping lower interface of this amphibolite complex could be formed by the displacement at the Franconian line thrust. It is one possible explanation for the increase of the vertical ray parameter. Corresponding reflections can be found in the 3-D seismic profiles. Close to the northeast of the drilling site the modeling results support indications for a second amphibolite complex which is responsible for a flattening of the vertical ray-parameter curve. The lower vertical ray-parameter values north-east of the hole exceeding 2000 m distance can be explained with a high velocity

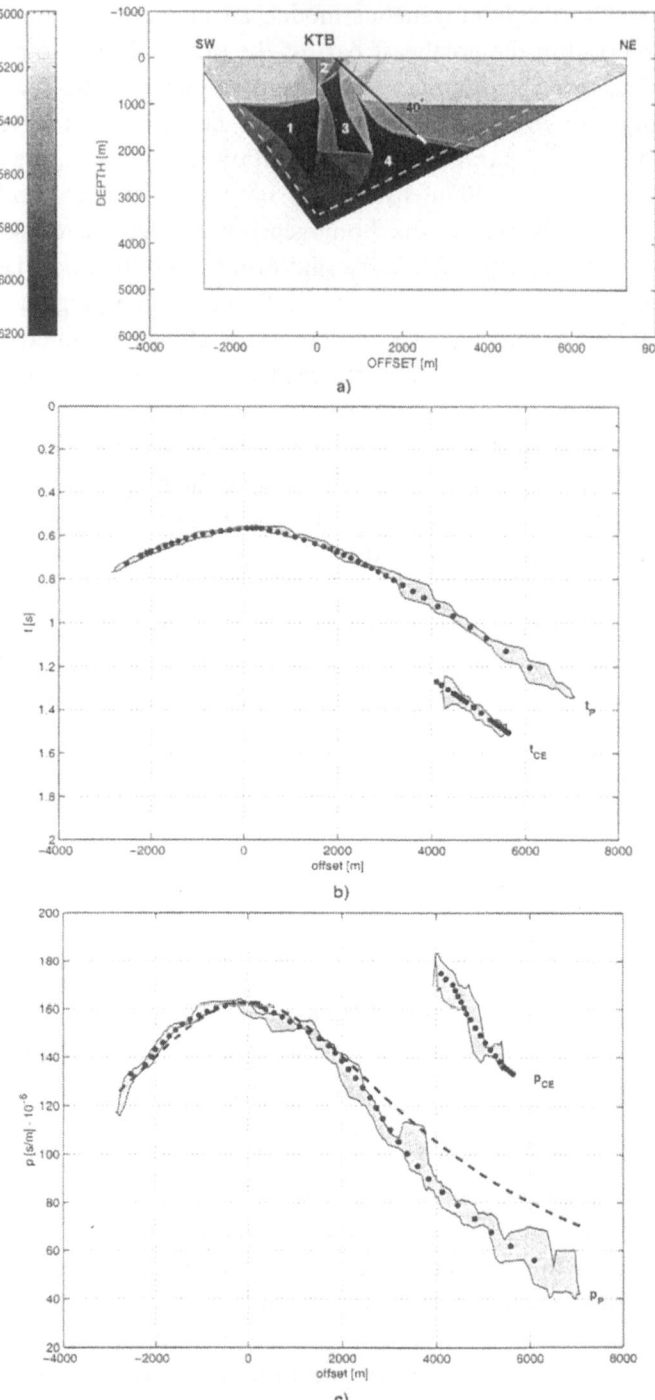

a)

b)

c)

body which is also required to model the gravity field. The travel-time effect of the slower granite complex is compensated by this body.

Modeling of the converted event CE allows a small region to be fixed where conversion must occur at the granite/metabasite boundary northeast of the borehole (see white line in Fig. 9a). Strong P-to-S-converted phases in the transmitted wave field have been observed in former investigations (LÜSCHEN *et al.*, 1996). The corresponding interfaces are related to fluid-filled cracks and low elastic material properties in fracture zones. The extrapolation of the conversion zone can be related to the surface trace of the Fichtelnaab fault zone (HARJES and JANIK, 1994; HARJES *et al.*, 1997). The fact that the observed conversion is strongest on the SH component cannot be explained by this 2-D modeling attempt. We consider it to be a 3-D effect that must be explained by 3-D survey data and 3-D modeling.

Conclusions

Modeling of vertical ray-parameter data provides an additional important tool to extract information from transmitted wave-field data. If there are three-component geophones as well as receiver/source arrays, the vertical ray parameter can be measured with high accuracy. The vertical ray parameter contains directional information as well as velocity information and thus allows modeling a velocity structure with lateral inhomogeneities. It is also possible to locate a P-to-S-conversion zone using vertical ray-parameter information.

Acknowledgements

The authors gratefully acknowledge the work of the various field crews. Special thanks are due to H. J. Dürbaum who coordinated the Integrated Seismic Survey Oberpfalz (ISO-89). We greatly benefited from discussions with M. Bliznetsov regarding the implementation of the CDR method. This study was supported by

Figure 9

Result of the combined modeling of slowness and travel-time data. (a) The obtained velocity distribution in the subsurface region that is illuminated by the transmitted part of the wave field. The numbers 1 to 4 denote high velocity complexes which can be related to amphibolite/metabasite units. The P-to-S converting boundary is marked in white. Linear extrapolation yields a 40° dipping interface which intersects the source profile at the surface trace of the Fichtelnaab fault zone. (b) Modeled travel times (dark dots) for the direct P wave (t_p) and the P-to-S-converted wave (t_{CE}). The gray shaded areas denote the standard deviation range obtained from the measured data. (c) Results of the slowness modeling (dark dots) for the direct P wave (p_p) and the P-to-S-converted wave (p_{CE}). The gray shaded areas denote the standard deviation range obtained from the measured data. For comparison the slowness values resulting from a homogeneous model (dashed line) are also given.

the German Science Foundation (Deutsche Forschungsgemeinschaft) and the Federal Ministry of Science and Technology (Bundesministerium für Forschung und Technologie) within the DEKORP and KTB programs.

REFERENCES

BLIZNETSOV, M., and JUHLIN, C. (1994), *Analysis of Wave Fields by the Common Excitation Array (CEA) Method*, J. Appl. Geophys. *32*, 245–256.

CASTEN, U., GÖTZE, H.-J., PLAUMANN, S., and SOFFEL, H. G. (1997), *Gravity Anomalies in the KTB Area and their Structural Interpretation with Special Regard to the Granites of the Northern Oberpfalz, Germany*, Geologische Rundschau *86*, 79–86.

CLAERBOUT, J. F. (1975), *Slant Stacks and Radial Traces*, SEP *5*, 1–12.

EMMERMANN, E., and WOHLENBERG, J. (eds.), *The German Continental Deep Drilling Program (KTB)* (Springer Verlag, Berlin 1989).

GAL'PERIN, E. I., *Vertical Seismic Profiling* (Soc. Expl. Geophys., Tulsa 1974).

HARJES, H.-P., JANIK, M., and KEMPER, M. (1990), *MSP—A Link Between KTB Borehole Data and Seismic Surface Measurements*, KTB Report *90-6b*, 137–155.

HARJES, H.-P., and JANIK, M. (1994), *Origin of Reflections from the Altenparkstein Fault Zone (KTB)*, KTB Report *94-2*, A97–A106.

HARJES, H.-P., BRAM, K., DÜRBAUM, H.-J., GEBRANDE, H., HIRSCHMANN, G., JANIK, M., KLÖCKNER, M., LÜSCHEN, E., RABBEL, W., SIMON, M., THOMAS, R., TORMANN, M., and WENZEL, F. (1997), *Origin and Nature of Crustal Reflections: Results from Integrated Seismic Measurements at the KTB Super-Deep Drilling Site*, J. Geophys. Res. *102*, B8, 18,267–18,288.

HARJES, H.-P., JANIK, M., MÜLLER, J., and BLIZNETSOV, M. (1998), *Imaging of Crustal Structures from Vertical Array Measurements*, Tectonophysics *286*, 185–192.

HLUCHY, P., KÖRBE, M., and THOMAS, R. (1992), *Preliminary Interpretation of the 3D-Seismic Survey at the KTB Location*, KTB Report *92*, 5, 31–52.

LÜSCHEN, E., BRAM, K., SÖLLNER, W., and SOBOLEV, S. (1996), *Nature of Seismic Reflections and Velocities from VSP-experiments and Borehole Measurements at the KTB Deep Drilling Site in SE-Germany*, Tectonophysics *264*, 309–326.

MÜLLER, J. (1997), *Richtungskontrollierte Stapelung und Migration von Bohrlochseismogrammen am Beispiel des MSP 2 Experiments in der KTB*, Thesis, Ruhr-University, Bochum, 99 pp.

RIABINKIN, L. A., *Fundamentals of resolving power of controlled directional reception l(CDR) of seismic waves*. In *Slant-stack Processing* (eds. Gardner, G. H. F., and Lu, L.), vol. 14 (Soc. Expl. Geophys., Tulsa 1991) pp. 36–60.

RIEBER, F. (1936), *A New Reflection System with Controlled Directional Sensitivity*, Geophysics *1*, 97–106.

RÖHR, C., KOHL, J., HACKER, W., KEYSSNER, S., MÜLLER, H., SIGMUND, J., STROH, A., and ZULAUF, G. (1990), *German Continental Deep Drilling Program (KTB)—Geological Survey of the Pilot Hole "KTB Oberpfalz VB"*, KTB Report *90-8*, B1–B55.

RÜHL, T., and HANITZSCH, C. (1992), *Average and Interval Velocities Derived from First Breaks of Vertical Seismic Profiles at the KTB Pilot Hole*, KTB Report *92-5*, 102–219.

WIEDERHOLD, H. (1992), *Interpretation of the Envelope-stacked 3D Seismic Data and its Migration—Another Approach*, KTB Report *92-5*, 67–114.

WÖHRL, T. (1981), *Geschwindigkeitsuntersuchungen am Falkenberger Granit*, Thesis, Ruhr-University, Bochum.

(Received January 9, 1998, revised September 11, 1998, accepted September 25, 1998)

Pure appl. geophys. 156 (1999) 319–344
0033–4553/99/010319–26 $ 1.50 + 0.20/0

⎥ Pure and Applied Geophysics

Elimination of Noise Caused by Spikes and Bursts in Vibroseis Data

ULRICH POLOM[1]

Abstract—Seismic recording systems without a telemetry system have often been affected by electromagnetic induced spikes or bursts, which lead to strong data distortions combined with the correlation process of the vibroseis method. Partial or total loss of the desired seismic information is possible if no automatic spike and burst reduction is available in the field prior to vertical stacking and correlation of the field record.

Currently, combined with the use of modern telemetry recording systems, the most common noise reduction methods in vibroseis techniques (e.g., spike and burst reduction, diversity stack) are already applied in the field to reduce noise in a very early state. The success of these automatic correction methods depends on the fundamental principles of the recording situation, the actual characteristic of the distorting noise and the parameter justification by the operator. Since field data are usually correlated and already vertical stacked in the field to minimize logistical and processing costs, no subsequent parameter corrections are possible to optimize the noise reduction after correlation and vertical stacking of a production record.

The noise reduction method described in this paper uses final recorded and stacked vibroseis field data at the correlated or uncorrelated stage of processing. The method eliminates signal artifacts caused by spikes or bursts combined with a standard convolution process. A modified correlation operator compresses the noise artifact in time using a single trace convolution process. After elimination of this compressed noise, re-application of the convolution process leads to a noise-corrected replacement of the input data. The efficiency of the method is shown with a synthetic data set and a real vibroseis field record. Furthermore, several thousand records from a 2-D deep seismic reflection project could be corrected with good results using this method.

Key words: Seismic data processing, vibroseis noise reduction.

Introduction

During the processing of deep crustal reflection seismic data from DEKORP 3 project (DÜRBAUM *et al.*, 1994), which has been measured by the vibroseis technique during 1990 using a modern SERCEL 368 (main line 2-D) recording system and additionally an older DFS V (cross line, 2-D and single-fold 3-D)

---

[1] DEKORP Processing Center, Institute for Geophysics, University of Clausthal, Arnold-Sommerfeld Str. 1, D-38678 Clausthal-Zellerfeld, Germany. E-mail: polom@ifg.tu-clausthal.de

recording system, noise problems occurred in the data sets measured by the DFS V recording system. Since this instrument has no telemetry system and no spike and burst reduction, considerable long-lasting noise was generated by the interaction of the correlation operator and electromagnetic bursts.

This interaction leads to anomalous signal appearances in the raw data after application of the correlation process, e.g., a spike in uncorrelated data reproduces the sweep signal in time-reversed order. Depending on the time position of a spike in uncorrelated data, this effect produces a distortion of the desired information in parts or even in the whole correlated trace. The situation becomes more complicated if several spikes occur at different times. They produce time-reversed sweeps which are overlaying themselves and the desired data.

The most common method to eliminate such disturbances is to suppress these spikes and bursts in an early recording state, where no vertical stack and no correlation have been applied to the field data. Using modern recording systems (e.g., SERCEL 368), including a telemetry system (i.e., digitization of the data is made in a box close to the receiver location and not in the recording truck) and a real time spike noise reduction, this disturbance is no longer an important problem in vibroseis data recording. Since DFS V system has no telemetry system, the electromagnetic events could be picked up by the main cable and were transmitted subsequently to most of the channels. In addition, the distortions had not been corrected prior to vertical stacking of the data. Therefore, most of the traces recorded were influenced by these distortions.

Since the cross-line data sets generated by DFS V system should be additional (low cost) information to the main line, especially for spatial dip analysis, only low-fold (8-fold) of single-fold coverage has been measured. The data were recorded using SEG-B multiplex data format and no correlation was applied in the field. Due to limited processing capabilities, the data could not be checked by field quality control.

No standard processing sequence was available to suppress this kind of noise in correlated data satisfactorily. Also, in uncorrelated data it is difficult to identify and correct the spikes and bursts using interactive or automatic processing methods. Therefore, the development of a specific processing sequence was necessary to separate and to eliminate such noise from the desired data. This method proved useful in obtaining stacked 2-D sections of the cross-lines and single-fold 3-D volumes for interpretation.

Description of the Method

1. Fundamentals

Based on the classic vibroseis convolution model (e.g., ANSTEY, 1970), the basic composition of the seismic reflected wave form can be represented in time domain by the equation

$$srw(t) = s(t)*r(t)*e(t), \tag{1}$$

where $srw(t)$: seismic reflected wave form (vibrogram), $s(t)$: source signal including recording filter ("Filtersweep"), $r(t)$: seismic event function, $e(t)$: earth filter (e.g., spherical divergence; absorption), * denotes convolution.

The vibroseis correlation process, i.e., convolution with the time-reversed source operator $s(-t)$, leads to the correlated seismic trace

$$srw_c(t) = \Phi(t)*r(t)*e(t), \tag{2}$$

where

$$\Phi(t) = s(t)*s(-t) \tag{3}$$

is the autocorrelation of the filtered sweep signal $s(t)$.

The typical source signal used in vibroseis technique is a frequency modulated sinoidal signal in which the instantaneous frequency varies with time.

A general representation of a non-tapered linear vibroseis sweep given by SERIFF and KIM (1970) is

$$s(t) = C_0 \sin 2\pi(f_0 + Qt)t, \tag{4}$$

where C_0: constant to control the amplitude, f_0: frequency to start from, Q: constant to control frequency variation.

Generally, the frequency range of interest is less than the recording frequency range limited by Nyquist frequency. Therefore, the observed frequency range in equation (2) is limited by the frequency range of the autocorrelation function $\Phi(t)$ of the sweep signal. The sweep time duration T is limited by the difference between total recording time and listening period (which is usually the same as the total time of the final correlogram).

2. Signal Distortion Caused by Spikes in Uncorrelated Data

Assuming that an external influence (e.g., electromagnetic induction from an electrical power line switch) generates an additional strong spike $sp(t)$ to the desired uncorrelated data from (1), i.e.,

$$srw(t) = s(t)*r(t)*e(t) + sp(t), \tag{5}$$

with $sp(t) = a(t_i)$ (t_i: time position of the spike; a: amplitude of the spike) and $sp(t) = 0$ elsewhere.

The processing stage of equation (5) is the most common domain of the spike and burst reduction possibilities used by modern recording systems. Most of these methods use energy for the spike identification. If the energy of the spike $sp(t)$ is much greater than the energy of the desired signal $s(t)*r(t)*e(t)$ beneath the spike, it will be identified and eliminated by the reduction software automatically. It is advantageous to do this prior to vertical stacking. However, as a disadvantage in most practical cases, the sampling dependent frequency range is present at this processing stage. This includes frequencies extending to the Nyquist frequency, including all non-desired frequencies outside the frequency range of the sweep $s(t)$. Therefore, the spike could be hidden by, e.g., strong monofrequent noise outside the frequency range of $s(t)$ but within the frequency range reaching the Nyquist frequency.

Application of the correlation process with the pilot sweep $s(t)$ produces two terms from equation (5).

$$srw_c(t) = srw_{c1}(t) + srw_{c2}(t). \tag{6}$$

The first term is the expected correlogram:

$$srw_{c1}(t) = \Phi(t)*r(t)*e(t), \tag{7}$$

the second term will be of the form:

$$srw_{c2}(t) = sp(t)*s(-t), \tag{8}$$

and describes the correlation artifact produced by the interaction of the spike $sp(t)$ and the correlation operator $s(-t)$. This process will generate a time reversed sweep artifact in the correlated signal.

Note that the artifact described in equation (8) is limited in bandwidth. The resulting bandwidth of the correlation artifact is less or equal to the bandwidth of the pilot sweep. In addition, the total time of the correlogram is usually limited by the listening period, which is mostly shorter than sweep time in practical cases. Therefore, in many cases only a part of the correlation operator is overlaying the desired correlogram.

3. The Influence of Spherical Divergence

To visualize the effect of spherical divergence in the following explanations, Figure 1 shows a synthetical example of a seismic event function $r(t)$ at the bottom of the figure. In addition, each single event is also plotted on a separate trace. The events of this initial model have been designed to the amplitude value of 1.

Usually, the signal energy or a raw seismic record (e.g., generated by dynamite source) is highest at times near the first arrival (direct or refracted waves), where the influence of $e(t)$ is small. Figure 2 shows the influence of a spherical divergence function $e(t)$ (e.g., $t^{-1.5}$ from 0 s to 1 s) on the trace ensemble from Figure 1.

In equation (5), the sweep $s(t)$ is convolved with the compound function $r(t)*e(t)$. Therefore, most of the energy of $srw(t)$ will appear at times beginning with time $t_s = 50$ ms, where $r(t)*e(t)$ has large amplitudes, and ending at time $t_e = 2050$ ms, which is the sum of start time t_s and the total sweep time duration T. Figure 3 shows the result of this convolution process, using a synthetically generated upsweep (10 Hz to 100 Hz range; 2 s duration and 100 ms cosine taper at both ends of the sweep) applied to the data from Figure 2. Since the listening period is usually shorter than the sweep time, sweeps from later events with less energy are mainly overlayed in most parts of the record by the energy of the "first arrival sweep."

Assuming that this recording stage has been influenced by a spike $sp(t)$ (see equation (5)) at time 1900 ms with nearly the same amplitude as the "first arrival sweep" (Fig. 4). The spike could not be identified in the trace containing all events, since it has been overlayed by the "first arrival sweep." Therefore, in uncorrelated data it is difficult to recognize a spike or burst with weak energy relative to the first arrival sweep, but with strong energy relative to the sweeps convolved with later events. Note that this is the common domain of the automatic

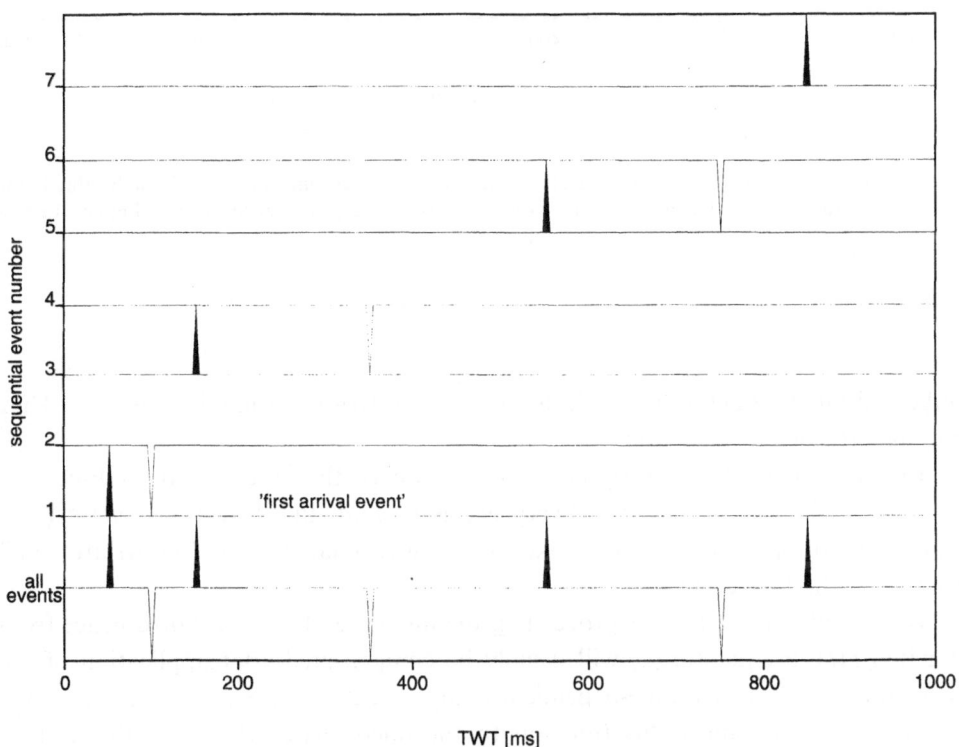

Figure 1

Synthetic seismic event function $r(t)$ (bottom trace). In addition, each individual event of the function has been placed in a separate trace. All of the events are designed to the power of one.

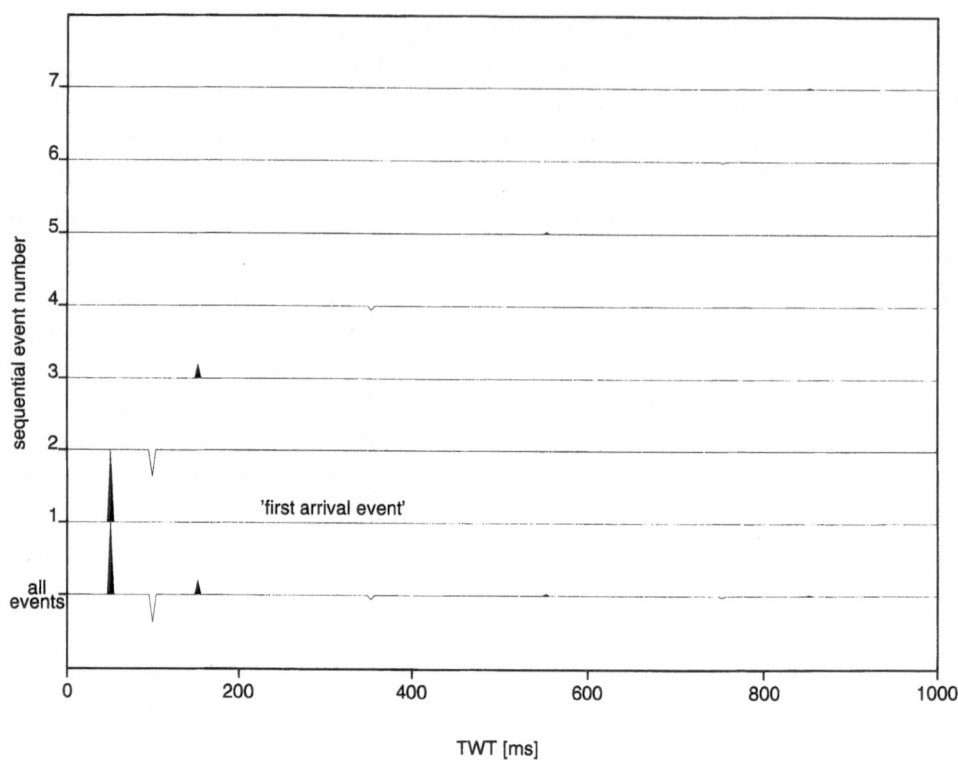

Figure 2
Data set from Figure 1 after application of a spherical divergence function $t^{-1.5}$. Usually, this is the recording characteristic of dynamite sources. Most of the trace energy is concentrated in times of the first three events.

spike and burst reduction possibilities used by modern recording instruments. They may ignore such a noise event.

Figure 5 shows the appropriate correlogram of the ensemble from Figure 4. Since the influence of spherical divergence is present, the distortions $sp(t)*s(-t)$ (equation (8)) generated by the spike in the uncorrelated data (Fig. 4) are small relative to the first events.

As a condition for further processing, the influence of spherical divergence from function $e(t)$ in equation (2) will usually be compensated after application of the correlation process by a time-dependent analytic scaling function $f(t)$. Note that it is not possible to apply this function to the uncorrelated data directly, as it is necessary to locate a seismic event at a fixed time position for application of $f(t)$. In contrast to correlated data, a seismic event in uncorrelated data expands over the time period of the sweep. Application of $f(t)$ to equation (6) leads to

$$srw_c(t)f(t) = \Phi(t)*r(t)*e(t)f(t) + sp(t)*s(-t)f(t). \qquad (9)$$

At this stage of processing, the influence of spherical divergence of $e(t)$ in the first term of equation (9) has been corrected by $f(t)$, but $f(t)$ also amplifies the distortion term $sp(t)*s(-t)$ in equation (9) considerably.

Application of the spherical divergence compensation function $t^{1.5}$ from 0 s to 1 s to the example in Figure 5 leads to Figure 6, which is equivalent to the situation from equation (9). With respect to the compensation function $f(t)$, the distortions at later times of the correlogram are amplified more than the distortions at the beginning of the trace. For the first time in the processing sequence, now it is possible to identify the distortion of the data. This kind of distortion usually would be eliminated from further processing by killing parts or even the whole trace. Processing low-fold or single-fold data sets with many distorted traces would lead to a considerable or total loss of subsurface information.

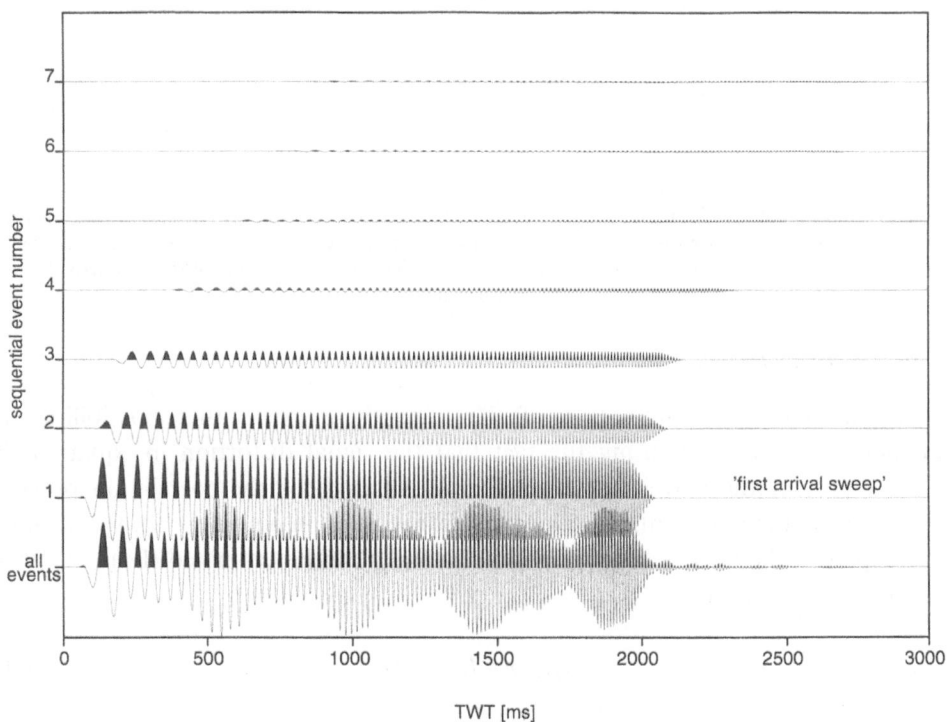

Figure 3

Convolution of the seismic event function (Fig. 2) with a linear upsweep (10 Hz–100 Hz, 2 s duration and 100 ms cosine taper at both ends of the sweep). Usually, this is the recording characteristic of vibroseis sources. Most parts of the synthetic vibrogram (bottom trace) are overlayed by the interacting energy of the 'first arrival sweep' and the sweeps from the following two events.

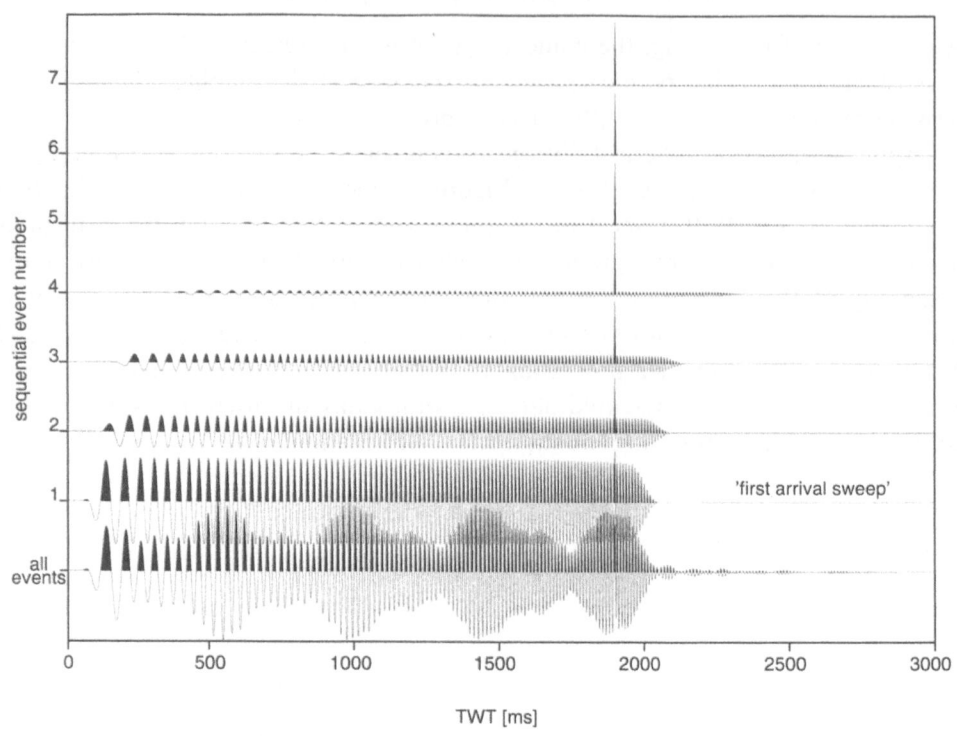

Figure 4

Data set from Figure 3 with an added spike at time 1900 ms. Due to their weaknesses, the sweeps from the later events are influenced in an increasing manner. It is hardly possible to identify the spike by its amplitude in the vibrogram (bottom trace) because it is similar to the amplitude of the first arrival sweep at this time.

4. Elimination of Distortion

Beyond the total elimination of the trace there exist other possibilities to eliminate the distortion, using the fact that the signal distortion in equation (9) forms a part of the correlation operator $s(-t)$ used for the correlation process.

Analyzing the bandwidth of the correlation artifact in equation (9), the position of the spike in the uncorrelated data (equation (5)) can be evaluated (POLOM, 1997) by the "times of frequency"

$$t_i(f_i) = f_i - f_0/G, \tag{10}$$

where f_i denotes the actual frequency at time t_v, f_0 is the initial frequency and $G = (f_e - f_0)/T$ is the gradient of the correlation operator used (f_e denotes the ending frequency of the sweep).

As an example, assume a spike-artifact distorted correlated trace from Figure 4 with 1 s duration, generated by a vibrogram of 3 s registration period using a 2 s linear upsweep with 10 Hz to 100 Hz frequency range ($G = 45$ Hz/s). The position of the spike in the uncorrelated data is not known, however it has produced a time-reversed linear sweep artifact ($G = -45$ Hz/s) from 95.5 Hz ($t = 0$ s) to 50.5 Hz ($t = 1$ s) in the correlogram.

Applying equation (10), this corresponds to a time t_i (95.5 Hz) $= 1.9$ s relative to $t = 0$ s and a time t_i (50.5 Hz) $= 0.9$ s relative to $t = 1$ s. Therefore, the correlation artifact should originate by a spike located in the uncorrelated data at 1.9 s. Elimination of the spike and subsequent correlation leads to a corrected correlogram without distortion term. The success of this method depends on the possibilities to determine the parameters of the correlation artifact exactly, for to get a good estimate of the spike position in the uncorrelated data.

Another way is to examine the correlogram from equation (9) for all possible correlation artifacts caused by spikes in uncorrelated data. Since all of these artifacts show the characteristics of time-reversed correlation operators, it is possible to compress them by a convolution process, using the time-reversed correlation operator $s(t)$, which is the same as the initial pilot sweep function. To

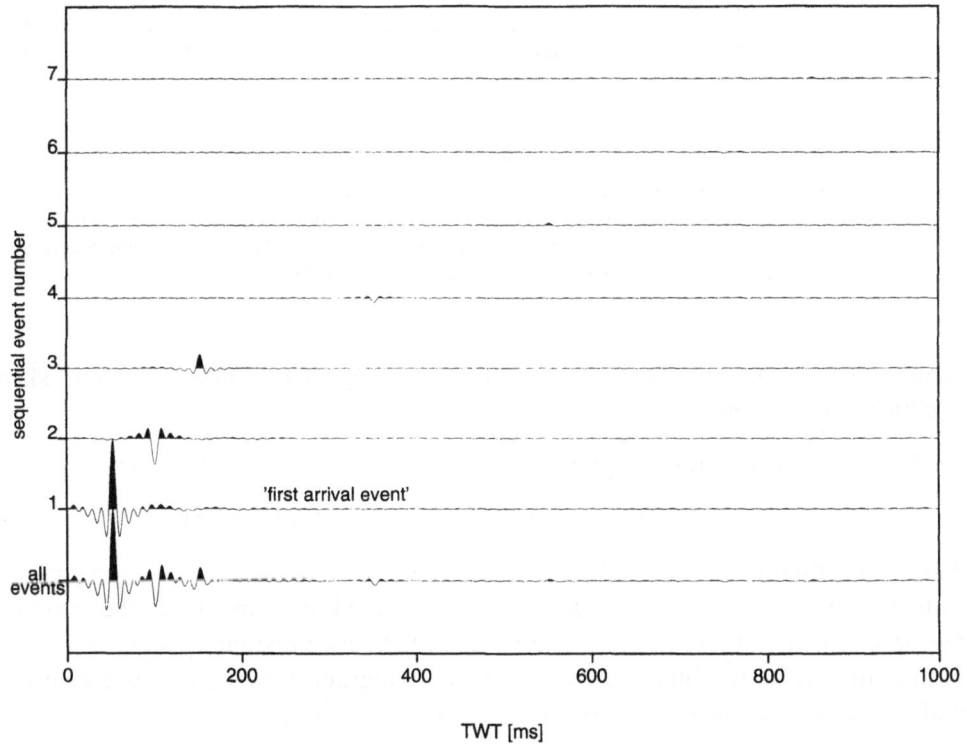

Figure 5
Result of correlation process applied to the data set from Figure 4. The energy of the correlogram is concentrated in the first three events, due to the influence of the spherical divergence. The energy of the correlation artifact seems to be minimal.

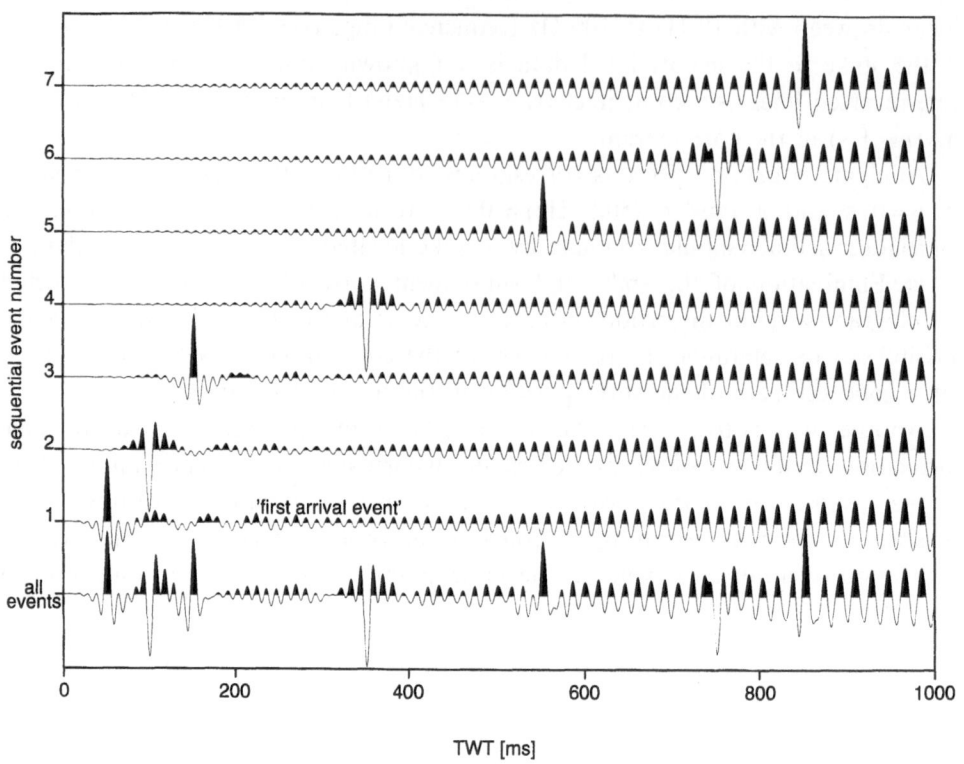

Figure 6
Data set from Figure 5 after correction of spherical divergence by $t^{1.5}$. This standard preparation used in seismic data processing leads to an amplification of the correlation artifact induced by the spike. Depending on the characteristic of the spherical divergence compensation function, the amplification of the artifact rises to later times of the correlogram.

distinguish this operator from the initial pilot sweep, it is denoted as $s'(t)$. Then equation (9) becomes.

$$srw_c(t)f(t)*s'(t) = \Phi(t)*r(t)*e(t)f(t)*s'(t) + sp(t)*s(-t)f(t)*s'(t)$$

$$= \Phi(t)*r(t)*e(t)f(t)*s'(t) + sp(t)f(t)*\Phi'(t). \qquad (11)$$

This transformation can be called "vibrogram representation" (VR) of the analytically scaled correlogram $srw_c(t)f(t)$, since it would be similar to the original vibrogram if no spherical divergence is present. Every previous existing spike has been amplified by the function $f(t)$ (in the correlogram domain) and is convolved with the autocorrelation function $\Phi'(t)$ (in the vibrogram domain). Beyond this, only the frequency range of interest defined by $s(t)$ is present in the data after application of the correlation process. Therefore, it is more favorable to identify a spike position in the VR than in the original vibrogram, because the VR has better signal-to-noise conditions relative to the spike, which is now "signal."

Figure 7 shows the appropriate VR of the example from Figure 6. Firstly, the amplitude of the "first arrival sweep" now is the same as the amplitude of the later events. Furthermore, the amplified distortion of the correlogram from Figure 6 has been compressed to the autocorrelation of the distortion with center position at 1900 ms. This compressed distortion energy can be identified clearly in the "all-events" trace (bottom of Fig. 7). Compared to Figure 4 there are substantially better conditions for an automatic spike elimination procedure.

After the identification, the portion of distortion energy from the second term of equation (11) can be eliminated. This can be done by application of surgical mute, automatic despiking, or similar data enhancement to the original vibrogram, but also to the VR. Presuming ideal correction conditions, the noise has been compressed to a very small time period in the VR. Therefore, the artifacts of the correction process in the VR are as nearly negligible as in the original vibrogram. The distortion term in equation (11) will be very small and can be neglected approximately. Then equation (11) becomes

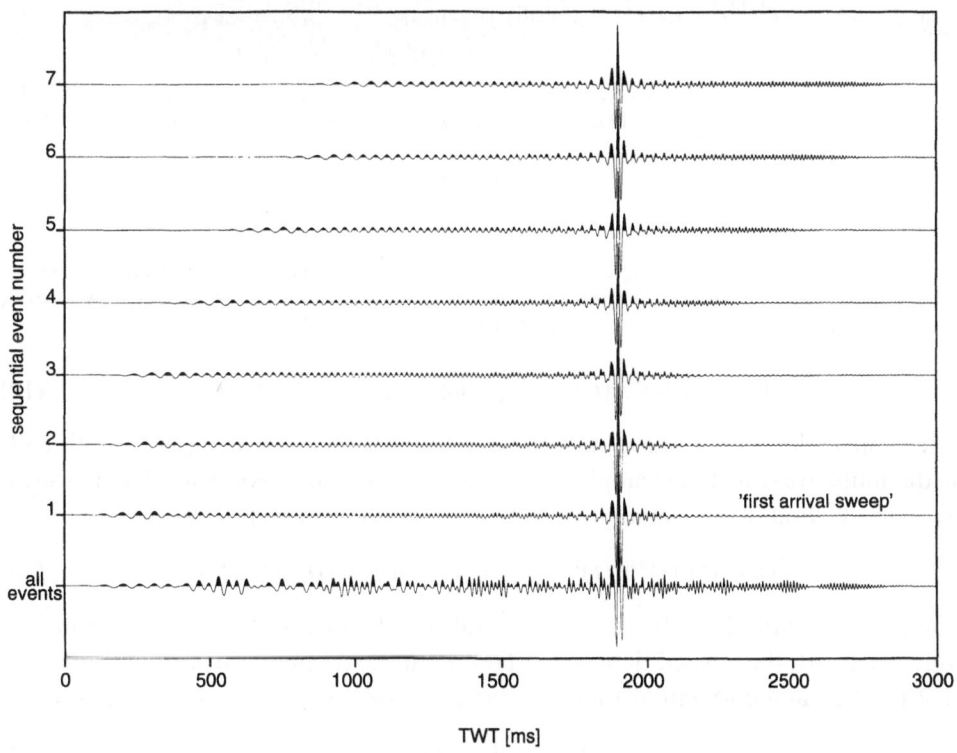

Figure 7

So-called "vibrogram representation" of the data set shown in Figure 6. The spike from the original vibrogram shown in Figure 4 has been convolved with the autocorrelation of the spike induced (and subsequently amplified) correlation artifact shown in Figure 6. Compared to Figure 3, the additional influence of spherical divergence is no longer present. Therefore, the spike position can clearly be identified in all traces.

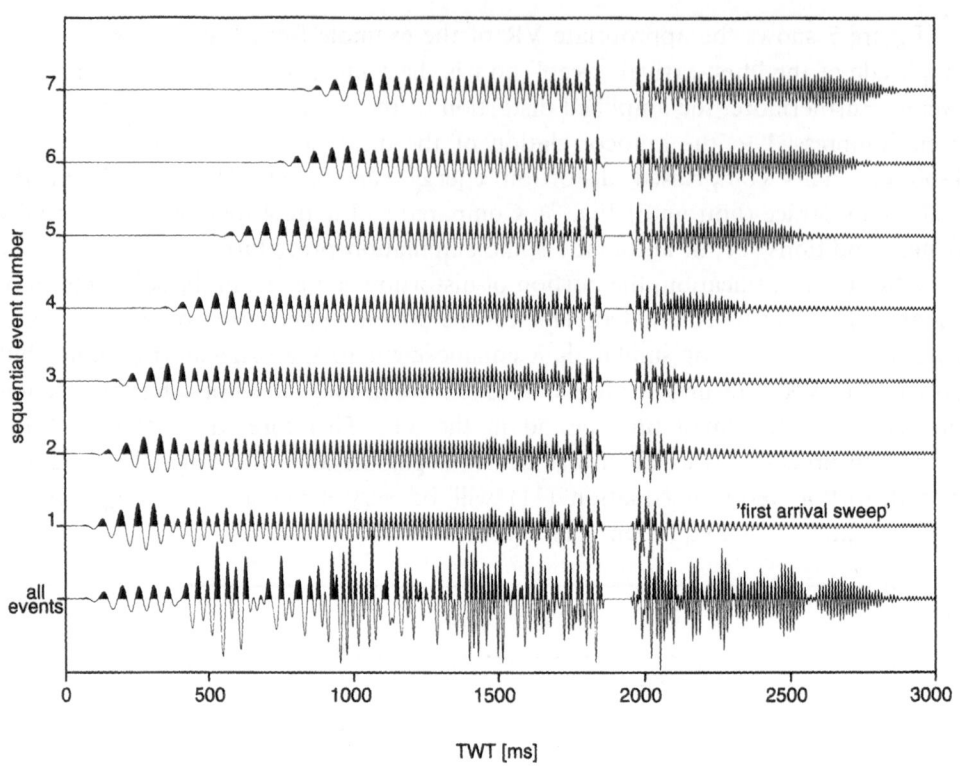

Figure 8
Vibrogram representation after surgical muting of the compressed noise artifact (from Fig. 7). To prevent undesired effects from the edges of the surgical mute window, it has been tapered by a cosine function.

$$[srw_c(t)f(t)*s'(t)]_{\text{corrected}} = \Phi(t)*r(t)*e(t)f(t)*s'(t). \tag{12}$$

After application of a surgical mute process (including a cosine taper at both ends of the mute area) to the example from Figure 7, the noise corrected VR is shown in Figure 8. Following convolution with the operator $s'(-t)$, leads to

$$[srw_c(t)f(t)*\Phi'(t)]_{\text{corrected}} = \Phi(t)*\Phi'(t)*r(t)*e(t)f(t). \tag{13}$$

Compared to equation (9), there is an additional function $\Phi'(t)$ in equation (13) convolved with the desired data term. To obtain the desired term without influence of $\Phi'(t)$, this autocorrelation function has to be designed to hold the equation

$$\Phi(t) = \Phi(t)*\Phi'(t). \tag{14}$$

Therefore, two conditions of the autocorrelation function $\Phi'(t)$ are required:

1. $\Phi'(t)$ has, at least, the same frequency range as the autocorrelation of the original source signal $\Phi(t)$.
2. The value of $|\Phi'(\omega)|$ has been designed to the power of one.

Assuming these conditions, $\Phi'(t)$ has the characteristic of a neutral element in a band-limited convolution process. Then equation (13) becomes:

$$[srw_c(t)f(t)\Phi'(t)]_{\text{corrected}} = \Phi(t)*r(t)*e(t)f(t), \qquad (15)$$

which is equivalent to the desired term from equation (2). Figure 9 shows the noise corrected correlogram after application of the correlation process to the data from Figure 8. Compared to Figure 6, the distortion has been mostly eliminated.

Using a conventional convolution process, the function $s'(t)$ has to be optimized (with respect to the conditions cited above) prior to application. Another method utilizes the pure phase shift operation for the convolution process presented by LI *et al.* (1995) to operate with $s'(t)$. Note that the correction process described does not depend on the characteristic of the sweep used. All equations described above hold for linear or nonlinear up- and downsweep sources.

However, in processing real seismic data, most of the noise correction methods known (e.g., deconvolution, *F-K*-filtering and many more methods) produce undesired effects. Therefore, Figure 10 shows the reference correlogram produced by correlation of the vibrogram data set from Figure 3 (without a spike) and

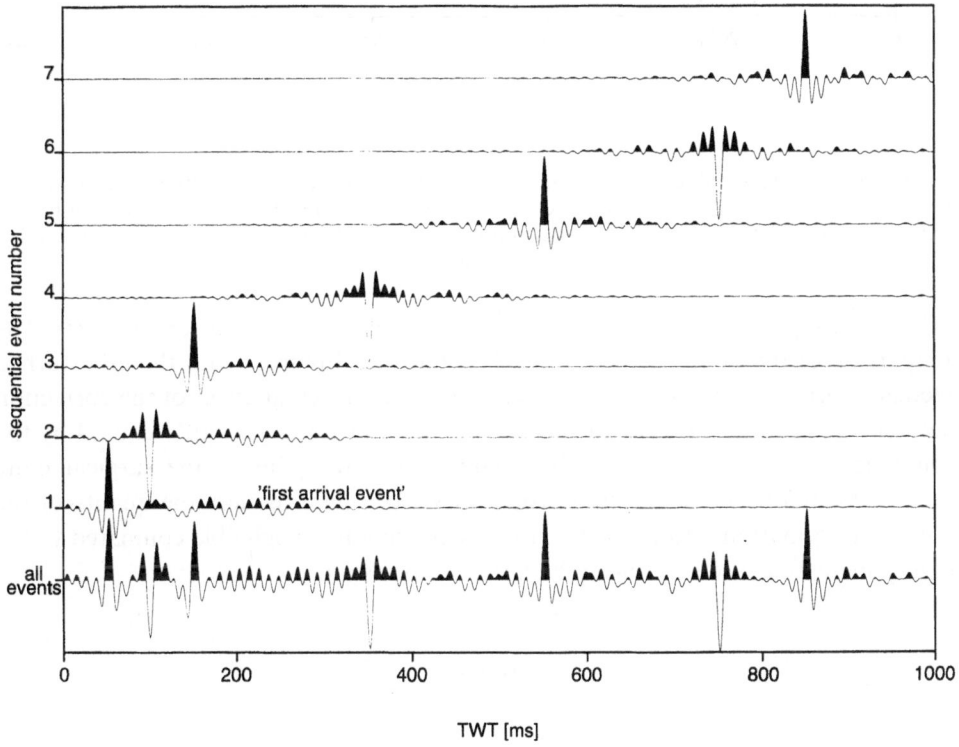

Figure 9
Noise corrected correlogram replacement after application of correlation process to the data from Figure 8. Comparison with Figure 6 reveals that the reverted sweep artifact has been totally eliminated.

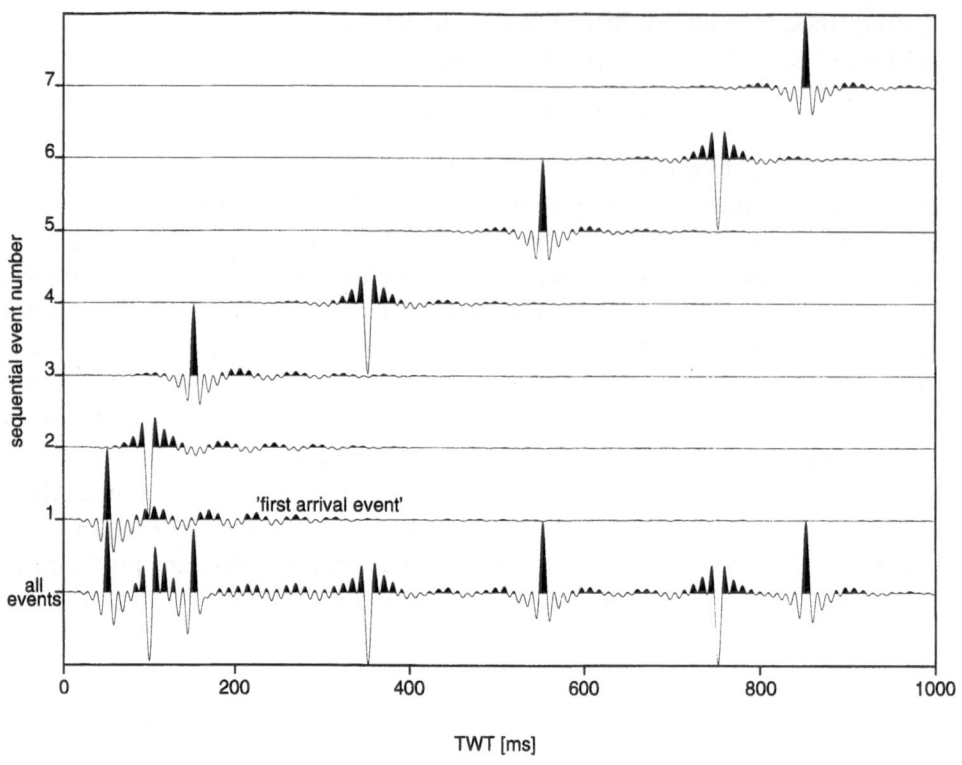

Figure 10

Result of correlation and subsequent analytical scaling of the vibrogram from Figure 3 (without a spike)
for reference. Comparison with Figure 9 shows that the only undesired effect of the correction process
used is a slightly higher noise level. This effect is primarily caused by a loss of signal due to the surgical
mute.

subsequent application of the spherical divergence compensation function $f(t)$. The
comparison of the noise corrected correlogram from Figure 7 with the undisturbed
reference correlogram (Fig. 10) yields an undesired marginal effect of the correction
method described as a rise of the noise level. This effect is primarily caused by the
simultaneous elimination of required data information during the surgical mute
process. Therefore, the correction process described here reduces the quality of the
required information. However, this quality reduction is negligible compared to the
quality increase caused by the spike-noise suppression.

Results

Application to Real Data Examples

The application of the correction method to a real seismic data set is demonstrated using a field record from DEKORP 3 deep reflection seismic project. Field

file 2024 (120 data channels, 13.7 s listen time, 5-fold vertical stack) was recorded in 1990 by a DFS V recording system. The elementary vibrograms of the record have been generated by five vibrators (12–48 Hz linear upsweep, 20 s sweep time, 200 ms taper). The data were stored using SEG-B multiplex tape data format and no correlation was applied in the field.

With respect to an empirical analytic gain function $f(t)$ ($f(t) = t^2$ to 3 s, then constant), Figure 11 shows 9 s of the correlated record. The amplitude representation of the data has been normalized by trace energy. The aim of the noise reduction technique described in this paper is the artifacts from channel 1 to channel 30 and from channel 70 to channel 90 initially. At least the artifact between channel 70 and channel 90 has an appearance similar to a time reversed sweep. The source of this artifact could easily be found in the original uncorrelated data shown in Figure 12, where a strong spike near 12.5 s impairs the desired data. The source of the other artifacts, however, could not be easily identified in the uncorrelated data of this record.

Therefore, the VR was used to obtain more information regarding possible data distortions. Figure 13 shows the appropriate time window of the VR compared to Figure 12. Since the VR contains only the frequency range of interest, determined by the frequency range of the sweep used, the distortions caused by 50 Hz and 100 Hz (which are probably picked up from electric power lines by the recording system) are successfully suppressed by the correlation process. Furthermore, the application of analytical scaling in the correlogram domain prior to VR transformation has corrected the influence of spherical divergence in the VR and has led to an amplification of all previous existing spikes and bursts. Thus, many more spike and burst positions could be determined in the VR data than in the original uncorrelated data shown in Figure 12. It should be noted that Figure 13 only illustrates one example of the distortions present in the VR of this record. There are more similarly compressed distortion elements at previous and later times of the VR.

During the entire project many records of the DFS V system have been affected by such distortions. It was nearly impossible to apply a manual editing for the correction process to all of these records. Therefore, an automatic detection and correction algorithm has been developed for the required operation, based on a single trace operation. Figure 14 shows the result of this automatic correction process, applied to the data example from Figure 13. Many spikes and bursts are correctly identified and eliminated. To prevent new artifacts caused by edge effects of the correction area, the transitions to the nondistorted data areas are controlled by cosine tapering. Note that there are other corrected time areas in the VR of this record not shown in this figure.

Subsequent application of an inverse VR transformation to the corrected VR leads to a noise corrected replacement of the data set. This final result is shown in Figure 15. The correction process improves the data quality in most parts of

recording channel

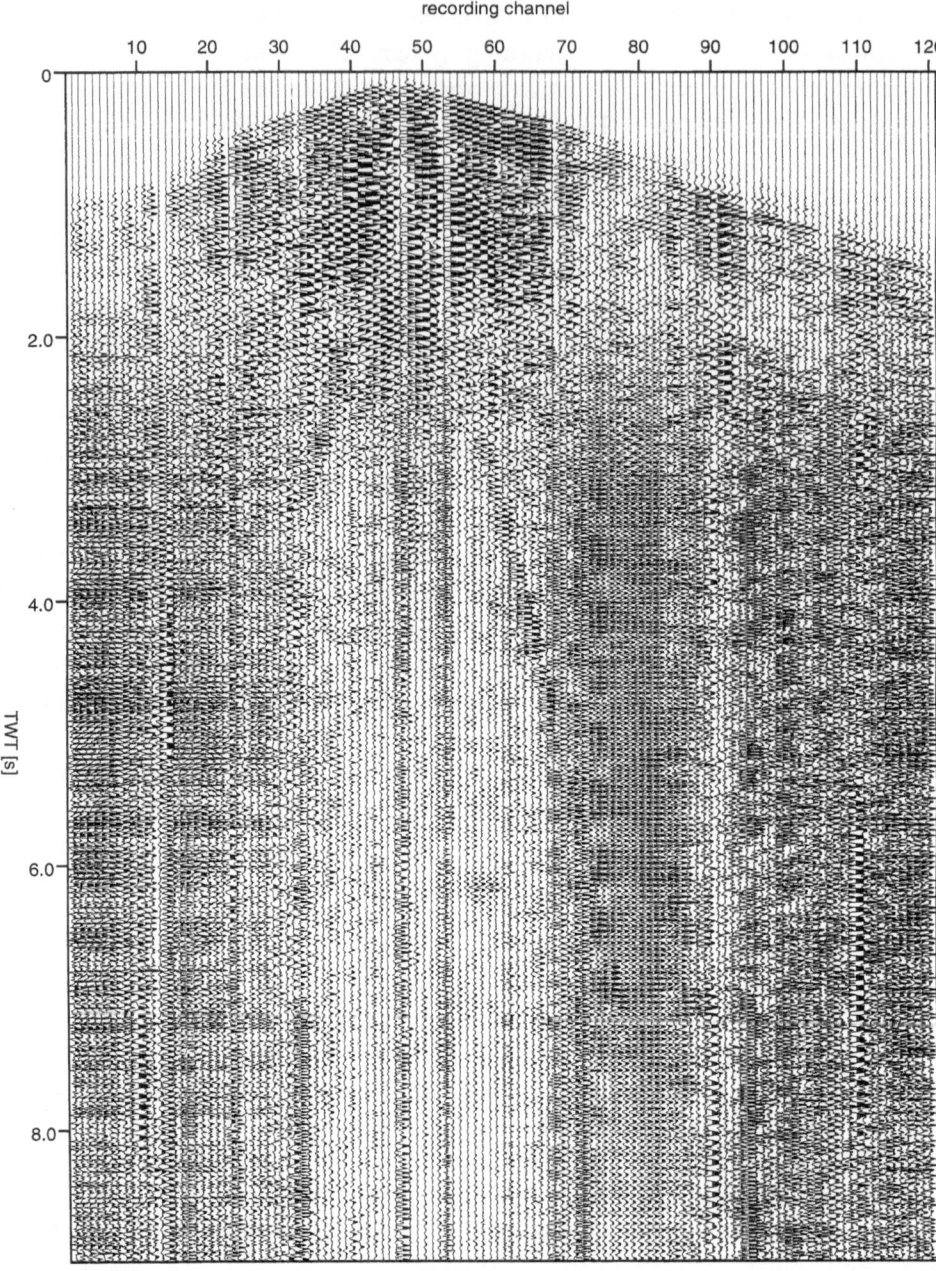

Figure 11
Field record 2024 (120 data channels, 80 m group spacing, 13.7 s listening period, 5-fold vertical stack) from the DEKORP 3 deep seismic project after application of correlation and analytical scaling (t^2 to 3 s, then constant). Traces are normalized by trace energy. The elementary vibrograms of this record were generated by five vibrators using 12–48 Hz linear upsweep, 20 s sweep duration and 200 ms taper. Strong artifacts from approximately channel 70 to channel 90 are impairing the seismic information of the data considerably. Also other traces seem to be affected by strong noise.

recording channel

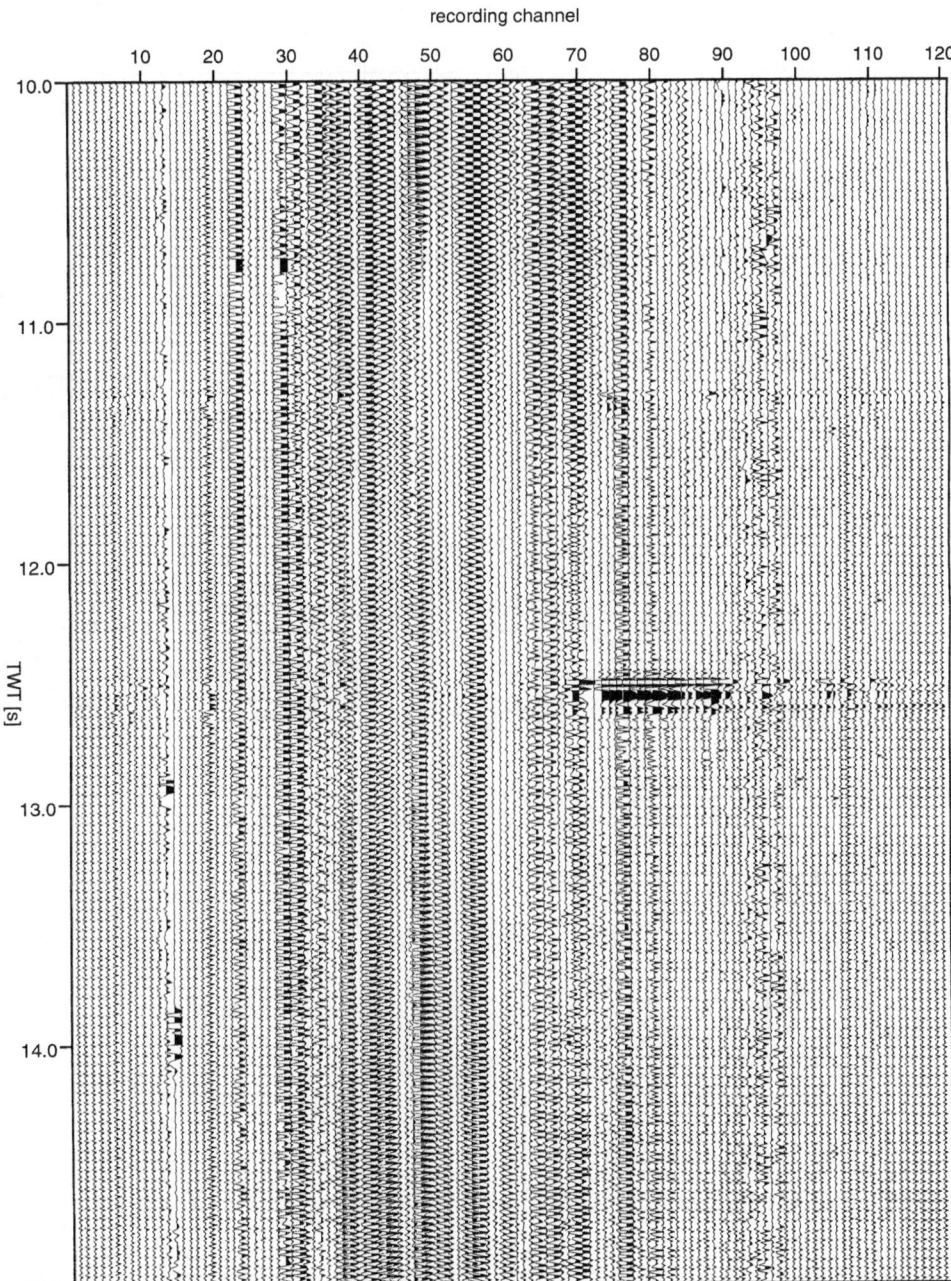

Figure 12

Part of the original vibrogram (normalized by trace energy) of the record shown in Figure 11. The strong spikes near 12.5 s TWT cause the reversed-sweep artifacts from channel 70 to channel 90 in Figure 11.

recording channel

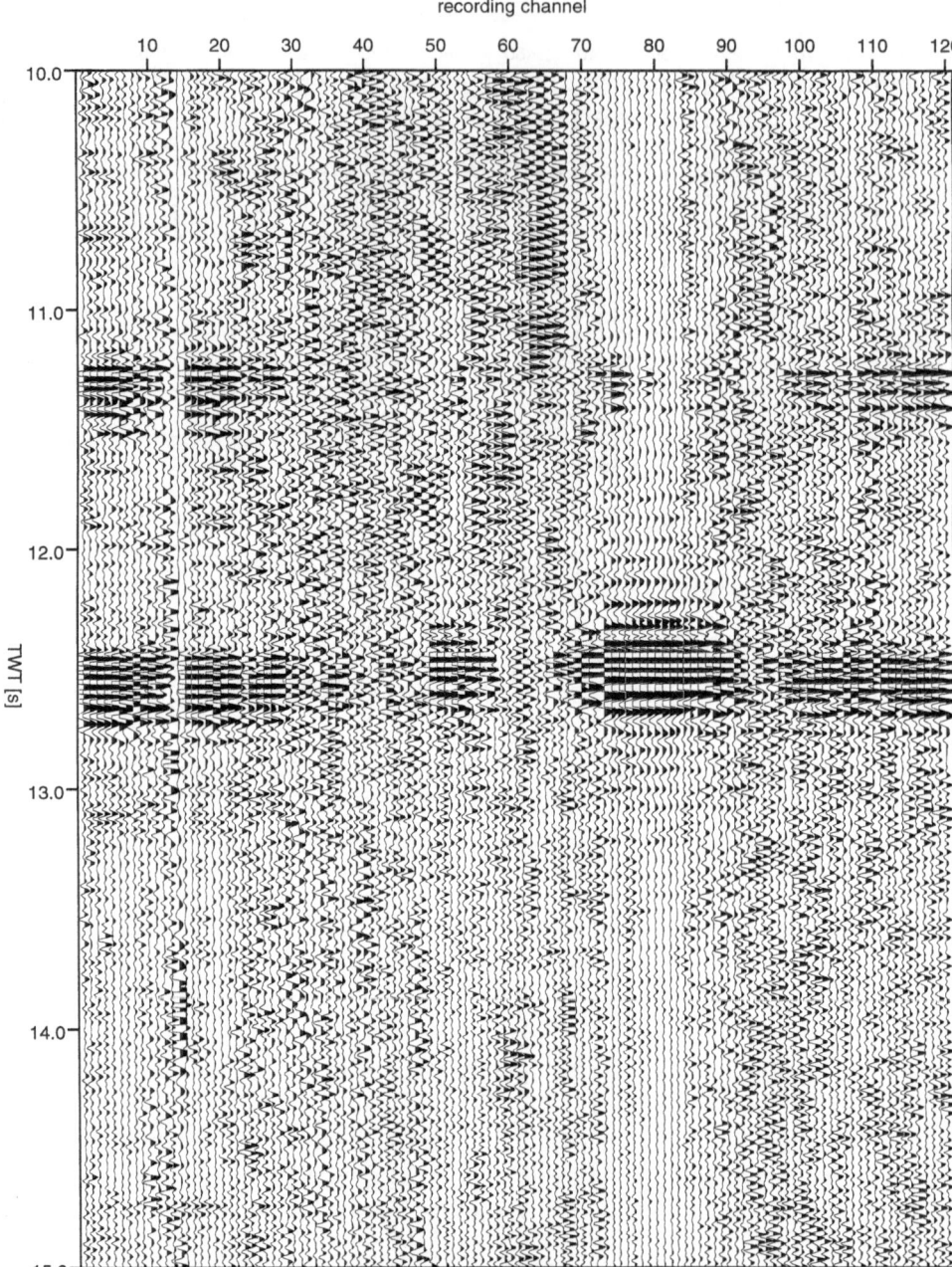

Figure 13

Part of the vibrogram representation (normalized by trace energy) of the record shown in Figure 11. Compared to Figure 12, many more distorting elements can be recognized in this figure. All of these elements are compressed time reversed sweep artifacts from the correlogram shown in Figure 11.

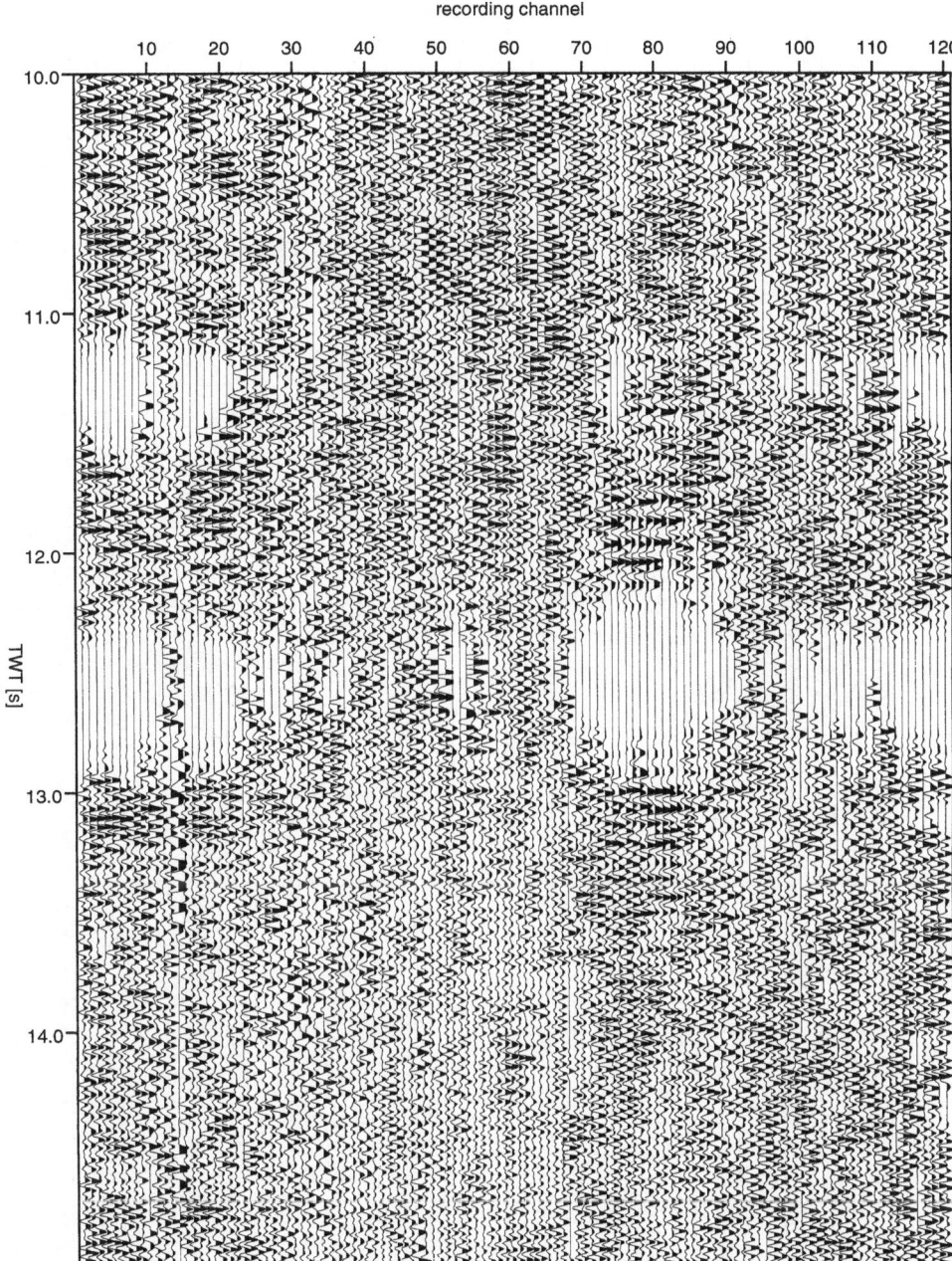

Figure 14
Result of an automatic detection and correction process (mute mode) applied to the data shown in
Figure 13. All of the automatic computed mute windows are tapered by a cosine function to prevent new
artifacts during further processing. Prior to plotting, the data has been normalized by trace energy.

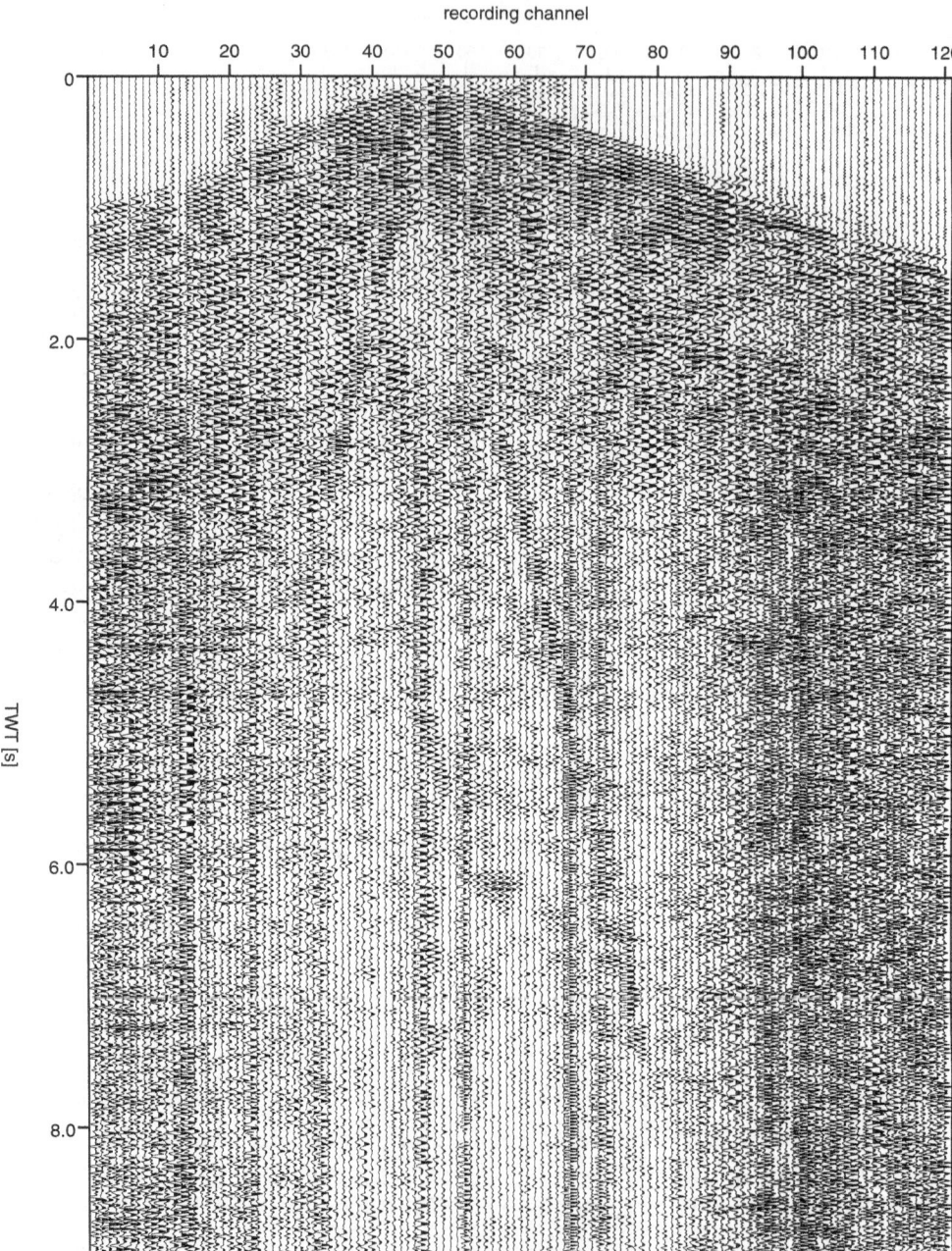

Figure 15
Final result (normalized by trace energy) of the correction process applied to the record shown in Figure 11. This result has been obtained by correlation of the automatically corrected vibrogram representation shown in Figure 14. Comparison with the original data (Figure 11) shows that most of the artifacts are eliminated.

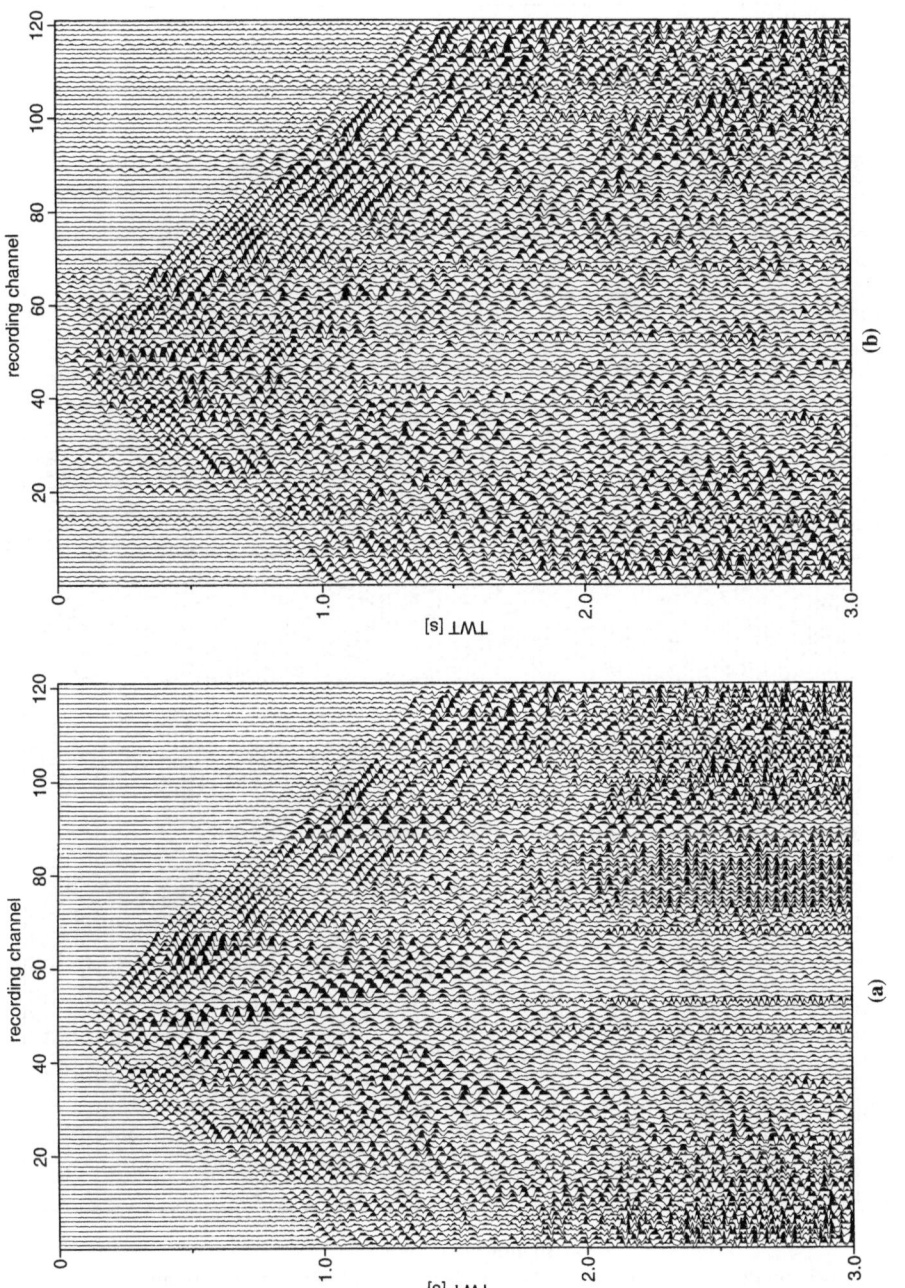

Figure 16

Zoom of the first three seconds (normalized by trace energy) of Figure 11 (16a) and Figure 15 (16b). As an additional effect of the correction process, the disturbing surface waves near the center of the record have also been diminished. An undesired rise of noise in times prior to the first arrival events is caused by an elimination of signal data information during the muting process shown in Figure 14.

recording channel

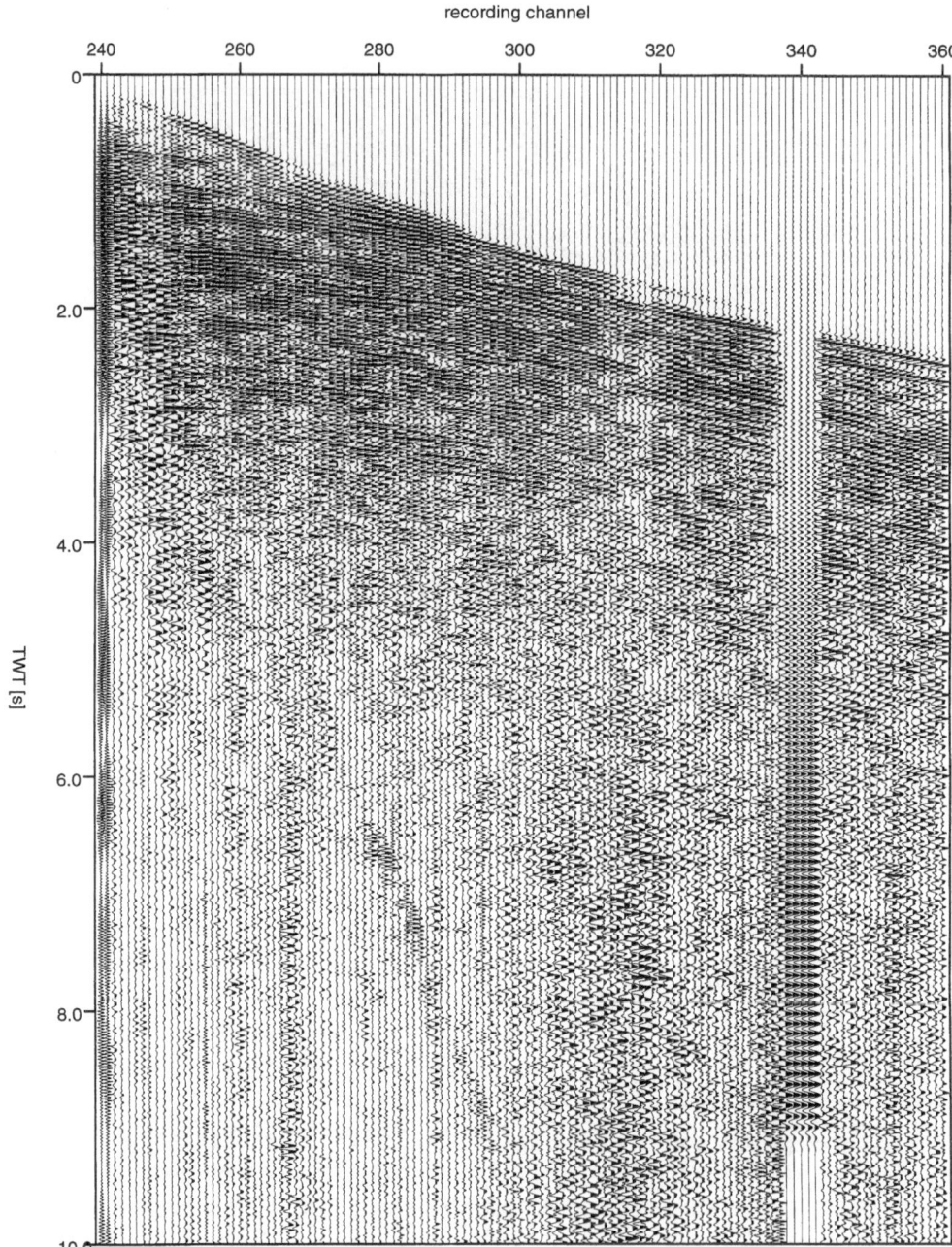

Figure 17
Part of the vibroseis production record 136 (360 channels in original, 60 m group spacing, 25 s sweep time, 16 s listening period, 10–50 Hz range, 8-fold diversity stack) from BASIN '96 deep seismic project recorded in 1996 by an OLYMPUS FRS* telemetry recording system. Correlation (in the field), analytical scaling and trace normalization has been applied to the data. A reversed sweep artifact strongly affects the traces from channel 337 to 342. In addition, other artifacts can be seen at channels 240 and 241.
* Trade Mark of SCHLUMBERGER.

the record. A comparison with Figure 11 shows no significant degradation of important reflection elements. The time-reversed sweep artifact from channel 70 to channel 90 has been totally eliminated.

Due to the correction of many distorted elements in this example, however, it was unavoidable to degrade parts of the record, especially at times before the first arrivals. Furthermore, due to the quantity of distorting elements, not all artifacts are eliminated totally by the automatic correction process. Nevertheless, these undesired effects are negligible compared to the quality increase from Figure 11 to Figure 15, especially in times from 2 s to 9 s TWT.

As a marginal effect of the automatic correction process using VR data, the strong surface wave energy near the shotpoint location (channel 40 to channel 60) has also been suppressed. A zoomed part of Figures 11 and 15 (Figs. 16a,b) exhibits this effect more clearly, which is caused by the frequency separation of the VR transformation when no spherical divergence is present. Generally, the surface wave energy contains low frequencies relative to the desired reflection signal energy, which contains a portion of high frequencies especially in the early stages of a seismogram. Therefore, by using upsweeps, the energy of the surface wave will be transformed to early times of the VR. The desired reflection energy from the same time in the seismogram will be transformed to later times in the VR. Thus, the energy of the reflection signal and the surface wave noise can be separated in time in the VR, where a subsequent individual correction is possible.

An additional possible application of the VR transformation is the utilization during vibroseis quality control in the field, even when using modern seismic recording instruments of high quality. These instruments are usually equipped with telemetric technique, spike and burst reduction possibilities and diversity stacking options. In addition, the actual production is under control by the instrument operator. Nevertheless, it is obviously possible to affect a production record by a spike or burst. For evidence, Figure 17 shows a part of a vibroseis production record (25 s sweep time, 16 s listening period, 8-fold diversity stack) from BASIN '96 deep seismic project (KRAWCZYK et al., 1997) recorded in 1996 by an OLYMPUS FRS* recording system.

The record contains a time reversed sweep artifact from channel 337 to channel 342. The appropriate VR of this record shown in Figure 18 verifies that the artifact is caused by a strong spike near 9.5 s TWT. Furthermore, a burst near 25 s TWT at channel 240 and 241 (near shotpoint) also leads to an artifact in the corresponding traces from Figure 17. Since this burst is near the end of the sweep time, it presumably has been generated by the vibrator itself.

---

* Trade Mark of SCHLUMBERGER.

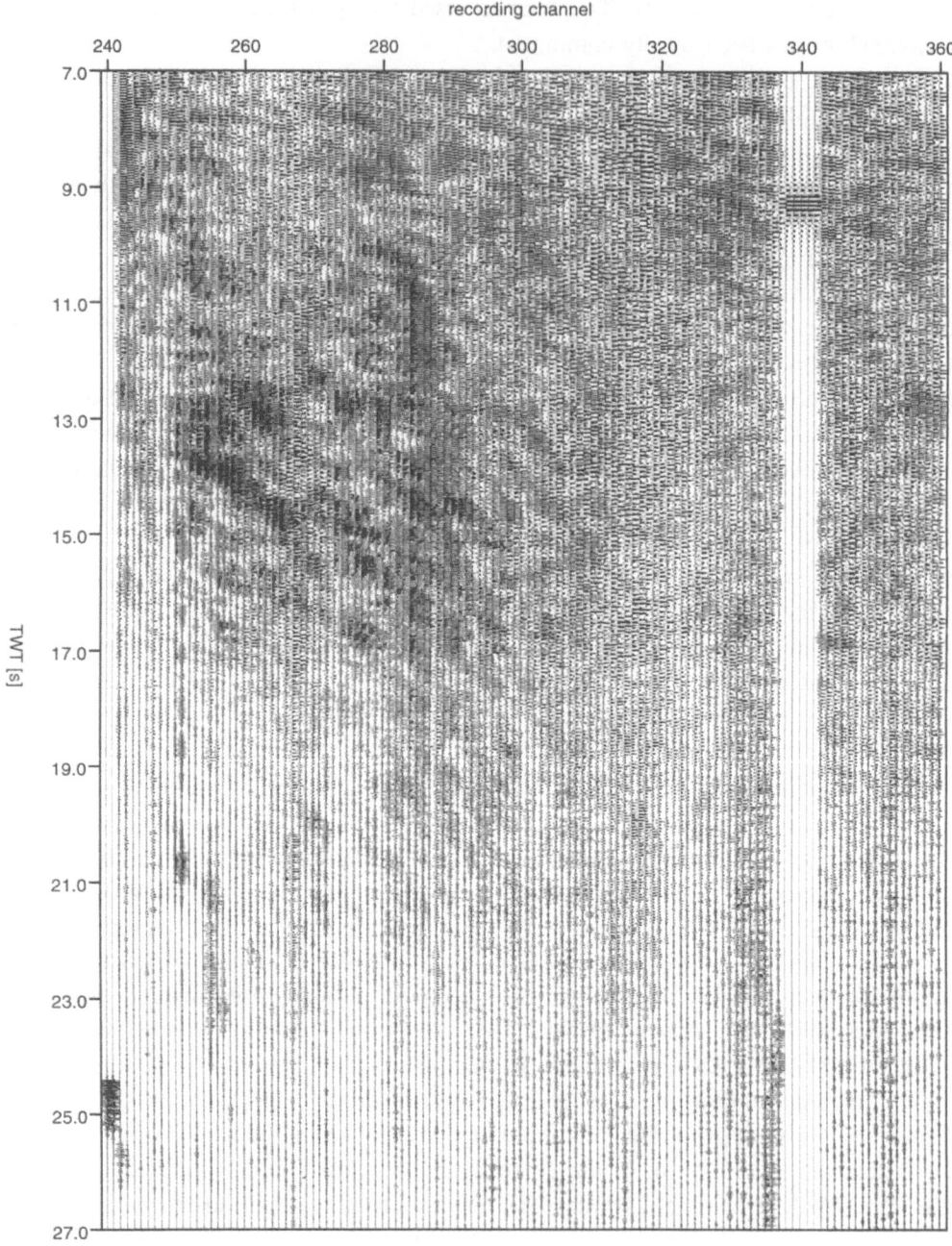

Figure 18
Part of the appropriate vibrogram representation (normalized by trace energy) of the data shown in
Figure 17. This data transformation indicates that strong spikes near 9.5 s TWT and bursts between 24.5
s and 25.5 s TWT are responsible for the two groups or artifacts in Figure 17.

Conclusions

The method described has been developed to correct strong distortions caused by spikes and bursts in vibroseis data recorded by a DFS V seismic recording system during the DEKORP 3 deep seismic project. In contrast to modern seismic recording systems, this instrument has no spike and burst reduction possibilities in the field. Combined with a 2-D main line, the DFS V data should be used to obtain additional cross-line informations from low-fold 2-D and single-fold 3-D seismic surveys in a crystalline area. Hence, the suppression of this noise was important to obtain useful cross-line and 3-D sections for spatial interpretation.

This method recovers hidden spikes and bursts in uncorrelated vibroseis data based on the appropriate correlogram. Analytical scaling in the correlogram domain (to correct the influence of spherical divergence) and a subsequent transformation similar to the inversion of correlation operation, leads to a vibrogram replacement in which the influence of spherical divergence has been corrected. Furthermore, the analytical scaling operation in the correlogram domain amplifies the artifacts caused by spikes and bursts in uncorrelated data and improves the detection of these distortions in the original vibrogram or in a vibrogram replacement. Subsequent interactive or automatic correction of the disturbances and repeated application of correlation operation leads to a correlogram (or correlogram replacement), where the data distortions are removed.

The advantages of the single trace based automatic correction process applied in the VR of the real data examples have allowed implementation as a standard processing sequence for all seismic lines recorded by the DFS V instrument, which had been affected by the artifacts from spikes and bursts in the uncorrelated data. During the processing of the vibroseis data from the DEKORP 3 deep seismic project, the automatic correction process was successfully adapted to several thousand records (each of 120 channels). Since the process operates on a single trace base, it is independent from the sort domain of the data.

The correction method also has been tested using high coveraged 2-D vibroseis data, although in this case the signal-to-noise improvement of the CDP stacking process was often more dominant. Good results have been obtained from single-fold up to 15-fold vibroseis data. Since it depends on the signal-to-noise ratio, the success of the correction process is more dominant at later times of a seismic section than in early times. The application to a more multi-coveraged 3-D vibroseis data set should be tested in the future.

Acknowledgements

I thank the DEKORP research group for the release of the field data and I am grateful to Dr. F. Keller for programming the important data input/output routines. Many thanks to Dr. M. Simon for his suggestful comments and the critical review of the manuscript. In particular, thanks to the colleagues from DEKORP Processing Center Clausthal for their support.

REFERENCES

ANSTEY, N. A., *Signal characteristics and instrument specifications*. In *Seismic Prospecting Instruments*, vol. 1 (Geb. Borntraeger, Berlin/Stuttgart 1970) 156 pp.

DÜRBAUM, H.-J., SCHMOLL, J., DOHR, G., REICHERT, C., WIEDERHOLT, H., STILLER, M., RYBERG, T., KAPP, I., HARTMANN, H.v., KLÖCKNER, M., KÖRBE. M., and THOMAS, R. (1994), *Profile DEKORP 3/MVE-90: Reflection Seismic Field Measurements and Data Processing*, Zeitschrift für Geologische Wissenschaften *22*, 631–646.

KRAWCZYK, C. M., LÜCK, E., STILLER, M., and DEKORP-BASIN '96 Research Group (1997), *The North German Basin-DEKORP—Campaign BASIN '96*, Annales Geophysicae *15*, C18.

LI, X. P., SÖLLNER, W., and HUBRAL, P. (1995), *Elimination of Harmonic Distortion in Vibroseis Data*, Geophysics *60*, 503–516.

POLOM, U. (1997), *Elimination of Source Generated Noise from Correlated Vibroseis Data*, Geophysical Prospecting *45*, 571–591.

SERIFF, A. J., and KIM, W. H. (1970), *The Effect of Harmonic Distortion in the Use of Vibratory Surface Sources*, Geophysics *35*, 234–246.

(Received October 1, 1997, revised April 18, 1998, accepted May 4, 1998)

To access this journal online:
http://www.birkhauser.ch

Pure appl. geophys. 156 (1999) 345–370
0033–4553/99/010345–26 $ 1.50 + 0.20/0

⌐ Pure and Applied Geophysics

How to Remedy Non-optimal Seismic Data by Seismic Processing

Jürgen Fertig,[1] Margit Thomas[1] and Rüdiger Thomas[1]

Abstract—Seismic data processing mostly takes into account the statistics inherent in the data to improve the data quality. Since some years the deterministic approach for processing shows many advantages. This approach takes into account e.g., the source signature, with the knowledge of its amplitude and phase behavior. The transformation of the signal into an optimized form is called wavelet processing. By this step an optimal input for deconvolution can be produced, which needs a minimum-delay signal to function well. The interpreter needs a signal which gives the optimum resolution, which is accomplished by the zero-phase transformation of the input signal. The combination of different input sources such as Vibroseis and Dynamite requires a phase adoption. All these procedures can be implemented via Two-Sided-Recursive (TSR-) filters. Spectral balancing can be accomplished very effectively in time domain after a minimum delay transform of the input signals. The DEKORP data suffer from a low signal/noise ratio, so that special methods for the suppression of coherent noise trains were developed. This can be done by subtractive coherency filtering. Multiple seismic reflections also can be suppressed by this method very effectively. All processing procedures developed during recent years are now fully integrated in commercial software operated by the processing center in Clausthal.

Key words: Wavelet processing, suppression of coherent noise.

Introduction

Seismic data are used today for a target-oriented interpretation of hydrocarbon deposits. Stratigraphic aspects play a dominant role which should provide an answer for depositional sequences. Reservoir characterization is now an important application for seismic methods. The analysis is carried out with stacked data. Unstacked data are used within an AVO analysis to show the change of signals due to different pore fills. In deep seismic sounding (projects: DEKORP, ECORS, etc.) should reveal hints regarding the evolution of deep parts of the earth's crust and upper mantle such as basin development in its temporal evolution. Low impedance contrasts in great depths produce a low signal-to-noise ratio.

The named application of seismic asks for a high reliability and resolution of the data. In former times statistical methods were used to make small details visible.

Institute of Geophysics, Technical University Clausthal, Arnold-Sommerfeld-Str. 1, D-38678 Clausthal-Zellerfeld, Germany.

Today a more deterministic path is taken to improve the data quality. The questions leading to the interpretation are from the physical and geological reasons of the outlook of seismic data. Therefore, it is necessary to know the influences of the wave path from the source to the receiver. At the source site normally we make an assumption about the source strength and coupling. Proceeding from the source to the receivers we have spherical divergence and a variation of the reflection coefficient with the angle of incidence. Absorption produces loss of higher frequencies. Peg-leg multiples from thin layers produce an interfering signal. At the receiver site we have to take into account different types of receivers such as geophones/hydrophones and the transfer characteristics of the recording instrument. Some of the named influences are known, others must be assumed in their influence on the seismic signature. To the extent we know the different sources of signal changes, they must be included into our processing scheme. Furthermore, several processing steps need optimized input data to permit the process to work in an optimal way. The processor's aim is to give an optimal seismogram (section) to the interpreter, with a high resolution both in time and distance.

The Phase of the Signal (Wavelet)

The seismic energy is produced by different sources: either impulsive or time-controlled sources (such as VIBROSEIS). Sometimes a combination of both source types is used. There is also a difference in the environment for the source; it can be onshore and offshore. The recording system can be a geophone, hydrophone, gimbal-phone, etc., or a mixture in which all have different transfer characteristics.

A great advantage for offshore sources is that we can pick up the source generated signal at some distance which can be used for further signal treatments. Figure 1a (McQUILLIN *et al.*, 1979) supplies a suite of marine source signals. All show different signatures of different time-length. However, the signal is known and as it is a scalar quantity, there is no change in the shape with distance from the source. Figure 1b (BORTFELD and FERTIG, 1983) shows the shape of a signal for the displacement within an elastic medium. A difference between the nearfield and the farfield must be pointed out. In the nearfield, the signal displays a one-sided character, in the farfield (more than five wavelengths from the source) a more or less time-differentiated signal's shape is obtained. Unfortunately, only a few records of onshore-signals are available; therefore a spike is assumed mostly as an input. Figure 2 (LINDSEY and MACURDA, 1985) shows clearly the different aspects for different wavelets which can give a completely false answer during the interpretation for a possible hydrocarbon's trap.

In the event the source-wavelet is unknown, the impulse response of the digital recording system is taken as the basic wavelet. This response has a definite frequency and phase spectrum. The phase spectrum is the important part as it is responsible for the signals shape and a given amplitude spectrum. The influence of this phase spectrum is given below.

For the data-processing and the interpretation, two distinct behaviors of the phase are of special interest. If the source signal is known, the phase can be changed by special filters. This process is called "WAVELET-PROCESSING." Minimum-delay (minimum-phase) wavelets are the optimum input for deconvolution processes, as many processing steps such as spectral balancing work optimal with minimum-delay signals; this type of a wavelet is also called "Processors Wavelet." The interpreter desires a signal which gives optimum resolution for a given bandwidth. The maximum amplitude should be at the time of the reflection, and its polarity and magnitude should show the right polarity and magnitude of the reflection coefficient. Also for subsequently referred interpretation methods, like "seismic attribute analysis," zero-phase wavelets are required. Due to these important aspects, the zero-phase wavelet is called "Interpreters Wavelet."

Figure 3 shows these general aspects. From bottom up a simulated geologic sequence of different "geologies" generated by random numbers is seen. The impedance contrasts are transformed to reflection coefficients with different ampli-

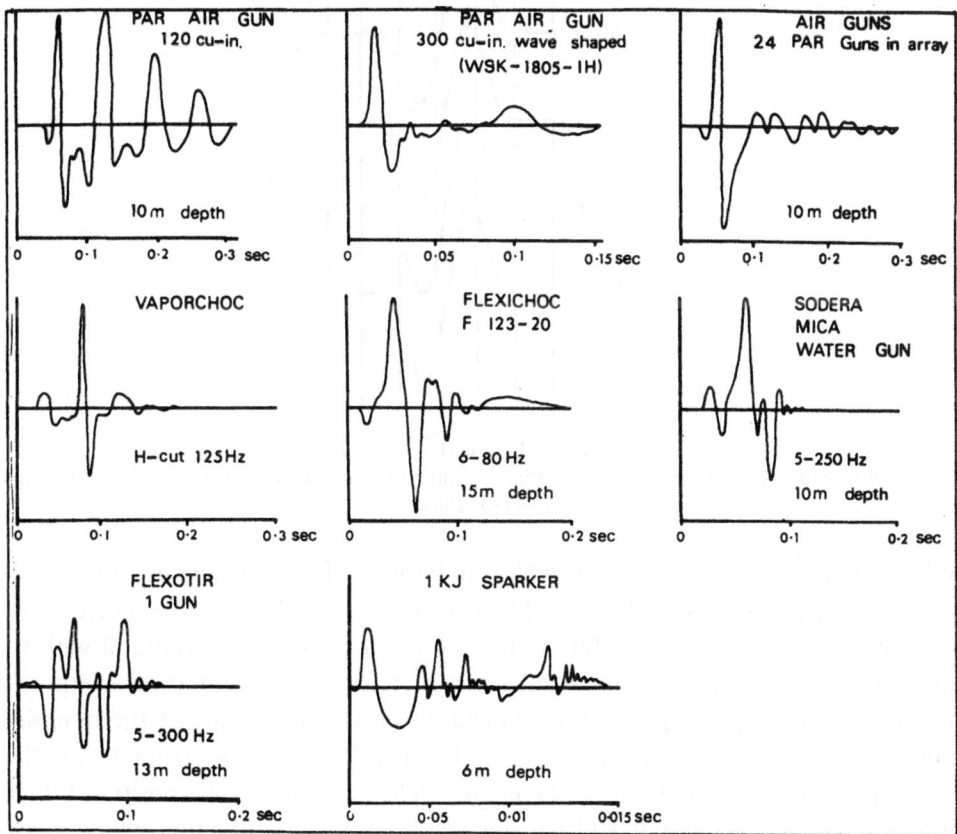

Figure 1a
Different seismic source wavelets (after McQUILLIN et al., 1979).

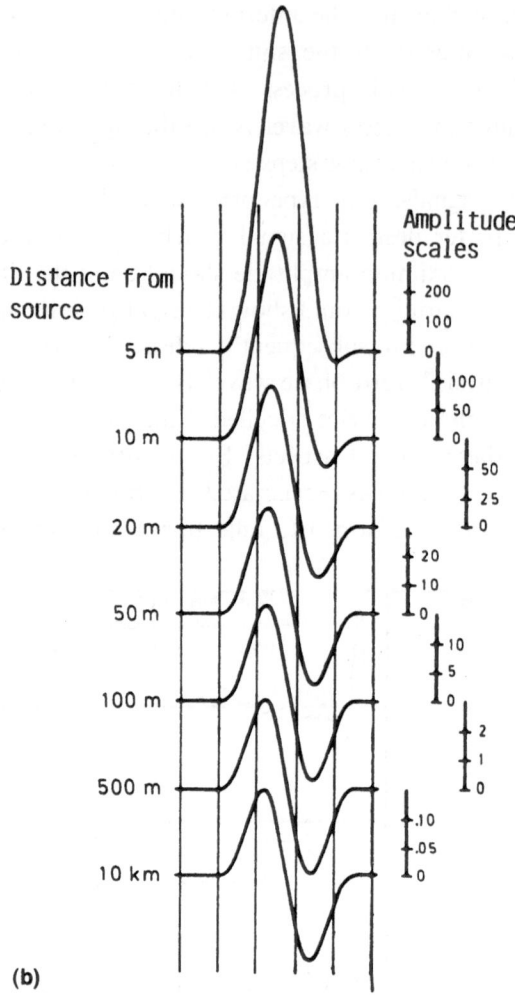

Figure 1b
Change of signature for different distances from a point source in an elastic medium (BORTFELD and FERTIG, 1983).

tudes and polarities. As an input wavelet, an impulse of a recording instrument with a 50 Hz notch-filter (DFS V) applied. The next trace shows the synthetic seismogram. To achieve more resolution, a conventional spike-deconvolution with the operator shown is applied which results in a more or less noisy trace. Sometimes a so-called optimum spike position is applied which first brings about the nonminimum-delay behavior of the input. The spike is shifted from the front to another position to produce minimum error energy. This optimum-spike-position-DECO produces a much better but still ringing result. Next, the minimum-phase correspondent of the wavelet is shown and the corresponding synthetic trace is given. Solely for demonstration purposes, the convolution of the reflection coefficients

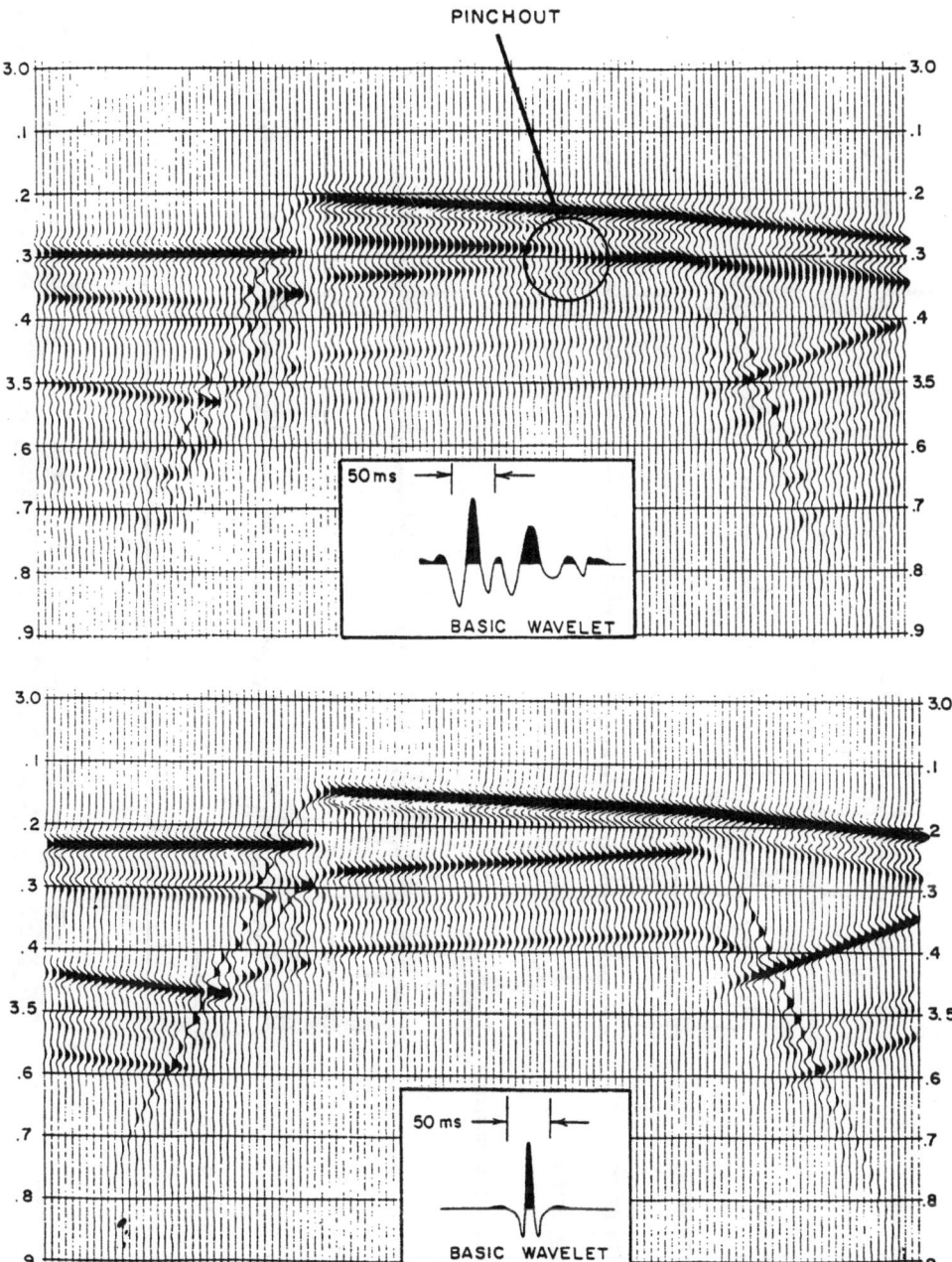

Figure 2
The influence of different source signatures on the shape of a seismic section. The non-symmetrical wavelet produces a false pinchout, simulating a hydrocarbon trap (LINDSEY and MACURDA, 1985).

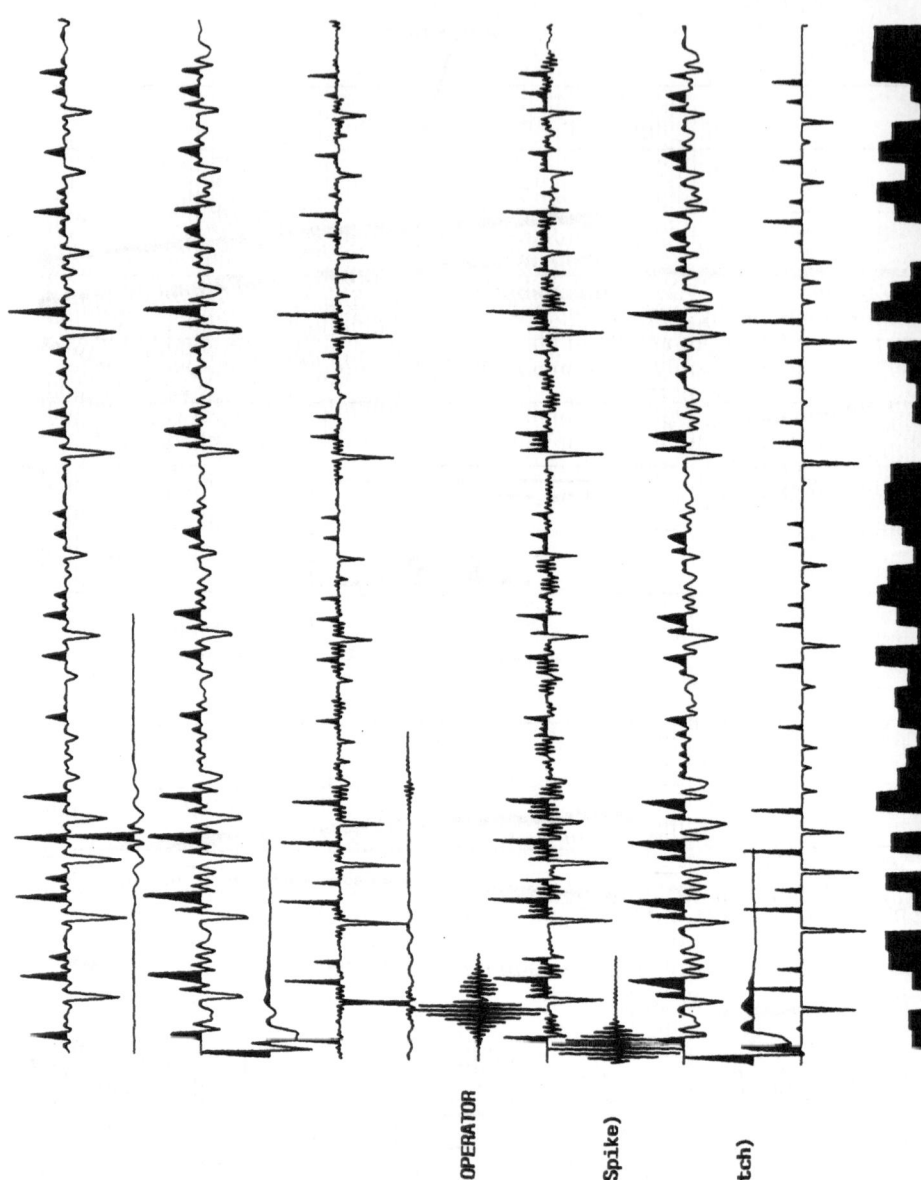

Synthetic Seismogram with
ZERO-PHASE Signal

ZERO-PHASE Signal

Synthetic Seismogram with
MINIMUM-PHASE Signal

MINIMUM-PHASE correspondent

Optimum SPIKE-POSITION DECON

Optimum SPIKE-POSITION DECON
for wavelet

Optimum SPIKE-POSITION DECON OPERATOR

Conventional DECON

Conventional DECON-Operator (Spike)

Synthetic Seismogram

Wavelet (DFS V with 50 Hz Notch)

Reflection coefficients

Geology (acoustic impedance)

with corresponding ZERO-PHASE signal is given. Here it can be seen that even without a deconvolution, the resolution is optimum and amplitudes show the right polarity and magnitude. This figure shows the benefits of wavelet processing for processing and interpretation.

General Wavelet Processing and the Adaption of Different Sources and Recording-instruments

If the input signal of a seismic recording is known, this signal can be manipulated in different and sometimes simple ways. One way is to use Two-Sided-Recursive (TSR) filters; a method presented by MARSCHALL (1977). Figure 4 left explains the general procedure. The representation is most elegant with the z transform and can be formulated in this sense: "Replace the known wavelet $W(z)$ by a desired wavelet $D(z)$." This formulation is given at the top. Unfortunately, this filter may not give a stable response as the inverse filter $1/W(z)$ is generally not stable (since $W(z)$ is mostly mixed phase). This can be accomplished by a well-known trick in signal processing: "Stabilization by correlation." By this procedure the nominator and denominator are multiplied by the time-reversed version of $W(z)$. The resulting nominator can be seen as the cross correlation between the desired and input wavelet. The denominator is the autocorrelation of the input wavelet. It can be seen that there is a strong correlation between conventional deconvolution and this waveshaping. To date, no advantage has been achieved. However, from the autocorrelation all phase information of the wavelet is lost, however the autocorrelation can be replaced by the corresponding autocorrelation of the minimum-delay equivalents $W_{min}(z)$ of $W(z)$. One step remains in which $W_{min}(z)$ must be ascertained from $W(z)$. This can be done in various ways or by different algorithms. A classical method is the Levinson-Algorithm: As shown by CLEARBOUT (1976) in a conventional deconvolution process via Levinson, the operator is the *inverse* minimum-delay equivalent of $W(z)$. The corresponding wavelet is stable and also minimum-delay, therefore the inverse of this decon-operator is stable and represents the wavelet $W_{min}(z)$ in the z domain. As the Levinson-algorithm is quite common in all processing packages, it is widely used. Another method is that of the

---

Figure 3

Aspects of wavelet processing. From bottom to top: blocked 'geology' expressed by acoustic impedance, corresponding reflection coefficients, the recording instrument's response as an input wavelet, synthetic seismogram, a conventional deconvolution-operator, result of spike deconvolution on the synthetic trace, an optimum spike-position operator as a means to compensate the phase behavior of the wavelet, results of this operation on wavelet and seismic trace. This should be compared with the usual decon, replacement of the wavelet by its minimum delay-equivalent and result in the shape of the trace, replacement of input wavelet by its zero-phase correspondent (the 'Interpreter's wavelet'), which gives the optimum resolution and appearance. The result is achieved by only a phase rotation without a change of frequency content.

Applications of TWO-SIDED FILTERS

$W(z)$ is an input wavelet of arbitrary phase.

1.)   Minimum-Delay Transformation of $W(z)$

$$D(z) : = W_{MIN}(z)$$

$$F(z) : = \frac{W_{MIN}(z)}{W(z)} = \frac{W_{MIN}(z) \cdot W(1/z)}{W(z) \, W(1/z)} =$$

$$= \frac{W_{MIN}(z) \cdot W(1/z)}{W_{MIN}(z) \cdot W_{MIN}(1/z)}$$

$$= W(1/z) \cdot \frac{1}{W_{MIN}(1/z)}$$

2.)   Zero-Delay Transformation of $W(z)$

$W_o(z)$ Zero-Phase equivalent of $W(z)$

$$D(z) : = W_o(z)$$

(same procedure as for 1.) $F(z) : = \dfrac{W_o(z) \cdot W(1/z)}{W_{MIN}(z)} \cdot \dfrac{1}{W_{MIN}(1/z)}$

3.)   Spike Transformation of $W(z)$
       ('Spike-Decon')

(same procedure as for 1.) $D(z) : = 1$

$$F(z) : = \frac{W(1/z)}{W_{MIN}(z)} \cdot \frac{1}{W_{MIN}(1/z)}$$

4.)   Other transforms

WAVELET PROCESSING

BY

TWO-SIDED FILTERS

$$F(z) = \frac{D(z)}{W(z)} = \frac{Desired\ wavelet}{Wavelet\ to\ be\ replaced}$$

Stabilization by correlation:

$$F(z) = \frac{D(z)}{W(z)} = \frac{D(z) \cdot W(1/z)}{W(z) \cdot W(1/z)} =: \frac{Crosscorrelation}{Autocorrelation}$$

As:

$$Autocorrelation:\ W(z) \cdot W(1/z) = W_{MIN}(z) \cdot W_{MIN}(1/z)$$

$W_{MIN}(z)$: min. delay equivalent of $W(z)$

$$F(z) = \frac{D(z) \cdot W(1/z)}{W_{MIN}(z)} \cdot \frac{1}{W_{MIN}(1/z)}$$

$$W_{MIN}(z)\ via\ \begin{cases} Levinson - Algorithm \\ Hilbert - Transform \\ Cepstrum - Analysis \end{cases}$$

$F(z)$ is a Two-Sided-Recursive Filter
(TSR-Filter)

Figure 4

Principles of wavelet processing by Two-Sided Filters. At the left, the general procedure as a replacement filter, with the important step of correlation to stabilize the procedure. The necessary minimum-delay equivalent can be found by employing different methods. At the right, applications of two-sided filters in wavelet-processing.

Hilbert-Transform method which is also described in Clearbout's book (CLEAR-BOUT, 1976); this method is ideal for long wavelets as the FFT can be used. Another kind of transform to a minimum-delay equivalent can be conducted via a cepstrum-analysis (TRIBOLET, 1979). The factorization of the autocorrelation-function also needs the time-reversed signal $W_{min}(1/z)$ which is easily determined. This signal is maximum-delayed, however applied in the right manner its inverse $W_{min}^{-1}(1/z)$ is stable again. The resulting filter $F(z)$ can be divided in two parts. The first part contains a nominator which causes no problems and a denominator which is stable; this part is called the feed-forward component. The second part is in principle an unstable filter; but applied as a feed-backward filter it delivers stable results. The application of parts of the filter in two directions explains the name two-sided; and by the existence of the denominator, both these components represent a recursive filter action, in all a Two-Sided-Recursive (TSR)-filter is obtained (FERTIG and THOMAS, 1995).

Applications of Two-sided-recursive Filters

In Figure 4 right some possible applications of TSR-filters are demonstrated. In a later chapter, the minimum-delay-transformation of an arbitrary wavelet is needed. After number # 1 in this figure it can be seen that the forward component is rather simple. As seen in the figure the backward component is always the same. The Zero-Delay-Transformation generates the most complicated part in the forward component when the Zero-Phase-Equivalent $W_0(z)$ of $W(z)$ is called for. Spike transformation can in principle be carried out, however this option in routine processing was not used.

The efficiency of wavelet-transform can be seen in Figure 5 (REDANZ, 1988). Here again a DFS-V-impulse response with a 50 Hz Notchfilter applied was used. The main intent here was to show the effect of wavelet-processing on the effective length of a wavelet. The effective wavelet-length L of a wavelet $W(t)$ is defined by BERKHOUT (1974) via:

$$L = \frac{\displaystyle\int_{-\infty}^{\infty} |t| W^2(t)\, dt}{\displaystyle\int_{-\infty}^{\infty} W^2(t)\, dt}.$$

The original signal has an effective length of about 10 time-units. The minimum-delay equivalent gives a length of about 4 time-units. The zero-phase wavelet with the same amplitude-spectrum shows the minimum length of only 1.4 time-units; typically the concentrated amplitude is at zero-time and the high concentration of the maximum explains the name "Interpreter's Wavelet" (see also FERTIG, 1997).

At the DEKORP Processing Center (DPC) in Clausthal the TSR-filters are applied in a different way by using the fast convolution. Figure 6 represents the general procedure of a replacement filter to transform the impulse response of a seismic recording system (on top) to its zero-phase equivalent (in the middle). In all applications, the minimum-delay wavelet shown in the top center is required.

The operator needed for zerophasing is gained by "impulsing." The TSR-filter is applied to a spike sitting in the middle of a time-series filled with zeros. Applying this operator to the wavelet produces the desired result of a short and symmetrical wavelet.

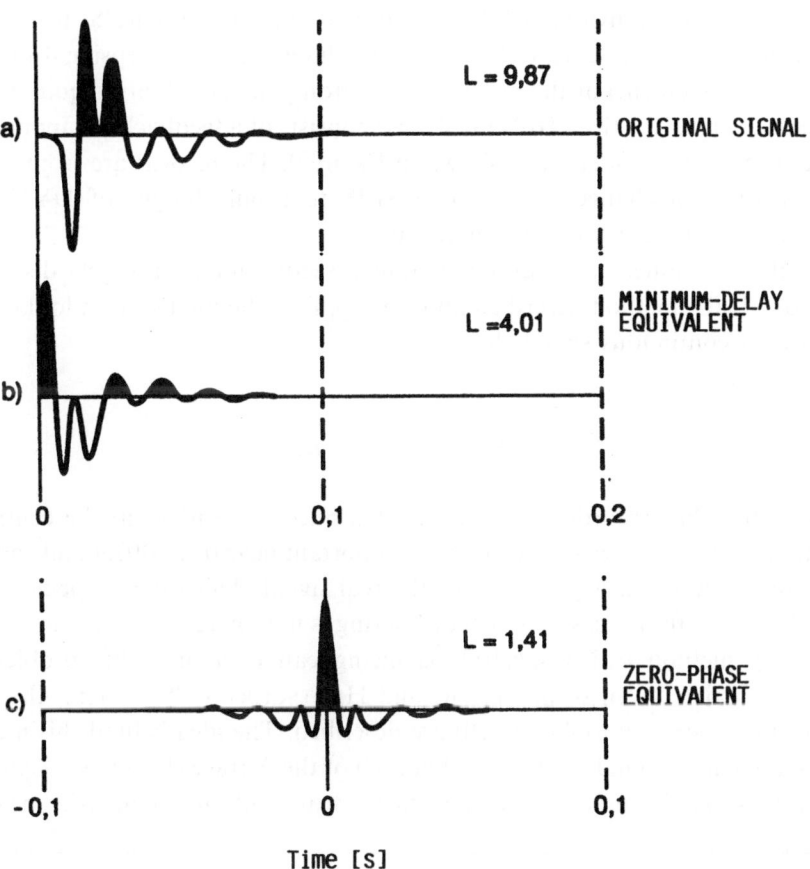

Figure 5
The effect of wave-shaping on the effective length L of the wavelet. The input (a) is a response of a DFS-V-recording instrument with a 50 Hz Notchfilter on. The minimum-delay (b) and zero-phase (c) correspondent were computed by TSR-filters.

Transition Zone Processing

By the same procedure as above some more problems can be solved:
— Adoption of different recording systems; comparison of "old" and "new" seismics. The filter operation is described by $F(z) = W_{new}(z)/W_{old}(z)$; the following steps remain the same as in Figure 4.
— Processing of combined onshore and offshore seismics. The offshore records are made by pressure-sensitive devices, the onshore records are derived by velocity-sensitive instruments.
— Due to different environmental conditions in land seismics, impulsive and VIBROSEIS systems must be combined. A possible processing flow is shown in Figure 7 (BRÖTZ *et al.*, 1987). The most important part is the minimum-delay (MD) transform of the individual wavelets. After this transformation, a spike-deconvolution can be applied, followed by a common bandpass filter. The general procedure of the individual MD transforms is shown in Figure 8. In both cases again the MD equivalents, either from the "filtersweep" (sweepsignal filtered by recording instrument) or the instrument recording filter itself are required. As can be seen, the filter for the VIBROSEIS case consists of a feedback-component only.

The success of this method is shown in Figure 9. The records are from southern Germany where a change from VIBROSEIS to small charges of DYNAMITE placed in small flushed holes was necessary.

If nothing is done, the resulting combined section looks strangely disturbed in the middle. If the method described above is applied, the final section looks smooth and shows a continuous signal shape.

Spectral Balancing

Sometimes shot records are covered by surface wave noise in the central part near the shot. Thus, these waves cover the important near trace offset and sometimes display the same frequency content as the real signal. This often happens in shear wave recordings. In this case, frequency filtering is not an adequate method for noise removal. A method called spectral balancing can overcome this problem. The general method is taken from FERTIG and HENTSCHKE (1987). Here, the general steps of this processing tool (Fig. 10) are described. The idea behind this method is the replacement of an individual signal in each of the N traces by a mean signal. Two mean values can be used: the arithmetical mean and the geometric mean. The

---

Figure 6
The general procedure for TSR-filtering at the DEKORP Processing Center (DPC) in Clausthal. The desired wavelet is a zero-phase correspondent of a DFS V-recording instrument. From top to bottom: input wavelet, its minimum-phase and desired zero-phase correspondent. The filter operator as a result of the application of the TSR-operations on a spike. Result of the application of this operator on the input wavelet.

DFS V . LC=8/18 . HC=128/70 . NOTCH=50 . SR= 2.00 MS. NSAMP= 251

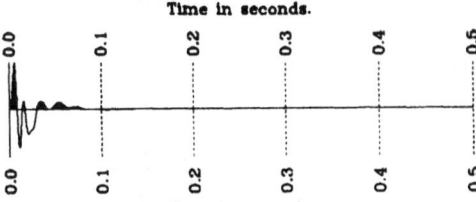

Taperwin.:0.000,0.500 Pow :1

DFS V Min.-Delay. LC=8/18 . HC=128/70 . NOTCH=50 . SR= 2.00 MS. NSAMP= 251

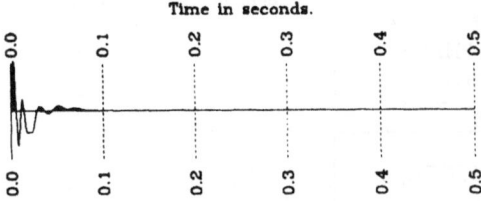

Taperwin.:0.000,0.500 Pow.:1

DFS V Zerophase . LC=8/18 . HC=128/70 . NOTCH=50 . SR= 2.00 MS. NSAMP= 251

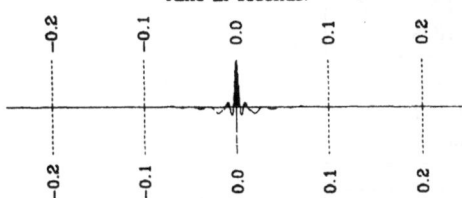

Taperwin.:−.250 , 0.0 / 0 0 , 0.250 Pow.:1

Operator after fwd. and bwd. recursion, SR = 2.00 MS, NSAMP = 501

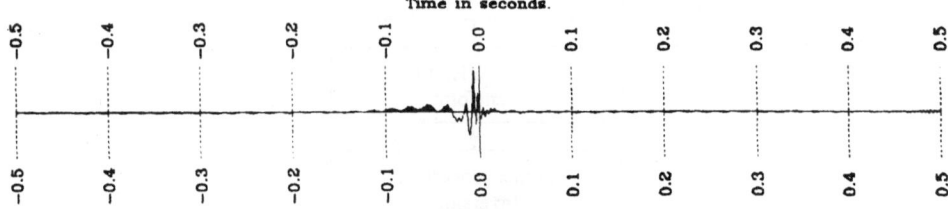

Result of Folding: Operator with wavelet, SR = 2.00 MS, NSAMP = 501

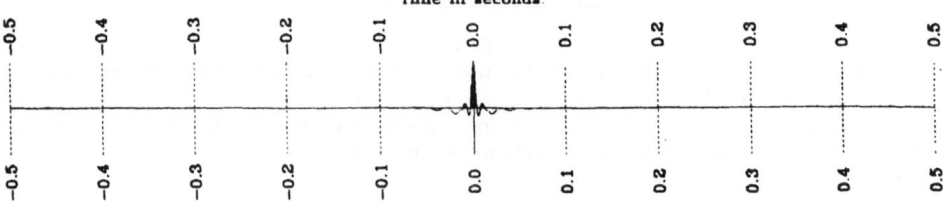

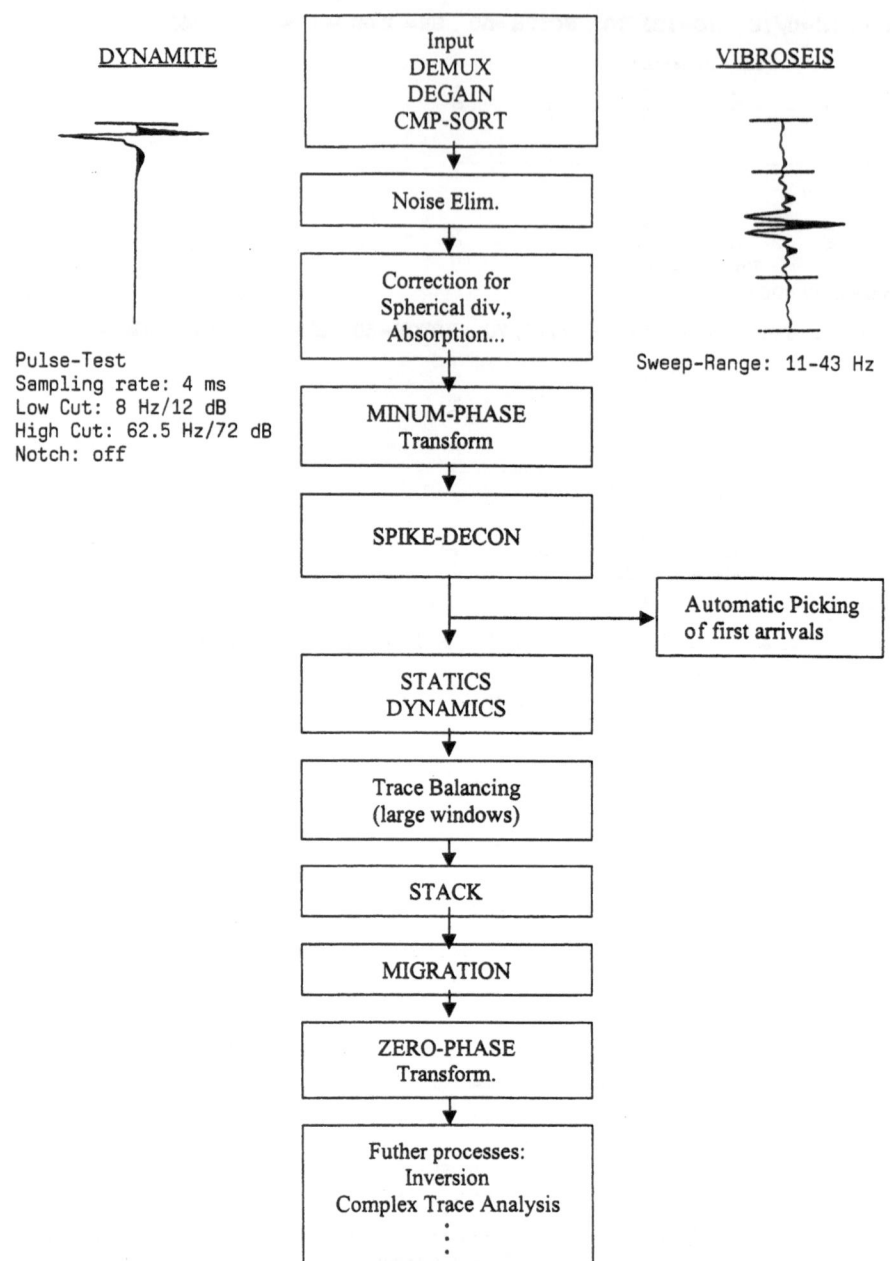

Figure 7

A generalized processing flow for the combination of explosive and Vibroseis records. The most important step here is the transformation of the basic wavelets into their minimum-phase correspondent ("Processor's wavelet") before spiking-deconvolution. After this step, both records can be combined and transformed to a common zero-phase section for further interpretation (BRÖTZ *et al.*, 1987).

MIN. DELAY TRANSFORM

OF

VIBROSEIS/DYNAMITE DATA

VIBROSEIS:

$AK(z)$: autocorrelation of filtersweep $FS(z)$
$W(z)$: instrument recording filter
$S(z)$: sweep
$FS(z) = S(z) \cdot W(z)$
$FS_{MIN}(z)$: min. delay equivalent of $FS(z)$

$$F(z) = \frac{FS_{MIN}(z)}{AK(z)} = \frac{FS_{MIN}(z)}{FS_{MIN} \cdot FS_{MIN}(1/z)}$$

$$\boxed{F(z) = \frac{1}{FS_{MIN}(1/z)}}$$

DYNAMITE:

$W(z)$: instrument recording filter
$W_{MIN}(z)$: min. delay of $W(z)$

$$F(z) = \frac{FS_{MIN}(z)}{W(z)} = \frac{FS_{MIN}(z) \cdot W(1/z)}{W_{MIN}(z) \cdot W_{MIN}(1/z)}$$

$$\boxed{F(z) = \frac{FS_{MIN}(z) \cdot W(1/z)}{W_{MIN}(z)} \frac{1}{W_{MIN}(1/z)}}$$

Replacement of individual wavelets of different records by the min. delay equivalent FS_{MIN} of the filtersweep.

Figure 8
Details of minimum-delay transform of Vibroseis and Dynamite data. Filtering of Vibroseis data corresponds only to a feed-backward operation. The filtersweep is the result of the sweep convolved with the instrument's response. Filtering of Dynamite data consists of both feedforward and feedbackward components.

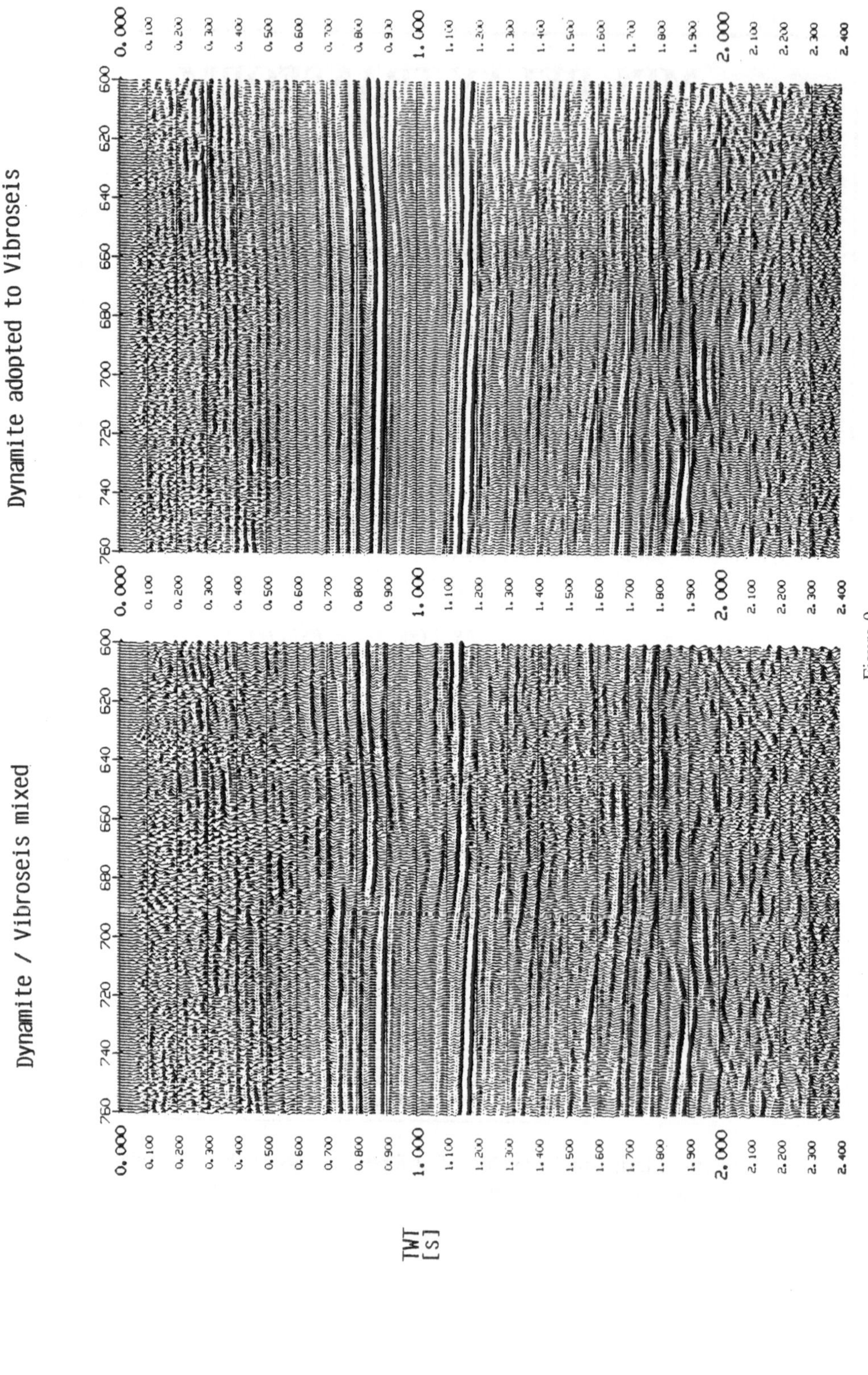

Dynamite adopted to Vibroseis

Dynamite / Vibroseis mixed

TWT
[s]

Figure 9

Results of combining Dynamite and Vibroseis data before (left) and after (right) waveshaping. The Dynamite data are gained by flushed holes filled with ca. 50 gr. of dynamite.

geometric mean is advantageous that spectral maxima of individual traces are reduced by a large amount; "noisy holes" will remain zero in all traces. This filter can be applied in frequency domain. In the time domain, this equates to convolving each trace with a symmetrical, rather long, filter of length LX.

As proposed and explained by TUFEKCIC *et al.* (1981) a one-sided filter can be applied to effect this. The first step is to compute the unit-distance prediction error filters $A_k(z)$ with the Levinson solution of Toeplitz equations of each individual trace. The procedure works in an optimal way if a MD transformation is first carried out. Next, the geometric-mean balancing filter is computed by averaging the logarithms of the prediction error filters of the individual traces. As the prediction filters are minimum-delay, the logarithms of the corresponding polynomials can be computed recursively by cepstral methods (TRIBOLET, 1979). The same procedure can be applied for the exponentation of the resulting mean to derive the geometric mean spectrum $GM(z)$. This results in short operators depending on the length of the prediction error wavelet. The main two advantages are:

—Everything can be done in time domain.
—The filter can be designed in a noise-free window (design-window) and applied in some other window (application-window) or for all traces.

Applications of spectral balancing are shown for two examples.

In Figure 11 the aim was to suppress the surface waves in the center of the record by this method. The design window was in the right part of the shot record outside the noise-covered area. The result of spectral balancing is shown on the right of Figure 11. It exhibits a balanced view.

A common method to generate shear waves by explosives is to apply converted *ps* waves (FERTIG and HENTSCHKE, 1987). The application of spectral balancing to this single record is given in Figure 12. The trace spacing was 30 m. Here particularly the effect of suppression high-frequency noise on individual traces can be seen. The right part of the figure is the result. Additional data analyses are given in the previously mentioned paper.

The above-mentioned method of spectral balancing is implemented as WIBL (Wiener Balancing) in a commercial processing routine and is very often used at the DPC Clausthal. Also for shallow seismics with strong noise, the method of balancing can considerably improve the data quality.

Methods for Suppression of Coherent Events in Seismic Data

Some types of seismic waves, such as surface waves and shear waves, are used as a desired signal in many areas, however these wave types have negative consequences for the data processing of *p* waves and for steep-angle seismics, especially if reflections appear only with weak amplitudes. In this case these waves are regarded as "noise." The situation is critical in particular, if the expected events in the target area are overlain by strongly developed "noise waves" and if no undisturbed primary reflections can be detected.

SPECTRAL BALANCING

<u>Filter:</u> $F_k(z)$ $Y_k(z) = X_k(z) \cdot F_k(z)$ $k = 1, ..., N$

OUTPUT = INPUT   FILTER

k = individual trace number
N = total number of traces in the gather
$X(z):$ = z-Transform of time-series x

$\overline{X}(z):$ = conjugate of X(z)

<u>ARITHMETIC MEAN:</u>
(AM)
$$F_k^a(z): = \frac{\frac{1}{N}\sum_{i=1}^N (X_i(z) \cdot \overline{X_i(1/z)})^{1/2}}{(X_k(z) \cdot \overline{X_k(1/z)})^{1/2}} = \frac{\frac{1}{N}\sum_{i=1}^N |X_i(z)|}{|X_k(z)|} = \frac{AM(z)}{|X_k(z)|}$$

<u>GEOMETRIC MEAN:</u>
(GM)
$$F_k^g(z): = \frac{\left[\prod_{i=1}^N |X_i(z)|\right]^{1/N}}{|X_k(z)|} = \frac{GM(z)}{|X_k(z)|}$$

$$GM(z) = EXP\left\{\frac{1}{N}\sum_{i=1}^N \ln|X_i(z)|\right\}$$

Fast Fourier Transform (FFT) produces symmetrical filter of length LX

<u>PREDICTION ERROR FILTER:</u> $A_k(z)$ as $X_k^{-1}(z)$ and
(UNIT DISTANCE)
$A_{ave}(z)$ as $GM^{-1}(z)$

$$\boxed{Y_k(z): = X_k(z) \cdot \frac{A_k(z)}{A_{ave}(z)} = A_k(z) \cdot A_{ave}^{-1}(z) \cdot X_k(z)}$$

$$\sum \ln X\overline{X} = \sum \ln\frac{1}{A\overline{A}} = -\sum \ln A - \sum \ln\overline{A}$$

Ln A(z): = CEPSTRUM of A (one-sided, causal)

EVERYTHING CAN BE DONE IN TIME DOMAIN

Figure 10
Procedure of spectral balancing. This can be accomplished by replacing the individual spectra of the
traces by the spectrum of the arithmetic or geometric means of the contributing traces within a record.
Here, the geometric mean is used. If the wavelets are minimum-delay all can be done in time-domain via
a cepstral analysis.

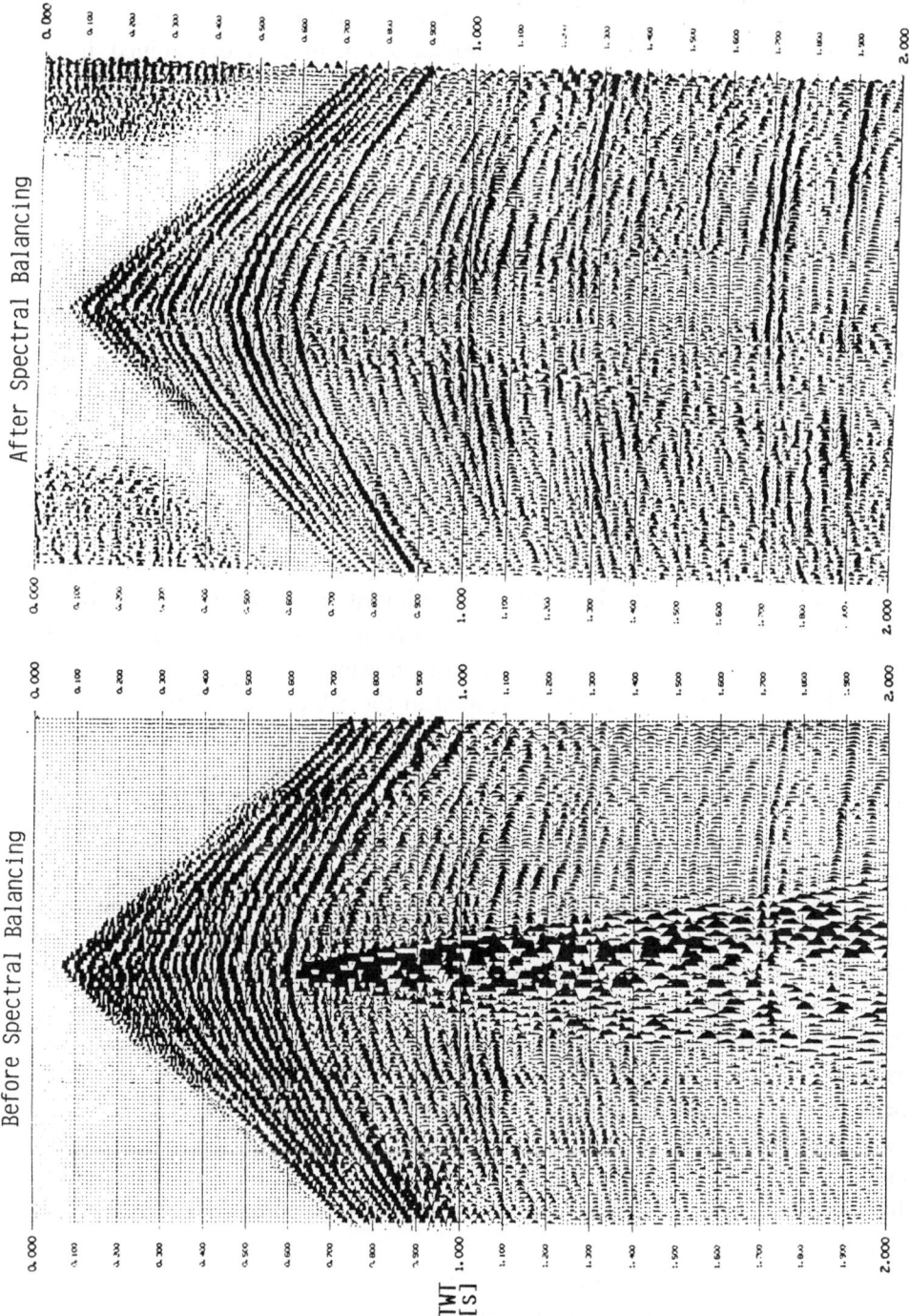

Figure 11

Results of spectral balancing on a shot record. The input is given on the left with surface waves with large amplitudes. At the right, the result after spectral balancing is given. The geometric mean spectrum was defined in the noise free parts of the input. The trace-spacing is 30 m.

In such cases, all processes which need an undisturbed reflection area for their application fail. By these processes an adjusted operator is determined in the undisturbed reflection. This operator is able to suppress the strong noise waves in the disturbed part of the seismogram.

Therefore, methods must be searched for that are independent of the knowledge of the desired seismic signal and that are orientated directly towards the clearly recognizable noise itself.

The process of muting, i.e., zeroing the traces in the disturbed time windows, is the fastest but also the most drastic method. The noise is eliminated completely, however signal components existing in this area are also lost for further processing.

The $(f\text{-}k)$ filter (MARCH and BAILEY, 1982; CHRISTIE *et al.*, 1983) is used with success if the noise components are mainly coherent with clearly defined inclinations or at least one clearly defined inclination area. This processing step reaches its limit if very steeply dipping "events" must be suppressed. The characterization "steep" must be considered differently. The frequency wave number $(f\text{-}k)$ filter is problematic if the inclination, e.g., expressed by dip in ms/trace, becomes very large, whereby some frequency components are moving into the area of the spatial aliasing. In that case not only the inclination, primarily readable in the seismogram must be eliminated, but also the apparent dip which is connected with the spatial aliasing must be removed. The aliased energy however, sometimes overlays the areas in which the reflection energy is located in the $f\text{-}k$ domain. This means that even here with a complete suppression, proportions of the reflection energy are lost.

To avoid the mentioned problem a different method was searched for, and at the DEKORP Processing Center in Clausthal the substractive coherency filtering provided in the program package of Seismograph Service Limited (SSL) has been chosen (KLÖCKNER and THOMAS, 1992). This was done since only weak reflection response was expected in the DEKORP deep seismic reflection data measured mainly in crystalline rocks.

This coherence filter has been applied for many years in the processing of deep seismic reflection data to improve the signal/noise ratio in stacked sections, to increase events correlating over several CMPs (even those with low amplitudes) and to weaken the non-coherent events (even those with strong amplitudes). This happens through the production and evaluation of test stackings of neighboring traces along dips which have to be determined in windows. That leads to an emphasis of these dips, along which the stacking amplitudes have their maximum. In this connection the so-called "coherence trace" is acquired and added to the initial trace.

With adoption of this procedure the power of the coherence filter can be controlled. At a subtractive coherence filter mode, the assessment factor must be chosen to be exactly -1 in order to subtract non-coherent events out of the seismogram. It makes no difference if the data set is available as a single-shot recording or if the process is applied on already stacked data, in which coherent noise amplitudes are apparent. This method can be used for primary, nonlinear, signal portions, for example for the suppressing of multiple reflections. In this case, it is ensured with the aid of normal moveout corrections that the multiple events in the single seismograms are corrected to horizontal. This means that the primary events are pictured overcorrected due to their higher stacking velocity. With the subtractive coherence filter the horizontal events, here the multiples, can be eliminated.

This method mainly has the advantage that it works to a great extent independently of the spatial aliasing. With this it can be used at every configuration of trace distance and dip.

The effect of spatial aliasing is based, in general, on a travel-time difference between two neighboring traces for a long signal (THOMAS, 1992). For horizontal horizons, a travel-time difference already appears in the single shot seismogram if the effect of the shot-geophone distance is not corrected. To make the spatial aliasing visible, the display in the $(f\text{-}k)$ area is especially suitable. Here, not only can the spatial aliasing be described and shown distinctly, but also the $(f\text{-}k)$ display of the results of other in time domain applied methods turns out to be an excellent interpretation tool.

This is most distinctly illustrated by a synthetically produced data set (Fig. 13/top). Shown are three different tilted events A, B and C (Dip: 0 ms/trace, 4 ms/trace and 10 ms/trace) in the $(x\text{-}t)$ and $(f\text{-}k)$ areas. In this example, all signals have the same amplitude. The frequency bandwidth of the signals was limited to 10–100 Hz. The term "event" is chosen, because in the $(f\text{-}k)$ domain it is not important to connect the content of the display together with a single seismic model. The event C, for example, can represent the travel time of a noise-train in the common shot gather, or the $T(0)$ travel time of a dipping horizon, or a noise-train in a zero offset section. Only the time-dip (travel-time difference from trace to trace) of the single event is important.

The events A and B show no spatial aliasing, while the frequency components of event C above 50 Hz are in the spatial aliasing domain. Under the assumption that event C should be eliminated, the area in the $(f\text{-}k)$ area affected by the spatial aliasing effects must be cut. However, these are just apparent dips with the inverse sign of real dips. If event B would dip in the other direction, proportions of event B would also be suppressed. If in the $(f\text{-}k)$ domain the "true" dip of 10 ms/trace is considered, only those parts affected by the spatial aliasing will be maintained (Fig. 13/center).

Figure 13 displays on the right the result after the subtractive coherency filtering. Here only the actual dip of 10 ms/trace was suppressed. Clearly it is recognized that these parts in the spatial aliasing area could be mostly removed without specifying their apparent dips, because this process is working independently on the spatial aliasing in the time-domain.

This processing step evidences good results not only with synthetic data. In Figure 14 (left) the record of a single shot is shown (trace distance: 30 m). Beneath the first event low velocity events, possibly shear waves, stand out which superim-

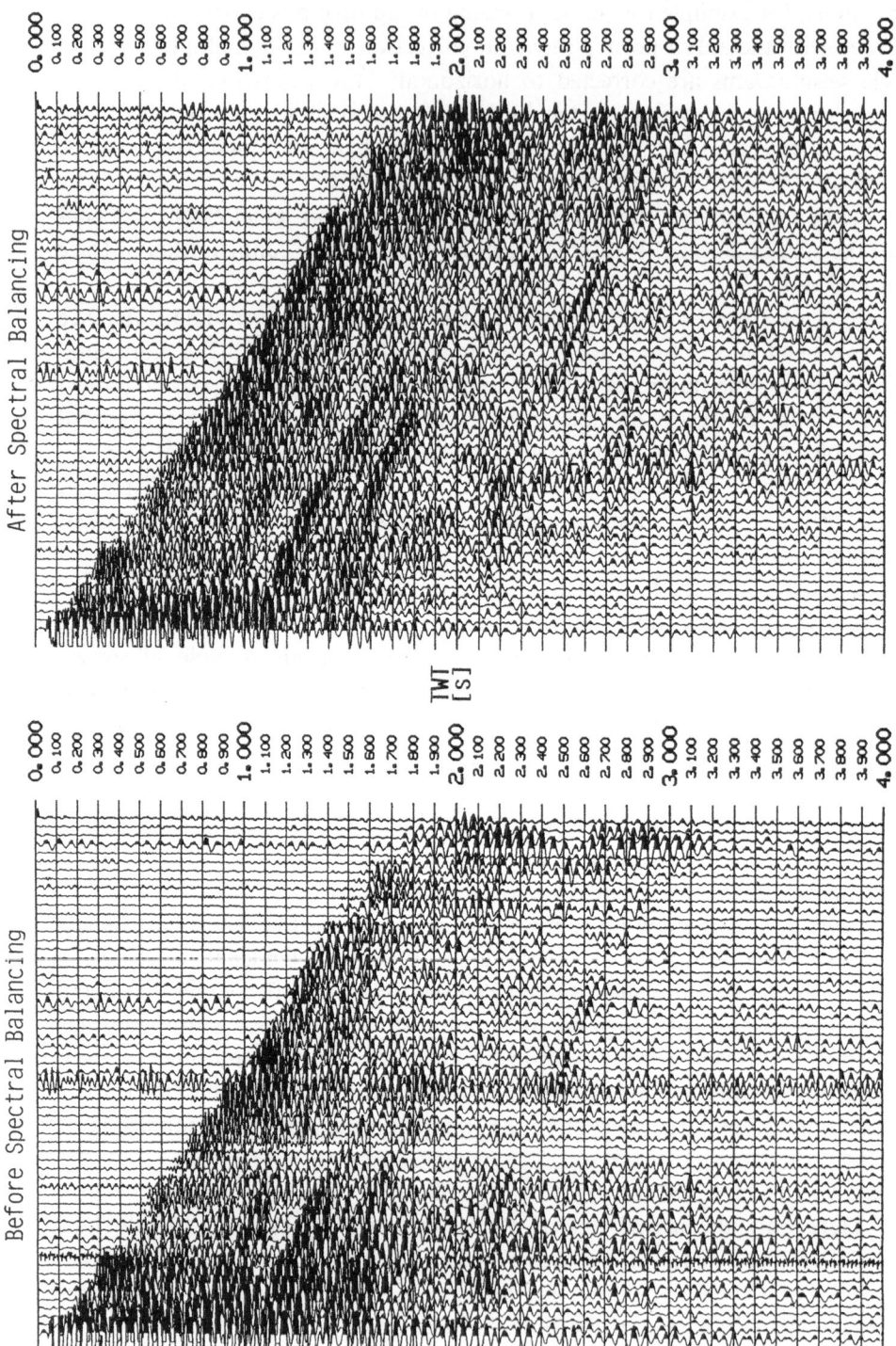

Figure 12

Horizontal components of *S* waves generated by explosives before (left) and after spectral balancing (right). The signals in the near-shot region have been improved; monofrequent noise has been reduced nearly completely. The trace spacing is 30 m.

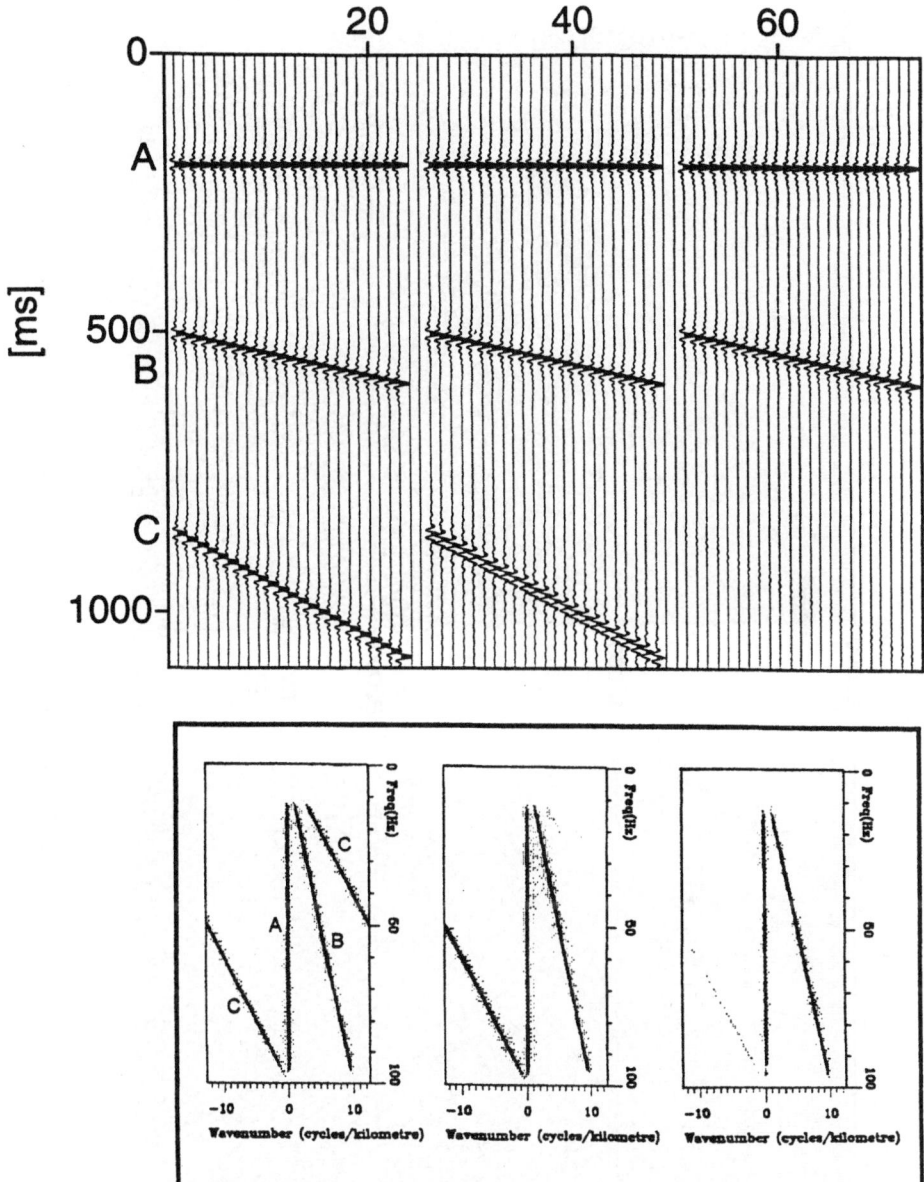

Figure 13
Three different dipping events (A: 0 ms/trace, B: 4 ms/trace, and C: 10 ms/trace) in the (x-t) area (top)
and in the (f-k) area (bottom). Left: unprocessed section. Middle: after application of the (f-k) filter
(inclination area 8–12 ms/trace). Right: subtractive coherency filtering.

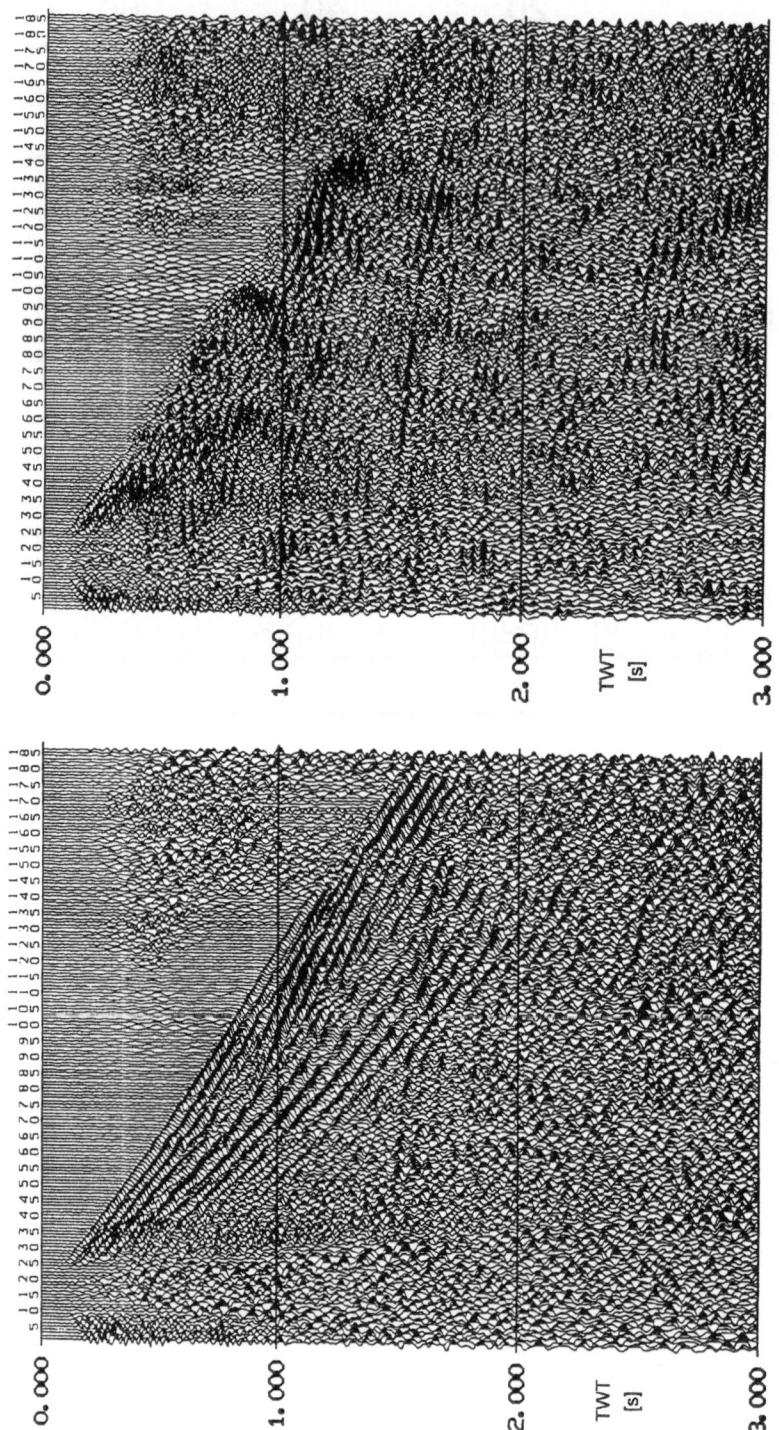

Figure 14

(left) Display of a single seismogram with Automatic Gain Control (AGC). (right) Result after the subtractive coherency filtering. Displayed with AGC.

pose the whole seismogram. A weak reflection event is recognized at 1.5 s TWT (Two Way travel-Time). Reflections can only be estimated at 2.5 s TWT. Figure 14 (right) shows the processing result after application of subtractive coherency filtering. In the time-domain down to 2 s TWT, clear reflections are now detected in areas that previously were superimposed by shear waves. Below 2 s TWT a complete band of reflections becomes visible.

Conclusions

In this article we wanted to demonstrate how difficult data can be improved by taking into account the best knowledge of all influences on the seismic signal. Different source characteristics from vibroseis and dynamite can be adopted by Two-Sided-Recursive Filtering. Conventional Decon-processes work optimal for minimum-delay signals, which can be produced by simple assumptions. To remove coherent events in seismic data, the substractive coherency filtering is a powerful tool. After all these prestack processes, a zero-phase-transform can be done again by recursive filtering, to provide an optimal input for the interpretations of seismic data.

Acknowledgement

Parts of this publication were supported by the "Bundesministerium für Bildung, Wissenschaft, Forschung und Technologie (BMBF)" under contact no. BEO 71/03GT 94054.

REFERENCES

BERKHOUT, A. J. (1974), *Related Properties of Minimum-phase and Zero-phase Time Functions*, Geophys. Prospect. *22*, 683–709.

BORTFELD, R. K., and FERTIG, J. (1983), *Über die Bedeutung der Scherwelle S* in der Schußseismik*, Erdöl und Erdgas *99*(4), 127–133.

BRÖTZ, R., MARSCHALL, R., and KNECHT, M. (1987), *Signal Adjustment of Vibroseis and Impulsive Source Data*, Geophys. Prospect. *35*, 739–767.

CRISTIE, P. A. F., HUGHES, V. J., and KENNETT, B. L. W. (1983), *Velocity Filtering of Seismic Reflection Data*, First Break *1*(3), 9–24.

CLEARBOUT, J. F., *Fundamentals of Geophysical Data Processing* (McGraw Hill 1976).

FERTIG, J., and HENTSCHKE, M. K. (1987), *Data Acquisition and Processing of Converted S Waves*, Geophys. Prospect. *35*, 148–166.

FERTIG, J., and THOMAS, R. (1995), *Optimales Preprocessing als Eingang für eine optimale Datenbearbeitung*. In *Von der Datengewinnung zum Lagerstättenmodell* (Dresen, L., Fertig, J., Jordan, F., Rüter, H., and Budach, W. eds.). 15. Mintrop-Seminar, Unikontakt, Ruhr-Universität Bochum, 79–103.

FERTIG, J., *Seismik*. In *Handbuch zur Erkundung des Untergrundes von Deponien und Altlasten* (Knödel, K., Krummel, H., and Lange, G. eds.) Band 3, Geophysik (Springer-Verlag, Berlin–Heidelberg 1997) pp. 405–446.

KLÖCKNER, M., and THOMAS, R. (1992), *Wesentliche Aspekte bei der Bearbeitung der Profile DEKORP 90-3A und 3B/MVE mit dem Schwerpunkt einer sorgfältigen Störwellenbeseitigung*, 5. Zwischenbericht, F + E-Projekt RG 8705-8, DEKORP 'Deutsches Kontinentales Reflexionsseismisches Programm,' 2. Phase, Institut für Geophysik der TU Clausthal (DPC Clausthal).

LINDSEY, J. P., and MACURDA, D. B. (1985), *Applied Seismic Stratigraphic Interpretation, Course Notes* (Geoquest Int., Houston, Texas).

MCQUILLIN, R., BACON, M., and BARCLAY, W., *An Introduction to Seismic Interpretation* (Graham and Trotman Limited 1979).

MARCH, D. W., and BAILEY, A. D. (1983), *Two-dimensional Transform and Seismic Processing*, First Break *1*(1), 9–21.

MARSCHALL, R. (1977), *Wavelet Processing by Means of Recursive Filters*, Paper presented at the SEG-Convention, Calgary.

REDANZ, M. (1988), *Waveletextraktion und Inversion von Reflexionsseismogrammen zur Ableitung von akustischen Impedanzen*, Berichte der Ruhr-Universität Bochum, Bochum.

THOMAS, R. (1992), *Untersuchungen über Ursachen und Wirkungen des räumlichen Aliasings in der seismischen Datenbearbeitung, insbesondere bei der Migration*, Dissertation, TU Clausthal.

TRIBOLET, J. M., *Seismic Application of Homomorphic Signal Processing*, (Prentice Hall 1979).

TUFEKCIC, D., CLEARBOUT, J., and RASPERIC, Z. (1981), *Spectral Balancing in the Time Domain*, Geophys. *466*, 1182–1188.

(Received October 1, 1997, revised May 19, 1998, accepted July 30, 1998)

 To access this journal online:
http://www.birkhauser-science.ch